ANIMAL BEHAVIOR

MECHANISMS • ECOLOGY • AND EVOLUTION

THIRD EDITION

ANIMAL BEHAVIOR

MECHANISMS • ECOLOGY • AND EVOLUTION

LEE C. DRICKAMER
Southern Illinois University at Carbondale

STEPHEN H. VESSEY
Bowling Green State University

Wm. C. Brown Publishers

Book Team

Editor *Kevin Kane*
Developmental Editor *Margaret J. Manders*
Production Editor *Daniel Rapp*
Designer *David C. Lansdon*
Art Editor *Janice M. Roerig*
Photo Editor *Laura Fuller*
Permissions Editor *Gail Wheatley*
Visuals Processor *Amy L. Saffran*

 Wm. C. Brown Publishers

President *G. Franklin Lewis*
Vice President, Publisher *George Wm. Bergquist*
Vice President, Publisher *Thomas E. Doran*
Vice President, Operations and Production *Beverly Kolz*
National Sales Manager *Virginia S. Moffat*
Advertising Manager *Ann M. Knepper*
Marketing Manager *Craig Marty*
Editor in Chief *Edward G. Jaffe*
Managing Editor, Production *Colleen A. Yonda*
Production Editorial Manager *Julie A. Kennedy*
Production Editorial Manager *Ann Fuerste*
Publishing Services Manager *Karen J. Slaght*
Manager of Visuals and Design *Faye M. Schilling*

Cover photo © Margot Conte/Animals Animals/Earth Scenes.

The credits section for this book begins on page 454, and is considered an extension of the copyright page.

Library of Congress Catalog Card Number: 90-86272

ISBN 0-697-07471-4

Printed in the United States of America by Wm. C. Brown Publishers, 2460 Kerper Boulevard, Dubuque, IA 52001

10 9 8 7 6 5 4 3 2 1

For Lancelot I, Callie, Real Trouble,
Madame Butt-Butt, Aphrodite, Rorschach,
Mehitabel, Electra, Lancelot II, Hopkins II,
Chelsie, Antigone, Pernod, and especially
Hopkins I, Max, and Guinevere (Winnie)

LCD

In memory of my father, Clifford H.
Vessey

SHV

BRIEF CONTENTS

CONTENTS

PART THREE   DEVELOPMENT AND LEARNING   151

PART FOUR   BEHAVIORAL ECOLOGY: SOCIAL PROCESSES   209

PREFACE

The study of animals and their fascinating behavior has been approached from a number of perspectives; ethology, comparative psychology, behavioral ecology, sociobiology, and others. It is our goal in this text to use evolutionary principles as a foundation for exposing students to a number of approaches, demonstrating that the varied viewpoints are complementary, not mutually exclusive. The subtitle, "Mechanisms, Ecology, and Evolution," reflects the broad perspective that dominates the book.

This text is suited for undergraduates taking their first course in animal behavior, but it has also been used in a variety of upper level undergraduate and graduate courses. Our approach—which focuses on concepts, processes, and methods—makes the book useful for all students, regardless of whether they're interested in the 'gee-whiz' performances of animals so often observed on television today, or in understanding the underlying mechanisms and evolutionary biology behind what is observed. The 20 chapters cover all of the topics encompassed by modern animal behavior. We have written each chapter so it is self-contained. Thus, instructors can arrange the material to suit their personal approach or the requirements of their particular course. The length of the book and the difficulty of the material are such that it should be completed handily in a semester course and also in the more intense instruction that often characterizes a term under the quarter-system.

Our presentation begins with basic material on the types of questions posed and answered by animal behaviorists throughout the history of the discipline.

While we have assumed some knowledge of basic biology, we do provide a bit of background material on evolutionary principles and related genetics to insure that all students have the same foundation. For each of the major parts of the book and the chapters within each part we follow a similar pattern. We first define and elaborate on the concepts and processes that underlie particular behavioral patterns. Then, through the use of appropriate research examples, we present the methods and techniques used by animal behaviorists. Through this approach, students are introduced to a variety of viewpoints that have contributed to the richness of the discipline.

The text is divided into six major parts. Part One covers the background for the study of animal behavior, including history, approaches, genetics, and evolution. In Part Two the mechanisms and processes that control behavior are presented and exemplified. Part Three is concerned with the development of behavior, learning, and motivation. In Part Four, the first of two parts on behavioral ecology, we deal with social processes: communication, aggression, reproduction, and parental care. Part Five, also on behavioral ecology, covers environmental processes: animal movement and orientation, habitat selection, feeding, and populations. The final section, Part Six, deals with the evolution of behavior patterns and the evolution of social systems. Throughout the book we make an attempt to integrate both proximate and ultimate perspectives, though the first portion of the book (Parts Two and Three) tends to emphasize the proximate issues (how) more and the latter portion of the book (Parts Four, Five, and Six) tends to emphasize the ultimate issues (why) more. We believe that this particular organization, working from events inside the organism through to those that are involved with the animal interacting with other animals and with its environment, provides a more complete picture of the discipline of animal behavior than other possible plans of organization.

CHANGES FOR THE THIRD EDITION

The basic goals and themes of the textbook remained the same through the first two editions, and, indeed, are the same for this third edition. Based on reactions to those first two editions, changes that are continually occurring in the study of animal behavior, and, with the help and advice of colleagues acknowledged below, we have endeavored to make further improvements in the organization and text material for this third edition. Among the major changes that should be highlighted are:

- The text has been divided into six parts rather than five as in earlier editions; we think that the current breakdown of chapters into six parts better reflects the nature of topics being studied by animal behaviorists today. (Previous Part Four on Behavioral Ecology has been divided into group chapters according to Social and Environmental Processes).

- The chapters in new Part Four have been rearranged. We feel that Communication (Chapter 11) is critical to all other social behavior and thus this chapter should lead off Part Four, followed by material on aggression, reproduction, and parental care.

- The chapter on Sexual Behavior and Reproduction in earlier editions has been split into two chapters on Sexual Reproduction (Chapter 13) and Mating Systems and Parental Care (Chapter 14).

- The chapters in what is now Part Five have also been rearranged. The flow is from how an animal finds its way about, to habitat selection, then feeding, and finally to behavior and population biology.

- New extended examples have been added to a number of chapters, providing students with a more in-depth examination of the processes by which animal behaviorists design and test experimental questions and the results obtained from such studies.

- All chapters have been updated with respect to current research; this is reflected in the content of the chapters, the revision of some Discussion Questions, and in the Suggested Readings at the end of each chapter.

- New illustrations and photos have been added or used to replace others providing for more effective understanding of concepts and research examples. Eight pages of color photos have been added, with illustrative material from all major chapters.

- An Instructor's Manual is available and will offer additional information on key chapter concepts, objective questions (helpful in preparing exams), suggestions for laboratory experiments and field trips, and a listing of audiovisual aids for each chapter.

ACKNOWLEDGMENTS

We would like to extend our sincere thanks to the numerous professional colleagues who have helped us in many ways and provided ideas, good discussions, and overall encouragement during the course of writing three editions of this book. We also would like to express our deep gratitude to the many undergraduate and graduate students at all colleges and universities, but especially our home institutions, who have used the book and provided careful input regarding its contents. Special thanks go to those who read all or parts of the manuscripts of the different editions at various stages and shared with us their most helpful criticisms and suggestions:

First Edition:

Terry Christenson
Tulane University

Victor DeGhett
SUNY College at Potsdam

Kenneth Able
SUNY at Albany

Robert Matthews
University of Georgia

Ronald Barfield
Rutgers, The State University

Irwin Bernstein
University of Georgia

David Barash
University of Washington

H. B. Graves
The Pennsylvania State University

Gordon Gallup
SUNY at Albany

Second Edition:

Donald Dewsbury
University of Florida

George H. Waring
Southern Illinois University

Patricia DeCoursey
University of South Carolina

Maureen Powers
Vanderbilt University

Arthur Coquelin
University of California at Los Angeles

Terry Christenson
Tulane University

Carl Bassi
Vanderbilt University

Third Edition:

Preston E. Garraghty
Vanderbilt University

Peter H. Klopfer
Duke University

Phil N. Lehner
Colorado State University

Max R. Terman
Tabor College

Donna Schroeder
College of St. Scholastica

Paul V. Cupp, Jr.
Eastern Kentucky University

Roger S. Sharpe
University of Nebraska–Omaha

Patricia Adair Gowaty
Clemson University

Charles Snowdon
University of Wisconsin–Madison

William J. Rowland
Indiana University

Thomas C. Rambo
Northern Kentucky University

Robert W. Matthews
University of Georgia

Jerry Lyons provided much needed encouragement and support when we started on this adventure many years ago; Jean-Francois Vilain deserves special mention for his patience, guidance, and reinforcement through the good times and bad of the first two editions. The manuscript editing and preparation for the first two editions were handled most capably by Robine Storm van Leeuwen; her touch is still evident in this third edition. For this edition we extend a warm thanks to Kevin Kane who helped to rescue our efforts and make this third version of the text possible and to Marge Manders who has been an excellent colleague, serving as our developmental editor. We thank Kristin Vessey and Karen Drickamer for their editorial advice and typing assistance. Both our wives deserve a special thanks for their continued patience, support, and encouragement as we put together each new edition of the book.

Lee C. Drickamer
Carbondale, Illinois

Stephen H. Vessey
Bowling Green, Ohio

PART ONE

THE STUDY OF ANIMAL BEHAVIOR

1

INTRODUCTION

Animals, including humans, are involved in a variety of complex, vital relationships with members of their own species, with members of other species, and with the physical environment. Our survival, like that of all animals, depends on our ability to procure food and shelter, to find mates and produce offspring, to protect ourselves from the elements, and especially in the early stages of human history, to avoid predators. Thus, we should not be surprised to learn that people have been interested in animal behavior for a long time. The relationship of the Pleistocene game hunters with their prey was just as intimate and as important as our relationship today with the insects and rodents that compete for our crops or spread disease.

WHY WE STUDY ANIMAL BEHAVIOR

Unlike other animals, we have demonstrated a desire for knowledge about the world around us that transcends our survival needs. Our curiosity about the living world is perhaps the main reason for our interest in animal behavior. By observing and experimenting with animals, we can learn about the relationships between animals and their environments, and about the internal processes that govern their behavior. Some researchers are interested in establishing general principles common to all behavior; they undertake comparative studies of different species and formulate models that explain the observed phenomena. Others are concerned with trying to understand our own species—

human brain mechanisms and behavioral biology and evolution. Still others desire to maintain and preserve the environment; they investigate behavioral processes of animals to conserve and protect endangered species. Many researchers use their knowledge about behavior to control economically costly animal pests.

PROCESSES IN THE STUDY OF ANIMAL BEHAVIOR

The return of migratory red-winged blackbirds (*Agelaius phoeniceus*) from winter haunts in warmer southern latitudes is a harbinger of spring in many areas of the United States (figure 1–1). The returning birds are also part of an annual cycle of events that has physiological, genetic, ecological, and evolutionary aspects. We can ask a variety of questions about this cycle, and we can test these questions experimentally. How do we go about it?

Part One of this book provides background information for examining the approaches and methods used to formulate and test questions in animal behavior. First, we examine the historical roots of the major approaches to the study of behavior (chapter 2). We then look at the methods problems involved in designing experiments to test hypotheses about behavior (chapter 3). Finally, we discuss some concepts of evolution and genetics to understand the dynamics of evolutionary processes as they have shaped behavior (chapter 4). In

FIGURE 1-1 Annual cycle of behavioral changes related to reproduction in red-winged blackbirds
The redwings' annual cycle provides us with many opportunities for studying the physiological, developmental, social, behavior-ecological, and evolutionary processes of animal behavior.

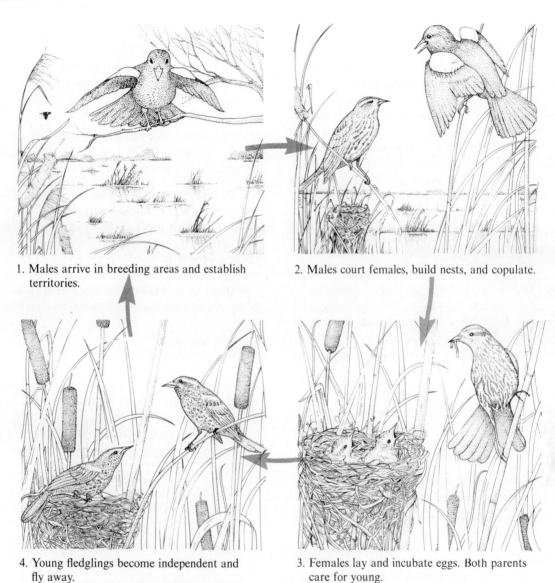

1. Males arrive in breeding areas and establish territories.

2. Males court females, build nests, and copulate.

4. Young fledglings become independent and fly away.

3. Females lay and incubate eggs. Both parents care for young.

5. (not pictured) Adult and young redwings migrate south by fall.

the major sections of the book that follow these introductory chapters, we will explore the various internal and external factors that regulate and influence animal behavior.

In writing this book we, the authors, have two major goals: first to present the key processes that affect animal behavior (internal mechanisms, development, social interactions, ecology, and evolution) and their significance; and second, to show how to formulate questions about behavior and to explore some of the methods used to investigate animal behavior. In the rest of this chapter, we make some observations about the behavior of redwinged blackbirds; we use these observations to question how and why the behavior patterns occur; and we see how our twin goals of learning about processes and learning how to ask questions will be realized. Also, we will examine briefly the importance of the evolutionary perspective in the study of behavior and the use of the comparative method, a powerful tool in the exploration of behavior.

MECHANISM PROCESSES

From mid-March to early April, male redwings arrive on the breeding area in small flocks. They establish individual territories in suitable habitats, usually marshy areas or wetlands. During migration, the social system of male redwings is characterized by gregarious flocking, but during the breeding season, the social system is characterized by individually defended territories.

- What neural and hormonal processes within the male redwings cause them to leave the wintering grounds and head northward to the summer breeding grounds?

- Which internal changes result in the behavioral shift from flocking to territoriality?

- What causes males to start singing and behaving aggressively?

The **mechanism processes** involve biochemistry and the functions of cells, tissues, and organs, and the external factors that affect behavior during the life of the individual animal. In Part Two we examine the internal mechanisms that control behavior: genetics, the nervous system, the hormonal system, and biological clocks.

DEVELOPMENTAL PROCESSES

When male redwings arrive at the summer breeding grounds and establish territories, they choose sites with particular characteristics, such as perches from which to display and sing. Female redwings arrive in mid-April and begin the reproductive cycle by selecting a mate or his territory.

- What processes in the male's past influence its choice of a territory site?

- How have the genetic inheritance and past experiences of each male interacted, resulting in the observed patterns used for territory establishment and defense?

- What experiences or learning events influence the female's selection of a particular mate or the specific characteristics of his territory?

Developmental processes involve the manner in which the genetics of the individual interacts with the environment to produce the physical and behavioral traits of the organism. These developmental events include a variety of learning experiences. In Part Three, we explore the factors in an organism's past experience and those in its present situation that shape observed behavior: development, learning, and motivation.

SOCIAL PROCESSES

Sometimes more than one female shares the same territory with a male mate. The female initiates nest-building activity. Courtship and copulation take place by early May. After completion of a nest, each female lays two to five eggs; the young birds hatch after an incubation period of fourteen to sixteen days.

- What behavior patterns are involved in male-female interactions after the females arrive and during the process of rearing the young?

- What forms of communication are used by the redwings during courtship?

- Why do some females share a male's territory with other females?

The study of **social processes** includes examining how animals communicate and interact. In Part Four we explore the behavior of individuals and groups. We examine communication, aggression, reproduction, mating systems, and parental care.

ECOLOGICAL PROCESSES

Young redwings are kept warm and are fed by the female for about six weeks while they grow, become fully feathered, and eventually learn to fly. They achieve full independence from their parents by early August. Throughout this period the male continues to defend the territory.

- Why do redwings prefer to make their nests in certain types of habitats?
- What foods do redwings consume, and what behavior patterns do they use to obtain these foods?
- What factors may account for the observed division of labor between males and females?

Behavioral ecology is the study of functional aspects of behavior. These processes arise from the interaction of animals with the living and nonliving environment (populations, ecological communities, and ecosystems). In Part Five we consider functional problems that relate behavior to the ecological context: orientation and migration, habitat selection, feeding relationships, and population dynamics.

EVOLUTIONARY PROCESSES

Adult red-winged blackbirds form mixed-sex flocks and fly south by about mid-September, and are followed several weeks later by flocks of younger birds.

- How has this particular species-specific social system developed over many hundreds of generations?
- Why has this particular social system developed for redwings instead of some other one, such as large nesting colonies?

Evolutionary processes result in behavior patterns that develop over time in a particular species through natural selection. Evolution by natural selection produces changes in morphology, physiology, and behavior. In Part Six we are concerned with evolutionary processes that have affected the historical development of individual behavior patterns, social behavior, and social organization.

This book divides the study of animal behavior into five processes; however, this division is only a convenient way to organize this study. The five processes are not separate, they are highly interrelated.

For example, we might ask, What behavior patterns are important for the interaction among male redwings during the establishment and maintenance of territories? This question is primarily concerned with social processes; in answering it, we would focus on the redwings' general social organization, the types of behavior patterns redwings exhibit, the sequence in which the behavior patterns occur, where in the territory the behavior patterns occur, and how the behavior patterns relate to the acquisition and defense of territory. A complete answer to our question involves all five processes. There is feedback between social processes and the physiological changes within individual birds: internal changes in hormone levels may produce shifts in behavior patterns. We must find out why redwings prefer certain types of habitats, and what factors determine territory size. We will also want to know how and why territoriality, and not some other system, has developed over many hundreds of generations.

PROXIMATE VERSUS ULTIMATE CAUSATION

Until recent decades, many animal behaviorists focused on the "how" questions: those involving the ways in which behavior is *directly* produced and controlled. We call these structural and mechanical influences **proximate** factors. The development of sociobiology and behavioral ecology has given a strong impetus to the investigation of "why" questions: those involving the **ultimate** factors that, through evolution, have produced the behavior. This approach has both advantages and disadvantages. Because the evolutionary perspective requires behaviorists to study animals in their natural contexts, it generates important questions about the functions of behavior. However, it has been criticized for overusing the concept of adaptation and for relying on "just so" explanations. The obvious risk of placing too much emphasis on evolutionary "just so" explanations is that the theories developed for observed behavior seem quite plausible, but cannot be properly tested using the scientific method. The fossil record provides bones and teeth and environmental evidence, but the behavior itself is not preserved and must often be inferred from sketchy and incomplete evidence. We should approach the study of animal behavior from the dual perspectives of proximate and ultimate factors: understanding how immediate factors affect behavior is just as important as explaining why behavior has evolved.

COMPARATIVE STUDIES

Comparative studies are often used to explore the ecology and evolution of behavior. Several brief examples will help illustrate the importance of the comparative method; we'll see this process in action many more times throughout the book.

Dewsbury (1972) examined the patterns of copulatory behavior for a wide variety of mammalian species. By comparing these copulatory patterns of males as they mate with females, we note significant differences (see figure 13–12). Males of the various species differed with respect to four major characteristics of their mating pattern: the occurrence of multiple ejaculations, the performance of multiple intromissions, thrusting during intromissions, and the presence of a copulatory lock at ejaculation. Dewsbury suggests that these differences may be significant in terms of restricting gene flow between populations. In chapter 4 we will examine further the underlying evolutionary processes leading to reproductive isolation, and in chapter 13 we look at the significance of pattern differences with regard to reproduction.

A second example concerns the ecology and social behavior of three species of marmots (Barash 1973a,b). Woodchucks (*Marmota monax*) inhabit fields at lower elevations, reproduce each year, are usually solitary and aggressive toward one another, and have juveniles that disperse at one year of age. Olympic marmots (*M. olympus*) inhabit meadows at higher elevation, reproduce every other year, are colonial and highly tolerant of conspecifics (members of the same species), and have juveniles that disperse at three years of age (figure 1–2). A third species, yellow-bellied marmots (*M. flaviventris*), live at intermediate elevations, generally reproduce annually but sometimes skip a year, are colonial but moderately aggressive, and have juveniles that disperse at two years of age.

These observed differences can be correlated with the length of growing season and the distribution of food. Short growing seasons at high elevations result in less food available to marmots of those habitats, and food tends to be clumped in its distribution. The longer growing seasons at lower elevations result in larger, more evenly distributed food supplies. Barash (1974) relates the behavior and ecology of these species by noting that aggressiveness is a key factor in the dispersal of juveniles. When less food is available, as with Olympic marmots, the young grow and mature slower, adults are more tolerant; consequently, a colonial social system is advantageous, probably because of the need for delaying dispersal of the young until they are capable of

FIGURE 1-2 A family grouping of Olympic marmots, a species that lives at higher elevations
Since the growing season is short, less food is available, thus juveniles grow more slowly and do not disperse until they are three years of age. The result is a colonial type of social system.
Source: Photo by David P. Barash.

independent existence. Where food is more plentiful, as in the woodchucks' environment, juveniles become large enough for independence after only one year, and adult aggressiveness is correspondingly more prevalent. Thus we see that the particular social system for a population is determined, in part, by the types and distribution of resources.

Animal behaviorists also use comparative studies to draw analogies between behavior patterns of diverse types of animals, including humans. We should be aware that analogies can be instructive, but they can also be misleading. We should rely on comparative studies only to produce tentative conclusions and to suggest hypotheses for further testing. We would be incorrect to conclude, for example, that because red-winged blackbirds compete for territorial space, humans also compete for territories in the same way or for the same reasons. Studies of red-winged blackbirds may suggest some of the functions and processes of territoriality — but only for redwings. We may then apply some of this knowledge to comparative studies of spatial relationships and their functions in humans or other animals. We may use our knowledge about other species to generate questions about human patterns of spacing, but we can answer questions about human behavior only by studying people directly.

Discussion Questions

1. For the animals listed below, think through the seasonal changes they experience and provide a series of questions about their behavior similar to those presented in the chapter for the red-winged blackbirds.
   (a) tree squirrels
   (b) turtles
   (c) spiders

2. For the questions you have posed above, indicate which of your queries are concerned with proximate issues and which are concerned with ultimate issues.

3. Examine figure 13–12 concerning copulatory behavior of mammals. From what you know at this early stage in your study of animal behavior, what sorts of observations would you need to make on birds to determine whether this same comparative approach would work for them as well as for mammals?

Suggested Readings

Beach, F. A. 1960. Experimental investigations of species-specific behavior. *American Psychologist* 15:1–18.
This easy-to-read article presents a second view, contrasting with and complementing that of Tinbergen, regarding the way we should approach the various questions to be asked about animal behavior.

Tinbergen, N. 1963. On aims and methods of ethology. *Zeitschrift fur Tierpsychologie* 20:410–29.
A classic exposition of the major framework for asking questions about animal behavior. Very readable, even for the beginning student.

2

HISTORY OF THE STUDY OF ANIMAL BEHAVIOR

*H*umans and their prehuman ancestors have left evidence—both deduced by us from archeological explorations and drawn and written by them—of their interest in the natural world. We know that some of this interest originated in need. Animals were a primary source of food, clothing, and materials for tools and shelter; thus, knowledge concerning their behavior was necessary for successful hunting. During the course of history, interest in animal behavior has also stemmed from human curiosity about the natural world. In this chapter we examine how and why people have studied animal behavior—from the early days of human existence, through the emergence of animal behavior as a scientific discipline in the nineteenth century, to the experimental and theoretical approaches of the present.

INTEREST IN ANIMAL BEHAVIOR

EARLY HUMANS

For many thousands of years, humans and their ancestors were hunters and meat-eaters. The early hominids and the first *Homo erectus* practiced a crude variety of hunting. Peking man, a form of *Homo erectus* 400,000 years ago, was an accomplished hunter and user of fire, and made tools from animal bones.

L. S. B. Leakey (1903–1972), an anthropologist known best for his discoveries of early hominid remains in Tanzania, proposed and tested a hunting strategy that was based on knowledge of animal behavior—a strategy that early hunters may have used to capture a rabbit or other small animal. Leakey suggested that, upon sighting the prey at about fifteen meters distance, the hunter should sprint directly toward the animal (a small animal often initially freezes in such a situation). Within two or three meters of the prey, the hunter should turn sharply either left or right, because the typical escape behavior of the prey is to make a sudden dash in one direction or the other. If both prey and hunter go to the left, the hunter is upon the animal and can grab it bare-handed (as Leakey demonstrated), or he might use a club or stone to strike it. If the hunter guesses incorrectly, he should stop, turn, and wait for the animal to stop. The process is then repeated and perhaps results in a successful capture.

Early *Homo sapiens* must have been keen observers of animal habits and characteristics. They needed to be familiar with the behavior of animals, not only to know where and how to hunt their prey, but also to protect themselves from potential predators. Hunters of the Upper Paleolithic (35,000 to 10,000 years ago) probably used fire to drive animals over cliffs or into cul-de-sacs where they could be slaughtered with rocks or clubs. A ravine with at least 100 mammoth carcasses has been located in Czechoslovakia, and the remains of thousands of horses that were stampeded over a cliff have been discovered in France.

FIGURE 2-1 Cave painting depicting man aiming arrow at deer

Cave paintings, such as this one found in Alpera, Spain, indicate that even in the early stages of human history people knew something about animal behavior, particularly about those habits and traits of animals that were important either as potential food sources or as possible predators on humans themselves.

Source: Photo by Hubertus Kanus/Photo Researchers, Inc.

Prehistoric cave paintings in France and Spain reveal other aspects of humankind's relationship to animals. These paintings realistically depict many types of game animals in ways that suggest close observation of the animals at various times in their life cycles. In addition, some of the drawings are symbolic representations of actual hunting scenes (figure 2–1). However, while early people were aware of the animals in their environment, their knowledge of animal behavior was probably limited to mostly practical concerns.

CLASSICAL WORLD

Interest in animal behavior in the classical world stemmed from curiosity about natural phenomena and a desire to record and categorize observations. For example, Aristotle (384–322 B.C.) wrote ten volumes on the natural history of animals, in which we note the first extensive use of the observational method. The following brief excerpts, translated from the original Greek, give us a flavor of what Aristotle's observations were like (the first two passages are true, the last is false) (Ley 1968, 36–37):

They say that the cuckoos in Hellice, when they are going to lay eggs, do not make a nest, but lay them in the nests of doves or pigeons, and do not sit, nor hatch, nor bring up their young; but when the young bird is born and has grown big, it casts out of the nest those with whom it has so far lived.

In Egypt they say there are some sandpipers that fly into the mouths of crocodiles and peck their teeth, picking out the small pieces of flesh that adhere to their teeth; the crocodiles like this and do them no harm.

The goats in Cephallaria apparently do not drink like other quadrupeds; but every other day turn their faces to the sea, open their mouths and inhale the air.

The Roman naturalist Pliny (A.D. 23–79) made extensive observations of the natural world. A quote from his *Natural History* provides some insight into the anthropomorphism (ascribing human characteristics or attributes to nonhumans) that characterized Roman perceptions of animal behavior (Nordenskiöld 1928, 55):

Amongst land animals, the elephant is the largest and the one whose intelligence comes nearest that of man, for he understands the language of his country, obeys commands, has a memory for training, takes delight in love and honour, and also possesses a rare thing even amongst men—honesty, self-control and a sense of justice; he also worships stars and venerates the sun and the moon.

We can see from these brief passages that early scholars were attempting to record what they observed in the world around them. Their perceptions of behavior were often colored by the lack of full knowledge about what was taking place, or by biases based on religion or philosophy. However, these early observations served as the basis for human understanding of the natural world for many centuries.

FOUNDATIONS OF ANIMAL BEHAVIOR

The rigorous scientific study of animal behavior did not begin until the latter part of the nineteenth century. We turn now to three major developments that contributed significantly to the study of behavior as it developed prior to 1900: (1) publication of the theory of evolution by natural selection, (2) development of a systematic comparative method, and (3) studies in genetics and inheritance.

FIGURE 2-2
Charles Darwin investigating the unique marine iguanas of the Galápagos Islands.

THEORY OF EVOLUTION BY NATURAL SELECTION

For several centuries, European ships made voyages of exploration and discovery to all parts of the globe. Often scientists were officially attached to the voyages, as Charles Darwin (1809–1882) himself was. These scientists and other crew members made observations of exotic fauna and flora and brought live and preserved specimens to zoos and laboratories in Europe, where scholars could observe, record, and speculate about the anatomy, behavior, and interrelationships of these newly discovered species. The following passage from Darwin's account (figure 2-2) of the marine iguana of the Galápagos Islands illustrates the kind of observations he made on animals in their natural setting (Darwin 1845, 336):

They inhabit burrows, which they sometimes make between fragments of lava, but more generally on level patches of the soft sandstone-like tuff. The holes do not appear to be very deep, and they enter the ground at a small angle; so that when walking over these lizard warrens, the soil is constantly giving way, much to the annoyance of the tired walker. This animal, when making its burrows, works alternatively the opposite sides of its body. One front leg for a short time scratches up the soil, and throws it towards the hind foot, which is well placed so as to heave it beyond the mouth of the hole. That side of the body being tired, the other side takes up the task, and so on alternatively.

Like all major scientific paradigms, the theory of evolution drew upon contributions by and suggestions from the work of other scientists. In 1798, Thomas Malthus (1766–1834), in his *Essay on the Principle of Population*, hypothesized that humans have the reproductive potential to rapidly overpopulate the world and outstrip the available food supply; that is, population increases geometrically, while the food supply increases arithmetically. The inevitable result is disease, famine, and war. Malthus's theory was an important influence on Darwin's thinking about the competition for survival among members of a species. A contemporary and friend of Darwin's, geologist Sir Charles Lyell (1797–1875) was among those who made observations of rock strata and successions of fossils that gave evidence of a process of continuous change in living material through time, an idea that was at odds with the biblical suggestion of the simultaneous creation of all living things. This evidence of geological change led others to the idea that species themselves were not fixed entities. The artificial selective breeding of domesticated stocks by English farmers provided additional support for the thinking of both Darwin and A. R. Wallace (1823–1913).

Wallace's voyage to the Malay archipelago, Darwin's travels on the *Beagle* to South America and the South Pacific, and their other studies and the intellectual influences of the time, led each man independently to formulate the theory of evolution by natural selection. The original theory states that although each animal species has a high reproductive potential, the number of animals of a species remains relatively constant over time. Thus, there is competition for survival. Variation in traits exists within animals of one species. Because some traits are more advantageous than others in the competition for survival, the operational process of natural selection occurs. Only those members of the species that are able to survive to produce more offspring contribute their characteristics to subsequent generations through their young.

Behavior, morphology, and physiology were all thought to be subject to the effects of natural selection. The following passage from *The Origin of Species* illustrates that Darwin clearly recognized the central role of animal behavior in determining the outcome of competition between animals (Darwin 1859, 94):

Amongst birds, the contest is often of a more peaceful character. All those who have attended to the subject, believe that there is the severest rivalry between the males of many species to attract, by singing, the females. The rock-thrush of Guiana, Birds of Paradise, and some others, congregate; and successive

FIGURE 2-3   Male bower bird in display
As in the passage quoted from Darwin, male birds of a variety of species display to attract females. The male bower bird builds a bower and adorns it with brightly colored objects.
Source: Photo by Gerald Borgia.

males display with the most elaborate care, and show off in the best manner, their gorgeous plumage [figure 2-3]; they likewise perform strange antics before the females, which, standing as spectators at last choose the most attractive partner.

Darwin concluded that species were not fixed entities. The theory of evolution by natural selection accounted for changes within a species through time, and also for the gradual appearance of new species. Recent developments in other biological fields—genetics in particular—have modified the theory of evolution by natural selection proposed by Darwin and Wallace. Today, some evolutionary biologists believe that evidence from the fossil record and genetic mechanisms supports the claim that rates of evolution vary through time (Stanley 1981). Change through evolution, in particular, the appearance of new species, may occur more rapidly during some time periods than at other times. We will explore the theory of evolution and the consequences of this theory for animal behavior in chapter 4.

COMPARATIVE METHOD

George John Romanes (1848–1894) is generally credited with formalizing the use of the **comparative method** in studying animal behavior. For Romanes, the comparative method involved studying animals to gain insights into the behavior of humans. Romanes sought to

support Darwin's theory with his proposal that mental processes evolve from lower to higher forms, and that there is a continuity of mental processes from one species to another. He argued that although people could really know only their own thoughts, they could infer the mental processes of animals, including other humans, from knowledge of their own. For Romanes, the similarities between the behavior of humans and that of other animals implied similar mental states and reasoning processses in humans and in nonhuman species. He suggested that a sequence could be constructed for the evolution of various emotional states in animals. Worms, which exhibit only surprise and fear, were placed lowest on this scale; insects were said to be capable of various social feelings and curiosity; fish showed play, jealousy, and anger; reptiles displayed affection; birds exhibited pride and terror; and finally, various mammals were credited with hate, cruelty, and shame.

Romanes's theory relied largely on inferences rather than on recorded facts; he made substantial use of anecdotes. A movement led by another Englishman, C. Lloyd Morgan (1852–1936), sought to counteract these faults by using the **observational method.** Morgan's basic tenent was that only data gathered by direct experiment and observation could be used to make generalizations and develop theories.

Morgan is probably best known for his "law of parsimony," which is now axiomatic in animal behavior studies, "In no case may we interpret an action as the outcome of the exercise of a higher psychical faculty if it can be interpreted as the outcome of the exercise of one which stands lower in the psychological scale" (Morgan 1896, 53). This statement, also called "Morgan's canon," has been interpreted to mean that in the analysis of behavior, we must seek out the simplest explanations for observed facts. Where possible, we should reduce complex hypotheses to their simplest terms to facilitate the clearest understanding of the mechanisms that control behavior.

THEORIES OF GENETICS AND INHERITANCE

The third development that greatly influenced research in animal behavior was the birth of the science of genetics and the development of modern theories of inheritance. In the 1860s, Gregor Mendel (1822–1884) reported his findings from breeding experiments on garden peas. These studies established key principles of the laws of inheritance of biological characteristics.

Present-day behavioral biology is based on the combination of evolutionary theory, which explains how traits can change through time, and genetics, which explains how traits are passed from one generation to another.

We now know that, like the morphological and physiological traits, an animal's behavior has a genetic component. Thus, behavior may change as a species evolves. This means that, as scientists, we can explore the genetic variation underlying various behavior patterns, just as others have investigated the effect of genetic inheritance on morphology and physiology. Behavior-genetic analysis had its beginnings in these early studies of inheritance and was then greatly expanded in the 1930s by the work of R. A. Fisher (1890–1962) and others. Modern behavior-genetic analysis (see Hirsch 1967; Fuller and Thompson 1978) is a powerful tool used by many animal behaviorists; of this we shall learn more in chapter 5.

EXPERIMENTAL APPROACHES

The ideas, methods, and theories established during the latter half of the nineteenth century form the foundation of today's experimental approaches to the study of animal behavior. (1) Comparative psychologists and physiologists have sought to determine the underlying causes of behavior—the control mechanisms. (2) Classical ethologists have been concerned primarily with the functional significance and evolution of behavior patterns, but have also developed explanations for behavior mechanisms, including drives, innate releasing mechanisms, and similar concepts. (3) Behavioral ecologists have explored the environmental context for behavior, and the ways in which animals interact with their living and nonliving environments. (4) Sociobiologists have applied the principles of evolutionary biology to the study of social behavior and organization in animals. We should now briefly examine the historical development of each approach. From these varied approaches to the study of behavior has come the modern synthetic view of animals living and behaving in their natural environment. Though we examine these approaches here as separate entities, bear in mind that they did not develop entirely independently of one another, and that in recent decades, they have become melded into a single discipline. The modern approach to the study of animal behavior contains elements of all of these approaches. As we can see by looking at the animal behavior courses offered at various colleges and universities and the titles of the textbooks used to teach such courses, those who work and teach in this area may call themselves ethologists, animal behaviorists, or comparative psychologists. However, they are all really pursuing the same goals using the same general theoretical frameworks, and practicing their craft using similar experimental techniques and methods.

STUDIES OF MECHANISMS

Comparative psychology is the study of different animals' behavior patterns in order to determine the general principles that explain their actions. Comparative psychology can best be understood by looking at the variety of approaches to behavior studies taken over the past century, which eventually led to comparative psychology's development. In today's world, comparative psychology has melded into the larger discipline that we call animal behavior or ethology.

PERCEPTUAL PSYCHOLOGY. Several distinct approaches to discovering the mechanisms underlying behavior emerged during the mid-nineteenth century. Researchers who were concerned with the mind/body dichotomy studied the relationships between physical and mental processes. Gustav Fechner (1801–1887), one of the founders of perceptual psychology and psychophysics, was interested in separating the processes of sensation (body) and perception (mind). His goal was the *objective* measurement of sensation (the reception of stimuli through the senses, such as sight and hearing), and the comparison of this direct measurement to *subjective* interpretation (perception) of the sensations.

PHYSIOLOGICAL PSYCHOLOGY. Modern physiological psychology developed from early attempts to relate behavior with the internal physiological properties and events of the organism. For example, Marie-Jean-Pierre Flourens (1794–1867) surgically removed portions of the brains of pigeons and recorded the resulting changes in the birds' behavior. Hermann von Helmholtz (1821–1894) studied the conduction speed of nerve impulses, and later, the physiology of vision. He ingeniously measured the speed of nerve conduction by experimenting on the frog motor neuron that triggers muscle contractions. First he stimulated the nerve at one point near the muscle, and then at a second point farther away from the muscle. The difference in amount of time elapsed between stimulus and muscular contraction in the two measurements is the conduction time for the distance between the two stimulus points. From this information, he calculated the speed of conduction.

Physiological psychology remains an important subdiscipline today, and work in this realm and in animal behavior are interconnected. Consider, for example, a study by Sperry et al. (1956), in which he surgically manipulated the position of the eyes in newts (*Notophtalmus viridescens*). Sperry removed the eyes and then replaced them so that they were upside down! Newts treated in this way behaved as if they saw the world upside down; they moved their eyes upward in response to the movement of an object downward in their visual field. This effect persisted even after several years. We learn from Sperry's work that in the visual system of the newt, the newt's optic neurons in the optic nerve travelling from the retina to the brain are labeled for spatial orientation. Thus, even though the eye has been rotated, the message sent to the brain along the nerve remains the same as if the eye were in the correct, normal orientation.

FUNCTIONALISM. By the late 1800s and early 1900s, Europe was no longer the exclusive center of behavioral studies, and individuals were conducting research investigations in comparative psychology at a number of laboratories in the United States. Two major new theoretical and experimental points of view arose during this period: functionalism and behaviorism. The **functionalists,** headed by John Dewey (1859–1952), studied the functions of the mind and how the mind operates, in contrast to studying how the mind is structured. Functionalists attempted to answer three major questions: (1) How does mental activity occur? (2) What does mental activity accomplish? and (3) Why does mental activity take place? Functionalism employed objective observation rather than introspection as its primary method.

The functionalist approach was the introduction into psychology of **adaptive behavior,** a notion prevalent in biology, that behavior functions in the animal's survival in its natural habitat. To these early psychologists, the concept of adaptive behavior implied that the response to a stimulus changes the sensory situation in such a way that the original conditions that produced the response are altered. For example, pain disappears when a sharp splinter is removed from the hand, and the original condition—the existence of a splinter—is also altered.

BEHAVIORISM. John B. Watson (1878–1958) was the principal founder of a new approach to the study of behavior, **behaviorism.** The basic tenet of behaviorists is that animal behavior consists of an animal's responses, reactions, or adjustments to stimuli or complexes of stimuli. Thus, most activities of an organism are products of its past experiences: the mind is a *tabula rasa* (a blank slate) at birth and cumulates all subsequent experiences. Psychology thus becomes the study of the ways in which past events affect behavior. To what degree can we predict and control behavior based on a knowledge of an animal's previous experiences? The methods utilized by Watson and his followers, for example B. F. Skinner (1904–1990), were strictly objective. Reports of subjective feelings or emotions were, by definition, not acceptable as scientific data. This restriction forced the behaviorists to study human behavior in much the same way they studied the behavior of any other animal, without benefiting from their subjects' verbal judgments or reports of feelings and perceptions. (It is noteworthy that Skinner's earliest papers dealt with innate aspects of behavior; studying the history of our discipline provides many insights and surprises!)

ANIMAL PSYCHOLOGY. Concurrent with the development of these viewpoints was the emphasis by Edward L. Thorndike (1874–1949) on the need for systematic, replicable experiments in comparative animal psychology. Thorndike used the puzzle box (figure 2–4) to perform a series of task-learning experiments, using cats as test subjects. A cat was placed in the box, which was fastened shut; the cat could open the door and obtain a reward placed outside the box by manipulating a shuttle-lever. From these experiments, Thorndike concluded that much of animal learning takes place by trial and error, and that rewards are a critical component of learning processes.

Animal psychology today is a diverse mixture of subdisciplines, both new and old. The study of comparative learning and learning theory is still quite important, as the work of Bitterman (1975), Seligman (1970), and Zolman (1982) exemplifies. Ecological aspects of learning have been examined by investigators

FIGURE 2-4   Thorndike puzzle box
A cat inside the cage can clearly see the reward, in this case a fish, placed outside. In order to obtain the reward, the cat must learn to manipulate a shuttle-lever system that raises the door of the cage.

like Kamil and Sargent (1981). The development of behavior is also a subject of continued investigation by researchers like Oppenheim (1982), who is examining aspects of neural development, Stern and Rogers (1988), who are studying the effects of early experience on adult behavior, and by groups like the one developed at the Wisconsin Regional Primate Laboratory by Harlow and colleagues (Suomi and Harlow 1977) that explored primate behavior development. The study of the physiological processes underlying behavior has also diverged into several pathways: the relationship of hormones and behavior (Lehrman 1965; Crews 1980; Goy et al. 1988); neural correlates of behavior (Hubel and Wiesel 1965; Brown et al. 1988) and brain chemistry; and psychopharmacology and behavior (Kelly et al. 1979). The cross-fertilization of genetics and behavior also produced a new subdiscipline called behavior genetics, which is concerned with the hereditary bases of behavior and how the interactions of genetics and environment affect behavior (Hirsch 1967; Oliverio 1983; Hood and Cairns 1988).

In 1950, Frank Beach, in his presidential address to the American Psychological Association, stressed that the discipline of comparative psychology was devoting too much attention to the white rat as a test subject, while ignoring many other types of available vertebrate organisms. Others, notably Lockard (1971) and

Hodos and Campbell (1969), have called our attention to the lack of an evolutionary perspective in comparative psychology, and to the incorrect use of the rat as a model for other organisms, especially humans. These critiques have stimulated more truly comparative investigations, for example, the work of Dewsbury (1972, 1975) on reproductive behavior in rodents. More attention has also been given to the natural context and actual field investigation of animals (Lockard 1971; Barash 1973, 1974).

COMPARISONS.   Discerning whether a particular study has been conducted by an ethologist or a comparative psychologist may be difficult at first. If we understand the historical differences between these two approaches, we can better appreciate the synthetic approach that characterizes the behavior studies of the past several decades.

Ethology was developed, largely in Europe, by researchers trained in biology. Ethologists traditionally observed a wide variety of animals in nature and conducted experiments under conditions that mirrored the natural setting as closely as possible. They concentrated their efforts on exploring questions of ultimate causation—the "why" questions of the evolution and function of behavior.

Comparative psychology originated primarily in America. Until the past several decades, most psychologists generally worked under controlled laboratory conditions. Much of their research was carried out on small rodents, particularly the domesticated rat. Comparative psychologists placed primary emphasis on proximate issues—the "how" questions of the physiological and developmental mechanisms underlying observed behavior patterns. Dewsbury's history of comparative psychology (1984) provides many details regarding the development of concepts and theories in this field, and many insights into the persons responsible for the experimental and theoretical work. Dewsbury defines comparative psychology as the attempt to make comparisons across species in order to develop principles of generality regarding animal behavior. He examines the course of development that characterizes this field since 1900 and notes the many myths that have been associated with what scientists and nonscientists alike have come to believe a comparative psychologist is.

Animal behavior is now a unified discipline with a broad synthetic approach: much of the research conducted by animal or comparative psychologists today is indistinguishable from that of other animal behaviorists with different backgrounds. These research endeavors include explorations of the genetic aspects of food-searching behavior in blowflies (McGuire and Tully 1986), the effects of aversive conditioning on learning behavior of honeybees (Abramson 1986), the role of hormonal factors in infanticidal behavior in rats (Brown 1986), and the role of the brain in budgerigars' interpreting acoustic information from contact calls (Brown et al. 1988).

STUDIES OF FUNCTION AND EVOLUTION

ETHOLOGY. The systematic study of the function and evolution of behavior, called **ethology,** is now a little over a century old. One of its most important principles is that behavioral traits, like anatomical and physiological traits, can be studied from the evolutionary viewpoint. For example, C. O. Whitman (1842–1910) made extensive observations of display patterns, which he termed instincts, in various species of pigeons. Whitman found that he could use displays (patterns of behavior exhibited by animals that function as communications signals) to classify animals according to similarities and differences in behavior. From its early beginnings, ethology developed into a separate science, with its own concepts and terminology, much as comparative psychology did. Today, as we noted previously, those working as ethologists are conducting the same sorts of studies as all others who study animal behavior.

The **ethogram,** an inventory of the behavior of a species, has been a starting point for many ethological studies. After making observations of an organism's behavior, ethologists then formulate specific questions about the adaptiveness and function of particular behavioral patterns. A student of Whitman's, Wallace Craig (1876–1954), defined two key categories of behavior patterns from his work with doves and pigeons. The first category includes the variable actions of an animal (such as its searching behavior to find food, a nest site, or a mate) and are called **appetitive behavior.** The second category includes stereotypical actions that are repeated without variation (such as the act of mating or the killing of prey) and are called **consummatory behavior.**

The ethological approach is used in another major area of inquiry: the determination of how key stimuli trigger specific behavior patterns. J. von Uexkull (1864–1944) demonstrated that animals perceive only limited portions of the total environment with their sense organs and central nervous systems. This sensory-perceptual world was termed the **Umwelt** by von Uexkull. Among the stimuli recorded by the sense organs, certain specific cues that ethologists call **sign stimuli** trigger particular stereotyped responses called **fixed action patterns** (FAPs). For example, the female three-spined stickleback fish's enlarged belly triggers courtship behavior in male sticklebacks (figure 2–5).

Credit for the synthesis of these early findings and for the further development of modern ethology belongs largely to two men, Konrad Lorenz (1903–1989) and Niko Tinbergen (1907–1988). Lorenz pioneered studies of genetically programmed behavior and investigated the importance of specific types of stimulation for young animals during critical periods of early development. Modern ethology's concern with four areas of inquiry—causation, development, evolution, and function of behavior—developed from a scheme proposed by Tinbergen (1963). (As psychologist Thomas McGill has noted, the first six letters of the alphabet can be used to remember these questions: *A*nimal *B*ehavior: *C*ausation, *D*evelopment, *E*volution, *F*unction.) Recognition for animal behavior as an independent discipline came in the fall of 1973, when the Nobel prize for physiology and medicine was awarded to three ethologists: Konrad Lorenz, Niko Tinbergen, and Karl von Frisch (1886–1982). Von Frisch had conducted research on animal sensory processes and made major contributions to the study of bee behavior and communication.

Modern ethology is characterized by varied types of investigations ranging from more traditional observational studies in natural environments (Geist 1971; Joerman et al. 1988) to experiments on the physiological bases of behavior (Bentley and Hoy 1974). The latter study is indistinguishable from those conducted by many physiological psychologists. Some ethologists work primarily with behavior genetics and the evolution of behavior (Manning 1971; Gerhardt 1979; Ukegbu and Huntingford 1988) or explore the relationships between hormones and behavior (Hinde 1965; Truman,

FIGURE 2-5   Courtship of male and female three-spined stickleback
The enlarged belly of the female three-spined stickleback fish (top) is a sign stimulus for the male of the species (bottom) to court and to entice the female to enter the nest he has built.

Fallon, and Wyatt 1976) or the nervous system and behavior (Nottebohm 1981; Rose et al. 1988). Others work on research problems in the field or in a laboratory setting that resembles the natural habitat. By employing experimental manipulation to test specific hypotheses, Kummer (1971) investigated the effects that transplantation of individuals had on the social behavior of baboons, Wickler (1972) studied the significance of color patterns in fish, and Gowaty and Wagner (1988) tested the aggressive behavior of eastern bluebirds.

Since the mid-1950s, the distinctions between ethology and comparative psychology have been slowly disappearing. Several events have opened communication between scientists of the two approaches. These events included the biennial meetings of the International Ethological Conference, many cross-visitations between researchers in Europe and America, and the publication of a number of international animal behavior journals. A common approach, which started with the notion of species-typical behavior, is beginning to emerge from this exchange of information. **Species-typical behavior** involves actions and displays that are broadly characteristic of a species, and that are performed in a similar manner by all its members. The autobiographical sketches of many leaders in animal

behavior (Dewsbury 1985) include a variety of perspectives, and provide excellent insight into the way the various approaches have independently developed, and how they have recently coalesced into a unified approach to the study of behavior.

At least one major long-standing controversy in animal behavior has been largely resolved during recent years. Early ethologists believed that much of an animal's behavior was instinctive or preprogrammed and was not affected to any great extent by experience. Psychologists claimed that learning and experience were the major determinants of behavior. Today, most animal behaviorists believe that neither of these viewpoints is entirely correct. Instead, as we shall see in chapter 9, the current focus is on the interaction of genotype, physiology, and experience, as the determinant of behavior, and on how the degree of genetic and experiential determination differs among animal species.

BEHAVIORAL ECOLOGY

In the past five decades a third approach to the study of animal behavior has emerged. **Behavioral ecology,** with origins in zoology, examines the ways in which animals interact with their environments, and the survival value of behavior (Morse 1980; Krebs and Davies

---

FIGURE 2-6   Factors affecting the habitat selection of a rabbit

Limitations and constraints that have come about through evolution determine the range of habitats and diets for the rabbit. Developmental and experiential factors influence the immediate choices of the animal.

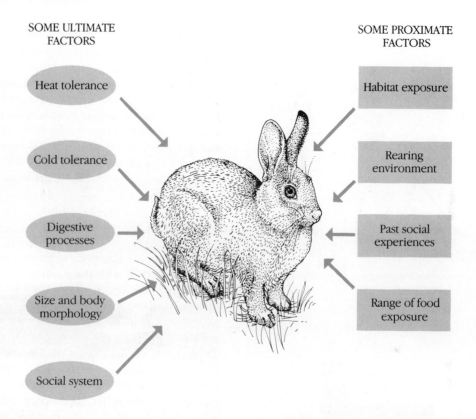

SOME ULTIMATE FACTORS

Heat tolerance

Cold tolerance

Digestive processes

Size and body morphology

Social system

SOME PROXIMATE FACTORS

Habitat exposure

Rearing environment

Past social experiences

Range of food exposure

1987). "Environment" as used here includes animals of the same species (conspecifics), other animals within the same ecological community, plants, and inorganic physical features of the habitat.

Behavioral ecologists are concerned with both ultimate and proximate questions about behavior. Suppose that we are interested in the behavior and habitat selection of a rabbit living on the edge of a large field (figure 2–6). Ultimate constraints that are operating in this instance would include the physiological tolerances of the rabbit: variables like temperature and moisture level restrict the rabbit to certain environments, and its digestive system can break down only certain types of foods, primarily vegetation. Proximate factors would include the rabbit's past experience with types of habitat and foods, which may lead to particular preferences for home site and diet. Ultimate factors establish the limits, and proximate factors affect the behavior of an animal within those limits.

Behavioral ecologists, trained primarily in zoology, ecology, and related fields, are also greatly influenced by the methods of comparative animal psychology. Behavioral ecologists often begin a field investigation and define questions about, for example, population regulation or predator-prey relations. Does the predator maximize its energy intake by utilizing some form of optimal foraging strategy? Certain aspects of the overall investigation (what are the most important features of the prey for predator recognition and detection?) may require experiments more systematic than those that can be done in the field setting. Thus, as behavioral ecologists, we might bring specific, testable hypotheses into the laboratory or controlled outdoor setting where the experiments are conducted. Attempts can then be made to relate laboratory findings to what is known about the animal in its natural field setting.

FIGURE 2-7   Shore crabs select mussels for food
(a) Shore crabs select those sizes of mussels that provide the
best rate of energy return more often than other sizes,
though (b) they do eat mussels of a variety of sizes.
Source: Data from R. W. Elner and R. N. Hughes, "Energy
Maximization in the Diet of the Shore Crab, *Carcinus Maenas*," in
Journal of Animal Ecology, 47:103–16. Copyright © 1978 by Blackwell
Scientific Publications Ltd, Oxford, England.

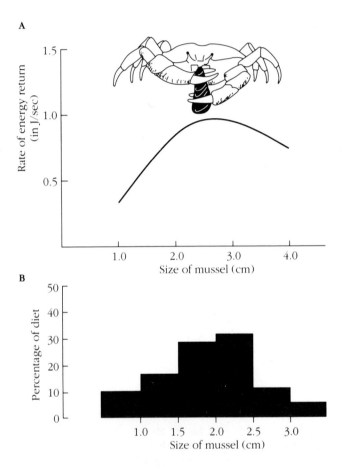

the time it takes to find food, and the ability of the predator to handle the prey, can influence prey selection.
We will explore these and other aspects of feeding in
chapter 17. We'll look at similar factors that influence
choices, and the selection of the most efficient pattern
for animal habitat selection in chapter 16.

Among the investigations in behavior ecology,
those of Emlen (1952a,b) on bird behavior and energy
budgets, Davis (1951) on population biology of rats, and
King (1955) on the relationship of prairie dog social behavior to habitat, were notable for the way they established topical areas for research work within the
developing discipline. In more recent years, topics that
have received particular attention by investigators include foraging strategy (Krebs 1978), the ecology of sex
and strategies of reproduction (Askenmo 1984; Clutton-
Brock 1988; Clutton-Brock et al. 1979), and social systems in relation to ecology (Christenson 1984).

SOCIOBIOLOGY

The most recent approach to animal behavior reached
maturity in 1975 with the publication of E. O. Wilson's
Sociobiology: The New Synthesis. **Sociobiology** applies the
principles of evolutionary biology to the study of social
behavior in animals. Sociobiology is a hybrid of behavioral biology (from the ethological perspective, with an
emphasis on ultimate questions), and the study of social
organization (with an ecological perspective) (Wittenberger 1981; Trivers 1985). As we shall see later, sociobiology, like the other approaches to animal behavior,
relies heavily on the comparative method. Diverse
groups of animals, living in a wide variety of habitats,
are examined to find similarities and differences in their
social systems. These examinations show if any general
patterns explain the social behavior of a species. Thus,
for example, we note that in some species of birds, young
born in one year may not breed the second year, but
help their parents rear the second year's brood. In other
instances, adult birds that have lost their mate (or their
clutch) may help close relatives rear young. These social
systems that involve helping at the nest occur in Florida
scrub jays (Woolfenden 1975), African white-fronted
bee-eaters (Emlen 1984), and acorn woodpeckers in the
western United States (Koenig, Mumme, and Pitelka
1984). Investigations of these and similar social systems
in birds reveal that nests with helpers are more successful—the number of young fledged is higher.
Helping behavior would appear to be using energy to
assist in rearing of offspring that are not the helper's

For an example of an investigation that used the
behavioral ecology approach, consider that done on the
prey-catching behavior of the shore crab (*Carcinus
maenas*). When these animals are given their choice of
what sized mussel to consume (figure 2–7), they select
the size that provides them with the highest rate of
energy return (Elner and Hughes 1978). Notice that although the crabs do eat mussels in a variety of sizes,
they may avoid the larger mussels because of the extra
time and energy needed to crack open the shells. The
wide size range that the crabs eat may represent a com-
promise: lots of time spent searching for just the right
sized mussel would be inefficient. Thus, we see that both

own progeny. How can such behavior evolve? Sociobiologists are interested in exactly that question and also what the advantages are for the individual birds if it helps versus the advantages if it does not help.

The various theories and concepts that constitute sociobiology have their roots in many earlier works. Among the most significant are the writings of Williams (1966) on natural selection and the concept of adaptation, Trivers (1971, 1972) on the evolutionary aspects of altruism and parental behavior, and Hamilton (1964, 1971) on the genetic theory underlying the evolution of social behavior. Studies conducted under the general heading of sociobiology include those on altruism in ground squirrels (Sherman 1977), on strategies for reproduction in damselflies (Waage 1979), on parental investment in wasps (Werren 1980), and on mate choice in American kestrels (Duncan and Bird 1989).

Since the mid-1970s sociobiology has had a significant influence on research in animal behavior. Faced with the challenge of devising new research questions and new methods to test aspects of sociobiological theory, investigators have re-examined older data in light of new predictions. One area of prediction and hypothesis—and the source of considerable controversy—is the application of sociobiology to *Homo sapiens*. Some sociobiologists argue that the principles used to investigate the social behavior of animals can be applied to investigate the social behavior of humans. Other individuals argue that sociobiology is merely a form of biological determinism. A complete resolution of this controversy is probably impossible.

SUMMARY

Archaeological evidence indicates that early humans had a practical working knowledge of the behavior of animals, particularly of those animals that were potential food sources or predators. By Greek and Roman times, writers like Aristotle and Pliny recorded extensive observations about and deductions from natural phenomena.

Three developments of the last half of the nineteenth century contributed to the emergence of animal behavior as a scientific discipline. First, Darwin and Wallace, each working from his own data and from ideas of previous investigators, independently put forth the theory of evolution by natural selection. Second, Romanes pioneered the development of the comparative method and used it initially to study mental evolution. Third, with Mendel's work on inheritance and the rediscovery and development of his findings at the turn of the century, modern theories of genetics and evolution emerged.

Derived from these diverse beginnings, four major approaches characterize current studies of animal behavior. Investigations of the mechanisms controlling behavior have historically been conducted primarily by *comparative animal psychologists and physiologists*. Although much of the early psychological research relied heavily on introspection and inference, these methods were later replaced by systematic, objective observations and replicable experiments. Modern animal psychologists explore such areas as physiological control of

behavior, sensation and perception, learning processes, and behavior genetics.

Ethology encompasses studies of the functional significance and evolution of behavior. Behavioral traits, like physical or physiological traits, are viewed in evolutionary terms and are thus subject to natural selection. Traditionally, ethologists have made many of their research observations in a natural setting. The research objectives and methods of modern ethologists range from observational studies and field experiments conducted to assess the function of behavior patterns, to investigations of the physiological bases of behavior.

Behavioral ecology generates studies that examine the biological relationships between an organism and its living and nonliving environment. The questions are asked from an ecological and sometimes evolutionary viewpoint, and investigations conducted in both field and laboratory settings utilize systematic, controlled experimentation.

Sociobiology has emerged as a new approach to the study of animal behavior in the past two decades. It applies principles of evolutionary biology to the study of social behavior in animals. Sociobiological research concentrates on field observations of social groups of animals in their natural environments, and the evolution of social behavior.

Discussion Questions

1. Select a type of bird, mammal, or insect that is easily visible during daylight hours and watch the animals of that species for several hours. Record as many as you can of the different behavior patterns, their frequencies, and patterns of occurrence.

2. Consider yourself a prehistoric *Homo sapiens.* What characteristics and habits of the other animals in your environment would you want to know?

3. You have been asked to study the behavioral biology of the zinger (*Zingus zingu*), a snake that lives in the Arizona desert. Formulate several questions about the zinger you would want to answer using each of the four major approaches to the study of animal behavior discussed in this chapter.

4. The animal's behavior, unlike the animal's skeleton, does not fossilize. What types of evidence would you look for in the fossil record to provide you with information concerning the behavior of animals?

Suggested Readings

Dewsbury, D. A. 1984. *Comparative Psychology in the Twentieth Century.* Stroudsburg, PA: Hutchinson Ross.
A thorough and scholarly rendering of the history of modern animal psychology. A nice blend of the individuals, experiments, and theories as they developed in this century.

Dewsbury, D. A. 1985. *Leaders in the Study of Animal Behavior.* Lewisburg, PA: Bucknell University Press.
A compilation of autobiographical sketches from those who have been the primary developers of the discipline of animal behavior, on both sides of the Atlantic Ocean. This is must reading for those who like to know how scientists get started in their field of endeavor and how ideas are developed and tested.

Klopfer, P. H. 1974. *Introduction to Animal Behavior: Ethology's First Century.* rev. ed. Englewood Cliffs, NJ: Prentice-Hall.
Written as a textbook and revised once, this short book gives a solid introduction to the ethological perspective on animal behavior.

Mortenson, F. J. 1975. *Animal Behavior: Theory and Research.* Monterey, CA: Brooks/Cole.
A compact treatment of the origins and development of the various themes that together make up much of modern animal behavior.

3

APPROACHES AND METHODS

For any observations of animal behavior (for example, the annual reproductive cycle of red-winged blackbirds [*Agelaius phoeniceus*] outlined in figure 1–1) we need to develop a framework in which to formulate testable hypotheses. What questions can we ask about the redwings' territoriality, social system, and habitat preference? What about the internal events that take place in males and females during the changing seasons? Which questions are the most significant? In the first section of this chapter, we examine in detail how the various approaches to the study of behavior have posed and answered experimental questions, and we consider the advantages and disadvantages of each approach.

Variation in how individual animals behave creates method problems for the investigator. For example, some red-winged blackbirds perform territorial boundary displays repeatedly in the face of an intruder (figure 3–1), while others display much less frequently under the same circumstances. Nest sites used by redwings may vary in their height above ground and the types of plant material used in construction. How can we design experimental procedures to account for this variation? In the second part of the chapter, we examine some general principles for conducting behavioral research.

Finally, what are the problems and pitfalls of experimental research in animal behavior? An awareness of the limitations of the various methods and techniques will help us to evaluate the results of behavioral

FIGURE 3-1   Territorial display by male red-winged blackbird
Redwings display when faced with an intruder attempting to enter the territory. The wingspread display is often associated with the conkaree call.
Source: Photo by Sarah Lenington.

investigations and to formulate better questions about behavior patterns we observe. The principles for research, and the problems and pitfalls of the experimental approach, are presented in the final section of the chapter. We believe that students of animal behavior must have some understanding of how to conduct research if they are to interpret what they are reading properly, and if they are to participate in field and laboratory exercises, for these principles are part of a basic course in animal behavior.

APPROACHES

ETHOLOGY

We can define **ethology** as the biology of behavior: the exploration of functional and evolutionary questions, and the mechanisms underlying *why* an animal exhibits certain behavior patterns under certain circumstances. The first step in studying an animal is the compilation of an **ethogram,** an inventory of the behavior performed by animals of the species under investigation. Behavior can be divided either into broad descriptive categories (e.g., courtship, nesting, sleeping, and feeding), or into more restricted units (e.g., specific patterns shown during various phases of courtship [figure 3–2]). The ethogram serves as a basis for posing questions about the adaptive value, ecological importance, and regulation of the various behavior patterns.

AN ETHOLOGICAL STUDY. By examining a study in detail, we can see how ethologists use observations from nature and field experiments to formulate and test questions. Niko Tinbergen studied the behavior of gulls for many years. Among the interesting problems he in-

---

FIGURE 3-2　Ethogram showing courtship pattern in orange chromide (*Etroplus maculatus*)
An ethogram can be compiled for all behaviors or for selected behaviors of a species.

a. *Charging:* an accelerated swim of one fish toward another

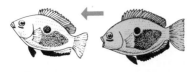

b. *Tail Beating:* an emphatic beating of the tail toward another fish

c. *Quivering:* a rapid, lateral, shivering movement that starts at the head and dies out as it passes posteriorly through the body

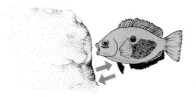

d. *Nipping:* an O-shaped mouth action that cleans out the (presumptive) spawning site

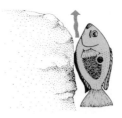

e. *Skimming:* the actual spawning movement whereby the fish places its ventral surface against the spawning site and meanders along it for a few seconds

vestigated was a curious phenomenon he first noted in his early observations of gulls (Tinbergen 1963; Tinbergen et al. 1962). After a young black-headed gull chick *(Larus ridibundus)* hatches and frees itself from the shell, the parents carefully pick up the remnant pieces of shell in their bills, fly off to some distant location, and drop them (figure 3–3), a behavior also seen in many other bird species. Why do parent birds remove the eggshells? What selection pressures could have led to the evolution of such a behavior?

Tinbergen and his colleagues considered several explanations of the behavior: The sharp edges of the shells might injure the chicks, or the shells might clutter the nest and hamper the parents' attempts to brood and feed chicks. These explanations seemed unlikely, because the eggshells of black-headed gulls are thin and easily crushed. The investigators noticed that the outer surface of the eggshells was mottled—but the inside surface was white. Although the exterior camouflages the eggs during incubation, the bright white interior may attract potential predators, such as crows *(Corvus corax)* or herring gulls *(Larus argentatus)*, which rely on visual cues for detecting prey. A more attractive hypothesis is that the parents remove the white eggshells to protect their black-headed young. The kittiwake *(Rissa tridactyla)*, a related species, does not remove the shells after chicks hatch; the chicks are white and thus not camouflaged at all. However, kittiwakes nest on cliffs in regions where these are few predators.

TESTABLE HYPOTHESES. A testable hypothesis arose from these observations: An eggshell left in a nest exposes the brood to a higher rate of predation. To test this hypothesis, Tinbergen and his coworkers performed field experiments in a large gullery in England. They scattered eggs of the black-headed gull over the dunes just outside the gullery. Some had the natural mottled coloration, and others were painted white. They watched the dune site from a blind and recorded the predation rates by crows and herring gulls for both egg types. The results, shown in table 3–1, clearly indicated that the white eggs were taken in greater numbers by both species of predators.

Tinbergen and his colleagues designed a second experiment as a further test of their hypothesis. They laid eggs out in the dune valley, and placed half-broken eggshells at varying distances from the whole eggs. The results revealed that the greater the distance between the whole egg and the half-broken eggshell, the lower the probability of detection of the whole egg (table 3–2). These and several additional experiments substantiated the idea that it is advantageous for adult gulls to remove the shells when the chicks hatch. Selection should favor those gulls that remove the shells and thus should eliminate a possible means for predators to detect the nest.

TABLE 3-1 Predation of black-headed gull eggs

	Eggs Painted White	Eggs with Natural Mottled Coloration
Taken by crow	14	8
Taken by herring gull	19	1
Taken by others	10	4
Total taken	43	13
Total not taken	26	55

Source: Data from Tinbergen (1963).

TABLE 3-2 Number of eggs taken and distance between whole gull eggs and broken shells

	15 cm	100 cm	200 cm
Eggs taken	63	48	32
Eggs not taken	87	102	118

Source: Data from Tinbergen (1963).

FIGURE 3-3 Young gull chick beside its shell fragments The chick's parents remove the shell fragments soon after hatching. Tinbergen and his colleagues showed that the white color of the inside of the shell, which normally shows when the shell is broken, attracts more predators than the mottled outside of the shell.

ADVANTAGES AND DISADVANTAGES. Ethologists proceed from observations usually made in a natural setting to the formulation of hypothetical explanations of the functions and the evolutionary development of the behavior. The best explanations are those that lend themselves to experimental testing, such as the reason for eggshell removal by black-headed gulls. One major advantage of the ethological approach is that it is based on an evolutionary perspective: The functions and adaptive significance of behavior can be discerned best under field conditions with an understanding of how evolution affects morphology, physiology, and behavior.

There are two important disadvantages of the ethological approach: First, working in the natural setting, ethologists cannot control the environment and lack knowledge of the observed animals' past history. Since factors like weather, habitat differences, or seasonal changes cannot be manipulated by the investigator, the functional significance of observed variations in behavior may be difficult to ascertain. (We should note, however, the investigator may turn certain of these situations to advantage by systematically gathering data under the different weather or habitat conditions.)

Second, providing a hypothetical evolutionary explanation for a behavior is often easy, but devising and conducting an experiment that will thoroughly and convincingly test the hypothesis can be difficult. Not all behavior patterns lend themselves to observation and analysis. This may be one reason we tend to look at reproduction when we are really interested in adaptation. For example, some African wild dogs (*Lycaon pictus*) prey on various ungulates (hoofed herd animals such as gazelles or zebras), and exhibit cooperative hunting behavior. We can hypothesize that this behavior evolved as a means by which a group of dogs could succeed in isolating, running down, and killing a larger prey. A single dog could achieve this feat rarely, if at all. Testing such a hypothesis directly, however, is impossible, because we cannot retrace the steps in the evolutionary sequence. Given the right set of observational data, we might conclude that single dogs or pairs of dogs were less successful hunters than larger groups, but these data would be only suggestive and would not provide a direct test of the original hypothesis about the evolution of group hunting.

COMPARATIVE PSYCHOLOGY

COMPARATIVE ANALYSES. **Comparative psychology** can be defined as the discipline devoted to comparative studies of behavior in animals and, in some instances, humans. The behavior patterns most often studied are those that involve learning and development. The primary emphasis is on "how" questions about the mechanisms that underlie observed behavior patterns. One way to begin research in comparative psychology is by identifying and characterizing the classes of behavior patterns in two or more species. Comparative analyses lead to the discovery of relationships between various types of behavior and species. A second way to begin is by selecting a species that is the most appropriate for investigation of a particular problem. Here, as several investigators have noted (Beach 1950; Hodos and Campbell 1969; Lockard 1971; Dewsbury 1984), the frequent use of the domestic rat has produced criticism of comparative psychology from both inside and outside the discipline.

A PSYCHOLOGICAL STUDY. By examining a specific research project, we can get a clearer look at the thinking and methods, and the advantages and disadvantages, of comparative psychology. Renner (1987) wanted to test whether the housing environment experiences of male rats would affect their exploratory behavior toward objects. Male rats were reared under one of two conditions; they were either group-housed in a large, 75 × 75 × 46 cm cage with various objects, some of which were changed daily; or individually housed in a small 30 × 15 × 15 cm cage with no objects. The former environment was termed the "Enriched Condition," and the latter, the "Impoverished Condition." The rats were reared in littermate pairs for ninety days, then one of each pair was randomly assigned to one of the two treatment conditions. The rats were housed as described for thirty days and then tested.

The test apparatus consisted of a startbox and four chambers arranged in a semicircle in front of the startbox, and each chamber contained an object for potential interaction. Among the measured variables were the average length of interaction between rat and object, and the number of different behavior patterns exhibited by the rats while interacting with the objects (figure 3–4). The results indicated that rats housed in the Enriched Condition interacted longer, and exhibited a

FIGURE 3-4   Male rat interactions with stimulus objects
(a) Mean length of bouts of interaction with the stimulus
objects in the arena, and (b) number of different behavior
patterns displayed in interacting with stimulus objects.

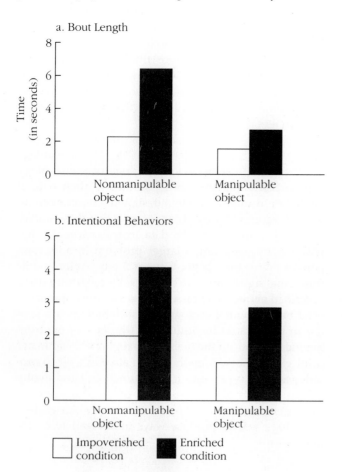

a. Bout Length

b. Intentional Behaviors

□ Impoverished condition   ■ Enriched condition

greater number of behavior patterns with the objects
than did rats from the Impoverished Condition. Both
the quality and quantity of exploratory behavior were
affected by the housing conditions during the previous
thirty days. This particular investigation illustrates an
attempt to elucidate some underlying mechanism to explain observed behavior—the "how" (proximate) questions about animal behavior.

ADVANTAGES AND DISADVANTAGES.   Many features of
this study characterize investigations in comparative
psychology and illustrate some of the advantages and
disadvantages of this approach. First, the experience and
environment of the subject animals can be controlled,
and the investigator can draw exact or nearly exact conclusions about the effects of conditions on behavior.
Second, one or two variables can be manipulated in a
systematic and replicable experimental design to ascertain their effects without the potentially confounding
influences of other uncontrolled variables. Finally, the
investigation is concerned primarily with development, in this instance, of a preference in the rats for
novel or familiar environments, and the proximate
mechanisms underlying this preference.

Certain drawbacks are inherent in the comparative
psychology approach. The animals, as those in the example, are often laboratory stock, housed in an environment that is simple compared with the natural
habitat. How does domestication of rats or other laboratory animals affect behavior, and what are the effects
of housing them in unnatural conditions? Comparisons
of wild and laboratory rats suggest that some behavior
patterns are affected but others are not (Price 1972, 1984;
Price and Huck 1976; Price and Belanger 1977; Boice
1972). For example, wild rats are more active than their
domestic counterparts, and laboratory strains exhibit
different learning patterns than do those captured in
the wild. Also, maternal behavior, including pup retrieval and nest construction, differs. In general, it appears that during the process of domestication, selection
may have produced a greater flexibility in behavior. Investigators must constantly be aware that laboratory
confinement or housing in unnatural conditions may
have changed their subject's behavior.

The comparative psychologists' approach may cause
them to not view their experimental methods and results in a complete context involving ecology and evolution. We should be concerned with both physiological
and evolutionary explanations for observed patterns of
behavior in a species. For example, in most studies,
knowing something about the natural environment of
a species and its recent phylogenetic history is necessary. A species that has narrow diet preferences may not
learn well when numerous types of foods are used as
reinforcement. In the study just discussed we might

want to question the ecological relevance of objects in the rat's natural world. As animal behaviorists, we must be constantly concerned about the *Umwelt*—the sensory-perceptual world of the species we are investigating.

Both the ethological and psychological perspectives have a great deal to contribute to the investigation and understanding of animal behavior. Because both "how" and "why" questions must be tested, an integrated approach that combines the methods of each discipline can provide the best total analysis of behavior.

BEHAVIORAL ECOLOGY AND SOCIOBIOLOGY

In the past two decades, new frameworks and approaches for animal behavior study have emerged. Behavioral ecology and sociobiology can be characterized in similar ways; they both involve a great amount of field work with some laboratory tests where appropriate; they both involve experimental manipulation under field conditions; and they are based in a reasoning process that includes both the ecological realm of the species under study, and concern with the way natural selection has shaped observed behavior. An example of a recent study in behavioral ecology will provide some insights into these approaches; we will learn considerably more about both behavioral ecology and sociobiology in the second half of this book.

A BEHAVIORAL ECOLOGY–SOCIOBIOLOGY STUDY. How do behavioral ecologists and sociobiologists approach a research problem? Hiebert et al. (1989) investigated the possible relationships between the size of the song repertoire, acquisition of breeding territory, and reproductive success in song sparrows *(Melospiza melodia).* They were interested in testing at least two questions: (1) Does the size of the song repertoire of the males, in any way, relate to establishment and maintenance of a territory? (2) Is there any relationship between the size of the song repertoire and the number of offspring successfully raised to independence? The study was conducted on Mandarte Island (6 ha) located 75 km south of Vancouver, British Columbia. This population of song sparrows had been studied for a number of years and most of the birds in the population were known individuals, marked with color bands on their legs. Hiebert and her colleagues visited Mandarte Island each spring from 1984 to 1987, and recorded the songs of male birds that

were establishing and residing on territories. They also mapped the size of male territories. Observations were also made of nests and eggs, and young were observed every two to five days during the reproductive season. Their data and ensuing analyses enabled them to conclude that males with larger song repertoires retained territories longer and had greater yearly and lifetime reproductive success (measured as the number of offspring raised to independence) than males with smaller song repertoires. Males with larger song repertoires were able to establish territories more rapidly, and their tenure on a territory was longer, than did males with smaller repertoires.

ADVANTAGES AND DISADVANTAGES. One advantage of studies that use behavioral ecology or sociobiology approaches is that animals are studied in their natural habitat or in an environment designed to mimic certain natural features. A second advantage is that these studies generally provide specific data to test particular hypotheses by dissecting a larger problem into its component pieces for a better view of the behavior and its functional significance (as did the song repertoire study described above). The disadvantages of the approaches used by behavioral ecology and sociobiology are similar to those faced by ethologists: the lack of environmental control and the inability to regulate the animal's prior experiences. Sometimes it is also difficult to provide testable hypotheses for the functional and evolutionary aspects of behavior. One of the challenges of the past decade well met by behavioral ecology and sociobiology was to produce ways to ask and test such questions.

DESIGN FEATURES IN ANIMAL BEHAVIOR STUDIES

If we are to conduct studies of animal behavior properly, it is necessary to have a basic understanding of the thought processes used to formulate hypotheses, and design experiments to test these hypotheses. We now examine some principles and problems associated with conducting experiments in animal behavior. To provide a basis for examining these principles, we will consider the design of an investigation of the effect of testosterone on aggressive behavior in male Mongolian gerbils *(Meriones unguiculatus)* (figure 3–5). Our test animals will be normal intact male gerbils, castrated males, and castrated males given testosterone replacement treatments.

FIGURE 3-5   Mongolian gerbil (*Meriones unguiculatus*)
These animals, native to the Gobi Desert and nearby regions in Mongolia and China, have become widely used test subjects in laboratory investigations of behavior. Here two unfamiliar males meet, and one sniffs the other, which freezes.
Source: Photo by Lee C. Drickamer.

FIGURE 3-6   Aggressive behavior patterns in small rodents
Classification of aggressive behavior patterns of small rodents enables scientists to observe and record similar behavior patterns in exactly the same way.

Source: Data from E. C. Grant and J. H. Mackintosh, "A Comparison of the Social Postures of Some Common Laboratory Rodents," *Behaviour* (1963) 21:246–59.

Hamster: nose–nose

Gerbil: left, aggressive; right, submissive

Rat: left, aggressive groom; right, crouch

Mouse: left attacking right

DEFINITIONS AND RECORDS

The first task is to agree on which of the many behavior patterns exhibited by gerbils we will consider to be "aggressive." We begin by observing several pairs of gerbils interacting. By watching and discussing the various behavior patterns exhibited, we may be able to decide which to record as aggression.

All observers must record similar behavior patterns in exactly the same way. (For an excellent discussion of the problems involved in describing behavior, read Drummond, 1981.) To ensure that our observations and data collection remain consistent, we need a permanent record of the behavior. Each investigator can refer to the record when making observations, so that definitions of behavior will not change during the time period of the study, and so that other scientists can use the visual or written definitions to repeat the experiments or to make comparisons between gerbils and other rodents.

We can make films, videotapes, photographs, drawings, and/or written definitions of the behavior patterns to be studied. **Films** or **videotapes** provide an unequivocal record of the patterns recorded as aggressive behavior. However, films have two disadvantages: (1) they are difficult to refer to during daily observations, and (2) other scientists cannot use our definitions without the film.

Photographs or **drawings** are easily referred to during experiments and can be published as part of the permanent record of research. Figure 3–6 illustrates some aggressive behavior patterns commonly observed in rodents.

In addtion to films, photographs, or drawings, we often provide **written definitions** of behavior patterns. In the past decade, two books have appeared that can help us formulate written definitions for behavior (see McFarland 1981; Immelman and Beer 1989). Both volumes contain definitions and descriptions for many behavior patterns, and also various processes and concepts in animal behavior. Thus, for example, in the study of gerbils we will record three patterns as aggressive behavior:

1. *Attack*: One gerbil bites, kicks, claws, or pushes the other, possibly inflicting physical harm.
2. *Chase*: One gerbil vigorously pursues another in order to catch and attack it.
3. *Aggressive Groom*: One gerbil mouths the neck, flanks, or anogenital region of the other.

As rigorous as these definitions may be, we still need to check the agreement between observers about the behavior patterns. Thus it is necessary to conduct a series of **interobserver reliability** tests. Two observers simultaneously watch several pairs of male gerbils interacting, and each observer maintains a separate record of the frequencies of the three kinds of aggressive behavior. Then we compute the correlation between the two sets of data; a high positive relationship between the two data records indicates good agreement between what the two observers have recorded for the behavior patterns. Thus, during the course of data collection, several different individuals can observe and record, and we will not have to worry that each of them is recording something different for the same behavior pattern.

When the results of the experiment are reported, all aspects of the methods used in conducting the investigation must be specified. Thus, another scientist can independently **replicate** our experiment and verify, contradict, or modify our conclusions.

DESIGN OF THE EXPERIMENT

HYPOTHESIS FORMULATION. Once the agreement is reached on what to consider aggressive behavior, our next step is to design the experiment. First, we must formulate specific hypotheses (sing., hypothesis), or questions for testing. Two types of hypotheses are involved in behavioral research: experimental hypotheses and null (statistical) hypotheses.

Experimental hypotheses are ideas that are developed by combining the reported investigations of other scientists with our own ideas. In our present example we may want to know:

1. Is aggression in adult male gerbils dependent upon the hormone testosterone?
2. Do male gerbils begin to fight before or after puberty?

These hypotheses lead us to a series of predictions. For example, if testosterone is critical for adult male gerbil aggression, then eliminating testosterone by castration should lessen or eliminate aggression in castrated males. Further, based on hypothesis 2 we can predict that male gerbils will not fight before the age of puberty, when production of testosterone increases.

It is extremely important to understand that animal behavior research is probabilistic. If two animals engage in a contest for a particular feeding site on a particular day, animal *A* may be more successful than animal *B*. But, on another occasion *B* may be more successful than *A*. The reasons for this reversal are not readily apparent or easily explained. In another example, consider the problem of measuring how fast a springbok (*Antidorcas marsupialis*) can run a hundred meters when fleeing from a cheetah (*Acinonyx jubatus*). Some antelope will run faster than others. To generalize about the speed of all antelope of a particular species, we need to average the speed of many animals and determine how much variation in speed can be expected among members of the test group. Clearly, an exact determination of fighting ability or running speed cannot be made, so animal behaviorists must deal with probabilities and predictions about behavior. Although we will not deal directly with statistics in this text, a thorough knowledge concerning statistics and experimental design is necessary for research in animal behavior. For now, a few basic principles can provide us with a background for understanding and critiquing animal behavior research.

The **null hypothesis** proposes that the frequencies of animals' behavior patterns do not differ significantly when the animals are given different experimental treatments. The null hypothesis cannot be proved; it can only be accepted or rejected using a statistical test procedure that provides a basis for interpreting answers to the *experimental* hypothesis. Our null hypothesis might be: There are no differences in the levels of aggression of normal intact male gerbils, of castrated male gerbils, or of castrated male gerbils given daily replacement injections of synthetic testosterone to replace the hormone normally secreted by the testes.

From the null hypothesis, we derive alternative outcomes for the experimental test and examine their interpretation. If indeed there were no significant statistical differences between all the treatment groups, our null hypothesis would be accepted. We would then return to experimental hypothesis 1 above and its predictions. The major prediction would be incorrect: no basis exists from the test data to claim that testosterone plays a critical role in male gerbil aggression. However, if the intact males and those receiving daily injections of synthetic testosterone exhibited statistically significantly higher levels of aggression than castrated males, we could reject the null hypothesis. Returning once again to the experimental hypothesis, we would conclude that the test revealed a key role for testosterone affecting aggression in male gerbils. Regardless of whether we accept or reject the null hypothesis, there may be important biological conclusions to be drawn from our experimental finding. (For expansion of these ideas and a discussion of the distinctions between hypotheses and deductions see Moore, 1984).

DESIGNATION OF VARIABLES. The next step in a research project is to establish the exact parameters to be investigated: the independent and dependent variables. **Independent variables** are the factors that the investigator has manipulated to define the treatment groups and conditions of the experiment. To test the first question regarding testosterone in male gerbils, at least three treatment groups are necessary: (1) normal intact adult male gerbils; (2) castrated adult male gerbils; and (3) castrated adult male gerbils given daily injections of synthetic testosterone. The outline of the experiment might look like that shown in table 3–3. In addition to establishing the treatment groups, the investigator must systematically control many other as-

pects of the experiment, including the ages and previous experiences of the gerbils, the length of the observation period, and the size of the arena in which the males interact. The investigator should hold these factors constant for all experimental test groups, so that the *only* manipulated factor in the design is the presence or absence of the hormone testosterone. In the present study, we will pair each gerbil with another male from the same treatment group in an open circular arena one meter in diameter for a ten-minute observation period.

The measures of behavior patterns that are observed and recorded are the **dependent variables.** In our experiment, the frequencies of the three aggressive behavior patterns will be the dependent variables.

CONTROL GROUPS. In addition to the design features discussed earlier, at least three other considerations arise in the course of setting up an experiment. Many experiments require control groups in order to provide a baseline for comparison with other experimental groups. A **control group** is an observed, unmanipulated set of test subjects, used so the investigator can see what animals do *without* the experimental condition. In the gerbil experiment, the intact males can provide data on normal gerbil aggression. The frequency of aggressive behavior by the pairs of castrated males and by the pairs of castrated males given testosterone injections can be compared with that of the pairs of normal intact males of the control group.

In other instances, a control group may serve to demonstrate that some portion of the experimental treatment procedure has not produced an unwanted effect that confounds the results. The castration surgery performed on some test subjects involves anesthesia and an incision in the scrotum to remove the testes; the surgery itself may have altered aggressive tendencies. Cas-

TABLE 3-3 Outline of experiment to test effects of testosterone on aggression in adult male gerbils
Groups 4 and 5 serve as controls for manipulations performed on gerbils in groups 2 and 3 to ascertain that the treatment, not the manipulation, affects aggressive behavior.

	Behavior Patterns (Frequencies)		
Treatment Groups	Attack	Chase	Aggressive Groom
1. Normal, intact males			
2. Castrated males			
3. Castrated males given testosterone injections			
4. Sham-operated intact males			
5. Castrated males injected with oil			

trated males' aggression levels may be different because of the decrease in blood testosterone levels, or because of the surgery. Thus we must add a fourth treatment group, sham-operated males (table 3-3, group 4). All surgical processes except actual removal of the testes are carried out on these males. If the surgical processes have no effect on aggression, then we would expect sham-operated males to exhibit the same amounts of aggressive behavior as do normal, intact males. Unless we have a control group, we will not be able to distinguish whether any observed behavioral effects in the testosterone-replacement therapy group are the result of the hormone or of the injection process. Thus we need to add a fifth treatment group of gerbils that are injected with peanut oil, which is the vehicle for the testosterone. For each manipulation in an experimental design, we need to determine what, if any, control groups are appropriate to the investigation.

Control groups are not necessary in all behavioral studies. For example, in research that simply describes the behavior patterns of various animal species, no control groups are needed. Nor would control groups be necessary in an assessment of a particular behavior pattern among groups of one species living in different habitats—for example, grooming behavior in groups of squirrel monkeys (*Saimiri* spp.). In this situation we would make comparisons between the monkeys in various habitats and draw conclusions about the relationship of grooming patterns to habitat features.

INDEPENDENT DATA POINTS.   The need to gather numerical pieces of information that are independent of each other is probably the most difficult problem of behavioral biologists and is based on both biological and statistical arguments. Behavioral data independence in our gerbil experiment can be considered in several ways. For any pair of gerbils, the behavior of one male will definitely influence the behavior of the other male. The principle of independent data points dictates that data from both males cannot be used as two distinct points in the same analysis. There are at least two mathematical solutions to this problem. We can combine the records of aggressive behavior for the two gerbils to produce one score for each pair. Or, data for "winners" and "losers" could be analyzed separately, but then we would need pre-established criteria for defining winners and losers.

FIGURE 3-7   Canada geese feeding on ground beside pond
The behavior of each bird may be influenced by actions of the other birds in the flock. Information gathered on the behavior of an individual bird generally cannot be considered independent of similar data on another bird in the same flock.
Source: Photo by R. Gates.

In the gerbil investigation, we will use each animal only once, because the aggressive performance of the male the second time it is tested is partly a function of its experiences in the first interaction. After being severely attacked in the first encounter, some males become submissive during all subsequent interactions, and erroneously low estimates of aggression frequencies would result. In contrast, other males might become hyperaggressive after winning the first encounter; reusing these males would result in erroneously high estimates of aggression frequencies.

Another aspect of the problem of independent data points is related to animals living in groups. For example, in an investigation of adult female goose feeding rates, the total number of females and males living in the flock could influence the feeding rate of each individual female (figure 3-7). The independent units for data collection and analysis in this example are flocks; an average feeding rate for all females in a flock would provide data for making comparisons between flocks of different sizes or flocks resident in different geographical locations or habitats. Assuming that each of the females in a single flock represents a separate data point would be incorrect: the behavior of each is not necessarily independent. The improper assessment of what constitutes an independent data point or observation for analysis is perhaps the most frequent problem with studies of animal behavior. Repeated use of the same animals for multiple tests without proper statistical analyses, or counting all members of an interacting group as independent of one another results in **pseudoreplication** and can lead to erroneous conclusions. If we wish to design and conduct an experiment properly, we must pay considerable attention to the difficult, but necessary, requirement of ensuring that the data points are independent.

The principle of independent data points is a derivative of one of the critical requirements of most statistical tests utilized by animal behaviorists—that is, each piece of information used in the analysis must be independent of every other piece of information. Special statistical tests have been developed to handle data analysis when the same test procedures are repeated on the same animals over time. These tests do not require complete independence of data points for information gathered on the same animal.

SAMPLE SIZE. A final factor in designing an experiment is the sample size—the number of animals that must be used in the investigation. Each experimental treatment group must contain enough animals to provide a complete and accurate assessment of the behavior. If the experiment is concerned with only qualitative descriptions of behavior, then the sample size should be large enough to permit an observer to record a thorough picture of the behavior, including estimates of average values for the dependent variables, and estimates of the amount of variation around the averages. In addition, most statistical tests require a certain minimum sample size for correct data analysis. Deciding the correct sample size to use for a particular experiment requires some practical experience with both the animal being studied and with the type of statistical test used in analyzing the results. Good animal behaviorists also take into consideration the fact that using too many animals may sometimes be wasting animal lives unnecessarily, which is expensive and can overextend our own limited time and energy.

VARIATION AND VARIANCE

Animals exhibit variation in appearance or performance of most morphological, physiological, and behavioral traits. For example, we noted that red-winged blackbirds exhibit variation in display frequency and in nest site selection.

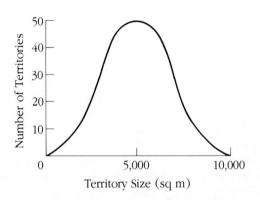

FIGURE 3-8   Territory size in red-winged blackbirds
Many measurable behavioral traits show some form of continuous variation. Territory size in redwings provides such an example. This graph illustrates territory size for a large sample of over 500 birds.

Variation may be either discrete or continuous. **Discrete** variation involves those measures that can take on only certain values. For example, clutch size (number of eggs from a pair in a season) in birds is a discrete measure; a species may lay 1, 2, 3, or up to 6 or more eggs per clutch, but they do not lay 1.5 or 3.6 eggs. Other measures are **continuous;** they can take on any value between some lower and upper limit. The nesting territories of red-winged blackbirds may vary from a few square meters to thousands of square meters (figure 3–8). Values assigned for different birds can be anywhere within this range, and are limited only by how refined we wish to make our measurement.

To assess the amount of variation that occurs for a particular behavior or trait in a sample, we use two principal measures, the mean and the variance. We can calculate the mean and variance for any group of values for a trait regardless of whether the variable is discrete or continuous. The **mean** (X) or central tendency, is the arithmetic average. Thus, from a sample of four clutches of eggs from yellow-shafted flickers we might get values of 2, 4, 6, and 8—with a mean of 5 eggs per clutch. In a second sample of four clutches the values of 4, 5, 5, and 6 eggs would also produce a mean of 5 eggs per clutch.

The variance is an estimate of the amount of deviation of the values in a sample around the mean of that sample; it is, by definition, the average squared deviation of the values from the mean of the sample. We calculate the variance S as,

$$S^2 = \Sigma(x_i - x)^2/N - 1,$$

where X represents the individual values in the sample, x_i is the mean, and N is the total number of values in the sample. So, for our first sample of clutches the variance S is 6.7, whereas for the second sample, the variance S would be 0.7. The means for the two samples are identical, but the variances are quite different; there is much more variation around the mean in the first sample.

We shall refer to the notions of mean and variance in both general and specific terms at a number of junctures throughout the book. Two other measures of the variation around the mean, the **standard deviation** and the **standard error of the mean,** are also often encountered in the animal behavior literature.

CAPTURING, MARKING, AND TRACKING ANIMALS AND ANIMAL SIGNS

Before discussing specific methods used for data collection in field settings, let us digress a moment to examine several related problems—issues that generally arise when we are conducting field studies, but that may also pertain to some laboratory work in animal behavior. These issues concern the methods of capturing animals, ways of marking animals for individual identification, procedures used for tracking animals, and signs left by animals that may assist us with our interpretation of their behavior. For additional techniques and details concerning the methods listed next, the student is urged to consult Lehner (1979) and Southwood (1978).

CAPTURING ANIMALS. To study many animals properly, it is necessary to capture them, even if only briefly. We may need to mark the animals for identification. We may need to make measurements of size, sex, or age. Or, we may wish to bring them into the laboratory for certain portions of our work. There are a variety of methods that we can employ (Eltringham 1978); the exact capture method we should use in a particular situation will depend upon the species and the nature of its habitat. The most common methods involve nets or traps. For many insect species we would use either sweep nets or dip nets. For birds we often use mist nets; nets of fine mesh set up using two poles, not unlike a volleyball net. Mist nets are often set up to intersect the flight of the birds, for example, across a clearing in the woods or along the path of a stream. Larger nets can be used with small cannon to capture flocks of birds at a feeding station or mammals baited to come to a particular site. Live traps for vertebrates and invertebrates are available in an enormous variety of shapes and forms, and can be used in the capture of almost all species of animals. Seine nets are often used to obtain samples of fish populations from a stream or lake. Pitfall traps, cans placed into the ground so their top rims are flush with the ground surface, are used for capturing a variety of invertebrates, reptiles, amphibians, and mammals. In aquatic habitats, electroshock can be used to stun large numbers of fish, which then float to the surface; they soon recover from their stunning experience. Finally, for many mammals, particularly larger beasts such as bears or ungulates, we employ a CO_2 cartridge gun and a tranquilizer dart. Animals shot in this manner are anesthetized for varying periods, and they either wake up naturally, or they are given an antidote when the investigator has finished tagging or measuring them.

ANIMAL IDENTIFICATION.   In much of animal behavior research, we want to be able to identify individual animals (see Stonehouse 1978). Our research questions might include: Which animals are dominant? Which females are mating with which males? Are young animals cared for only by their mother, or do other members of a group share in this process? One way we can identify individual animals is by their natural marks (Pennycuik 1978). Examples of the use of this technique include coat color patterns in zebras (*Equus burchelli*) (Petersen 1972) and giraffes (*Giraffa camelopardalis*) (Foster 1966); differences in bill patterns in Bewick's swans (*Cygnus columbianus*) (Scott 1978); and variation in physical characteristics, natural mutilations, and scars in primates (Ingram 1978), lions (*Panthera leo*) (Pennycuik and Rudnai 1970), and bottlenose porpoises (Wursig and Wursig 1977).

A second technique for individual identification involves tagging animals; the nature and location of such tags will vary with the species (see Stonehouse 1978 for details). Among the various tagging techniques used are metal and plastic leg bands in birds (Spencer 1978; Patterson 1978); fin clips or punches and metal finger-

Powder tracking

Small rodents and other animals can be tracked by covering their fur with fluorescent powder (chapter 3) and then following the trail they leave with a blacklight to provide information concerning their movements (chapter 15) and habitat use (chapter 16).

Source: Photo by Lee C. Drickamer.

PLATE 1

Male stickleback fish (*Gasterosteus spp.*)
Color patterns on the flanks of these fish increase in intensity during territorial boundary disputes (chapter 12). The change in color is under hormonal (chapter 7) and neural control (chapter 6).
Source: Photo by J. J. Tulley, courtesy of F. A. Huntingford.

PLATE 2

Ground squirrel family (*Spermophilus richardsonii*)
Ground squirrels have an annual cycle involving hibernation (chapter 8) and 4–5 months when they are awake and above ground during which time they reproduce (chapter 12).
Source: Photo by Gail R. Michener.

PLATE 3

ling tags for fish (Laird 1978); toe clips, ear punches, dye marks, and tattoos for a variety of mammal species (Lane-Petter 1978); and small dots of paint or dye on invertebrates (Southwood 1978).

Another animal marking technique involves radio-isotopes, either in collars or bands, or in subcutaneous implants (Linn 1978). Because of potential problems with radioactivity and possible effects both on the animals bearing the radioisotope label and the environment, the use of this technique has been curtailed considerably. More recently, Sheridan and Tamarin (1988) have developed the use of radionuclides for identification of individual meadow voles (*Microtus pennsylvanicus*) in order to study longevity and reproductive success; radionuclides are safer and not harmful to the animals carrying the label.

TRACKING ANIMALS. Having marked animals for individual identification, we now might ask: How can the animal be followed over time to study its behavior? We can choose from among several methods. Probably the most popular of these techniques involves radiotelemetry. An appropriate animal is captured and fitted with a collar containing a radiotransmitter and battery or implanted with a similar device. Once released, the transmitter will send out signals that can be detected by a radio receiver and antenna. This sort of technology has been adapted for use with animals ranging in size from small birds, lizards, and rodents, to whales and elephants. The size of the transmitter package and the range over which the receiver can pick up the signal vary with the size of the animal and the needs of the investigator. By following an animal for many days, it is possible to obtain an accurate picture of the area that it uses and the portions of the day when it is active. Radio signals can now be detected by satellites circling the earth; information from the satellite is then fed to a computer at a ground station and animal movements can be analyzed. This technique has been used effectively with caribou (*Rangifer tarandus*) in Alaska (Miller et al. 1975).

For some animals, particularly large mammals or birds in flight, we may use vehicles moving along the ground or airplanes from overhead to follow movement patterns. For many aquatic animals, including many vertebrates (especially fish and whales) and a variety of invertebrates, we can use boats and diving gear to observe both from the water's surface and from below. In all of these situations, photographs or videotapes may aid in gathering information and enhance the processes of identifying the individual and scoring the behavior patterns.

Another technique for following animals has been developed recently and has promise for use with a variety of small animals; to date much of its use has been with rodents (Kaufman 1989). The procedure, sometimes called the "Shake-and-Bake" method, involves using a finely powdered fluorescent dye. The powder can be created in a variety of colors, e.g., magenta, lime green, blue, etc. The animal that is to be tracked is captured and placed into a plastic bag with the powder. After gently shaking the bag for a few seconds, the pelage of the animal becomes covered with the dye. Once the animal is released, it will deposit a trail of dye for four to eight hours wherever it goes. On a night after the animal has been allowed to put down the dye trail, we go out with a special black light that causes the dye to fluoresce, and the trail of the animal can be followed quite easily. In this way, we obtain a record of the travels of the animal; the technique can be used on animals that are either nocturnal or diurnal with equal facility.

ANIMAL SIGNS. Finally, are there any traces, signs, or constructions animals leave that can help us to interpret their behavior? Examples of animal signs that we might expect to find include tracks, fecal material, egg shells, and animal remains. Almost all animals leave tracks of some kind, except those in aquatic habitats. Impressions of feet can be found in mud, sand, or other types of surfaces. We may use tracks to determine the direction of movement of an individual or group, the composition and size of a group, and depending upon whether there are age or sex differences in the nature of the tracks, something about the age and sex of the individuals. Examination of fecal material can tell us something of the diet of an animal. Egg shells may help us identify nest sites and indicate which species are nesting in a particular habitat. The remains of animals might hold clues to the cause of death, or could even tell the tale of an act of predation. Other, possibly more active, signs we might discover would be nests, egg cases and spider webs. Nests of birds can provide considerable information about the site selection and construction of the nest, and when eggs or young are present, about the development of the nest. Spider webs are instructive with respect to their tactics for prey capture: the variety of locations and forms of web construction can be used to deduce information about their prey.

FIELD METHODS FOR DATA COLLECTION

The design features for studies of animal behavior apply equally in laboratory and field situations. One of the early pioneers in conducting animal behavior research, T. C. Schnierla (1950), used an experimental strategy that combined field investigation with appropriate laboratory tests of specific points. He was also a strong proponent of the comparative method and carried out behavioral observations under varied field conditions. Many of the pioneering ethologists, including both Tinbergen and Lorenz, conducted large portions of their research under field conditions.

In recent years, field studies of many animal species have focused considerable attention on research strategies and methods of collecting data under field conditions. Many of the techniques for behavioral observation that were first devised for use with primates have found broad application with many other animal species. We will mention topics like focal-animal sampling and instantaneous sampling only briefly; these and other related topics are discussed in detail by Hinde (1973), Altmann (1974), Dunbar (1976), and Lehner (1979).

Focal-animal sampling (Altmann 1974), involves recording all of the actions and interactions of one particular animal during a prescribed time period. Using this technique, an observer may watch a large number of animals, recording the behavior of each for a short period (e.g., five or ten minutes per animal), or the observer may record the behavior of fewer focal animals, each being watched over a longer time period (hours per animal). As an example of the use of this technique, consider the study conducted by Bercovitch (1986) on dominance rank and mating activity in male olive baboons (*Papio anubis*). The question tested concerned the relationship between male dominance rank and access to sexually receptive females for mating. Intense observations were made on only one or two baboons each day. The females that were the subjects of the focal sampling were in behavioral estrus, exhibiting consort relationships with males. Using focal animal sampling, it was possible to record all of the male partners of the sexually receptive females. When the data were analyzed using information from only adult males that interacted with the receptive females, there was no clear relationship, between dominance status and mating activity. However, if adolescent males were included in the analysis, then a relationship emerged, with dominant males engaging in more sexual activity with receptive females. One advantage of this observation technique is that intense samples can be gathered with a focus on particular animals.

A counterpoint to focal-animal sampling involves scan-sampling; the technique is sometimes also called instantaneous sampling. An observer using this technique watches each animal for only a few seconds at periodic intervals and records the activity(s) that the animal is performing only at the specific time marks indicated by the sampling scheme. The intervals between samples of the behavior of each individual can vary, but generally, they are a few minutes to a half hour. As an example of the use of this technique, consider a study of feeding behavior of two captive groups of lemurs (Ganzhorn 1986): the ring-tailed lemur (*Lemur catta*) and the roughed lemur (*L. fulvus*) housed in semi-natural environments. Scan sampling was conducted on the groups of lemurs using a thirty-minute interval between records of individual activity. The species and part of the plant being eaten were recorded when the lemur was eating—at the time the data was to be recorded. By recording the data in this fashion, the observer was able to make many records on all the lemurs in both groups during all seasons of the year. The analyses of these feeding data indicated that roughed lemurs spent less time eating during the day, had longer feeding bouts, and consumed more fruit than ringtails. One advantage of this technique is that it allows the observer to sample widely across all animals in the group, and across a wide range of behavior patterns,

There are other sampling schemes, e.g., one-zero sampling, sequence sampling (Altmann 1974), that are beyond our present scope. Also not covered are other aspects of correct procedure design for collecting and analyzing data; these may be of interest to some students (see also Hinde 1973; Dunbar 1976).

We should note two final important points. First, despite the absence of rigid laboratory-type control, field studies should not be considered less accurate than laboratory investigations. Second, the same constraints and methods considerations apply whether we are working in the field or in the laboratory.

OBSERVATIONAL METHODS

In addition to the topics covered elsewhere in this chapter and at other appropriate points in the book, several selected issues should be discussed briefly in conjunction with the use of observational methods for studying animal behavior. Most of these issues have particular relevance to investigations conducted in the field, but many points also pertain to laboratory conditions. (For a thorough treatment of these and related topics see Lehner, 1979).

OBSERVING VERSUS WATCHING.

Most of us watch animals on a daily basis—pets, wildlife, or simply other humans. We may notice when something about the behavior of a dog is different than usual or when a particular bird species flocks to the feeder in greater numbers. These visual and mental notes constitute "watching." **Observing** animal behavior involves systematic recording (writing, tape recording, or filming) of the activities of particular animals, usually with specific test questions governing the nature of the data recorded.

Making thorough, proper observations requires that the observer maintain an awareness of some critical considerations. Two of these—the problem of equal observability of all of the subjects under study and the establishment of interobserver reliability—are discussed elsewhere in this chapter. The latter is important both in terms of the agreement between the data recorded by two or more people on the same animals, and may act as a check against possible observer sex or age bias. One observer may be recording a disproportionate number of some actions for a particular age or sex.

Another key consideration for conducting observational research is scheduling the sessions to coincide with the time of day (and year) when the behavior we want to study is likely to occur with reasonable frequency, but not such that a distorted picture is obtained. For example, it may be erroneous to select only the peak periods for the occurrence of a particular class of behavior, such as aggression, as this would give a biased impression; but it would be equally unwise to study the same animals only during their resting period.

Proper observational research requires certain equipment. Generally, binoculars or spotting scopes are needed to identify animals and to see some aspects of their behavior accurately. For some species, it may be necessary to construct a blind or other unobtrusive vantage point in order to record the animals' activities without disrupting their normal patterns. We must always be aware that in both field and laboratory settings, the mere presence of an observer can influence the subjects' activities.

Finally, some form of reliable, accurate scheme of data collection must be devised. Construction of proper data sheets, ways of encoding behavior, and ways to keep track of the identities of subjects require practical experience. Lehner (1979) devotes a section of his book to this subject.

IDENTIFYING SUBJECTS. At least two major problems should be mentioned in conjunction with marking animal subjects. First, some of the techniques just outlined may alter the behavior of the subject, or may influence the appearance or actions of the subject, resulting in changes in the reactions of other animals to the subject (see Burley, Krantzberg, and Radman 1982). This could indirectly influence changes in the subject's behavior. Second, some investigators have adopted the practice by which animals are identified and recorded in their data by names, such as "Swifty," "Old One," and the like. This practice should be avoided. When we name an animal based on its physical appearance or some behavioral trait, we attribute to it certain qualities that will almost certainly result in subsequent observation bias. Even the practice of randomly assigning human names (such as "Ralph" and Sue") to animals can result in the association of particular traits with names; this can also bias what we do and do not record.

NATURAL VARIATION VERSUS MANIPULATION. We often begin a behavioral study by making general observations of the subjects. When we have decided on the hypotheses to be tested and the general approach, we arrive at several critical questions: Can the hypotheses be investigated without any manipulation—that is, can we rely on natural variation? Do we need to consider the possibilities for manipulating either the subject(s) or the environment? For some studies it will be necessary only to devise a usable scheme for recording the data. For instance, if we were interested in comparing the frequencies of a particular bird species' foraging behavior in deciduous versus conifer forests, no manipulations would be required, as the natural variation in the two types of habitats provides us with the basis for a testable hypothesis.

If, however, the hypothesis being tested involves the effects of some social variable—for example, flock composition on foraging behavior—it may be necessary to manipulate the number and sexes of birds in the foraging flocks by netting and removing some birds. Alternatively, we might be interested in whether provisioning the habitats with supplementary food would alter the foraging rates of the birds in either habitat. In this instance, we would be manipulating the environment.

FIGURE 3-9   Bat in flight using large ears for echolocation
Animals live in perceptual worlds that differ from that of humans. To understand behaviors of animals properly, we must also investigate their varying perceptual worlds.
Source: New York Zoological Society Photo.

METHOD PITFALLS

Several potential pitfalls may entrap the unwary investigator; we become aware of most of them through experience. Three pitfalls deserve special attention: (1) perceptual worlds, (2) correlation versus causation, and (3) differential observability of animals.

PERCEPTUAL WORLDS.   One important task of animal behaviorists is to gain a thorough understanding of the perceptual world—the *Umwelt,* of each animal. Every animal lives in a different perceptual world. All too frequently, animal behaviorists assume that the animals they are studying live in the world of our human perceptions; this situation is only rarely true. For example, bats and many rodents emit sounds at very high frequencies, well beyond the range of human hearing (Griffin 1958; Noirot 1972). These high-frequency sounds are important, in bats for navigation and catching prey (figure 3-9 and figure 3-10), and in mice for communication between the young pups and the mother. Or, consider the visual world of insects—which is quite different than our own; bees and many other

---

FIGURE 3-10   Bat's system of echolocation
Bats navigate by echolocation whereby they emit high energy sounds that bounce off objects in the environment and return as echoes to their sensitive hearing systems.

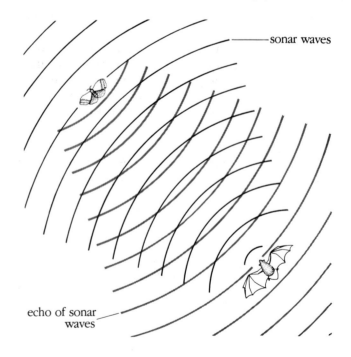

sonar waves

echo of sonar waves

insects have eyes consisting of hundreds of individual cells, called *omatidia*, each of which sends signals to the insect's brain. Its visual world thus appears as a series of points, sort of like a photograph that has been taken with very large-grained film. To understand the behavior patterns of animals and to test various behavioral phenomena properly, we must know something about animals' sensory systems and how organisms interpret various stimuli in their environments. We will examine sensation, perception, and the *Umwelt* in more detail in chapter 6.

CORRELATION VERSUS CAUSATION. When explaining why a particular behavior occurs, we must be aware of a second problem area—confusing *correlation* with *causation*. For instance, if we are investigating songbird reproduction and find that the birds always mate at the same time of the year soon after the beginning of heavy spring rains, we would probably find a good correlation between the onset of rains and subsequent mating behavior. However, it would be incorrect to infer that rainfall itself is a direct *cause* of mating behavior. Increased rainfall may lead to changes in both the quality and quantity of vegetation available to the birds for food. Changes in diet may, in turn, produce hormonal changes, which may then lead to behavioral changes. In addition, other factors may affect the mating behavior of the birds, including external factors (such as number of hours of daylight, availability of nesting materials, or social relations with other birds), and internal factors (such as biological clocks [see chapter 8] or previous mating experience).

Thus, explaining the timing of the onset of songbird mating behavior involves considerably more than simple correlation between environmental and behavioral changes. A series of experimental tests might establish the links in the chain between an external process, such as the increase in rainfall, and the internal events that resulted in behavioral changes. Correlation is an important tool for structuring our thinking and for formulating hypotheses about the causes of behavior, but we must not confuse correlation with causation. Any of several possible sequences can explain a particular behavior. Each step in any hypothetical sequence must be analyzed experimentally to confirm the causal connections between events in the sequence.

DIFFERENTIAL OBSERVABILITY OF ANIMALS. A third potential pitfall for investigators conducting studies of animal behavior centers on how long and for what period of time they observe different animals within the group. For example, if investigators observe juveniles more frequently than other group members, the conclusions of the investigation could be significantly biased. A good illustration of this point is the relationship between social-dominance status and mating frequency of adult male rats at a garbage dump (table 3–4). To test this relationship, we need the following information: (1) the social rank of each adult male rat, (2) the observed frequency of mating by each male, and (3) the number and length of times each male was observed.

We can determine social rank by watching the males at the dump for several weeks. The top-ranking male *B* is the rat that can defeat all six of the other adult male rats living at the dump. The next highest male *D* can defeat the remaining five, and so forth down to the lowest-ranking rat *M*, who loses all his fights. We must also record how many times we see each male copulating with an adult female (column 3). On the basis of the frequency of copulation and rank of the male rats, we may conclude that higher social status among males is associated with a greater amount of mating activity.

TABLE 3-4 Social dominance and mating in adult male rats at a garbage dump

Male	Social Rank	Observed Matings Per Month	Times Observed Per Month (hr)	(Corrected) Number of Matings Per Month Per Times Observed (hr)
B	1	18	85	0.21
D	2	15	80	0.19
K	3	14	68	0.21
G	4	10	41	0.24
C	5	7	30	0.23
J	6	5	26	0.19
M	7	4	22	0.18

However, this may not be true; what if the lower-ranking male rats were not seen as often? The lower-ranking males may be observed less frequently, and recorded as mating less frequently, but they could be engaging in frequent mating behavior in areas not visible to the observer.

To complete the analysis, we need to know the number of times and length of time each male rat is seen in a standard time period—say, one month. If all the males are seen about the same number of times and for the same total amount of time, then the correlation between higher social status and more mating activity is valid. If, however, as table 3–4 shows, males are seen more frequently or less frequently depending on social rank, then we must make a correction for observability. The data in column 5 of table 3–4 have been corrected for the *observability* of each male and are expressed as the number of matings per month per time observed. In this case we note that there are no differences in the mating rates of males of different ranks. While this finding does not mean that the original conclusion is invalid, the corrected results do indicate that further study will be necessary to determine what the lower-ranking, less-visible males are doing when the observer cannot see them.

USING ANIMALS IN RESEARCH

One issue that has become particularly important in the past decade for those who conduct research on animals involves the care and well-being of the subjects. This is particularly true for laboratory animals, though there are genuine concerns for the safety and health of animals in field pens, and under natural, free-ranging conditions as well.

A major area of importance involving wild animals is the study of endangered or threatened species. The federal governments of the United States and in Canada, many of the individual states and provinces, and many other countries have passed laws governing the sale and importation of species that are endangered or threatened in the wild. Scientists who wish to conduct research on these animals must comply with the laws, and when they publish their findings, they generally are asked to certify that they have obeyed all applicable laws. Ironically, only intensive and extensive study of the behavioral and reproductive biology of these endangered species, whether they be mammals, birds, reptiles or other taxa, can give us any hope of protecting the appropriate habitat, of providing proper breeding conditions, and of ensuring the continued reproduction of these important animals.

For those who work in the laboratory, at least two types of issues are important: matters concerning the actual care of the animals, and matters concerning experimental research. Among the former are caging, lighting, food, water, and general health. These issues have been addressed by the "Guidelines for the care and use of animals in research" (1986) put forth by the National Institute of Health. Experimental research can range from simple observations of animals housed under specific conditions that, although not involving surgery, may cause the animal a certain amount of pain or distress; to manipulations that do involve surgery, such as removal of the testes, or brain operations. Guidelines put forward by groups like the Animal Behavior Society and the Association for the Study of Animal Behaviour (1986), and others (American Society of Mammalogists 1987; American Ornithologist's Union 1988), now serve to direct investigators in these animal welfare matters. Proper procedures must be followed; they often involve approval of the protocols by a committee of fellow scientists and a veterinarian. Investigators are urged to consider whether the procedures are necessary, what the significance is of the answers that will be obtained, and how can the number of animals used be minimized, consonant with the objectives of the experiment. It is the responsibility of everyone working on animals to be aware of these rules and guidelines, and to make every effort to comply on all issues. We think it is important to note that scientists have taken the lead on these matters, putting forth their own sets of guidelines, serving on committees that formulate national guidelines, and serving on thoughtful review boards for scientific proposals.

Again, the irony here involves the enormous benefits to be gained from research on animal behavior. A number of fields benefit from the investigations conducted on animals, including human medicine, conservation biology, wildlife biology, and anthropology, to name but a few. The benefits range from preservation of species, to gaining a deeper understanding of the way in which all animals, including our own species, interact with their living and nonliving environment.

SUMMARY

An ethological study of animal behavior traditionally begins with an *ethogram,* an inventory of an animal's behavior, and seeks to answer "why" questions by determining the functions and evolution of behavior, as is exemplified by Tinbergen's studies of the significance of eggshell removal in gulls. The advantages of the *ethological* approach are that animals are studied in their natural habitat where functional relationships are more readily discerned, and problems can be examined from an evolutionary perspective. However, the ethological approach sacrifices control over environmental parameters and over knowledge of an animal's history. In addition, devising a good test of the evolutionary significance of a behavior pattern is often difficult.

A study of animal behavior using the *comparative psychology* approach begins with the classification and comparison of the behavior of different species in order to discover relationships between them. This approach is exemplified by Renner's exploration of the effects of experience on object interactions in rats. Experimenters have a high degree of control over both test subjects and environmental conditions; they can manipulate one or more variables while they hold all others constant. However, the breadth and strength of the conclusions of these studies are limited by the heavy reliance upon laboratory stocks for test subjects, by the influence of domestication on laboratory animals, and by an overemphasis on methodology to the detriment of a more complete view of the context and functional importance of a behavior pattern.

Behavioral ecology and *sociobiology* attempt to dissect the ecological and social aspects of animal behavior from a functional and evolutionary perspective. These approaches have the advantage of studying the animal in a natural or natural-like habitat. Furthermore, when applied correctly, the resulting dissection of behavior patterns provides a bridge between the functional perspective and the more mechanistic perspective we have used to characterize comparative psychology. In addition to sacrificing some degree of control over the animals, studies using the frameworks of behavioral ecology or sociobiology sometimes suffer from oversimplification of past evolutionary processes that have resulted in the species' presently observed behavior patterns. There is a tendency to formulate "just-so" theories that often prove difficult to test experimentally.

Because animal behavior deals with probabilistic phenomena, correct experiment design is of great importance. The investigator must define the behavior patterns being investigated; must state the *experimental hypotheses,* or questions about observed biological phenomena, and the *null hypotheses,* or statements about expected differences between experimental treatment groups; and must designate the *independent variables,* or factors that are controlled and manipulated by the investigator, and the *dependent variables,* or behavior patterns that are measured as outcomes of the experiment.

In addition, the investigator must determine if *control groups* are necessary, either to provide a foundation for comparison or to demonstrate that the factors being manipulated, and not the manipulative processes themselves, are responsible for the observed results; must insure the *independence of data points;* must use sufficient numbers (*sample size*) of animals in the experiment to provide proper analytical results; and must keep accurate and complete records of experimental procedures so that experiments can be replicated. Field investigations of animal behavior operate under the same design constraints and method requirements as do laboratory experiments.

Behavioral scientists must be aware of the need to understand the perceptual world of the animal being investigated and of the problem of differences in observability in animals of different ages, sex or social status. Correlations between environmental factors and behavior should not be interpreted as causes of behavior.

In recent decades, animal behaviorists, like all biological scientists, have become more cognizant of issues surrounding the health and welfare of the animals with which they work. Under both field and laboratory conditions, there are a variety of issues pertaining to the care and safety of research animals that are now governed by various guidelines and procedures. These procedures are designed to aid and facilitate research, not impede progress. There are enormous benefits to be gained from animal behavior research in human health and medicine, in conservation biology, and in our general understanding of biological phenomena.

DISCUSSION QUESTIONS

1. Animal behaviorists ask questions concerning causation, development, evolution, and function. Some species of gulls pick up shellfish, fly up a short distance in the air, and drop them on rocks to break them open. Describe the specific series of questions you might ask in each of these four areas about this behavior, and suggest some of the methods you would use to answer these questions. What specific hypotheses would you test? What experimental procedures would you use to test these hypotheses?

2. The effects of early feeding experiences on later food preferences were studied in guinea pigs. The animals were grouped five per cage, and each of the three groups was fed only one type of food—A, B, or C—for three weeks. Then the guinea pigs were tested individually to determine their food preferences in a "cafeteria" situation, with all three food types present. The results are shown in the table that follows. What can you say about the relative effects of early feeding experiences in guinea pigs? What principles of experimental design and methodology have been violated in conducting this experiment?

		Food Eaten In Cafeteria Test (g/24 hr)		
		Type A	Type B	Type C
	Type A	9.6	2.8	5.7
Rearing	Type B	3.8	6.2	7.8
food	Type C	1.4	.6	16.2

3. Suppose you were asked to conduct a study assessing the diet of red foxes living in a farmland area. You have sufficient funds and foxes, and a natural setting in which to conduct both observation and manipulative experiments. What types of tests would you conduct to answer the question about diet? What experimental problems would you expect to encounter and how would you solve them?

4. To conduct sound experiments in animal behavior, an understanding of proper methods and experimental design is necessary. What other reasons can you cite for acquiring a firm knowledge of these principles for work in animal behavior?

5. Many research animals are housed in zoos throughout the world. What special concerns about caged animal behavior pertaining to health and welfare would you have if you were an administrator or veterinarian at a zoo?

Suggested Readings

Altmann, J. 1974. Observational study of behavior: sampling methods. *Behaviour* 49:227–67.
The key paper summarizing focal-animal sampling, scan-sampling and other techniques for the observational study of animals, under field or laboratory conditions.

Beach, F. A. 1960. Experimental investigations of species-specific behavior. *Amer. Psychol.* 15:1–18.
Beach defines species-specific behavior and discusses a total framework for questions in comparative psychology. He was one of the pioneers in developing comparative psychology and animal behavior.

Colgan, P. W., ed. 1978. *Quantitative Ethology.* New York: Wiley.
An edited volume of articles dealing with the design experiment and analysis of behavior. Written more for the advanced student than the beginner.

Dewsbury, D. A. 1973. Comparative psychologists and their quest for uniformity. *Ann. N.Y. Acad. Sci.* 223:147–67.
A sound criticism of the approaches and methods of comparative psychology. Presents and discusses specific problems and suggests alternative patterns of thinking.

Drummond, H. 1981. The nature and description of behavior patterns. In *Perspectives in Etholoy*, eds. P. P. Bateson and P. H. Klopfer. 4:1–34. New York: Plenum Press.
An excellent book chapter on the entire processes of observing and describing behavior patterns of animals.

Hirsch, J. ed. 1987. *Journal of Comparative Psychology* 101 (3).
This issue, marking the 100th anniversary of the journal, is dedicated to a review of comparative psychology, past, present, and future. There is excellent breadth and depth, and much of the material should be understandable to animal behaviorists with only a moderate background.

Immelmann, K., and C. Beer. 1989. *A Dictionary of Ethology*. Cambridge, MA: Harvard University Press.
Not really a book for reading, but rather a fine reference source for definitions of behavior patterns, as well as concepts and procesess related to behavior. Carefully written with succinct definitions, but not so terse that the beginning student cannot gain a great deal of knowledge from using the book.

Krebs, J. R., and N. B. Davies. 1987. *An Introduction to Behavioural Ecology*. 2d ed. Oxford, England: Blackwell Scientific Publications, Ltd.
The only real textbook introduction to the various aspects of the field of behavioral ecology. Well written with excellent examples and good explanations of some difficult concepts.

Lehner, P. N. 1979. *Handbook of Ethological Methods*. New York: Garland.
Excellent introduction to the variety of methods and approaches to the study of animal behavior. Useful for the advanced student as well as the novice.

Tinbergen, N. 1963. On aims and methods of ethology. *Zeit. Tierpsychol.* 20:410–33.
Written by a Nobel laureate and founding father of modern ethology. Presents the four-question framework used for the study of behavior, with comments on the methods used by ethologists.

Trivers, R. 1985. *Social Evolution*. Reading, MA: Benjamin Cummings.
A very readable text on all aspects of sociobiology. One real strength of this book is the fine set of examples chosen to illustrate the various concepts and processes that comprise the basis for research in sociobiology.

4

GENETICS AND EVOLUTIONARY PROCESSES

In order to understand and interpret animal behavior, we must be familiar with certain aspects of genetics and with the theory of evolution. Like animal morphology and physiology, animal behavior has, over time, undergone changes that adapt the animals to their various habitats and that contribute to reproductive success. We can use our knowledge of evolutionary processes to formulate tentative explanations of the ultimate causation of observed behavior patterns and to design experiments to test the functional significance of the behavior.

Since modern evolutionary theory is based on some principles of genetics, we examine basic genetics in the initial sections of this chapter. We then take a close look at evolutionary theory and explore the relationships between genetics and evolution under the topic of population genetics. To find out how the process of evolution works, we examine the ways that evolutionary processes result in adaptive changes—within species as well as in the production of new species. We next examine genetic variation, kinship, and units of selection in the evolutionary process. We conclude with some cautionary comments on the use of evolutionary explanations for behavioral phenomena.

GENETICS

SPECIES AND GENE POOLS

In biological terms, a **species** consists of animals that share similar characteristics and habits and that interbreed freely with one another (Mayr 1970), but that are reproductively isolated from other species in nature—that is, members of one species cannot or do not breed with members of another species. Biological species are dynamic entities that change over time, and thus are the primary units of study for evolutionary biologists.

Each individual possesses a number of **genes,** or basic units of heredity. Genes are located on chromosomes, and they establish the fundamental framework for the development of anatomical, physiological, and behavioral traits. The complete set of genes for any organism is called its **genotype;** the actual physical, physiological, and behavioral traits that we see expressed are called the **phenotype. The gene pool** of a species consists of a hypothetical collection of all genes of all reproductive individuals.

We can employ the species concept in the categorization of living organisms. In the classification scheme utilized by zoologists, the species is usually the smallest recognized unit. Although some species, like dogs or horses, are subdivided into subspecies, races, or breeds,

TABLE 4-1   Classification for some animals
The basic unit of animal classification is the species. Each species may be further classified in a hierarchy of categories based on characteristics shared with other animals and on patterns of evolutionary ancestry.

Common Name	Kingdom	Phylum	Class	Order	Family	Genus	Species
rhesus monkey	Animalia	Chordata	Mammalia	Primates	Cercopithecidae	*Macaca*	*mulatta*
Norway rat	Animalia	Chordata	Mammalia	Rodentia	Muridae	*Rattus*	*norvegicus*
red-winged blackbird	Animalia	Chordata	Aves	Passeriformes	Icteridae	*Agelaius*	*phoeniceus*
three-spined stickleback	Animalia	Chordata	Osteichthyes	Gasterosteiformes	Gasterosteidae	*Gasterosteus*	*aculeatus*
migratory locust	Animalia	Arthropoda	Insecta	Orthoptera	Acrididae	*Schistocera*	*gregoria*

generally all members of these subunits can interbreed (Mayr 1963). The basis of each succeeding higher order of classification (table 4–1) is designated by the degree of evolutionary and genetic relatedness, so animals of different species that share many characteristics and a recent common ancestry are classified within the same genus (group of related species). The closer together two species are classified, the greater the percentage of overlap between the two species' gene pools.

BASIC PRINCIPLES

A **karyotype** is a diagrammatic representation of the contents of a cell nucleus, including the chromosome number, size, and form. Each somatic cell of most organisms is **diploid,** and contains $2n$ chromosomes, arranged in n homologous pairs, one derived from each parent (figure 4–1). Since each sexually produced offspring must have $2n$ chromosomes, the **gametes** (ovum and sperm) are **haploid;** they each have a single set of unpaired chromosomes, a total of $1n$. Changes in chromosomes numbers during the life cycle of a mouse are diagramed in figure 4–2.

Genes are arranged along chromosomes at particular positions, or **loci.** Alternative forms of a gene, called **alleles,** may be present at a given locus. Although many alternative forms of a single gene can exist, an individual can have only two possible alleles of any gene at a particular locus, one on each chromosome. An animal is **homozygous** for a given locus when both alleles at that locus are identical (for example, *vv* or $++$ alleles in the mouse). When the alleles at the same locus

are different (for example $+v$), a **heterozygous** condition results. Since literally thousands of pairs of alleles are involved in a normal fertilization, the task of analyzing the genetic bases of behavior is quite complex. Here we use a convention established for denoting the origin of alleles; the *wild* type (the normal condition found in the general population) is designated by '+' and various *mutant* types are designated by letters such as '*v* '.

Populations of animals that are homozygous for all or most of their loci, such as inbred strains of mice, are referred to as genetically **homogeneous;** all members of such a population have essentially the same set of genes, or genotype. In contrast, populations of animals that are heterozygous for many loci are referred to as genetically **heterogeneous;** members of such a population show variations in the alleles.

Some alleles at a particular locus are **dominant** and express an effect when present on only one of the paired chromosomes. In contrast, other alleles are **recessive** and express an effect only when *both* chromosomes have the recessive allele. Another possible condition is **partial dominance,** in which two alleles interact to produce a genetic expression in the phenotype somewhere between the effects of the two alleles.

For a clearer picture of the dominant or recessive effects of genes, let us consider the following example. Suppose we are dealing with just one locus that controls whether a mouse is normal or a "waltzer" (one that exhibits aberrant circling and erratic movement patterns). There are only two possible alleles, $+$ (normal, dominant) and *v* (abnormal, recessive) for the locus in

FIGURE 4-1   Human karyotype

There are 23 pairs of homologous chromosomes in each body cell of the human ($2n = 46$).

Source: Photo by J. S. Yoon, Bowling Green State University.

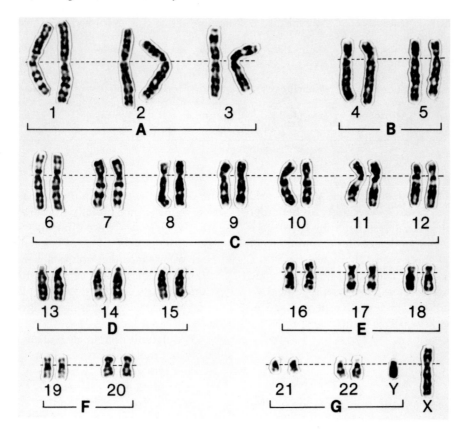

question (see table 4–2). If a normal male mouse of genotype $+v$ mates with a normal female of genotype $+v$, then one-fourth of the offspring will be $++$, one-fourth will be vv (both homozygous conditions), and one-half of the progeny will be heterozygous, $+v$. Since these genes are expressed in the phenotype, three-fourths of the progeny will be normal and one-fourth will be waltzers, because wild $+$ alleles are dominant to recessive v alleles.

What are the links between these hereditary units we have labeled "genes" and the behavior of animals we observe? How do genes affect the biochemical and physiological processes within the cells and organs of an animal? Biochemical research indicates that specific gene DNA sequences code for the structure and synthesis of particular proteins. Proteins are found in cell walls, in the structure of organelles within cells, and in

the liquid medium within cells, where they are critical features of many ongoing chemical reactions. Some proteins are also enzymes, catalysts in the cell's biochemical processes. Thus, genes are linked to behavior via biochemical processes that regulate the structure and function of the tissues and organs (see figure 5–9).

EVOLUTIONARY PROCESSES

Today, the **synthetic theory,** which is based on the principle of reproductive success, is the most widely accepted evolutionary theory (see Mayr 1963; Stebbins 1966; Huxley 1974). The basic tenet of this theory is that the more successful members of a species are those that pass on their genes, and thus their characteristics, to more members of the next generation.

FIGURE 4-2   Mouse life cycle chromosome changes

Sperm 1N = 20   Ovum 1N = 20

Zygote 2N = 40

Embryo 2N = 40

Adult 2N = 40

Sperm 1N = 20   Ovum 1N = 20

RELATIVE GENETIC FITNESS

Relative genetic fitness is a critical concept in the synthetic theory of evolution by natural selection; it refers to the probability that a particular genotype will be present in the offspring of the species being examined. Fitness values range from 0 to 1. When this number is multiplied by the frequency of a genotype in the present reproducing generation, the resulting product indicates the expected frequency of that genotype in the next generation. For example, a fitness value of 1.0 means that the genotype will be present in the population of the subsequent generation at the same rate as it is present in the population of the producing generation.

We should note here that many concepts in genetics have underlying assumptions, not all of which always hold true. The concept of fitness is based on the assumptions that the population is of sufficiently large

TABLE 4-2   Possible combinations of two alleles, + and *v*, at one particular locus in the mouse, when a +*v* female mates with a +*v* male
The + allele is dominant to the *v* allele.

		Female Parent	
		+	*v*
Male Parent	+	++	+*v*
	v	+*v*	*vv*

	Genotype		Phenotype
1	++	=	normal
2	+*v*	=	normal
1	*vv*	=	waltzer

size and that there is no differential movement of genotypes into or out of the population. However, even with its limitations, the concept of fitness is useful, since it describes a particular genotype's relative ability to continue in succeeding generations.

SURVIVAL

To pass on its genes, an animal first must survive by finding food and shelter and by avoiding predation and disease. Then the animal must successfully mate, and the offspring must survive to reproduce. Much of what we study in animal behavior is concerned with how animals play the two parts of the evolutionary "game": survival and reproduction. What we discover is that genetically based physiological and behavioral traits favorable for survival and reproduction are more likely to be passed on to future generations than those that are unfavorable. Less favorable traits that put the animal at a disadvantage will slowly disappear from the population, because the genotype fails to continue in succeeding generations.

The phenotype is the physical manifestation of the genetic blueprint. Beginning with conception, a series of developmental stages occurs in the organism that involve the interaction of the genetic information with the environment, and result in the phenotypic manifestation. The phenotype is the end result of **epigenesis.** Epigenesis is viewed as the complex unfolding of an organism's morphology, physiology, and behavior through the interaction of the organism's intrinsic genetic program with extrinsic stimulation and experience (see chapter 9).

DARWIN AND BIOLOGICAL EVOLUTION

Biological evolution through natural selection can be defined as a change in gene frequencies within a population that alters the gene pool from one generation to the next. Thus, biological evolution encompasses both small directional changes that result in minor modifications (**adaptation**) and longer-term directional

changes that result in new forms (**speciation**). Although Darwin knew little about genes, he was able to assemble the important elements of the process of biological evolution. Examining the manner in which Darwin derived his conclusions should be instructive for our understanding of this process. The following account, summarized in figure 4–3, attempts to reconstruct the stages by which Darwin arrived at the theory of natural selection (from Mayr 1977). Darwin derived the initial facts from Malthus's 1798 essay on population.

- *Fact 1: All species are capable of overproducing.* For example, one pair of houseflies, if unchecked, could produce more than six trillion progeny in

FIGURE 4-3   Components of natural selection theory

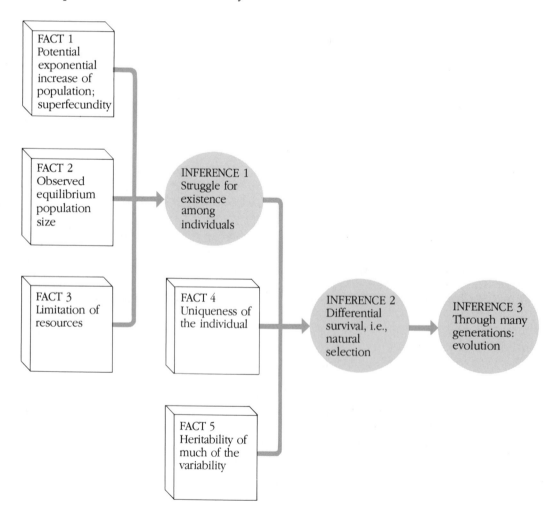

one year. Even elephants, which are not particularly fast breeders, could produce nineteen million descendants in 750 years from one pair.

- *Fact 2: Populations of species tend to remain relatively stable over time.* The death rate tends to equal the birth rate, and most fluctuations are either temporary or repeated cyclically. For example, close monitoring of a population of white-footed mice (*Peromyscus leucopus*) over a number of years shows that their number fluctuates annually, decreasing to between five and twenty in a two-hectare plot in winter and increasing to between thirty and one hundred in the summer after spring breeding (Vessey 1987). Although these may seem like wide fluctuations, they are in fact small, given the range of possibilities.

- *Fact 3: Limitation of resources.* Populations do not go on increasing indefinitely: because resources are limited, eventually a population runs out of something. We usually think of food as most important, but other requirements, such as shelter or nesting space, may limit the population. Many species, particularly invertebrates, are limited by local weather conditions. For example, storms or seasonal shifts in temperature can eliminate many members of a population.

  - *Inference 1: Struggle for existence among individuals.* Exponential population growth, combined with a fixed supply of resources, results in a struggle for existence among individuals. Although this idea was not new with Darwin—others had suggested that such a struggle is a beneficial feedback device to maintain the balance of nature—he emphasized that much of the competition is between members of the same species. Competition refers not only to predator-prey interactions but also to a struggle among animals needing the same limited resources. As we shall see, most competition is subtle and more aggressive animals do not necessarily leave more offspring. In the mouse populations mentioned above, only one or two pups from a litter of four or five survive to the weaning age of three to four weeks. Should they survive to weaning age, their life expectancy is just ten weeks, even though the laboratory animals of this species can live one to two years or longer.

- *Fact 4: Uniqueness of the individual.* From his studies of animal breeding, Darwin confirmed what others had observed: Every animal in a herd is different from every other, and extreme care must be used to select the sires and dams from which to breed the next generation.

- *Fact 5: Heritability of much individual variation.* Although Darwin's understanding of genetics was meager, he assumed correctly that the potential for variation is always present, and that it is renewed in every generation.

  - *Inference 2: Differential survival, or natural selection.* Excessive fertility and individual variation come together to set the stage for evolution. Some individuals have traits that enable them to outcompete their conspecifics, and these individuals are more likely to survive and reproduce.

  - *Inference 3: Through many generations—evolution.* The natural selection of individuals with heritable qualities continued over many generations leads to gradual change in the appearance of individuals—representing a change in gene frequencies in the population.

POPULATION GENETICS

Population genetics deals with two aspects of evolution: the sources of and the dynamics of variations. More specifically, if we are working on population genetics, then we will be making inquiries into (1) the basic units of genetic variation at the level of the gene and chromosome and (2) the causes of changes in the frequency of these units' occurrence in populations.

MUTATION AND RECOMBINATION

There are two main sources of genetic variation: mutation and recombination through sexual reproduction. **Mutation** is a heritable change in the genetic material in the form of an alteration in the sequence of nucleotide pairs in the DNA molecule. Most spontaneous mutations are copying errors that occur in the course of DNA replication. Mutations can occur as minute chemical changes that result in the substitution of one nucleotide for another in the DNA molecule (*point* *mutations,* additions or deletions of one or a few nucleotide base pairs) or as major structural changes in the chromosome (*chromosomal aberrations*). The latter can occur in a number of ways, such as loss of a piece of chromosome (*deletion*); breakage of a piece, which then flips end over end and reattaches (*inversion*); or transfer of a piece from one chromosome to another (*translocation*).

Recombination occurs in sexually reproducing species when homologous chromosome pairs cross over, so that part of the maternal chromosome exchanges with part of the paternal chromosome. Further variability is introduced at each generation because the chromosomes sort randomly and independently during cell division, producing new combinations of genetic material. The haploid gametes contain some maternal and some paternal chromosomes, plus some that are mixtures (due to crossing over).

We estimate that ten thousand gene loci, many bearing multiple alleles, exist in humans. The total number of possible diploid genotypes is therefore astronomical.

CAUSES OF EVOLUTION

HARDY–WEINBERG FORMULA. What about the actual process of evolution via genetic changes? Does the scrambling of genes through segregation and recombination cause evolution? In other words, does one allele increase in the following generation at the expense of another? The **Hardy-Weinberg formula** allows us to sample a population and determine whether a set of alleles is at equilibrium or whether the relative frequencies change as a result of sexual reproduction.

In order to apply the Hardy-Weinberg formula, we assume that the population is large, that it is **panmictic** (each individual has an equal opportunity to mate with any member of the opposite sex, and the genes in the population are mixed randomly). (We should note that completely random mating probably does not exist in any species in nature.) In addition, we assume that gene frequencies do not change and that evolution is not occurring at that locus. We use the Hardy-Weinberg formula to generate expected frequencies of alleles (with

FIGURE 4-4   Hardy-Weinberg equilibrium
When all alleles have an equal chance of being transmitted from generation to generation, their relative frequencies can be described by the Hardy-Weinberg formula. A two-by-two matrix illustrates the possible combinations of offspring from sperm and ova. Under these conditions alleles $_A$ and $_a$ are at a particular locus with frequency $p_{(A)}$ and $q_{(a)}$.

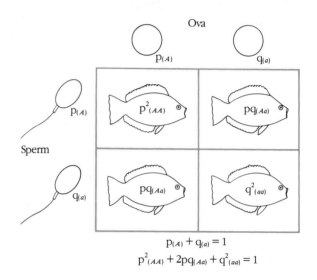

$$p_{(A)} + q_{(a)} = 1$$
$$p^2_{(AA)} + 2pq_{(Aa)} + q^2_{(aa)} = 1$$

our assumption that gene frequencies don't change). Then, if the observed frequencies don't fit the values expected from Hardy-Weinberg, we reject that assumption.

Let's consider an example. Suppose that there are only two alleles, *A* and *a*, at a given locus; each gamete carries either *A* or *a*. If *p* is the relative frequency of *A*, and *q* is the relative frequency of *a*, then $p + q = 1$. Will sexual reproduction itself alter *p* and *q*? What will be the frequencies of the diploid genotypes *AA* and *aa* in the next generation?

If *p* is the frequency of the *A* allele, it is also the probability that a gamete will contain *A*; similarly, *q* is the probability that a gamete will contain *a*. Because the probability of the union of independent events is equal to the product of their separate probabilities, the probability that the two *A* gametes will come together is $p \times p$ or p^2. Likewise, the frequency of *Aa* is $2pq$, and that of *aa* is q^2; the frequencies must total 1 (see the matrix of all possible combinations in figure 4–4). The equation $p^2 + 2pq + q^2 = 1$ is known as the Hardy-Weinberg formula.

Let us use some real numbers to see how the Hardy-Weinberg formula predicts gene frequencies, and tests whether the population is at equilibrium for the alleles. Suppose we mix 6,000 *AA* individuals with 4,000 *aa* individuals. The initial diploid frequency of *AA* is 0.60, of *Aa* is 00.0, and of *aa* is 0.40. Thus the value of *p* is 0.6, and that of *q* is 0.4. In the next generation and every generation thereafter, the frequency of *AA* is p^2 (0.36); of *Aa* is $2pq$ (0.48); and of *aa* is q^2 (0.16). The frequency of the *A* allele is thus $0.36(p^2) = 1/2(0.48)(2pq) = 0.36 + 0.24 = 0.60$, the starting frequency. In other words, the gene frequencies do not change. Thus we learn that segregation and recombination do not cause evolution.

Use of the Hardy-Weinberg formula enables us to sample a population at one point in time and determine whether or not it is at equilibrium for a set of alleles, without having to take repeated samples across several generations. The following example illustrates this point. The Quinault Indians, a Northwest Coast group, were studied by Hulse (1971) for the incidence of the MN blood group. Of a sample of 201 individuals, 77 were homozygous for the *M* allele (*MM*), 101 were heterozygous (*MN*), and 23 were homozygous for the *N* allele (*NN*). Is the MN trait under selective pressure?

To find the expected frequencies in the next generation, we first calculate *p* and *q* and then substitute those numbers into the Hardy-Weinberg formula to get the proportion of each phenotype. Finally, we calculate the expected numbers and compare them with the original sample. In this case $p = 0.63$ and $q = 0.37$. The proportions of blood type in the next generation turned out to be 39.7% MM, 46.6% MN, and 13.7% NN. Converting these proportions to 201 individuals (for comparison's sake), we get 80 with MM, 94 with MN, and 28 with NN—close to the existing numbers. For each set of values of *p* and *q*, there is only one expected combination of homo- and heterozygotes that should persist indefinitely if the conditions of Hardy-Weinberg equilibrium are met. We have therefore demonstrated that the population in this instance is panmictic and that no selection for the MN blood group trait is occurring.

FORCES FOR CHANGE. If sexual reproduction does not cause evolution, what does? One force for genetic change is **mutation pressure.** Our discussion of mutations has indicated that if allele *A* mutates to allele *a* in an individual, then *p* and *q* will change in the population. The problem is that mutations are relatively rare, on the order of 10^{-5} per gene per replication, or less. In fact, Wilson and Bossert (1971) have calculated that it would take about ten thousand generations to reduce the frequency of a particular allele to one-third its original value.

A second, more rapid way to change gene frequencies is by **gene flow.** Animals that are moving from one population to another may have gene frequencies that are different from those in the new population (see figure 4–5). Gene flow results in evolution, provided that the new phenotypes are adaptive (are fit), and the new alleles are incorporated into the gene pool of the population. A third force for change is **natural selection,** which may be defined as the differential change in relative frequency of genotypes due to differences in the ability of their phenotypes to obtain representation in the next generation.

A fourth force to consider is **genetic drift,** the alteration of gene frequencies through sampling error. Genetic drift operates to some degree in all populations, but is significant only when populations are small (from ten to one hundred individuals). In a small population, gene frequencies can change greatly because each chance mating combination has a big impact on the total population. In the same way, the batting average of a baseball player who hits once for three times at bat is .333; if he gets a hit on his next at bat, his average will jump to .500. If a player's average of .333 is based on 300 at bats, his next hit will increase his average to only .336.

The ultimate fate of any allele is that it is either lost ($p = 0$) or fixed ($p = 1$) in the population. Heterozygosity tends to be lost as a result of genetic drift. A special case of drift is the **founder effect,** a situation in which new populations are started by small numbers of individuals, for example, one in which a pair of birds colonizes an island. These individuals can carry only a fraction of the genetic variability of the parental population; thus the new population is likely to differ from the old. In a second example, the starting population of white-footed mice in an isolated woodlot before spring breeding is sometimes only one or two pairs. This event forms a genetic bottleneck that results in widely fluctuating allele frequencies. The mice that survive the winter to begin breeding the following spring will carry only a subset of all of the possible alleles that were present in the population the previous fall during the period of peak numbers.

FIGURE 4-5   Evolution by gene flow
Population *B*, with a low frequency of the white allele, gains members from population *A*, which has a high frequency of the white allele. The frequencies of the white allele would be expected to increase in future generations of population *B*.

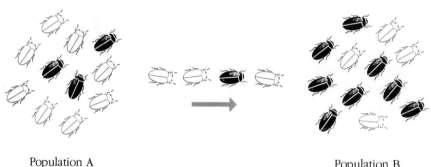

Population A   Population B

FIGURE 4-6   Stabilizing, directional, and disruptive selection
Each type of selection can occur during evolution. Each type is shown with the original population ("before selection"). The expected result ("after selection") is graphed in terms of the frequencies of phenotypes within a population.

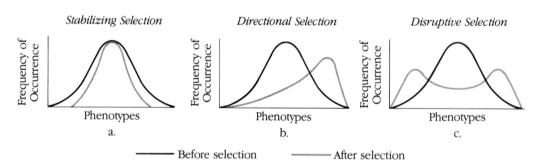

SELECTION, ADAPTATION, AND SPECIATION

NATURAL SELECTION

We have defined natural selection as a change in the relative frequency of genotypes that derives from the differential ability of their phenotypes in gaining representation in the subsequent generation. Three major types of selection can occur with respect to genotype/phenotype variations: (1) stabilizing selection, (2) directional selection, and (3) disruptive selection (figure 4–6).

STABILIZING SELECTION. Consider stabilizing selection; we find that the average or normal individuals in the population have an advantage over the extreme variants in terms of reproductive success. **Stabilizing selection** occurs when the environmental conditions that are most important for survival of a particular population remain relatively constant over a long period of time (figure 4–6a). This type of selection does not favor a specific optimum genotype; rather, the norm is represented by a number of variable genotypes that are well adapted to the existing set of environmental conditions.

DIRECTIONAL SELECTION. In **directional selection,** a regular shift in the adaptive characteristics of a species takes place. When there is a progressive shift in some aspect of the environment, or when part of a population migrates into a new area with different environmental conditions, directional selection can occur (figure 4–6b). The Darwin's finches of the Galápagos Islands provide an example of this type of selection. The ground finches and cactus-feeding finches (*Geospiza* spp.) have evolved from the ancestral fringillid type that colonized the islands. Some species in this group have large stout beaks and consume a diet of large, hard seeds. Another group in the islands has species with more refined, pointed beaks useful for obtaining nectar from flowers. Peter Grant and his colleagues (Grant, 1981; Grant and Abbott, 1980; Grant and Grant, 1980) outlined the manner in which these differences in beak morphology could be the result of competition for food sources. This competition has resulted in directional selection for specific beak characteristics adapted for different feeding strategies and food habits.

Another example of directional selection is the artificial selection used in the domestication and breeding of animals, like cattle and dogs, and in the development of specific strains of laboratory animals, like mice and fruit flies. We shall say more about artificial selection in the discussion of behavior genetics in chapter 5.

DISRUPTIVE SELECTION. When individuals of a species formerly in a homogeneous environment are later subjected to different selection pressures in different parts of that species' geographical range, **disruptive selection** may occur (figure 4–6c). Different variants of a trait or behavior may be more adapted for particular conditions at different locations within the geographical range of the species. Disruptive selection—or, as it is sometimes labeled, **diversifying selection**—is relatively rare in nature. One possible example of disruptive selection has been reported for an African butterfly species (*Papilio dardanus*). The color pattern of this butterfly is characteristic for a particular locale, and in each case, matches the pattern of another butterfly species in that locale that is distasteful to predators. This mimicry protects the *Papilio dardanus* from predation. Intermediate color patterns do not occur; if they did, they would not match those of any particular local species. Thus we can postulate that disruptive selection has produced a series of discrete types (Clarke and Sheppard 1960), each matching a local form, as a means of avoiding predation.

ADAPTATION

The term **adaptation** has two meanings. In some contexts, adaptation refers to all of the traits and characteristics that enable an organism to survive and reproduce. Thus we may speak of an animal as being adapted for life under a particular set of ecological conditions. For example, the bills of many bird species are adapted for obtaining a variety of food types, ranging from insects to seeds to fish (figure 4–7).

Adaptation is also used in another way to refer to the evolutionary selection that results in specializations of anatomy, physiology, and behavior. In this sense, adaptations are temporary stopping points in long-term evolutionary sequences; observing the actual processes is much more difficult. Most environmental changes are relatively gradual, and thus their impact on adaptation may be expected to be gradual. An adaptation that enhances fitness today is not necessarily adaptive for some generations in the future.

When discussing adaptation as an evolutionary process we need to define several critical terms:

- The **ecological niche** of an organism is its location and functional role in the ecological system; niche refers to traits like nest sites, food habits, and seasonal and daily activity rhythms. Because of the large number of characteristics we use to define the animal's living space and traits, the niche is referred to as an *n*-dimensional space that includes all of the animal's habits and its place in the ecological system (Hutchinson 1957).

- **Competition** occurs when some resources necessary for survival and reproduction are available in limited supply and are needed by animals of different species (*interspecific competition*) or by individuals of the same species (*intraspecific competition*).

- The **environment** of an organism consists of its living and nonliving surroundings. Thus, we consider living animals of the same and different species, vegetation, soil, geological formation, and climate components of the environment of a terrestrial organism.

ENVIRONMENTAL ADAPTATIONS. Through the study of evolution, we have determined that two major selective forces, environment and competition, influence the evolution of adaptations in animals. The effects of these two forces cannot be readily separated, but we can make some useful distinctions. General environmental

FIGURE 4-7   Bill adaptations in six species of birds
The bills of (a) the falcon, (b) the kingfisher, (c) the anhinga, (d) the pelican, (e) the chickadee, and (f) the flycatcher, illustrate some of the variation in bill types adapted for each species' specialized techniques for obtaining and consuming food. Falcons use hooked bills to tear flesh; kingfishers use long slender bills to grab and hold fish; anhingas spear fish with their long slender bills; pelicans use their large bills and the pouch below it to dip into the water to net fish; chickadees have small short bills for gleaning insects from vegetation and for eating seeds; and flycatchers gather insects as they fly out from and back to their perches with their mouths wide open.

FIGURE 4-8   Salamander
Salamanders are widely distributed and occupy both aquatic and terrestrial habitats. Where two closely related species occupy similar habitats, behavioral changes can result to reduce the overlap of the niches they occupy.
Source: Photo by R. Brandon.

eggs, the female moves slightly upstream and repeats the body movements. Dislodged sand and gravel settle over the eggs, which are thus more secure against currents and better concealed from potential predators.

COMPETITIVE ADAPTATIONS. Adaptations can also result from competition between individuals of different species. An axiom of ecology states that no two species can occupy *exactly* the same niche in the same geographical area. Animals that compete for the same resources (e.g., food, nest sites) have some degree of niche overlap; the severity of competition between individuals of two species varies with the degree of niche overlap. Two possible outcomes of severe competition are (1) local extinction of the individuals of one species or (2) the development of some form of coexistence through a change in the habits of the individuals of one or both species. To reduce competition and niche overlap, the individuals of one (or both) species may develop adaptations that alter patterns of resource utilization.

Let's consider a specific example. Two related salamander species, *Plethodon hoffmani* and *P. punctatus* (figure 4–8), share the same moist environment in the

adaptations are the traits and habits of an animal that have probably developed in response to climate or to various ecological features of the habitat. For example, females of several species of salmon that are native to the Pacific Northwest and neighboring parts of Canada and Alaska have spawning grounds on the rocky bottoms of glacial streams. Before the female drops her eggs at a selected site, she lies on her side and rapidly moves her body and tail to dislodge sand and smaller stones from among the rocks. The dropping eggs are fertilized by sperm from a male, and then fall into the cracks and crevices created by the female's actions. After laying the

FIGURE 4-9   Character displacement in two species of cricket frogs
Acris gryllus and *A. crepitans* can be found in the southeastern areas of the United States. Where the distributions are nonoverlapping, calls from the two species are similar; but in zones of overlap, the calls are sharply differentiated.

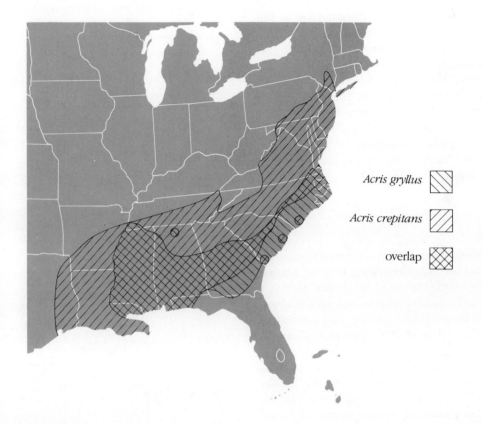

Appalachian Mountains. The two species share several characteristics—for example, both are nocturnal—so we may ask whether they compete for certain critical resources that may be available in limited supply, and if so, to what extent. Fraser (1976) demonstrated that adults of these two species have dissimilar utilization patterns of critical resources of food and habitat; the degree of niche overlap is lessened and competition reduced. The dissimilar patterns of resource utilization may be based in part on different adult body sizes and thus different resource requirements. There is, however, strong potential competition between adults of the smaller species (*P. hoffmani*) and the similarly sized juveniles and subadults of the larger species (*P. punctatus*). Staggered feeding times and occupation of different habitats probably act to lessen the degree of niche overlap and min-

imize the extent of this competition. We can see from this example how adaptations may have evolved in two species that occupy the same area and that may, at some time in the past, have had a higher degree of niche overlap.

Interspecific competition may also manifest itself in a phenomenon called **character displacement** (Brown and Wilson 1956), which can occur when two closely related species with similar niche requirements overlap geographically. In the zone of overlap, the two species show sharper differentiation for some traits than each does in the nonoverlapping zone. For example, two species of cricket frogs (*Acris*) (figure 4–9) have quite similar calls in eastern Texas and in southern Georgia, areas of nonoverlapping distribution (Blair 1958, 1974). However, the calls are sharply differentiated where the species overlap in central Mississippi. Because of this differentiation, when both species are present, each can identify conspecifics easier.

We should note that these examples are of static rather than dynamic situations. Whether two coexisting species evolved separately and have come to have overlapping distributions, or whether they evolved different body sizes and patterns of resource utilization as a result of sharing the same area at some time in the past, is difficult or impossible to determine.

In addition to competition; predation, disease, and mutualism can also affect the processes of adaptation through evolution by natural selection. Predation, which involves adaptations of both the predator organism and its potential prey, will be examined in some detail in chapter 17. The role of disease's effect on individuals and on populations is discussed in chapter 18. Mutualism is considered when we discuss the evolution of social behavior in chapter 20.

SPECIATION

How are new species formed? What processes are involved in speciation? The formation of new animal forms through evolution is called **speciation.** In order for populations of a single species to diverge and eventually to evolve into separate species, two conditions must be met: (1) a barrier of some type must form between populations of a species, and (2) a reproductive-isolating mechanism must develop while the populations are separated, so that when and if the barrier disappears, any separation in genotypes that has developed will not be eliminated by the exchange of genes. Although not observable directly, the evolution of new species occurs in several ways (Mayr 1963; Huxley 1974; Futuyma 1987); we consider two of them here.

GEOGRAPHIC ISOLATION. Speciation by **geographic isolation** (sometimes also called **allopatric speciation**) is a gradual process involving several steps or stages. As figure 4–10 shows, a beetle species may initially be divided into several subspecies. If the subspecies *A* and *B* are then separated by a barrier such as the mountain range, evolutionary changes in the now independent populations may develop so that individuals in subspecies *A* may emerge from pupae in early summer, while those in *B* may emerge in late summer. When the mountain range wears down, perhaps over thousands of years, and the individuals from the two groups are

FIGURE 4-10   Speciation by geographic isolation in beetles

Two subspecies, A and B, of a beetle have slightly different color patterns (Stage I) with an intergrade zone. During a period of geographical isolation, subspecies A and B (Stage II) develop separate breeding seasons so that if they should again become free to intermingle (Stage III), through a breakdown of the barrier, the two forms have become reproductively isolated and can now be considered separate species. Stage III is not necessary for speciation, but it provides a nice natural test.

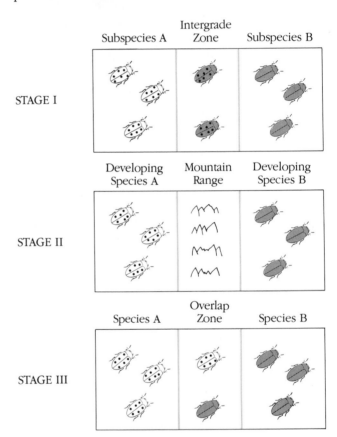

again free to intermingle, there will no longer be any gene flow between the two groups because of the difference in their annual cycles. We can now say that separate species *A* and *B* have evolved. Geographic isolation is believed by most investigators to have been the most common cause of speciation among animals.

Not all evolutionary biologists agree about the existence of a second form of speciation, **sympatric speciation.** In this process two forms diverge from a single species into separate species even though they inhabit the same geographical region. All members of certain insect species, for example, live out their entire life cycles on one plant species. A mutation that causes some individuals to transfer to and remain permanently on another plant species could result in two distinct species that are isolated from each other by virtue of their exclusive use of different plant species. In this instance, the selection of a particular plant species is a key factor in reproductive isolation.

REPRODUCTIVE ISOLATION. How do species remain distinct? What are the mechanisms that evolve to maintain the separateness of species? The importance of species-isolating mechanisms lies in the most efficient use of energy for reproduction. When no viable progeny are produced, there is a high cost in genetic fitness for an animal to expend energy and gametes on nonproductive matings. Evolution by natural selection has produced mechanisms to help ensure efficient mating activities.

Species-isolating mechanisms are generally of two kinds: *prezygotic* and *postzygotic*. **Prezygotic** mechanisms occur before mating and zygote formation. **Postzygotic** mechanisms occur during the development of the zygote or during the life of a viable hybrid produced by a mating between members of different species. All of these isolating mechanisms are outcomes of the speciation process; isolating mechanisms are not a *cause* of speciation.

There are several kinds of prezygotic-isolating mechanisms. (1) In **habitat isolation,** members of related species do not interact to mate with one another because of preferences for different habitat types (see chapter 18). (2) **Seasonal** or **temporal isolation** involves mating activity at different times of the year or different times of the day and eliminates the possibility of matings between members of related species (see chapter 8). (3) **Behavioral isolation** occurs when members of different species select mates and choose conspecific partners on the basis of coloration patterns and sometimes quite complex courtship displays (see chapter 13). (4) When members of two species have such different genital structures that mating is not possible, the effect is called **mechanical isolation.** (5) In some cases copulation may take place, but due to chemical and structural incompatibility, no fertilization takes place, and thus no zygote is formed.

Postzygotic mechanisms entail a genetic disharmony in the hybrids. The term **genetic disharmony** refers to a broad class of phenomena that may lead to inviability of the hybrids, sterility of any hybrids that are produced, or the production of weak or sterile second-generation hybrids. Also, some hybrids may fail to mate successfully because they exhibit courtship patterns that are different from either parent species, which leads to a kind of behavioral isolation.

GENETIC VARIATION, KINSHIP AND UNITS OF SELECTION

POLYMORPHISM

We might expect that alternative alleles at a locus would disappear through selection. In fact, as an example, as many as 30 percent of the gene loci in fruit flies are **polymorphic**—that is, there are two or more alleles for that locus present in the gene pool for the population. So, we may ask: What maintains this polymorphism? One of several possible explanations is **heterozygote superiority,** in some environments the heterozygote is more fit than either of the homozygotes. An example of this phenomenon occurs in humans; heterozygosity for the sickle-cell trait provides protection from malaria. In parts of Africa and India and in some Mediterranean countries, the prevalence of malaria as a cause of mortality is sufficient to have produced a frequency of the sickle-cell gene in excess of 0.1, even though in the homozygous state, this gene is usually lethal.

Another mechanism for producing balanced polymorphism is **cyclical selection.** The population numbers of some species of animals, such as meadow voles (*Microtus pennsylvanicus*), undergo drastic cyclical fluctuations. Krebs and his colleagues (1973) have argued that one allele is favored during earlier phases of population growth, and another is favored during the peak and decline phases of the population cycle. Others (Crow and Kimura 1970) have argued that this polymorphism is due to selectively neutral genes that are changing in frequency through genetic drift and are of no adaptive value to the individuals possessing them.

MEASUREMENT OF GENETIC RELATIONSHIP

In social groups of most species, the members are related and thus may be inbred to some extent. On the one hand, inbreeding reduces heterozygosity, diminishes the adaptability and benefits of heterozygote superiority, and increases the possibility of the accumulation of deleterious recessive alleles. On the other hand, the closer the genetic relationships of group members, the more intricate the social bonds and degree of cooperation between the group members, as we shall see later.

Several measures of relationships are used in population genetics. For now, we only need to discuss one of them. The **coefficient of relationship,** designated by *r*, is the proportion of genes in two individuals that are identical because of common descent. For example, in parent-offspring relationships, only one-half of the genes from one parent go to an offspring, because the gametes are haploid; thus $r = 1/2$. The *r* for full siblings is, on the average, also 1/2 because they share one-fourth of the genes from each parent. A general rule for computing *r* is to count the number of paths back to the common ancestor and to the related individual. The number of paths is the power to which 1/2 is raised to give *r*. If there is more than one common ancestor, the paths must be traced separately and their probabilities added together to obtain the coefficient of relationship. Figure 4–11 gives examples of the calculation of *r*.

UNITS OF SELECTION

Up to this point we have assumed that the unit of selection is the individual. It is possible, however, to argue that there are other units of selection, all the way from genes themselves up to entire ecosystems. The notion that the gene is the unit of selection is relatively recent (Williams 1966; Dawkins 1976). The individual may be thought of as a unique and temporary vehicle for pieces of DNA, which replicate themselves and are potentially immortal.

We can consider interdemic selection and kin selection to be at two ends of a continuum of relatedness above the level of the individual. By **interdemic selection** we mean selection operating at the level of the **deme,** which is a local population whose members breed randomly. **Kin selection** is defined as selection acting through close relatives. Although all members of a species are related to some extent, we assume that the deme is large enough and outbred enough to ensure a low

FIGURE 4-11   Coefficients of relationship
To compute the proportion of genes that are identical in two individuals because of common descent, count the number of paths connecting those individuals by way of a common ancestor. This number is the power to which ½ is raised. Thus for half-siblings there are two paths, so ½ is raised to the second power. If there is more than one common ancestor, the process must be done separately for each common ancestor and the probabilities summed.

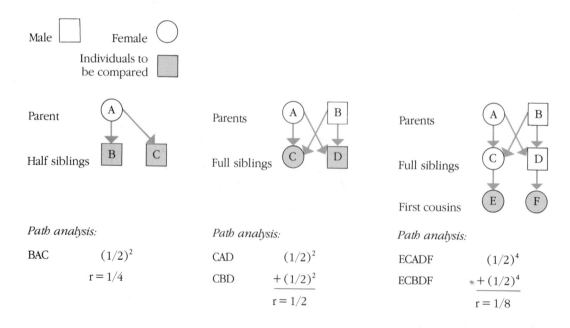

average coefficient of relationship. Instead of thinking that individuals or genes survive and reproduce or disappear, we can think of entire demes or populations that survive and colonize new areas or become extinct. In the latter case, of course, the entire gene pool is lost. The theory of **group selection** (Wynne-Edwards 1962)—that entire demes may be the units of selection—holds that a trait will evolve if reproductive restraint confers reproduction advantages on the entire group, and if selfishness with regard to reproducing is disadvantageous for the group. Wynne-Edwards developed his theory to explain a situation in which animals apparently behave in a manner that is beneficial to the group but that might reduce that individual's own fitness. Evolutionary biologists have been reluctant to invoke group selection, because it is likely to be much slower than individual selection. Thus, Wynne-Edwards's ideas, discussed further in chapter 20, have been generally rejected.

Smaller units within the deme are frequently composed of related individuals who share a certain amount of their genetic material through common descent. We can think of these kin units as extensions of individuals. Kin selection is thus an extension of the selfish-gene idea. We would expect units of closely related individuals to be cooperative among themselves and to be competitive with other nonrelated units of close relatives.

Individuals may reduce their own fitness by acts that increase the fitness of others (**altruism**). Hamilton (1964) showed how such behavior could evolve via kin selection. He introduced the concept of **inclusive fitness,** which is the sum of direct and indirect fitness. **Direct fitness** is determined by the reproductive success of one's own offspring. **Indirect fitness** is determined by the reproductive success of relatives other than one's own offspring. For altruistic genes to spread by kin selection, the ratio of the recipient's benefit (b) to the altruist's cost (c) must be greater than the reciprocal of the coefficient of relationship (r) (Hamilton 1964). That is,

$$\frac{b}{c} > \frac{1}{r}$$

In siblings who share one-half their genes, $r = 1/2$, and therefore b/c must exceed 2 for altruistic genes to spread. In other words, if an individual more than doubles the fitness of a sibling through an altruistic act that causes

that individual to leave no offspring, genes promoting that behavior could spread through the population. The more distant the relative, the lower is r, and the higher the benefit-to-cost ratio must be. These aspects of selection are discussed in greater detail in chapter 20.

EVOLUTIONARY STABLE STRATEGIES

In recent years, behavioral ecologists have applied new mathematical models to problems of behavior and evolution through natural selection. One such model is **optimality theory,** based on the idea that individuals that are maximally efficient at certain activities—foraging or seeking mates, for example—should have a selection advantage to ensure the greatest propagation of their genes in subsequent generations (MacArthur 1965; Maynard Smith 1976).

The approach presents at least one problem, however. What if there are several possible "best" solutions to a situation? Here we move into the realm of contingent probabilities, where the optimum behavior of an individual in a particular situation depends on the behavior of other individuals. In situations that involve multiple variables, behavioral ecologists and sociobiologists employ the concept of **evolutionary stable strategy** (ESS), a strategy that if adopted by most members of a population cannot be bettered by an alternative strategy (Maynard Smith 1976). To clarify what we mean by an ESS, let's consider an example.

The common yellow dung fly (*Scatophaga stercoraria*) is found in pastures where cattle are grazing. The males and females meet and mate at pats of fresh cow dung; the female oviposits, and the larvae develop in the cow pat. At fresh cow pats, the ratio of mature reproducing males to females is 4 or 5 to 1. What is the best strategy (ESS) a male can follow to achieve a maximum frequency of successful matings? As we might expect, the strategy is multifaceted. Consider first the question of where the males search for females around a fresh cow pat. The expected (based on the mathematics of ESS theory) and observed mating captures of females by males in various zones around a fresh cow pat are graphed in figure 4–12 (Parker 1974, 1978). A male may mate by capturing a female or by taking over

FIGURE 4-12   **Captures of male *Scatophaga* at fresh cow pats**

Comparisons are shown between predicted and observed captures in zones on and around the cattle droppings. The predicted curve is based on an ESS (see text). Zone A = the surface of the cow dropping; Zone B = zone nearest the dropping on the grass to a distance of 20 cm; Zone C = a similar grassy band in the area from 20 to 40 cm from the edge of the cow pie; Zone D = 40 to 60 cm region; and Zone E = 60 to 80 cm from the dropping.

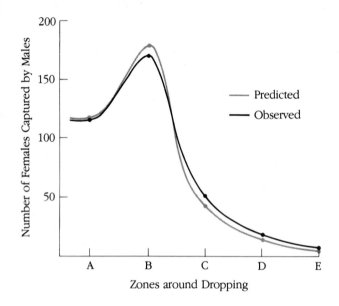

PROBLEMS WITH EVOLUTIONARY EXPLANATIONS

Although evolutionary theory is a useful tool for developing explanations for observed behavior patterns, it should not be used indiscriminately. We should note the following cautionary comments about the evolutionary perspective:

First, we must take care that our reasoning is not circular. New species evolve in several ways. We may argue correctly that species-isolating mechanisms have developed in the course of evolution to ensure efficient use of energy and gametes in mating. It would be incorrect, however, to claim that two species evolved where previously there was but a single species because the species developed species-isolating mechanisms. The isolating mechanisms evolved during speciation and should not be presented as the cause of speciation.

Second, the recording of behavioral phenomena presents a "static" picture on the evolutionary time scale. To visualize the past events and dynamic processes that have influenced the present adaptations of a species is difficult. In discussing the evolution of behavior patterns, we must guard against overextending our hypothetical explanations of the behavior pattern's functional significance and the natural selection pressure that led to each aspect of that complex behavioral sequence. We should provide evolutionary explanations only if we can support them with data, or if we can use them to devise experiments to test a hypothesis. In short, we should not feel compelled to provide a complete evolutionary explanation for each existing behavior pattern.

Third, in discussing the evolution of behavior, we must bear in mind that the environment is a passive force, not an active force. The environment itself does not impose a strategy for evolution on an organism. The adaptations that characterize an animal species represent temporary stopping points in evolution and are the results of the action of natural selection upon the phenotypes within that species. The environment provides the setting for evolution, and thus the parameters of the environment are critical determinants of the nature of the adaptations that evolve.

one that has been captured by another male. The decision factors are the length of time it takes to mate, how much time it takes to dislodge a competitor, the age of the dung pat, and when to move on in search of another fresh cow pat.

From the example, we see how natural selection might operate on several related patterns simultaneously. The behavior in this example can at least theoretically be described by an evolutionary stable strategy. The use of ESS theory is important to behavioral ecologists, both as a modeling tool and as a stimulus to formulate testable questions.

SUMMARY

The primary units of study for evolutionary biologists are *species*, or groups of animals that share similar characteristics and habits and that interbreed freely with one another, but that are reproductively isolated from all other species. The classification of animal types is hierarchical and is based on commonality of ancestry and similarity of characteristics.

Each individual of a species possesses *genes*, the basic units of heredity. The complete set of genes for an organism is its *genotype;* and the hypothetical collection of genes within a species is the *gene pool.* For each *locus* on a chromosome, one of various *alleles*, or alternative forms of a gene, may be present. These alleles can be *dominant* or *recessive* or can exhibit *partial dominance.*

Biological evolution through natural selection is measured by the *fitness* of the trait in the organism or group, and by changes in *gene frequencies.* Traits that have a genetic basis and that favor survival and reproduction will be present in higher frequencies in subsequent generations. The theory of evolution and the heritability of variation accounts for the differences in some animals' rates of survival and reproduction in the face of limited resources.

Population genetics provides us with the *Hardy-Weinberg formula,* a useful tool that enables us to sample at a point in time, rather than over several generations, to determine whether or not there is an *equilibrium* for certain sets of alleles. Among the forces for genetic change are *mutation pressure, gene flow, genetic drift,* and *natural selection.* The latter, natural selection, can be subdivided into *stabilizing selection, directional selection,* and *disruptive selection.*

Adaptation is the evolutionary process by which changes occur over time in an animal species so that members of the species are well suited for life in that particular environment. Adaptations occur as populations respond (through natural selection) to the *environment* (an organism's living and nonliving surroundings), to *interspecific competition,* and to *niche differentiation.*

Speciation, the evolutionary process by which new forms appear, can occur only when populations of a species are separated, and when isolating mechanisms develop during the period of separation. After the separating barrier is removed, the species maintain their distinct identities through *pre-* and *postzygotic-isolating mechanisms. Genetic variation,* or *polymorphism,* is maintained through several mechanisms, including heterozygote superiority and balanced polymorphism.

Although when we examine evolutionary theory, we assume that the unit of selection is the individual; several theories have proposed other units, ranging from whole ecosystems or demes at one end of the spectrum, to genes (DNA) at the other. Evolutionary theorists also examine *group selection* and *kin selection,* both of which are based on *altruistic acts,* or acts performed by individuals that benefit other individuals or the group even though their own fitness may be reduced.

The concept of *evolutionary stable strategies (ESS)* has been developed by behavioral ecologists to apply to situations in which the optimum behavior of an individual animal is a function of multiple variables.

Evolutionary theory is useful for explaining observed behavior. It has its limitations, however—we are taking a static look at behavior when in fact evolution is a dynamic process.

Discussion Questions

1. Animal species must maintain a certain degree of flexibility to be able to change in response to environmental changes. At the same time, due to competition and other factors, these species must attain a certain degree of specialization. What are the dynamics of "walking the fine line" between flexibility and specialization?

2. Explain how random events can affect the stages of the evolutionary processes discussed in this chapter.

3. How might natural selection operate to facilitate the evolution of species-isolating mechanisms?

4. Outline the major steps in adaptation and speciation.

5. What is the role of environment in evolution?

6. Charles Darwin married his first cousin, Emma Wedgewood. What is the coefficient of relationship between the Darwins' children?

Suggested Readings

Barash, D. P. 1983. *Sociobiology and Behavior.* 2d ed. New York: Elsevier North-Holland.
Condensed treatment of the new hypotheses and concepts of sociobiology. Very readable style, excellent introductory chapters and solid use of examples throughout.

Dawkins, R. 1976. *The Selfish Gene.* New York: Oxford University Press.
Written for scientists and nonscientists. Presents an original and thought-provoking viewpoint of the role of genes and DNA in evolution.

Williams, G. C. 1966. *Adaptation and Natural Selection.* Princeton: Princeton University Press.
Readable, somewhat philosophical treatise on aspects of modern evolutionary thought. Still very current as Williams's ideas continue to receive greater attention.

Wilson, E. O. and W. H. Bossert. 1971. *A Primer of Population Biology.* Oxford, England: Blackwell Scientific Publications, Ltd.
An excellent introduction to population genetics and related topics.

PART TWO

MECHANISMS OF BEHAVIOR

5

BEHAVIOR GENETICS

In part 2 we explore the concepts and methods associated with behavioral mechanisms, in particular, those involved in the reception of information from internal and external sources, and those involved in the production of observed behavior. Because critical information affecting the morphology, physiology, and behavior of an organism is contained in its genotype, we begin with behavior genetics. Chapters 6 and 7 explore the roles that the nervous and hormonal systems play in the reception of external cues and the generation of behavioral responses. Chapter 8 examines the phenomenon of the biological clock, or mechanisms of behavior control present in many organisms as patterns or cycles of activity—for example, daily rhythms.

Historically, genetics and animal behavior followed separate paths of scientific development, but around 1960, a synthesis of these two disciplines began the path to modern behavior-genetic analysis. Beginning as early as the late nineteenth century and continuing into the 1930s, many psychologists and zoologists were questioning the effects of inheritance on behavior, notably Robert Tryon, Sewall Wright, and R. A. Fisher; these early investigators provided the groundwork for techniques and theories that have been more fully expounded since the 1960s. In this chapter, we examine several aspects of behavior genetics, beginning with the kinds of questions asked by behavior geneticists. We then discuss the methods used in behavior-genetic investigations and the significance of the findings of this research. We conclude with some comments on behavior genetics and evolution.

QUESTIONS IN BEHAVIOR GENETICS

We should note at the outset that behavior genetics is an expanding subfield of animal behavior not yet fully defined. Further development of the field may change the research objectives, but questions about the role of genes in influencing and controlling behavior will continue to be posed (Hirsch 1967).

QUESTIONS OF CONTROL AND DEVELOPMENT. To use the behavior-genetic approach, we begin by asking: Do individuals within a species differ regarding some particular behavioral trait? Do these intraspecific variations have a possible or probable genetic basis? What is the mechanism for the behavioral trait's inheritance? In some cases we may be able to ask, How many genes are responsible for influencing a particular behavior pattern? or How does gene action affect the development of various patterns? What proportion of the variation is attributable to genetic influences and what proportion is attributable to environmental influences? (Gene-environment interactions during epigenesis development will be treated more fully in chapter 9.)

QUESTIONS OF POPULATION AND EVOLUTION. We begin research in this area with the determination of the frequencies of known behavior-influencing genes in a population. Is the behavior in question adapted to the species' present environment? Do phylogenetic comparisons of related species reveal similarities or differences in genotype and behavior that may be used to determine the evolutionary history of certain behavior patterns? What forces or factors may be influencing changes in gene frequencies over time?

METHODS OF ASSESSING GENETIC DETERMINATION

Behavior geneticists use two basic approaches in their research: (1) they hold the environment constant in order to explore the effects of genetics, or (2) they hold genetics constant in order to study the effects of the environment. The starting point for investigations is a simple formula:

$$V_T = V_G + V_E + V_I$$

where V_T represents the total phenotypic variation observed in a population for a particular trait; V_G represents the genotypic component of the total variation; V_E represents the environmental component; and V_I represents the variation attributable to interactions between genotypic and environmental factors ($V_I = V_G \times V_E$). In other words, the phenotypic variation is the sum of genotypic, environmental, and interactive forces. (For a brief discussion of variation and the measurement of sample variance, see chapter 3.)

INBREEDING

One way to study the effects of environmental parameters on behavior is to hold the genetic component constant by using homogeneous strains of animals. One method of achieving genetic homozygosity (a homogeneous population of animals with the same genotype) is by inbreeding, or using brother-sister matings for many generations. Because the animals all have the same genotype in an inbred strain, we know that $V_G = 0$, so $V_I = V_G \times V_E = 0$, $V_T = V_E$. In mice, after about thirty generations of inbreeding, some 98 to 100 percent of the allele pairs are homozygous. During the inbreeding process, many recessive alleles that are lethal or otherwise detrimental to successful reproduction attain the homozygous condition. Thus, to obtain a viable inbred strain of mice, we must begin with multiple brother-sister matings because many lines will die out before a high degree of homozygosity is achieved.

If our experimental subjects have a common genetic background, we can manipulate various aspects of the environment to determine the relative importance of the external parameters influencing the behavior. Consider an experiment in which mice from each of four inbred strains C57/Bl/1, C3H/Bi, DBA/8, and JK, were divided into two test groups: one group would be exposed to a noxious stimulus for two minutes when four days of age, and the other would not be exposed (Lindzey et al. 1963). At four days of age, the rats were placed in a metal tub; the noxious stimulus was a two-minute activation of a doorbell attached to the side of the tub. Control mice were placed into the tub for a similar period of time, but were not exposed to the bell. All the mice were tested at thirty days of age by placing them in the same washtub where the exposure to the noxious stimulus had occurred. During a two-minute test period repeated each day for ten consecutive days, the amount of urination and defecation and the total movement of the mouse across the various sections of the washtub were recorded. Mice exposed to the noxious stimulus defecated and urinated more and were less active than the control mice not exposed to the trauma of the bell sounding. Though there were some distinct strain differences for the dependent variables, these were consistent between control and experimental mice within each strain: Some strains consistently exhibited lower mobility scores; for such strains, both the control and experimental mice exhibited low mobility, even though the two treatments did result in differences between experimentals and controls within strains.

The relatively clear-cut conclusions reached in this study are not always possible in many behavior-genetic analyses. Often the performance of test animals falls between the two extremes, complicating the interpretation of results. We should note that the conclusions from studies using inbred strains are limited to the particular strain and to the specific variables measured in an investigation. Numerous additional factors, such as differences in methods or procedures between laboratories, and the age, sex, and previous experience of test subjects, may affect the behavior and thus the interpretations and conclusions of such studies.

STRAIN DIFFERENCES

As an alternative to the foregoing procedure, we can compare and assess the effects of heredity on behavior by systematically examining two or more genetically homogeneous inbred strains of the same species maintained under the same environmental conditions. In this instance we are holding V_E constant in order to assess genetic effects and thus $V_T = V_E$.

The performance of four strains of inbred mice was studied using a water maze (Upchurch and Wehner 1988). The water maze consisted of a circular pool of water 122 cm in diameter and 32 cm deep (after Morris 1981). The water was made opaque using powdered paint that does not stain or adversely affect the mice. An escape platform made of clear Plexiglas (10.5 cm³) was placed 0.5 cm below the water surface; the location of the platform could be moved about the pool. Test mice from all four strains were reared under identical housing conditions, so that any differences in behavioral performance could be attributed to their genetic differences. In the experimental tests, mice were required to use visual cues to locate and swim to the escape platform. Mice that could not make it to the escape platform were "rescued" after 60 seconds. Two of the strains tested, C3H and BALB, were not able to perform the required task; mice of both strains lack good visual acuity. Mice of the other two strains, C57 and DBA, succeeded in locating the platform, though they differed in one significant way. The C57 mice could locate the platform either by a visual cue designating the location of the platform, or by the platform's location remaining the same from trial to trial so that its position could be learned. DBA mice could also use the visual cue to find the platform, but with no cues they did not remember the platform's location. The difference between these two strains behavior may be related to the difference in place-learning capacity of their brains.

Lynch and Hegmann (1972) used an experimental design that held environment stable and varied inherited traits to assess how variations in genotype affect a particular behavior pattern. They compared the nest-building activity of house mice (*Mus domesticus*) from five different inbred strains reared in the same environment. Over five consecutive days, they measured the amount of cotton used by each mouse in each strain to make its nest in its cage. The combined result for both sexes indicated that the strains differed significantly in the average daily amount of material used in nest construction: strain 1 used 1.3 g; strain 2 used 1.3 g; strain 3 used 0.7 g; strain 4 used 1.1 g; and strain 5 used 0.7 g (figure 5-1).

The results demonstrate genetic variance in nest-building behavior in house mice. The existence of such genetic variation is a necessary, but not a sufficient, condition for a behavioral response to selection pressure.

FIGURE 5-1 House mouse nests in laboratory
The difference in size of nests constructed by mice can have a genetic basis. These two nests were built by house mice selected for construction of large and small nests.
Source: Photo by Lee C. Drickamer.

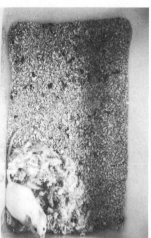

Studies of inherited behavior differences between strains point out which behavioral traits are genetically labile, and thus possibly affected by selection pressure.

SELECTIVE BREEDING

Domesticated pets, livestock, and many ornamental and agricultural plants are the products of selective breeding (figure 5-2). In behavior genetics research, artificial selection techniques show whether a trait is under some degree of genetic control. If it is, then it can be affected by selection pressures. Artificial selection is also used to estimate the degree to which genes are expressed in phenotypes.

ARTIFICIAL SELECTION TECHNIQUES. Artificial selection is based on the assumption that some degree of genetic variability exists for the trait in question. Animals at the distribution extremes of a large sample population are then bred through several generations, and the strength of expression is tested.

Consider the following example. Hirsch and Boudreau (1958) tested fruit flies (*Drosophila melanogaster*) to see whether their reactions to light were under genetic control. They used a Y-maze with one arm lit and one

FIGURE 5-2   Dog breeds

The (a) dachshund, (b) collie, and (c) German shepherd have been bred to serve different functions. The dachshund is used to hunt animals that live in dens or burrows; the collie is a herd dog, used primarily with sheep; and the German shepherd serves as a guard dog, rescue dog, and police dog.

Source: Photo (a) by P.A.E. Davis/Photo Researchers, Inc. Photo (b) by Mary Eleanor Browning/Photo Researchers, Inc. Photo (c) by Jeanne White/Photo Researchers, Inc.

A

B

C

FIGURE 5-3   Distributions of light approach scores of fruit flies

Graph (a) measures the responses of the initial population of flies, and graphs (b) and (c) measure the high and low response lines after 2 and 29 generations of artificial selection.

Source: J. Hirsch and J. C. Boudreau, "Studies in Experimental Behavior Genetics," in *Journal of Comparative & Physiological Psychology*, 51:647–51, 1958. Copyright © 1958 by the American Psychological Association.

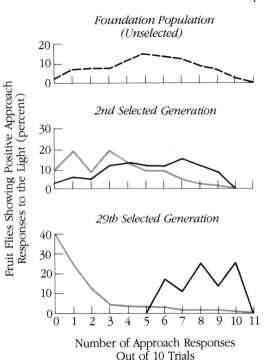

—— High response line

—— Low response line

*Foundation Population
(Unselected)*

2nd Selected Generation

29th Selected Generation

Number of Approach Responses
Out of 10 Trials

Fruit Flies Showing Positive Approach Responses to the Light (percent)

arm dark to determine the percentage of flies that approached the light source in ten trials (figure 5–3 a). Flies that made the greatest number of approaches (high line) and those that made the fewest approaches (low line) were then bred separately. Progeny of high-line flies were tested in the apparatus, and individuals that made the greatest number of approaches (high line) and those that made the fewest approaches (low line) were again bred separately. Progeny of second-generation high-line flies were tested in the apparatus, and individuals that made the greatest number of approaches to light were bred to produce the next generation. The same procedure was followed for the low line; flies making the

fewest approaches to light were selected to serve as parents for each succeeding generation. For both high and low lines, the procedure was carried out for a total of twenty-nine generations. Figure 5–3b and c show the frequency distributions of responses to light for the high and low lines after two generations and after twenty-nine. By the last generation, little overlap remained between the responses of flies from the two lines. Hirsch and Boudreau concluded that the photic responses of this species of fruit fly are under some degree of genetic control.

Another example of artificial selection helps us to understand the processes involved further. It too combines this selection technique with inbreeding, but mates cousins or nieces. Pfaffenberger and colleagues (1976) used this combined technique to investigate the production of guide dogs for the blind. The experiments involved several lines of German shepherds. Among the favorable traits achieved or enhanced by artificial selection and inbreeding were reduced walking speed, improved overall temperament, and greater uniformity of response to various training procedures. In both of the foregoing examples, one key question arises: How can the degree of genetic control of the specific trait be measured?

MEASUREMENT OF HERITABILITY. **Heritability** (h_b^2), used in a broad sense, is an estimate of the **degree of genetic determination** (°GD) of any particular trait. Heritability is a population characteristic; it is *not* a term that is applied to an individual's inheritance of a specific trait. In this broad sense, heritability of a particular trait in a specific population sample is defined as

$$h_b^2 = {}^{\circ}GD = \frac{V_G}{V_G + V_E}.$$

That is, the degree of genetic determination of a trait is the ratio of genetic variation to genetic plus environmental variation (Fuller and Thompson 1978).

One way to assess the degree of genetic determination is to use inbred strains and some genetic crosses. In a hypothetical study of wheel-running activity (table 5–1), two strains of mice (*Mus domesticus*), A and B, are tested, along with a strain of mice from crosses of A and B, called the F_1. Each strain (A, B, and F_1) is homogeneous; hence any variation in wheel-running activity within the strain must be attributed to V_E. If we make a cross of F_1 mice, producing individuals (F_2) that are no longer homogeneous, the variance in wheel-running activity then includes both genetic and environmental variation: $V_G + V_E = V_T$. We can estimate

TABLE 5-1   Calculation of degree of genetic determination (°GD) for wheel-running activity in mice[1]

Mouse Strain	Mean Number of Wheel Revolutions Per 24 Hours	Variance
A	1,107	112
B	5,680	418
F_1(A $\times$ B)	5,235	325
F_2(F_1 $\times$ F_1)	4,745	465

$$V_E = \frac{112 + 418 + 325}{3} = 285$$

$$V_G = 465 - 285 = 180$$

$$°GD = V_G/V_T = 180/465 = 0.39$$

1. F_1 is a cross of two inbred strains, A and B.

the degree of genetic determination (°GD) as V_G/V_T. In this example, °GD works out to 0.39; that is, 39 percent of the variation in wheel-running activity of these mice can be attributed to genetic influences.

In other instances, we may be interested in predicting the efficiency of selection (Fuller and Thompson 1978). In such cases what we measure is called the realized heritability, or h_r^2. The major method of estimating h_r^2 utilizes artificial selection. We can, for example, consider selection for larger litter size in hamsters (figure 5–4). The base stock of hamsters has a mean litter size of 8.4 pups. The **selection differential, S,** is defined as the difference between the mean litter size produced by the selected breeding parents and the mean litter size of the base stock. Here the selected male and female parents have a mean litter size of 13.6 pups, so $S = 13.6 - 8.4 = 5.2$.

The **response to selection, R,** is defined as the difference between the mean litter size of the offspring generation produced by selective mating and the mean litter size of the base stock. If the offspring generation has a mean litter size of 10.1 pups, then $R = 10.1 - 8.4 = 1.7$. The ratio of the response to selection to the selection differential (R/S) is a measure of the h_r^2. In our example $R/S = 1.7/5.2 = 0.33$, so about one-third of the variation in litter size for this particular stock of hamsters can be attributed to genetic factors. Clearly, if a small value for S produces a large response for R, then the heritability estimate will be high; conversely, if a large selection differential produces only a slight shift in R, then the h_r^2 estimate will be low.

FIGURE 5-4   Measurement of realized heritability In this example h_r^2 was calculated using artificial selection and computations of selection differential (S) and response to selection (R) for litter size in hamsters.

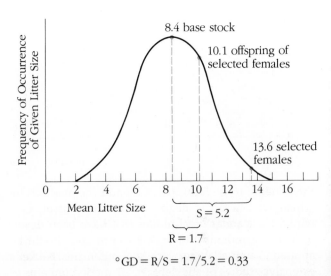

$$°GD = R/S = 1.7/5.2 = 0.33$$

As an alternative to this approach, heritability can be estimated from the regression of the response to selection on the cumulative selection differential (Bakker 1986). Much research has been conducted on behavior genetics in rodents and fruit flies, while little has been done on many other organisms. Bakker (1986) made extensive studies of the genetics underlying aggressive behavior in stickleback fish (*Gasterosteus aculeatus*). As part of his effort, Bakker selected both male and female sticklebacks to determine the degree of genetic determination for aggressiveness of juveniles. He selected in both directions, for more aggressiveness in one direction and less aggressiveness in the opposite direction. His results (figure 5–5) revealed that within two generations, both male and female sticklebacks could be selected when juveniles for higher or lower levels of aggression. Using the regression of the response to selection on the cumulative selection differential, he determined that h^2 for the aggressiveness trait in male sticklebacks was 0.51, and 0.64 for females. Note from the figure, however, that the aggression response shifted back toward the control levels for both sexes in the third generation. This situation illustrates the complexity of

FIGURE 5-5

Three generations of selection for level of aggressiveness in male and female sticklebacks. The dependent variable is a mean juvenile aggression score.

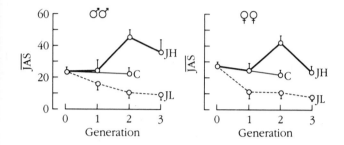

the genetics underlying behavior and the methods by which we assess the degree of genetic control. Certainly, the standard techniques that have been developed using rodents and fruit flies can and should be applied to a much wider variety of organisms. Bakker's extensive studies of sticklebacks are a good impetus to move in that direction.

Estimates of realized heritability can be compared for various traits; however, such comparisons have inherent problems. As we noted earlier, conclusions in behavior genetics are limited by constraints, including the population and strain(s), the type of test, the environmental conditions. Another, perhaps less problematic use for estimates of realized heritability is the prediction of response to selection. If we have calculated h_r^2 and the selection differential is known, the response to selection for a particular trait can be predicted.

Another way to calculate heritability involves utilizing the resemblance between relatives (Falconer 1981; Hartl 1988). Genetic relationships can be used to measure heritability by measuring the trait's covariance between parents and offspring, between full sibs, or between other genetic relatives. This method involves calculating the slope (b) of a regression line for the value of the trait in parents against the value for the same trait in the progeny (or between sibs etc.); we are interested in the covariance between the trait in the relatives. Consider, for example, the age of puberty in female mice (figure 5–6). The slope of the regression line for mothers and daughters is b = 0.20. This regression procedure is similar to that used for calculating realized heritability using parents and offspring. Because in this case we are calculating the regression using only one parent, only one-half of the genes from any one parent are passed on to the offspring. We must calculate h² as 2b, or 0.40 in the example above.

FIGURE 5-6   Parent-offspring resemblance.

The age of puberty, as measured by first vaginal estrus, is compared for mothers and daughters in house mice. The slope of the regression line serves as the basis for calculating heritability.

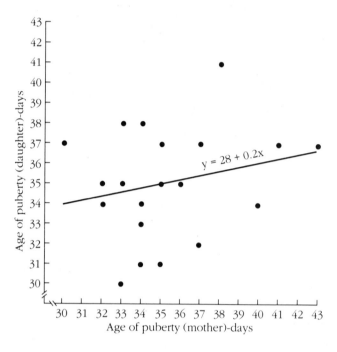

The field of quantitative behavior genetics has received extensive treatment in recent decades. Roberts (1967), Fuller and Thompson (1978), Falconer (1981) and Hartl (1988) provide detailed discussions of measurement of heritability and related topics.

CROSS-FOSTERING

We can transfer neonatal animals from the parent female to another female of the same species or strain, or to a female of a different species or strain (if she will successfully rear the young). We can then compare genetically similar animals with different rearing environments, and assess the relative importance of the effects of genotype versus the effects of maternal-care environment on certain behavioral traits. This technique of cross-fostering helps us differentiate species-specific behavior from environmentally influenced behavior. If genetically similar animals reared under different conditions (by either a biological parent or by a foster parent) exhibit similar behavior, then we can conclude that genetic control of that behavior is fairly

FIGURE 5-7 Selection for high and low rates of defecation in rats exposed to a novel, stressful situation When pups born to mothers of each line are cross-fostered to mothers of the other line and are later tested for defecation rates, the observed rates correspond to those of their genetic parents.

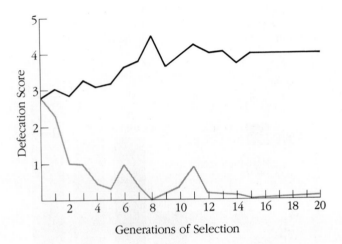

rigid. However, if the behavior differs significantly, we can conclude that it is strongly influenced by environment. Two types of cross-foster transfers are possible, within a species and between species. Examples of both should help us to understand this process.

TRANSFERS WITHIN A SPECIES. Broadhurst (1965) conducted a study of whether the defecation rates of rats subjected to novel or stressful environments were affected by postnatal maternal influences. He used selective breeding to obtain two distinct lines of rats; one group exhibiting high rates of defecation, a sign of "emotionality," and the other exhibiting low rates (figure 5-7). To test whether the greater "emotionality" of the mothers in one group led to behavioral differences in their progeny, he bred rats of both groups, and at the time of birth, cross-fostered some pups of each subline to mothers of the opposite subline. Young pups that had been cross-fostered were later tested. The rates of defecation of cross-fostered animals were the same as the rates of defecation of their genetic parents. Therefore, we can conclude that the different defecation rates of the two sublines can be attributed to genetic factors rather than to postnatal environmental factors.

TRANSFERS BETWEEN SPECIES. In a different type of cross-fostering study, Quadagno and Banks (1970) tested the effects of early environment on certain behavior patterns of two rodent species—an inbred strain of the house mouse (*Mus domesticus*) and the pygmy mouse (*Baiomys taylori*). Pups of both species were reared under one of four treatment conditions: (1) *Mus* pups by *Mus* dams (control), (2) *Mus* pups by *Baiomys* dams, (3) *Baiomys* pups by *Mus* dams, (4) *Baiomys* pups by *Baiomys* dams (control). Each litter consisted of three pups. In normal litters, all three pups and the dam were the same species; in cross-fostered litters, one pup was transferred from the other species, the other two pups and the dam were the same species. Thus, the environmental stimulation included not only maternal influences, but also possible sibling influences. Note that any possible confounding effects from having litters of different sizes were eliminated.

For the first twenty-one days of life, the mice were reared in the litters as described, and then each mouse was housed individually until testing began at sixty-five days of age. Each mouse was given a series of social and behavioral tests. The tests produced several significant results, only a few of which are mentioned here. First, general-activity scores of cross-fostered *Baiomys* were higher than for control *Baiomys*, and general-activity scores of cross-fostered *Mus* were lower than for control *Mus*. Some environmental factor, possibly handling by the dams, or activity of sibs from the foster species, must account for this difference. Second, in paired encounters involving combinations of conspecific-reared and cross-fostered mice, individuals that had been cross-fostered exhibited more positive affiliative responses to members of the other species. Again, an environmental factor probably accounts for this effect. Third, mating tests of mice from the different treatment groups showed that mice had not lost their ability to mate with members of their own species. Sexual behavior is apparently less susceptible to the effects of early experience with heterospecifics and is under greater genetic influence than are some other behavior patterns.

One problem that we should note concerning the use of mammals in the study of behavior genetics concerns the fact that mammals go through a period of development in utero before they are born. Many other animals hatch from eggs. In the case of mammals, the period of development prior to birth is difficult or impossible to control. For eggs, we can provide some measure of control by housing the eggs under constant conditions, from the time they are laid until they hatch. Thus, we need to sound a note of caution concerning the many studies that have used mammals as subjects for behavior-genetic analysis, particularly those in which there is some possibility that the prenatal environment may influence the expression of the traits we are studying.

MUTATIONS

Another technique that we can employ involves mutant strains, in which a behavioral effect can be traced to a particular allele where a single gene affects morphology and behavior. For example, consider the mating behavior of fruit flies (*Drosophila*) that have a single mutant allele. Mutant-type adults have yellow bodies; wild-type females have gray bodies. Male yellow mutant flies, when mated with wild-type females, have reduced reproductive success, compared to the success of wild-type male to wild-type female matings (Bastock and Manning 1955; Bastock 1956). We can trace this fertility decrease to distinct differences in the mating behavior patterns of the mutant flies. When the males court females, they engage in wing-vibration displays (courtship behavior). The pattern of wing vibration is altered in mutant flies; the bouts of vibration are shorter and occur less frequently. The effect of these changes is reduced stimulation for the courted female. Further tests confirm that the difference in mating success is due to the difference in mating behavior and not to other factors, such as variations in body color or scent.

Individual genes or groups of genes that act as a unit do not necessarily exert a single effect, but may have multiple effects; this phenomenon is called **pleiotropism**. In fruit flies, the yellow mutant allele affects at least one morphological characteristic—body color, and one behavior trait—courtship display.

MODES OF INHERITANCE

Another question in which we as behavior geneticists would be interested concerns analyzing the mode of inheritance of a behavioral trait. When two animals that show distinct differences for a behavioral trait are mated, how is the behavior of the offspring affected? Consider the wheel-running activity of mice, diagrammed in figure 5–8. Suppose that individuals of one strain of mice run an average of 100 m/hr, and mice of a second strain run only 20 m/hr. If the progeny produced by mating males with females of the opposite strain show virtually the same wheel-running speed as either parent strain—either 20 m/h or 100 m/hr—then there is a **dominant**

FIGURE 5-8   Modes of inheritance for wheel-running activity in mice
When two strains of mice, low (P_1) and high (P_2), which exhibit significant differences in wheel-running behavior, are mated, three modes of inheritance can be exhibited: dominance—progeny exhibit behavior like one parent or the other; intermediate—progeny exhibit slow wheel-running activity in between levels recorded for the two parent strains; and overdominance (hybrid vigor)—progeny exhibit more wheel-running activity than either parental type.

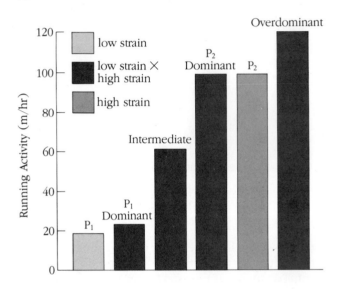

mode of inheritance. A behavioral response by the progeny that falls somewhere in between that of both parents (e.g., 60 m/hr) indicates an **intermediate** mode of inheritance. In some instances, the wheel-running speed of the progeny may be greater than that exhibited by either parent strain (e.g., 120 m/hr); this indicates **overdominance**, or **hybrid vigor**.

Schröder and Sund (1984) investigated the inheritance and learning of a water escape behavior in mice. The apparatus consisted of a small tank of water with a wire mesh platform just above the water level at one end of the tank, similar to the apparatus we discussed in the section on strain differences. Mice were given a series of five trials, one per day for five days, to measure the time it took to escape from the water onto the platform. Two parental strains, C57Bl and Balb/c, were chosen because they differed widely in their ability to learn this task (figure 5–9). When mice of the two strains were cross-bred to produce the F_1 generation and these progeny were tested, they exhibited significantly faster learning of the water escape response than either parental strain. This is an example of overdominance.

FIGURE 5-9   Learning curves for a water escape response for two strains of mice and their F₁ hybrid cross. The results show a faster learning response for the F₁ stock than for either parental strain.

Source: Data taken from Figure 3 on page 227 of J. H. Schröder and M. Sund. 1984. Inheritance of water-escape performance and water-escape learning in mice. *Behavior Genetics* 14:221–33.

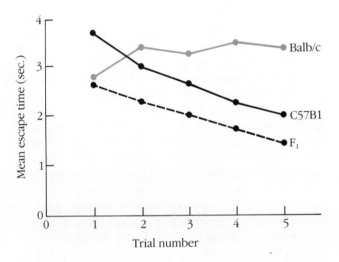

TWIN STUDIES

One topic that has long been of interest to behavior geneticists concerns twins in our own species. Identical, or **monozygotic**, twins develop from the splitting of a single fertilized egg; fraternal, or **dizygotic**, twins develop from two separate fertilized eggs. Because monozygotic twins have the same genotype, they provide scientists with a unique research opportunity. Identical twins are more alike in their characteristics and behavior traits than are fraternal twins. For example, correlations of intelligence test scores between ordinary siblings or fraternal twins are about 0.50; test-score correlations for monozygotic twins are between 0.73 and 0.90. (Erlenmeyer-Kimling and Jarvik 1963). Correlations between test scores of unrelated individuals range from 0.00 to 0.30. When analyzing data like these, we must be aware that rearing environments for identical and fraternal twins are probably not the same. The differences in test scores must be interpreted in light of the possible variations in rearing environments, the composition and size of the same sample population, and the type of intelligence test used for gathering the data.

Cavalli-Sforza and Bodmer (1971) and Ehrman and Parsons (1981), using data from Sheilds (1962) and Newman et al. (1937), have summarized several studies on monozygotic and dizygotic human twins. Values for monozygotic twins reared together or apart and for dizygotic twins were tabulated for hereditary and environmental influences on various traits in humans. Across all three types of twins, and within each type, the heritability values for height are higher than are those for weight. The heritability values are somewhat higher for weight than those for measures of IQ or personality. All of these heredity component values were positive; they ranged from about 0.30 to over 0.80. In contrast, estimates of the environmental component for these same traits were more variable, with a mixture of negative and positive values (total range from about −0.60 to +0.80). Ehrman and Parsons (1981) note that variations in both the hereditary and environmental components may be due to the nature of the tests and the differences in the sample populations. Another potential problem encountered when interpreting data from twin studies is that environmental conditions are not actually the same for monozygotic and dizygotic twins (Smith 1965). In general, monozygotic twins are treated more similarly than are dizygotic twins—by parents, other relatives, and peers.

The heritability of IQ has been the subject of some controversy in recent decades. Jensen (1967, 1973), Shockley (1969), and others have attempted to present data to support racial differences in IQ and to argue that heredity plays a more dominant role in perpetuating these differences than does environment. Hirsch and his colleagues (Hirsch 1975, 1981; Hirsch, McGuire, and Vetta 1980) have provided sound explanations of some of the fallacies and other problems inherent in the arguments of Jensen, Shockley, and others. A quote from Hirsch's 1981 article provides the best summary of this viewpoint (p. 33):

The key to "establishing the relative roles of heredity and environment" has been believed erroneously to be the heritability estimate. But heritability estimates cannot be made for human intelligence measurements, because the heritability coefficient is undefined in the presence of either correlation or interaction between genotype and environment, both of which occur for human intelligence. When correlation exists, either (1) between genetic and environmental contributions to trait expression, or (2) between environmental contributions to trait expression in both members for a parent-child or sib pair, heritability is not defined. Furthermore, when heritability can be defined, for example in well-controlled plant and animal breeding experiments, it has no relevance to measured differences in average values of trait expression between different populations; heritability estimates throw no light upon intergroup comparisons! Also, heritability estimates provide no information about ontogeny and are thus irrelevant to the formulation of public policy on education and social conditions.

GENES TO BEHAVIOR

We may wonder: How are the processes taking place at the level of the gene and DNA translated into physiology and behavior? Since genes do not code directly for behavior, what links are there between the genotype of the organism and the behavioral component of its phenotype? One way in which the cellular processes involving DNA can affect the behavior of developing organisms is during **epigenesis,** the interaction of the genetic program and the experiences and environment of the organism. We will encounter epigenesis in more detail in chapter 10. Biochemical events involving genes can also affect behavior in adults through the turning on and off of specific genes or gene complexes, which leads to the synthesis of particular proteins. A schematic representation of the major components in an organism's system and their feedback relationships is shown in figure 5–10. The model illustrates the pathways that connect genes and behavior in both developing and adult organisms.

Some investigators have used **mosaics,** organisms whose tissues are of two or more genetically different kinds, to test aspects of the gene-to-behavior sequence; mosaics permit us to observe both anatomical and behavioral anomalies combined in the same animal. Hotta and Benzer (1972, 1973) exposed fruit flies to chemicals that increased the frequency of mutations and produced various "abnormal" types of flies. They then used genetic techniques in mating the flies to produce mosaics that had some normal and some mutant tissues— for example, one normal wild-type red eye and one mutant white eye (Benzer 1973)—as well as some affected behaviors—for example, courtship.

Benzer then sought to determine what tissue parts must be mutant for the abnormal behavior to be expressed. He tested many fruit flies for the presence or absence of various mutant parts, analyzed the resulting statistics, and made a map of the early embryonic fly on which he located the **foci,** the groups of cells that differentiate into specific structures and organs. We can use these maps to trace the effects of particular mutant genes on behavior through the structures and physiological processes they affect.

Consider a second example. Investigators in another laboratory (Thiessen and Yahr 1970; Thiessen, Yahr, and Owen 1973), explored the relationship between genetics, testosterone production, and territorial marking in Mongolian gerbils (*Meriones unguiculatus*). Intact male gerbils normally mark their environment by rubbing the ventral sebaceous gland on surfaces. This behavior is greatly diminished in castrated males, but increases when they are given exogenous testosterone. The question tested by these investigators was: Does hormone (testosterone) alter the metabolism of neurons in the preoptic area of the brain? If the hormone induced changes in RNA synthesis (by turning genes on and off), the affected protein synthesis might in turn affect other neural processes and lead to effects on territorial marking behavior. The researchers gave some castrated gerbils treatment with testosterone only, and gave other gerbils treatment with testosterone plus actinomycin D, a compound that inhibits RNA synthesis. The gerbils receiving only the hormone exhibited increased levels of marking behavior, as predicted from previous work. However, for gerbils given the hormone and the actinomycin D, levels of marking behavior remained about the same as for castrated gerbils. Though the evidence here is only suggestive, it does provide some insight into how investigators can begin exploring the relationships between gene action and behavior.

GENETICS AND EVOLUTION

GENE FREQUENCY

We have looked at the ways genes affect behavior, but there is another aspect to the gene-behavior relationship. That is, behavior may significantly affect the frequency and expression of certain genes in a population (Ford 1964; King 1967; Oliverio 1983). Changes in the gene pool, one of the outcomes of natural selection, can in turn affect behavior.

For example, variations in courtship behavior (the processes of mate selection and synchronization of activities that lead to mating) may alter gene frequencies. Merrell (1949, 1953) used wild-type and mutant fruit fly strains to test this possibility. Four mutant strains, designated yellow, cut, raspberry, and forked, had sex-linked recessive alleles, expressed only in the male

FIGURE 5-10   Model illustrating relationship between genes and environment in control of behavior

Conclusions: Relationships between genes and behavior: (1) Genes do not make *traits;* many steps occur between DNA segments and behavioral capacities of an animal. (2) Genotype (= genome, the sum of genetic information) is a molecular code for regulated construction of proteins. (3) Raw materials obtained from the environment, such as protein building blocks, are broken down to incorporate the amino acids into species-specific proteins. (4) Therefore, the key role of genes is enzyme production which in turn regulates entrance and exit of substances into cells. (5) Enzymes regulate reactions and thus control development.

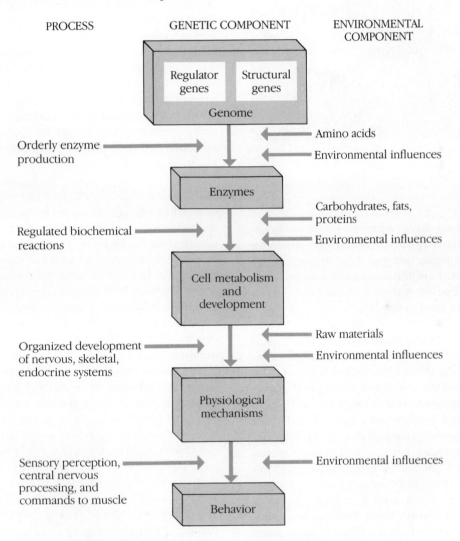

phenotypes and detected by examination of external characteristics. He started populations of flies with a mutant gene frequency of 0.5. Thus for each generation of flies, a departure from random mating would be indicated by an increase or decrease of the mutant gene in males. After several generations, the number of raspberry, cut, and yellow mutants decreased; the number of forked mutants showed no significant shift. Other tests demonstrated that although fertility or viability did not differ for any of the mutant types, mating success differed significantly. Thus the tests implicated some aspect of courtship and mating behavior as the critical factor leading to differences in productivity, and hence to shifts in gene frequencies.

FIGURE 5-11 Deermice (*Peromyscus leucopus*) in natural habitat

Source: Photography by Hal Harrison from Grant Heilman.

ADAPTATIONS

Other investigations have demonstrated that gene-controlled behavior may vary between groups of animals of the same species. Some of these differences are adaptive to habitat characteristics and others are adaptive to stages of population density fluctuations. Let us consider a series of brief examples.

HABITAT ADAPTATIONS. Prairie deermice (*Peromyscus maniculatus bairdi*) of the midwestern United States inhabit only fields, and are never caught in forested areas (figure 5–11). In contrast, woodland deermice (*P. m. gracilis*) inhabit only forested areas in south-central Canada and the northern and northeastern regions of the United States. Mice of these subspecies exhibit several behavior patterns and preferences that are predictable from the habitats they occupy. Woodland deermice prefer a temperature of 29.1 C; grassland deermice prefer a temperature of 25.8 C, in agreement with their relatively cooler habitat (Olgivie and Stinson 1966). That is, open grasslands are generally cooler than woodland habitats at the same latitude, and the behavior of mice from the two subspecies reflects this relative difference. In addition, the two subspecies have different reactions to sand. In a one-day test, grassland mice removed a median of 5.9 lb of sand from a hopper, in contrast to the 0.1 lb that woodland mice removed (King and Weisman 1964).

Again this result seems logical, because grassland deermice are terrestrial and live in burrows in the ground; woodland deermice are semi-arboreal, and sometimes nest in trees.

Dewsbury and colleagues (Dewsbury et al. 1980; Webster et al. 1981) have explored a variety of behavior patterns across a wide range of rodents. In particular, they determined that some species spend no time climbing, while others are off the ground between one-third and one-half of the time. Those taxa that climbed more are those that we would associate more with arboreal habitats in nature. Those rodent taxa that burrow or live in subterranean nest sites engaged in more sand and peat digging than did rodents that had habitats involving little or no nesting or food finding below the ground surface. These cross-species comparisons made while holding the test situation constant permit us to make generalizations about particular behavioral adaptations and their utilization in different types of environments.

POPULATION FLUCTUATIONS. The population density of meadow voles (*Microtus pennsylvanicus*) rises and falls in fluctuations that last three or four years (Chitty and Chitty 1962; Krebs 1966). During the low phase of the density fluctuation in a particular area, voles are difficult to find. In contrast, at the time of peak populations, there may be over two hundred voles per hectare. Trapping data indicate that many animals emigrate during the increase phase of the population fluctuation; at peak density, however, few animals emigrate, and losses in numbers must be attributed to deaths. Blood protein production is regulated by specific gene loci, and the blood proteins can be measured by electrophoresis. **Electrophoresis** is a technique in which substances are separated from one another on the basis of their electrical charges and molecular weights.

One such locus, transferin (Tf), has two possible alleles, designated Tf^C and Tf^E (Myers and Krebs 1971; Krebs et al. 1973). Comparison of the frequencies of heterozygous and homozygous alleles in female voles from dispersing and nondispersing populations reveals a striking difference (figure 5–12). The Tf^C/Tf^E allele combination is significantly more frequent in the dispersing population. Additional data indicate that 89 percent of the loss of heterozygous females from populations during the increase phase is attributable to emigration (female voles dispersing from the local population). We should keep in mind that in this instance no direct gene-to-behavior sequence has been

FIGURE 5-12 **Transferrin genotypes in female meadow voles**
Genotypes of dispersing females are compared with those of resident females in the autumn during the increasing phase. C, C/E, and E represent the three transferrin genotypes.

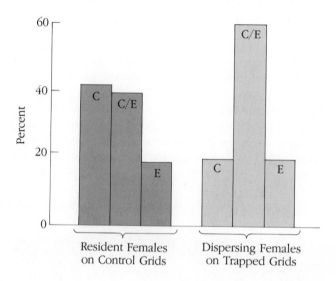

clearly delineated; however, changes in gene frequencies have been correlated with changes in population density. The foregoing example also illustrates how rapidly the genotypic makeup of populations can change.

MOLECULAR TECHNIQUES, GENETICS AND BEHAVIOR

It is noteworthy that in recent decades several techniques based on genetics have been developed and are proving to be of great value to those who study animal behavior. These techniques involve the measurement of proteins or DNA to assess some of the genotypes of free-living or penned animals. They use electrophoresis in the same way used above with transferrin. The blood or tissue proteins, separated by electrophoresis, can aid the determination of the parents of any particular progeny. By knowing more about the genetic relationships between animals, critical hypotheses regarding social organization and social behavior can be tested.

To illustrate this point further, consider several examples. Melnick et al. (1984) were interested in determining whether inbreeding occurred in populations of rhesus monkeys (*Macaca mulatta*). Seven groups containing a total of 292 monkeys were studied. Blood samples from more than 80 percent of the animals were tested using electrophoresis. The data developed from the patterns of similarities and differences in blood proteins permitted the investigators to derive two major conclusions: (1) There was a high degree of gene flow between groups; this is probably related to the male rhesus monkeys' emigration from their natal groups at about puberty, usually to join another group. Thus, the genetic data are in agreement with the observations of behavior. There was a very low frequency of consanguineous matings—that is, very little inbreeding.

Another example of the use of the protein electrophoresis technique involves the issue of whether female deermice (*Peromyscus maniculatus*) mate with more than one male when they are in estrus, and whether the litter that is born contains pups that were fathered by more than one male (Birdsall and Nash 1973). Pregnant female mice were live-trapped in nature, and blood samples were taken from them and from all their progeny born in the subsequent litter. All of the possible alleles at three separate loci were mapped using the electrophoresis technique. For some 10 percent of the litters, the data on genotypes indicate that the female was inseminated by more than one male. This genetic finding again confirms the information known about the mating behavior of these mice.

A more recent technique, called DNA fingerprinting, is similar to the protein electrophoresis, in that the molecules are separated from one another by electric charge and molecular weight. The technique allows scientists to determine the biological parents of a particular individual with a very high degree of accuracy. The technique involves purifying DNA from tissues obtained from the test subjects, breaking up the DNA into fragments using compounds called *restriction enzymes* that split the DNA molecule at specific locations, and separating the various DNA fragments according to size with electrophoresis. For any one individual, the pattern of DNA fragments is inherited in a Mendelian fashion, one-half from each parent. Thus, it should be possible to match up the pattern of DNA fragments for an unknown test subject with those of all possible parents and assess which ones are the biological parents. DNA fingerprinting requires tissue samples from all animals in a population, as does protein electrophoresis, to be fully useful in assigning paternity and maternity. The reason that DNA fingerprinting is often considered better than protein electrophoresis is the variety of DNA fragment patterns is far more numerous than the numbers of different alleles at particular loci; thus the accuracy of the test is much improved.

Hedge sparrows (*Prunella modularis*) live in a social system that involves a variety of mating possibilities (Davies and Lundberg, 1984); these include monogamous pairs, polygyny with a male and several females, polyandry with a female and several males, and various multiple combinations of males and females. Burke et al. (1989) used DNA fingerprinting to see whether males can discriminate their own young from others in broods of more than one male parent. By observing the birds in the Botanic Garden at Cambridge University in England throughout the reproductive period (courtship, mating, and rearing the young to fledging), and also collecting tissue samples from the birds for DNA fingerprinting, Burke and coworkers were able to determine the male's investment in offspring that may or may not be his own biological progeny. As it turns out, males apparently do not discriminate between progeny that are their own and progeny of other males in broods with multiple male parents. However, males do spend more time assisting with the feeding of offspring where they had had greater access to the female parent during the mating period. Access to the female is a good predictor of paternity. These molecular techniques will continue to supply much needed genetic information so that investigators can combine their knowledge about behavior with what they learn about genetic relationships.

SUMMARY

Behavior genetics is concerned with two categories of questions; first, the role and relative importance of inheritance in development and regulation of behavior, and second, the role of genetics in the evolutionary process. The basic formula used to express the genotypic and environmental contributions to phenotype variation is $V_T = V_E + V_G + V_I$. In other words, the phenotypic variation is the sum of genotypic, environmental forces, and the interaction of genotype and environment. While the assumption that $V_I = 0$ is often made, we should keep in mind that the gene-environment interaction represented by V_I is a critical parameter in most situations.

We can use a variety of methods to assess the relative importance and interaction of genetic and environmental parameters in regulating behavior. *Inbreeding*, or brother-sister matings for many generations, is used to develop homogeneous strains of laboratory animals. This permits us to hold genetic factors constant while we manipulate environmental variables, or to hold environmental factors constant while we examine inherited behavior differences between different strains or species. By using the *variance* values for a trait from two parental strains, and the F_1 and the F_2 crosses, we can obtain an estimate of the *heritability* (h_b^2) for a trait (heritability in the *broad* sense or °GD). We can calculate an estimate of *realized heritability* (h_r^2), which is useful in predicting the effectiveness of selection, by computing the ratio of response (R) to the selection differential (S). Heritability can also be estimated by using parent-offspring correlations. *Cross-fostering*, or the transfer of neonatal animals from the parent female to another female of the same species or strain or to a female of a different species or strain, is used to determine which behavior patterns are species-specific and which are affected by the maternal environment.

Another area of investigation in behavioral genetics is the determination of the mode of inheritance—which can be either a *dominant* mode, an *intermediate* mode, or an *overdominant* mode (also called *hybrid vigor*). In addition, studies of *mutant* strains are used to determine the processes that take place at the level of the gene, and the results of these processes that we see translated into morphology, physiology, and behavior.

Behavior geneticists are concerned with the ways behavior affects the frequency and expression of genes and how changes in the gene pool can affect behavior as well as morphology. Lastly, genetic techniques such as *protein electrophoresis* and *DNA fingerprinting* can be used to ascertain genetic relationships between animals and relate that information to their behavior.

Discussion Questions

1. How would you use the techniques of behavior genetics to study fitness as measured by reproductive success?

2. What effect would plasticity (flexibility) of phenotype have on the effectiveness of artificial selection for a trait?

3. Stamm (1954) investigated the genetics of food-hoarding behavior in two strains of rats. The two parents strains were Irish (mean hoarding score = 5.8) and black-hooded (mean hoarding score = 12.5). A cross between the two parental strains produced a new F_1 strain of rats with a mean hoarding score of 13.2, which was not significantly different from that of the black-hooded parental strain. What is the mode of inheritance of this hoarding trait in these rat strains? Explain your reasoning in arriving at this conclusion.

4. Two populations of fish of the same species inhabit neighboring lakes that have identical environmental conditions, including fauna and flora. The population in one lake is genetically homozygous; the population in the second lake is heterozygous. Chemical pollution from agricultural practices in nearby fields enters both lakes and results in rapidly shifting, less stable conditions, producing changes in the fauna and flora. What can you predict about the ability of the two populations of fish to survive and maintain their numbers in each pond, given the environmental changes? Explain your prediction(s).

5. Among the techniques for studying the genetics of animal behavior that we discussed in this chapter were use of inbred strains, cross-fostering, and artificial selection. What are the advantages and disadvantages of each of these approaches? For each technique, give a brief example situation in which you would use them to test a question about behavior genetics.

6. Recall that the formula for calculating realized heritability is $h_r^2 = R/S$, with R representing the response to selection and S the selection differential. Below are the values needed to calculate R, S, and h_r^2 for an experiment involving measurement of the amount of sand dug by mice from five strains—A, B, C, D, and E—each of which was selected from one generation for a tendency to dig sand. The data given are the mean grams of sand dug from a trough in twenty-four hours; ten mice were tested for each mean shown. After calculating the h_r^2 for each strain, answer the following questions: What can you conclude about the h_r^2 in the mice? What restrictions would you put on these conclusions?

			Strain		
	A	*B*	*C*	*D*	*E*
Original population	62.6	41.5	10.8	17.6	31.5
Selected mice used for breeding	70.2	45.8	14.2	21.6	36.8
Offspring generation	63.5	44.6	11.2	20.8	33.8

Suggested Readings

Craig, J. V. 1981. *Domestic Animal Behavior.* Englewood Cliffs, NJ: Prentice Hall.
Covers the relatively new area of applied animal behavior, including domestic pets and livestock. Of particular interest to students concerned with veterinary medicine, animal science, and zoological parks.

Ehrman, L., and P. A. Parsons. 1981. *Behavior Genetics and Evolution.* New York: McGraw Hill.
A general textbook, probably written at a slightly lower level than the next entry, but with solid coverage of topics involving human behavior genetics and genetics and evolution.

Fuller, J. L., and W. R. Thompson. 1978. *Foundations of Behavior Genetics.* St. Louis: Mosby.
A general textbook for a course in behavior genetics. Excellent treatment of a range of topics with in-depth coverage of many items only mentioned in our chapter.

Manning, A. 1976. The place of genetics in the study of behaviour. In *Growing Points in Ethology,* ed. P. P. G. Bateson and R. A. Hinde. Cambridge: Cambridge University Press.
Historical context and definition of the role of behavior genetic analysis in the larger framework of behavior analysis are the strong points of this book chapter. Covers the types of research strategies emphasized by those using genetics as a tool for the study of behavior.

6

THE NERVOUS SYSTEM AND BEHAVIOR

*E*very living organism is continuously bombarded by environmental stimuli. To survive, animals must have sensory receptors to receive information, some type of nervous system to sort out and interpret stimuli, and effector (motor) and endocrine systems to produce behavioral responses to appropriate stimuli. In this chapter we examine how the nervous system takes in sensory information and generates behavior. Chapter 7 explores the roles of endocrine systems in mediating behavior. Behavior genetics, behavioral endocrinology, and neuroethology are areas of animal behavior concerned with its mechanisms and proximate causation that have emerged in the past several decades; however, neuroethology has captured the center stage in the 1980s. The far-reaching developments in neuroscience have enabled those who are interested in behavior to enhance greatly their knowledge of nervous system control of behavior by using the new information and techniques generated from these discoveries. In this chapter, we deal with the basics of the nervous system's effect on behavior; complete coverage of neuroethology necessitates an entire book (see Camhi 1984 in Suggested Reading List).

We look first at the basic units of the nervous system, the neurons, and how they transmit impulses. Next we consider the systems of sensation, perception, and the filtering mechanisms that sort relevant from less important stimuli. We then explore different types of nervous systems, their evolution, and their relationships to different types of animals' behavior capacities. Next, we examine research techniques and the kinds of findings

on the nervous system and behavior that they generate. Finally, we consider two in-depth examples of the relationships between the nervous system and behavior: cricket song and sound communication in frogs.

THE NERVOUS SYSTEM

The nervous system carries messages throughout an organism by means of the **neuron,** a basic structural unit that shows many variations on a similar theme (figure 6–1). Typical neurons consist of a **soma** (cell body) with a nucleus, an **axon** with threadlike **synaptic processes,** and **dendrites** emanating from the soma. The position of the cell body and the numbers and kinds of dendrites vary with nerve cell type.

COMMUNICATION OF IMPULSES

Communication within the nervous system is by the passage of an electrical impulse. There are two stages: (1) movement of the impulse among the neuron and (2) transmission of the impulse from one neuron to another. Within the neuron, the electrical impulse is conducted from the dendrites to the end of the axon by the movement of ions (electrically charged atoms) across the cell membrane. The impulse, initiated when a stimulus reaches dendrites at one end of the neuron, results in a

FIGURE 6-1   Three types of neurons
Neurons are the basic structural building blocks of nervous systems. Shown here are a generalized neuron, a receptor or sensory cell, and a brain cell from a vertebrate.

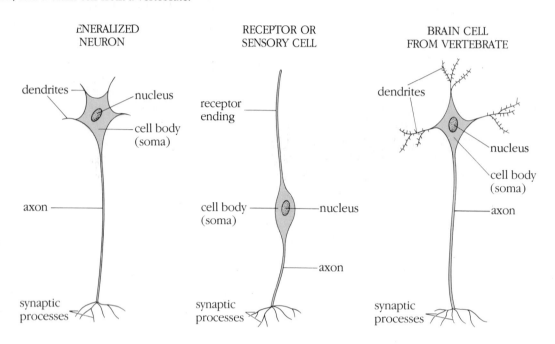

change in **membrane permeability** of sodium (Na⁺), potassium (K⁺) and chloride (Cl⁻) ions. This leads to shifts in ion concentration, which then changes the electrical potential between the inside and the outside surfaces of the neuron's cell membrane, and thus conduction of the impulse. This local change in membrane ion balance also results in a similar change in neighboring regions, and in this manner, the impulse is propagated along the neuron. The normal resting balance of charged ions is then restored in the **refractory period,** a brief period after the passage of an impulse when another stimulus cannot produce an impulse.

An impulse is transmitted from the axon endings of one cell to the dendrites of the next cell at the **synapse** (cell junction). An impulse causes synaptic **neurotransmitters** (chemical messengers) to be released into the **synaptic cleft** or **gap** (intracellular space). These neurotransmitters affect the permeability of dendrite membranes and thus initiate an impulse that is conducted by the processes just described. In this way an impulse passes from neuron to neuron. **Neuromodulators** are chemicals released by neurons that act at synapses to facilitate or inhibit impulse transmission, affecting a population of neurons without initiating action potentials and without synaptic transmission.

PROPERTIES OF NERVE CELLS

The activities of nerve cells are unique (see Hodgkin 1971 for a review of neural impulse conduction). First, each neuron either fires or does not fire: the production of an impulse is an **all-or-none** phenomenon. There are few gradations in the intensity of nerve impulses; moreover, the propagation process is virtually identical for all neurons. Transmission along neurons in most organisms is **unidirectional:** impulses cross synaptic junctions in only one direction. These properties of neurons reduce ambiguity, because the same simple message is carried by all types of nerve cells.

If nerve impulses are all similar, how do neurons transmit any information about the nature of the stimulus or its strength? Different types of environmental information are picked up by different sensory receptors and carried along separate neural paths to particular areas within the central nervous system for decoding and interpretation. The separate but integrated systems for different types of information (for example, information for the eyes or the ears) prevent confusion about whether the stimulus was originally photic (relating to light) or auditory (relating to sound). The strength of stimulation may be communicated in at least two ways: (1) the same neuron may fire repeatedly with brief refractory periods, or (2) several neurons carrying the same type of information may fire simultaneously. Thus, sensory information may be encoded by the individual neuron's firing frequency and by the number of neurons firing.

SENSATION AND PERCEPTION

SENSORY EQUIPMENT

The term **sensation** refers to the process of transducing environmental stimulation or energy (for example, sound, light, heat, mechanical forces) into electrical impulses. The events involved in the sensation of external stimuli are peripheral, occurring at the body surface. To receive this environmental information and to respond to the demands of the particular environment, animals have evolved a variety of sense organs that are often specialized. Sensory receptors within muscles and organs provide information on muscle tension, body position, and internal conditions. Here we will be concerned primarily with those receptors that receive external cues, often called **exteroreceptors.** Examples of exteroreceptive sensory systems include the electric organ and sensory system of certain fish living in murky waters; the echolocation system used for navigation and hunting by bats, which involves the production of high frequency pulses and the reception of the returning echoes with specialized ears; and the specialized system for sensing geomagnetic forces useful in orientation by pigeons, as some evidence suggests (Walcott et al. 1979).

The electric fish (*Gymnarchus niloticus*), inhabits muddy rivers in parts of tropical Africa, where visual navigation would be difficult. Special organs near the rear of the fish (figure 6–2) emit a continuous stream of very weak electrical impulses, and porelike structures on the head contain sensory receptors that are stimulated by extremely small changes in the electrical field around the fish, set up by the fish's electrical impulses (Hopkins 1974; Machin and Lissman 1960). The sensory receptors detect moving or stationary objects in the environment as distortions in this electrical field. The electric fish can thus navigate in its murky environment with little difficulty. We should note that neither the stimulus nor the machinery for picking up these types of electrical stimuli are part of the human sensory world.

FIGURE 6-2   Impulses of electric fish
Electric fish (*Gymnarchus niloticus*) use weak electrical impulses emitted by a specialized tissue in the tail region to find their way about in murky water. The electrical field around the fish, indicated by lines in this figure, is continually monitored by pore organs located around the head region. An object in the water disturbs the symmetry of the field and thus can be detected by the fish.

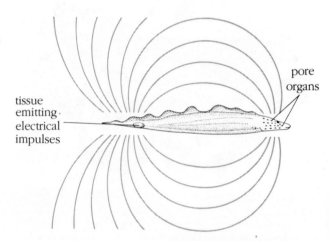

Silkworm moths (*Bombyx mori*) are stimulated by chemical sensory cues as are humans, but the quality of the stimulus and the acuity of the moth's chemical sense organs are clearly not within human sensory capabilities. The female silkworm moth has a special way of communicating her availability as a reproductive partner; a specialized gland on her abdomen emits a chemical substance called bombykol (Beroza and Knipling 1972). A male moth (figure 6–3) flying downwind from the female can detect this specific odor even when the concentration is as low as several molecules per million. Fine hairs on the male's featherlike antennae are stereochemically specialized to detect the odor. Once the male detects the scent, he flies upwind toward its source to locate and mate with the female. Males of other moth species and most other animals are apparently not affected by the chemical attractant and do not respond to its presence in the air. These species-specific odor cues (called **pheromones**) can be used as bait to capture particular insects when eliminating pests or when capturing animals for study.

FIGURE 6-3   Male silkworm moth

The male moth, here the giant silkworm moth *Hyalophora euryalis*, has specialized receptors on its antennae for detecting the chemical attractant bombykol.

Source: Photo by E. S. Ross.

SENSORY CAPACITIES

Within each modality in which an organism senses and transduces incoming stimuli, there are several aspects of the stimuli encoded into the neural message. For vision, these aspects may include color, polarization, image or pattern, and movement. Mechanical senses include pressure sensors on the body surface (tactile sensations), substrate-borne vibrations, hearing (both airborne and waterborne oscillations, including bat ultrasound and marine mammal sonar), organs of balance that register gravity and movement (semicircular canals), lateral line systems that detect displacements of water in fish, and internal proprioceptors that monitor the position and stretch of muscles and other tissues. Some animals possess electrical senses (fish that use electrical discharges). Lastly, the chemical senses involve detection of waterborne molecules (taste) or airborne molecules (smell).

These sensory systems and their capacities will vary with each animal species we study. To understand how the animal takes in and processes information from its environment, it is important to understand how its sensory systems work. It is equally important to understand how evolution has shaped its sensory systems, and the advantages and disadvantages of such systems in the animal's environment.

FIGURE 6-4   Auditory sensitivity of the turtle

Each curve is based on data from four turtles.

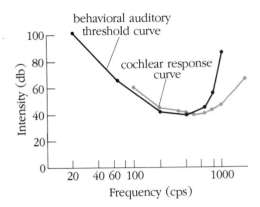

PERCEPTION SYSTEMS

Once information has been coded into electrical impulses, it may be conveyed throughout the animal's nervous system. **Perception** is the analysis and interpretation of sensory information. This decoding process is the function of the **ganglia,** bundles of neurons found in many invertebrates, or in the **central nervous system** in vertebrates. In the simplest nervous systems, little or no decoding takes place; a stimulus-response system is "wired" into the basic plan of the animal. The way an animal perceives a particular stimulus is a function of the type of sensory information it receives, the structure of its nervous system, and the past experiences permanently encoded in its nervous system.

When we investigate the sensory and perceptual worlds of diverse types of animals, we need to make both physiological and behavioral measurements. If we place electrodes in nerve tracts, which conduct impulses from a sensory end organ toward decoding centers, we can record whether neural impulses are initiated when we provide stimulation. For example, when we place electrodes in the cochlear nerve of a turtle and play pure tones near the turtle's ear, we obtain a response record for auditory thresholds (figure 6-4). The turtle (*Pseudemys scripta*) is maximally sensitive to airborne sound in the range of 200 to 600 cycles per second (Gulick and Zwick 1966).

FIGURE 6-5   Tactile discrimination in octopodes

The proportion of grooved surface in the cylinders decreases from left to right. Beside each object the number indicates the total percentage of the surface that is grooved. Octopodes have difficulty distinguishing between objects with the same or nearly the same total percentage of grooved surface (a, f, and g) regardless of the pattern of the grooves. They can, however, distinguish cylinders with different amounts of grooved surface (a and b).

Proportion of Grooved to Flat Surface

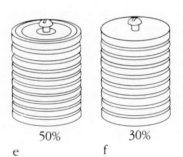

30%	20%	14%	0%
a	b	c	d

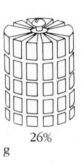

50%	30%
e	f

26%

g

An experimental assessment of perception must be designed so that the behavioral task an animal performs indicates the stimulus's reception and its interpretation by the animal. Octopodes can be trained using food rewards and mild electric shocks to choose cylinders with varying groove patterns on the basis of total grooved area (figure 6–5). However, they cannot distinguish the patterns of the grooves (Wells and Wells 1957). Octopodes can learn to distinguish cylinders *a* and *b* but cannot discriminate between *a, f,* and *g*. Techniques used to explore the sensory and perceptual capacities of various animals must be carefully designed to take into account the animal's response capabilities and behavior patterns.

An animal species can thus be characterized by the sensory system it has for receiving input from the environment, and by the nervous system structures it has

FIGURE 6-6   Tympanic membranes in moths
The tympanic membrane of noctuid moths is located in the thoracic region (a) and is connected to the central nervous system by two neurons, A₁ and A₂ (b).

POSITION OF
TYMPANIC MEMBRANE

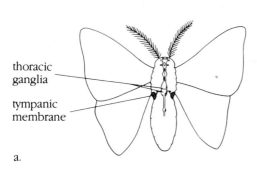

CUTAWAY VIEW OF
TYMPANIC MEMBRANE

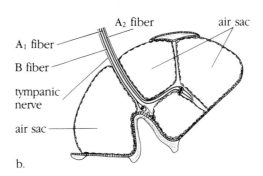

for analyzing and interpreting the information received by the sensory system. Together, the sensory and perceptual machinery determine what has been called the *Umwelt* of the animal (see chapter 3). To repeat, when investigating the behavior of any organism, we must thoroughly understand both its sensory system and its perceptual world.

SENSORY FILTERS

Ethologists developed the concept of **sensory filters,** in connection with sensory and perceptual processes. As we noted at the beginning of the chapter, every animal is continuously bombarded by an enormous number of diverse stimuli. Receiving, processing, and responding to all of these stimuli would be impossible. One of the primary functions of the sensory and nervous systems is the selection of only the pertinent stimuli. To accomplish this selectivity, animals have evolved two major types of information-filtering systems, one peripheral and the other central.

PERIPHERAL FILTERS.   Peripheral filters function at the level of the sensory receptors. Each animal species can receive information from a limited number of sensory modes. The sensitivity range of a particular mode is a limiting factor in the reception of environmental stimuli. Thus, not all animals can sense ultraviolet light as bees can; and few organisms can transduce weak

electrical current into neural impulses as can the electric fish. The limitations of the peripheral sensory system act as an initial screen, and only a limited number of the stimuli that impinge on an organism result in messages sent to decoding centers in its nervous system.

Certain species of noctuid moths (underwing moths of the genus *Cataocala*) have special hearing receptors that consist of tympanic membranes located on each side of the body in the thoracic region (figure 6–6a) (Roeder and Treat 1961; Roeder 1970). Two neurons, A₁ and A₂ connect each "ear" to the central nervous system (figure 6-6b). These moths can detect predatory bats and evade capture by them.

Roeder placed small electrodes into each of these neurons and recorded the nerve impulses produced as a result of bat vocalizations and artificially created sounds, which he directed toward the moth. By varying the frequency and intensity of these sounds and recording the nerve impulses, he was able to draw several conclusions about peripheral sensory filters. First, when sound frequency is varied, there is no change in pattern from either neuron: the moth cannot detect differences in frequency. When the sounds are soft, the A₁ neuron responds; the A₂ does not respond. When the sounds are loud, both neurons respond, but the A₁ fires more often than it does with soft sounds. Thus some aspects of incoming sounds are filtered (they do not reach the central nervous system), and different patterns of firing in the two neurons encode the sound stimulus for sounds that are passed on to the central nervous system.

FIGURE 6-7   Eggs used to test retrieval behavior in herring gulls

The eggs shown here, which vary in size, shape, color, or color pattern were used by Tinbergen to test the stimulus qualities that are most important in eliciting egg retrieval behavior in herring gulls. Tinbergen found that size and shape were more critical than other factors.

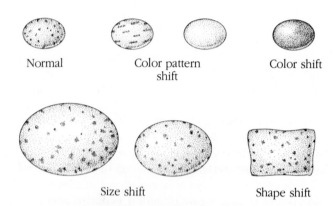

CENTRAL-FILTERING PROCESSES.   Central-filtering processes (called **innate releasing mechanisms [IRMs]** by ethologists) within the nervous system sort out incoming information, select relevant or important stimuli for further action, and eventually produce a particular response. Evidence for the existence of these processes comes from several sources.

When we present an animal with a variety of stimuli, all of which are detected by its sense organs, it responds selectively. That is, its behavior is based on only one or a few of the stimuli presented. For example, when an egg rolls out of its nest, a brooding herring gull (*Larus argentatus*) retrieves it with its bill, pulling it back into the nest. Tinbergen (1953) tested herring gulls to determine which stimulus qualities of the egg—its size, shape, color, or pattern (figure 6–7)—might trigger the retrieving response.

Tinbergen measured the bird's preferences by placing pairs of model eggs on the edge of the nest and recording which egg the bird chose to retrieve first. Then, using eggs that varied in size, shape, color, and color pattern, he found that for retrieval, the most critical sensory cues were egg size and shape. The gull's visual system is quite capable of receiving incoming stimuli from the eggs and transducing these stimuli into messages sent to the central nervous system, but the gull's behavioral response was directed only at specific stimuli. Thus, Tinbergen concluded that some type of central-filtering process must be in operation during the

decoding of the visual input. This central-filtering process affects the gull's preference for particular stimulus qualities of the egg.

Animals exhibit different responses to the same stimulus; the response depends upon their internal conditions (see Motivation in chapter 10). Our everyday experience demonstrates how the central filtering process operates on the auditory mode: In a crowded college dormitory, a stereo is playing the latest popular music at a high volume and many conversations are taking place, and two students are talking seriously about campus politics. Although they are receiving a large quantity of auditory input, they can hear one another and can concentrate on the topic at hand, because central-filtering processes act on the incoming nerve impulses from the ear and sort out only the relevant stimuli for further processing.

EVOLUTION OF THE NERVOUS SYSTEM

The topic of evolutionary changes in the structures and properties of animal nervous systems and the behavior changes associated with them is vast and complex. In this section we will elucidate the ties between the nervous system, evolution, and behavior. Our generalizations are, of necessity, rather broad.

SIMPLE SYSTEMS

Unicellular organisms do not possess a formal structure that could be called a nervous system, but they do exhibit the general phenomenon characteristic of most cells, called **irritability,** the wavelike passage of a stimulus from one point of a cell to another. Cells respond in a nonspecific fashion to external stimuli. In the process of evolution, irritability became channeled and refined, and specialized tissues and structures developed for receiving and transmitting information within the organism. Changes in the nervous system were critical for the evolution of multicellular animals, and for complex, flexible, and integrated behavior sequences.

The **nerve net** (figure 6–8) represents one of the earliest specialized neural organizations that conducts messages between cells. It is characteristic of coelenterates, such as the sea anemones (e.g., *Stephanauge*), the hydra (e.g., *Obelia*), and the jellyfish (e.g., *Aurelia*). However, some organisms with a nerve net possess another form of nervous system as well, and many higher animal

FIGURE 6-8   Nerve net from sea anemone
The many neuronal fibers in the sea anemone appear to be randomly oriented, but even this degree of organization allows for the important property of irritability.

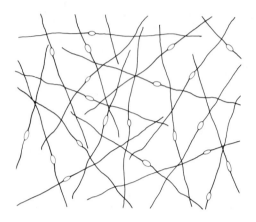

species retain some form of nerve net—for example, in the gut wall of vertebrates where peristaltic contraction involves diffuse conduction. The nerve net appears to be a random arrangement of nerve fibers. Synapses occur at junctions between the neurons, but unlike synapses in the nervous systems of other organisms, those in coelenterate nerve nets are nonpolarized and conduct impulses in all directions. An impulse that begins at one point passes to most of the other neurons in the net because each neuron is through conducting. Only later did a system evolve that allows impulses to travel in only one direction along discrete, closed channels.

The nerve net regulates simple behavior, such as postural changes and feeding responses, and is a good mechanism for localized responses: distances between receptors and effectors are short, and responses need not be mediated through a central nervous system. Starfish (class Asteroidea), for example, have small pincers on their backs that help keep small organisms from settling there. The actions of these pincers are controlled by local nerve nets.

RADIALLY SYMMETRICAL NERVOUS SYSTEM

Two trends marked the evolution of slightly more complex nervous systems: First, different types of neurons became specialized to serve different functions; and second, arrangements of neurons became more ordered, forming nerve tracts and integrative centers. The first major development after that of the nerve net was the radially symmetrical nervous system characteristic of starfish (figure 6–9) and sea urchins. Starfish have a *circular nerve ring* in the central portion of the body, with **nerve tracts** (collections of axons) extending into each arm. Within each arm is a network of sensory and motor neurons, called a **plexus.** The radially symmetrical nervous system permits more flexible and more complex behavior than does the nerve net system.

Some starfish responses are reflexive and involve only peripheral sensory and motor neurons; other responses involve integrated behavior mediated through the central ring. Most important, coordinated movements of the five arms are partly controlled by impulses that travel through *sensory neurons* and *interneurons* to *motor neurons.* These pathways necessarily involve the central ring. Thus, the distance between sensory receptors and muscle effectors is increased in radially symmetrical nervous systems. Alternation of excitation and inhibition (the processes of stimulation and stimulation blocking of muscle tissue by motor neurons) produces extension and retraction of the arms, and results in locomotion. Starfish are capable of righting themselves if placed upside down, and of wrapping themselves around a bivalve shellfish (e.g., a clam) to open the shell and obtain food; the interneurons and central ring mediate the coordinated movements necessary to perform these tasks.

Animals with radially symmetrical nervous systems also evolved more specialized types of sensory receptors, including systems for receiving contact, taste, and general chemical signals. The combination of refined system for sensing the environment and coordinated responses means that starfish and related organisms possess a greater degree of flexibility and diversity of behavioral responses than do simpler forms. However, their nervous systems' total capacities for adaptation to their environment are still relatively limited.

BILATERALLY SYMMETRICAL NERVOUS SYSTEM

The bilaterally symmetrical nervous system, characteristic of worms, mollusks, arthropods (including insects and crustaceans), and vertebrates, is more advanced than the radially symmetrical system. A bilaterally symmetrical organism generally possesses a distinct head and tail and has a segmented body (figure 6–10). We can observe several evolutionary developments in the nervous systems of these animals.

FIGURE 6-9 Starfish nervous system
The starfish is characterized by a radially symmetrical body plan. Its nervous system, as shown here, contains a central ring and nerve tracts with sensory and motor fibers that connect the central ring with each arm.
Source: Photo by Runk/Schoenberger from Grant Heilman.

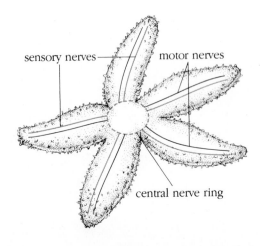

The first development is the refinement and diversification of sensory organs and systems. For instance, sensory receptors are very important in some insects for their spatial distribution and mate finding. Male house crickets (*Acheta domesticus*) use sounds produced by stridulation (the rubbing of specialized structures on the wings to produce a chirping sound) to establish and maintain distinct territories (exclusive use of an area) or to attract females (Heiligenberg 1966; Bentley 1969; Bentley and Hoy 1970). Specialized sensory receptors for registering the chirping sounds are located in the cricket's head region. We shall see this system in more detail in the final section of the chapter.

Most insects have evolved highly complex visual receptors called **compound eyes,** which have no lenses and are composed of many separate light receptor units called **ommatidia** (figure 6–11). Each ommatidium is stimulated by a point of the object being viewed, and an overall image is a mosaic of many such points. Octopodes, a group of mollusks, have evolved a visual receptor with a lens, similar in some respects to the eyes of vertebrates. Other examples of specialized sensory receptors are the antennae used by some moths to detect bat vocalizations, and the taste receptors of blowflies.

FIGURE 6-10 Earthworm nervous system
Earthworms are bilaterally symmetrical organisms, with bodies divided into many segments. The earthworm nervous system has several important features: the concentration of neurons in ganglia toward the anterior or head region, the ventral fiber tract, and sensory and motor neurons that connect to the ventral nerve cord within each body segment.

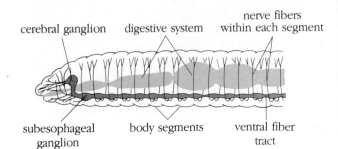

FIGURE 6-11   Compound eye and ommatidium of insect
The compound eye of many insects is comprised of many individual photo-receptive cells called ommatidia. Each ommatidium "sees" one point of the visual object, and together these points from the ommatidia form a mosaic image.

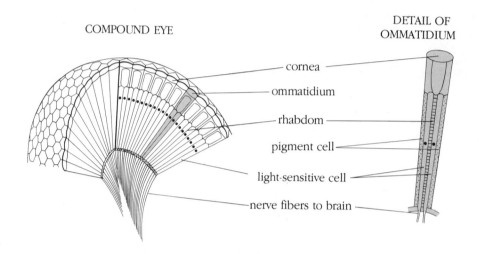

COMPOUND EYE

DETAIL OF OMMATIDIUM

cornea

ommatidium

rhabdom

pigment cell

light-sensitive cell

nerve fibers to brain

The second development in bilaterally symmetrical organisms is the formation of an articulated skeleton—an external (in invertebrates) or internal (in vertebrates) support structure. The evolution of skeletal structures was accompanied by changes in effector (muscle) systems. Concurrently, patterns of muscle innervation and control were evolving, making possible coordinated movements of various body parts and more rapid and refined responses by the organism.

The third and perhaps the most important evolutionary development seen in bilaterally symmetrical organisms is the **centralization** of neural processes and behavior control. Two separate but related changes are involved in centralization. In earthworms (*Lumbricus terrestris*), nerve tracts coalesced into a distinct, ventral nerve cord (see figure 6–10). Within each body segment, sensory neurons from receptors enter the nerve cord, and effector neurons exit the nerve cord to go to muscle systems. With this system some local control is preserved, but the possibility for activity coordination of body segments is increased.

Many invertebrates' neural tissues are concentrated in ganglia in anterior body segments. Since an organism with a head and a tail most frequently ambulates anteriorly, a great deal of incoming information

originates in front of it. Thus, having the necessary neural machinery near the source of stimulation—in the head rather than in the tail region—is advantageous. Successful monitoring of the external environment requires more types and numbers of sensory receptors, and integration of this incoming information requires more ganglia; consequently, more neural tissue is concentrated in the anterior segments.

GENERALIZED VERTEBRATE NERVOUS SYSTEM

A sequence of forms, which included generalized vertebrate nervous systems, developed from bilaterally symmetrical organisms. The evolutionary changes in vertebrate nervous systems were characterized by (1) further concentration and enlargement of the central nervous system, (2) development of many interconnections between nerve cells, and (3) **encephalization,** which encompassed changes in size and configuration of the anterior regions of the nervous system, which led to the evolution of the brain. A number of these changes can be seen in figure 6–12.

Many behavioral phenomena characteristic of vertebrates are intimately tied to these changes in the nervous system. First, the complexity of behavior patterns

FIGURE 6-12   Vertebrate nervous systems
Nervous systems of vertebrates are characterized by
centralization, encephalization, and the development of
many interconnections within the nervous system.
Differences among vertebrate species in the size of the brain
and cranial capacity are evident here in the drawings of
reptile, bird, cat, and human nervous systems.

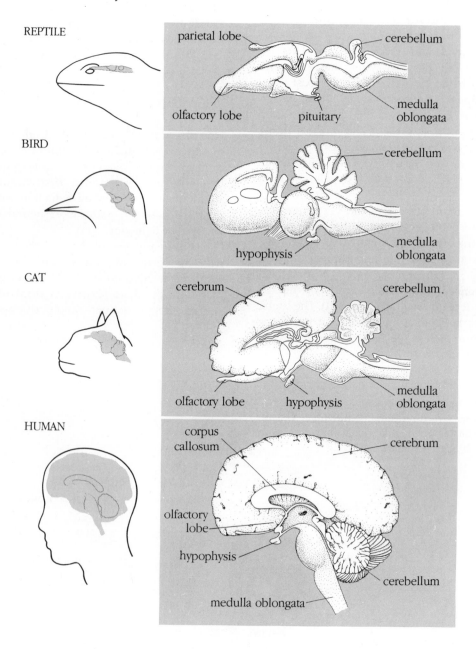

and the general flexibility of behavioral responses of vertebrates exceeds that of other animal groups. Second, because of changes in the structure of the nervous system and in the morphology of neurons themselves, vertebrates' responses are generally faster than those of invertebrates, though some invertebrate responses occur as fast as any vertebrate response (e.g., escape responses of cockroaches). A third consequence, a result of the formation of true brains through encephalization (the enlargement of anterior portions of the nervous system), is a greater capacity for information storage. This feature of the vertebrate nervous system has enormous ramifications for behavior: the retention of past experiences may influence future behavior. Finally, a further result of encephalization is the capacity to form associations between past, present, and possible future events. The ability to predict behavioral responses in particular situations, and the ability to deal with stimuli in the abstract, reaches its fullest development in humans.

Another important feature of the vertebrate nervous system is the evolution of the **limbic system.** This is comprised of the septum, the cingulate gyrus, the hypothalamus, the amygdala, the hippocampus, and the fornix (figure 6–13). The limbic system is sometimes called the **rhinencephalon** because many of the system's functions depend on olfaction (smell). The limbic system mediates behavioral functions that pertain to needs gratification, such as sex, feeding, and actions related to "emotions."

METHODS OF INVESTIGATING THE NERVOUS SYSTEM

Investigators have experimented with different techniques to explain the neural mechanisms that control behavior. In each of the following subsections, we look at an experimental technique used to investigate the relationships between the nervous system and behavior, and discuss the results of research using the technique. For some experiments, several different techniques were used to explore these relationships.

RECORDING NEURAL ACTIVITY

One direct way we can discover relationships between neural mechanisms and behavior is by implanting electrodes in an organism's nervous system and recording the neural impulses directly. By recording impulses and simultaneously observing behavior, we can draw correlations between patterns of neural discharges and patterns of behavior.

Lettvin and his colleagues (1959) performed an intriguing series of studies utilizing the electrode-recording technique (see also Maturana et al. 1960). They placed frogs (*Rana pipiens*) in front of a gray hemisphere, 35 cm in diameter and positioned in front of the

---

FIGURE 6-13   Primate limbic system
A variety of separate structures comprise the limbic system (grey areas) in primates. The monkey brain shown here is representative of higher vertebrates.

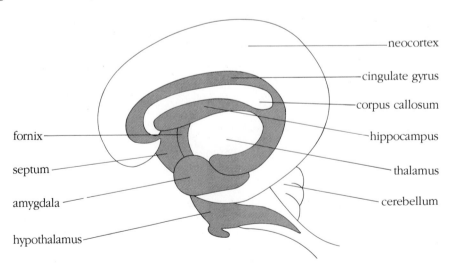

frog's eye. They prepared each frog so electrode recordings could be made from the ganglion cells associated with the frog's retina. Using magnets inside the hemisphere, they moved objects around the hemisphere that the frog could see, and simultaneously recorded the neural impulses that passed through the frog's ganglia. Lettvin and his associates were able to identify five major types of ganglion cells:

1. Moving-edge detectors, which respond to any moving edge that passes across the frog's visual field;
2. Sustained contrast detectors, which fire repeatedly when a moving object passes into the visual field and stops there;
3. Net-dimming detectors, which respond when the general level of illumination, particularly at the center of the visual field, is decreased;
4. Convexity detectors, which fire when any small object passes across the visual field and when any dark body that has a convex leading edge passes through the field of view;
5. Darkness detectors, which fire more frequently as the level of illumination is decreased.

Each of these types of detectors can presumably be related to the habitat and behavior of the frogs. Thus, the convexity detectors have become known as "bug detectors," because they are probably involved in detection and capture of prey.

In related studies on toads' recognition of prey, Ewert (1980, 1984, 1985) demonstrated that wormlike stimuli, represented by rectangles moving past the toad in the direction of the long axis, produced the greatest number of turning and prey capture movements. Attempts to relate these behavioral responses directly to response patterns of ganglion cells in the retina have generally proven unsuccessful. However, further tests revealed that the information carried from retinal cells is conducted to at least five areas in the toad's brain. One of these areas, a subpopulation of cells called T5(2), is of particular interest. These cells exhibit selective firing activity when wormlike stimuli are presented to the toad—there is a good correlation between the pattern of neuronal firing in this group of cells and the behavioral response profile when the length, velocity, and other attributes of the stimulus presented to the toad are varied.

Are the T5(2) cells necessary for prey recognition? Are they alone sufficient for prey recognition? Given our current techniques, only a partial answer to these questions is possible. If the T5(2) cells are destroyed on one side of the brain, stimuli moving on the opposite side of the toad's visual field do not evoke prey-capture behavior (Ewert 1980). As in other vertebrates, much of the visual information from each eye is transmitted to the optic area on the opposite side of the brain. Data from frogs indicate that ablations such as the destruction of the T5(2) cells on one side of the brain do not result in blindness, nor do they result in any related motor incapacity (Ingle 1973, 1976; Comer and Grostein 1981). But our ablation techniques are a bit too crude to target only the T5(2) cells, and other cells in this region may also have been destroyed. Thus, we can tentatively conclude only that some cells—possibly the T5(2) cells—are necessary for prey recognition.

A possible test of the T5(2) cells' sufficiency to recognize prey uses an electrode with a small electric current to stimulate cells in this subpopulation. When this is done in the absence of visual stimuli, the toads respond as if they had sighted a worm. Thus, some tectal neurons are at least sufficient to trigger the orienting response. Again, we must caution that more than just T5(2) cells may be involved—the electric current may have also stimulated some neighboring cells. Only more refined techniques and additional experiments will completely answer our questions.

A third series of experiments, conducted primarily with the mud puppy (*Necturus maculosus*), reveals more about retinal response to moving objects (Dowling 1976; Werblin and Dowling 1969). When tiny spots of light are projected on the retina, recordings show ganglion cells are either excited (fire) or are inhibited. For each ganglion cell, there is a corresponding region of the retinal receptor cells called its **receptive field**. In general, the receptors associated with each retinal ganglion cell—its receptive field—are arranged in pairs of concentric circles (figure 6–14a). When light shines on the inner circle of receptors, they stimulate the ganglion cell to fire. The inner circle is termed **on-center**. If light is directed at regions of the outer circle as well, the ganglion cell's response is less than when light shines on the on-center region alone. Thus, the outer portion of the field has receptors that exert an inhibitory effect, and is termed **off-surround**. These ganglion cells, therefore, are called **on-center, off-surround cells**. In addition, all vertebrates examined thus far have some retinal ganglion cells which act in the opposite way; they are called **off-center, on-surround cells**. When the receptive fields of neighboring ganglion cells are mapped, a series of overlapping circles results (figure 6–14b). Moving stimuli, like the mock bugs or worms presented to frogs or toads, would thus be "seen" by their photic images' passing across the retina as a series of excitations and inhibitions of the ganglion cells.

FIGURE 6-14   Receptive fields of mud puppies
(a) When light is shown on the retina, the activity of ganglion cells is excited if the light strikes on-center and inhibited if it shines on the off-surround neurons. (b) Mapping a series of receptive fields results in an overlapping pattern, with fields of the same or different sizes, depending on the organism.

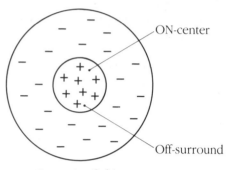

Receptive field
of one cell
(a)

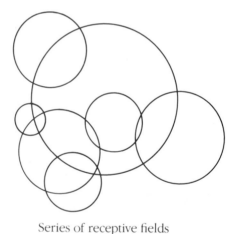

Series of receptive fields

(b)

FIGURE 6-15   Blowfly feeding
When the blowfly is hungry, sensory mechanisms on its feet detect the presence of sugars, and the proboscis is extended to obtain nourishment.

TRANSECTION

When a blowfly (*Phormia regina*) is hungry, chemoreceptors on its legs are stimulated by potential food sources it encounters. It extends its proboscis, thereby stimulating its oral taste receptors, and it begins to feed. But when does a blowfly stop eating? When it is satiated?

In a fascinating series of experiments, Dethier and his associates (Dethier 1962, 1976; Dethier and Bodenstein 1958; Dethier and Gelperin 1967) used the technique of transection to analyze the feeding behavior of the blowfly (figure 6–15). They **transected** (cut) certain nerves to observe the effects on a particular behavior pattern—food intake—and to identify the neural events that are involved in the determination of satiety.

Dethier and his colleagues tested several unsuccessful hypotheses before they reached their conclusions. They first thought hunger and satiety might be directly related to blood sugar levels, so they injected hungry flies with sugar solutions, but the flies were not satiated with the sugar solutions. When a blowfly feeds, ingested material travels first to a blind sac called a crop, where it is stored before it is digested. To test whether hunger is related to the crop fullness, they performed microsurgery to tie off the crops of a number of flies. This procedure prevents food from accumulating. But the flies remained hungry after this operation and continued to feed. More determined than ever to solve the problem, Dethier and his coworkers hypothesized that hunger and satiety might be directly controlled by the fullness of the hindgut, the major portion of the fly's alimentary canal. They tested this idea by delicately giving the flies "enemas," thus filling the hindgut. Again, the flies remained hungry.

Dethier and his colleagues then considered the remaining portion of the digestive system—the foregut, where the initial step in digestion takes place. By using very refined, miniaturized surgical techniques (involving a paraffin block to hold the fly, forceps, and a razor blade) and a dissecting microscope, they were able to transect a small nerve that connects the foregut and the brain. They hoped to find out whether this nerve carries the message of foregut fullness, which in turn stops feeding behavior. When hungry blowflies with this nerve transection were given an opportunity to feed, they too began to eat—but they did not stop eating. From these data and related information, Dethier and his colleagues concluded that as long as there is food in the blowfly's foregut, stretch receptors located there send messages to the brain that inhibit further eating behavior. Severing the nerve makes the fly permanently hungry. Additional research revealed that stretch receptors located in the body wall also send messages to the brain that act to inhibit further feeding when the fly is satiated.

Consider a second example. The feeding responses of garter snakes (*Thamnophis sirtalis*) have been tested using the transection technique (Burghardt 1966, 1969; Burghardt and Pruitt 1975). Garter snakes respond to pieces of cotton that contain extracts of prey items (e.g., earthworms or slugs) with tongue flicks and attacks. Do garter snakes detect the prey via the olfactory system or via the vomeronasal organ, a separate set of chemosensitive receptors located in the nasal cavity of some vertebrates? To test this question, Halpern and Frumin (1979) performed the following operations on the snakes: (1) transection of the bilateral olfactory nerves, (2) transection of the bilateral vomeronasal nerves, or (3) a sham operation which included all aspects of the transection surgery except actually cutting the nerve. The results indicated that while 100 percent of the snakes in all three groups responded with attacks prior to surgery, only 10 percent of snakes that had the vomeronasal nerve cut responded after surgery. Response levels remained at 100 percent for snakes in the other two test groups. The vomeronasal organ must be important for this type of food discrimination in the garter snake.

NEURAL STIMULATION

Another method we can use to investigate neural control of behavior is **electrode stimulation**—that is, by placing electrodes directly into specific areas of the peripheral nervous system or the brain and passing an electrical current through the electrodes. If the stimulation evokes a recognizable sequence of behavior that is normally part of that species' repertoire, then the electrode is probably in or near a nerve pathway or a brain area that controls or generates the observed behavior.

Willows (1967) used this technique on the marine mollusk *Tritonia*. These organisms have several major ganglia as their major decoding center, instead of a brain.

Electric current passed through electrodes to a ganglion and within individual neurons in a ganglion (figure 6–16a) may evoke particular behavior patterns. By providing electrical stimulation to specific nerve cells, Willows elicited turning and swimming movements (figure 6–16b). Others (Brazier 1968) have used this technique to map the neural pathways and the parts of the brain responsible for generating certain motor patterns in species ranging from insects to mammals.

Electrical stimulation of a rat's (*Rattus norvegicus*) lateral hypothalamus through implanted electrodes elicits behavior patterns such as eating, drinking, aggression, and grooming (Panksepp 1971; Valenstein 1969; Valenstein, Cox, and Kakolewski 1970), even when careful examination confirms that the electrodes are in virtually identical locations. To further explore this issue, Bachus and Valenstein (1979) placed electrodes in the hypothalami of a group of rats. They then preselected rats that exhibited drinking behavior when electrical stimulation was provided. Bachus and Valenstein used electrolytic lesions through the same electrode in order to destroy tissues around the electrode's tip. As a further test, they gave some rats considerable prelesion stimulation experience, and gave other rats minimal experience. After the lesioning, they again gave electrical stimulation and found that the stimulation evoked drinking behavior in spite of the tissue damage. Bachus and Valenstein concluded that individual differences in drinking behavior in response to electrical stimulation in the hypothalamus through implanted electrodes cannot be attributed solely to variation in the specific location of electrodes within the hypothalamus.

TELEMETRY. Stimulation by radio signals, called **telemetry,** is a variation on the electrically evoked response technique. Telemetry's advantage is that it permits test animals to move about freely without restraint, even with the electrodes in place. One of the first investigators to utilize this technique, Jose Delgado, employed it to investigate aggressive behavior in rhesus monkeys (*Macaca mulatta*) (Delgado 1963, 1966). He determined the **dominance hierarchy** (rank ordering of animals from most dominant to most subordinate) by observing four rhesus monkeys in a room equipped with perching platforms. He temporarily removed the top-ranking monkey and implanted electrodes in the areas of its brain known to elicit and inhibit aggression. He attached a radio-receiver pack to the monkey's back. When he sent a radio signal to the pack, it produced five seconds of electrical stimulation to a

FIGURE 6-16   Testing *Tritonia* through stimulation
Use of stimulation through electrodes to evoke distinct
behavior patterns is one method for mapping which nerves
and decoding centers affect particular behavior patterns. An
apparatus (a) used to suspend the marine nudibranch
Tritonia in sea water permits placement of electrodes into
specific ganglia or neurons. Turning and swimming
movements (b) can be produced when specific neurons are
stimulated.

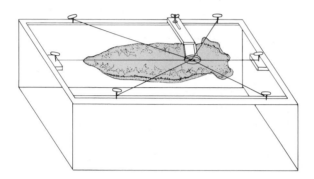

Tritonia suspended in apparatus for electrode stimulation

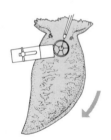

Stimulation
causing left
turning

Stimulation
causing right
turning

Stimulation
causing
swimming

selected area of the monkey's brain through a designated electrode. Stimulation of one area of the brain,
the posteroventral nucleus, caused the dominant monkey's aggression to increase. (A lower-ranking monkey
learned to press a lever at the side of the room that activated the radiostimulator to send an impulse to another brain area and inhibited aggression by the
dominant male.) This technique also has been used effectively to study the neural control of social aggression
in primates (Herndon, Perachio, and McCoy 1979).

SELF-STIMULATION. A second variation of the electrode stimulation technique permits the subject animal
to provide **self-stimulation** by pressing a lever or some
other device. Rats will learn to push a lever to stimulate
electrodes implanted in specific areas of the brain (Olds
1956). When electrodes are placed in the amygdala or
septum, the rats press the lever frequently, which suggests that the stimulation is rewarding or pleasurable.
When electrodes are implanted in other locations, the
rats press the lever very infrequently; the stimulus effect
is probably less pleasurable or even punishing.

The self-stimulation technique has been used to study aspects of the neural bases of behavior; for example, male rhesus monkeys' responses to fear and tension. Maxim (1977) prepared monkeys by implanting electrodes at self-stimulation sites in the hypothalamus. When he presented the monkeys with a snake, a fear stimulus, they pressed the bar for self-stimulation more frequently. When he paired an implanted male with a nonimplanted male, the implanted monkey pressed the bar more frequently in response to the nonimplanted male's strong dominance behavior. More bar presses were also recorded when either the implanted or nonimplanted male was about to engage in intense play-wrestling. Bar pressing was not related to other activities, such as eating, drinking, or playing with objects. In this instance, self-stimulation appears to have a fear-reducing effect: the rate of bar pressing was related to the potential threat or the exhibition of another monkey's dominance behavior.

Research on brain neurochemistry has uncovered a class of compounds called **endorphins** (Watkins and Mayer 1982). These chemicals appear to be the body's own opiates, which exert a pain-relieving effect. Endorphins have been related to several behavioral and physiological phenomena, including hibernation and water balance; and they may also function as neurotransmitters. Acupuncture treatment apparently induces the release of endorphins in the brain, which then have an anesthetic-like effect on the patient. The self-stimulation effects outlined above may involve the release of endorphins.

LESIONS AND SPLIT-BRAIN TECHNIQUES

Using electrical currents, chemicals, or microknives to destroy specific brain areas, or to cause **lesions,** is more drastic than the other techniques discussed thus far. If lesion treatment results in behavioral modification or deficit, we can draw tentative conclusions regarding the capacities of the remaining brain tissue and about the functions of the destroyed tissue. Lesions in different regions of rhesus monkeys' brains produce different social contact behavior (Girgis 1971). If tissue is destroyed in the medial portion of the amygdaloid nucleus, monkeys avoid social contact through flight or withdrawal. Lesions in the lateral regions of the amygdaloid nucleus make the monkeys placid and nonaggressive. In an attempt to produce a model of how brains operate as decoding and integrative centers, Arbib (1972) reviewed considerable literature on the use of lesions to explain the functioning of neural pathways and of specific regions of the brain.

The lesion technique was used early by Lashley (1929) primarily in rats, to study the effects of brain damage on behavior. He noted that cortical lesions produced deficiencies in tasks such as visual discrimination learning. Rats with varying degrees of ablation of the cortex exhibited functional losses that affected their ability to learn new tasks; some even lost previously learned discriminations. However, the passage of time often led to recovery—animals again could learn and retain the visual discriminations of cube sizes or black-and-white patterns. These observations led Lashley to his hypotheses of **mass action** and **equipotentiality.** Simply stated, this means that different portions of the cortex have equivalent capacities, and one region can substitute for another in the performance of various sensory, motor, and association tasks.

Beginning after World War II and extending into the 1970s, work by Roger Sperry (1961, 1974) challenged the Lashley hypothesis. Studies using rats, cats, and humans have given us a different perspective on how the brain functions. In particular, these investigations involved human patients and related animal models in which some or all of the connections between the two cortical hemispheres were severed. A variety of functions are affected by such surgery. Functions relating to language (speech, writing), and calculations are localized in the left cerebral hemisphere, where the right half of the visual field projects (figure 6–17). Functions relating to spatial abilities and many nonverbal capacities are localized in the right cerebral hemisphere, where the left half of the visual field projects. Following surgery that severs the intercerebral connections (those that connect the two brain hemispheres with each other) normal transfer between hemispheres does not occur. Objects touched by one hand cannot be located using the other hand, and subjects cannot recognize an object they have just been shown if that object is then presented to the other half of the visual field. Sperry was a co-recipient of the 1981 Nobel Prize in Physiology or Medicine for his pioneering work in neurobiology.

The foregoing studies and much additional information support the notion that there are indeed areas of the brain cortex with specialized functions. Hebb (1949) proposed that groups of cell assemblies comprising thousands of cells, may be the sites of localized function where information is stored and retrieved. It is possible that there may be some degree of equipotentiality within the cells of such assemblies. In chapter 10, we examine further how memory traces are made and stored in the brain.

FIGURE 6-17   Split-brain patients
When the connections between the two cerebral
hemispheres are severed, neuroanatomical observations and
postoperative testing demonstrate that various functions are
localized in one side of the brain or the other.

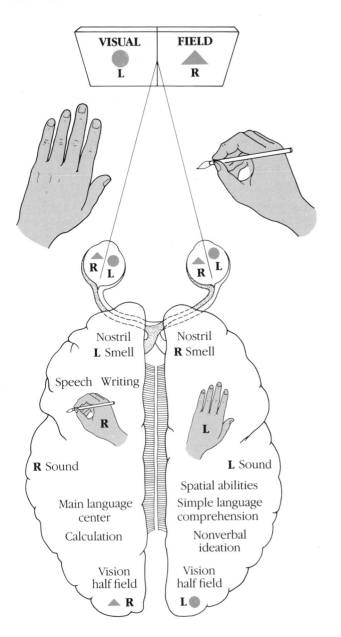

An important study of neural tissue lesions or ablations was reported by Anand and Brobeck (1951) in their
experiments on the effects of hypothalamic lesions on
food intake in rats. When they made a lesion in the ventromedial hypothalamus, the rat would not stop eating,
a condition called **hyperphagia,** and became obese. In
contrast, when they made a lesion in the lateral hypothalamus, the rat reduced or ceased its eating activity,
a condition called **hypophagia,** or **aphagia.** Thus the
medial hypothalamus contains a "feeding" center, and
the lateral hypothalamus has a "satiety" center. Investigations utilizing electrical stimulation confirmed these
findings. Stimulation of the ventromedial region of the
hypothalamus stops feeding behavior; stimulation of the
lateral region of the hypothalamus initiates feeding behavior. Powley and Keesey (1970) provided additional
data and an alternative interpretation of the effects of
hypothalamic lesions on feeding behavior in rats. They
concluded that a lesion in the lateral hypothalamus
lowers the set point around which weight balance is
maintained. Aphagia after surgery results from the rat's
attempt to reduce its body weight to the new set point,
after which normal regulation follows. The aphagia is
not the result of the surgical recovery period, as previous investigators claimed. Powley and Keesey caused
presurgical weight reduction in male rats by starving
them briefly before lesioning. Rats that underwent
prelesion starvation exhibited brief hyperphagia after
surgery and then regulated body weight normally
around the new set point. Weight-control mechanisms
appear to be active in rats immediately after surgery;
the rats are not simply undergoing weight loss as a result
of surgical recovery or destruction of a control center.

As we shall see in more detail in chapter 7, hormones trigger the neural activity that underlies maternal behavior in recently parturient females (Terkel
and Rosenblatt 1972; Zarrow, Gandelman, and Denenberg 1971), but maintenance of the maternal behavior
also has a nonhormonal basis (Numan, Leon, and Moltz
1972; Rosenblatt 1970). So which areas of the brain
maintain maternal behavior? When Numan (1974) made
lesions in the medial preoptic area (MPOA), data from
observations of maternal behavior in both MPOA-
lesioned and sham-operated rats (table 6-1) indicated
that the maternal behavior of the lesioned rats was
drastically impaired. The MPOA and/or the connections to the anterior hypothalamus that were destroyed
by the lesion are, therefore, important for normal maternal behavior.

TABLE 6-1   Mean values for three measures of maternal behavior in MPOA-lesioned and sham-operated female rats

	MPOA-lesioned female rats		Sham-operated female rats	
	Preoperative	*Postoperative*	*Preoperative*	*Postoperative*
Nursing (sec) (30-min test)	1,551	4	1,578	1,323
Retrieving (percentage)	100	0	100	89
Nest building (percentage)	96	0	94	76

Source: Data from Numan (1974).

Experiments on rats, dogs, and other animals have shown that MPOA lesions in males result in a cessation of sexual behavior (Hart 1974; Giantonio, Lund, and Gerall 1970). No investigators have reported a return of sexual activity in male animals with MPOA lesions. However, Twiggs, Popolow, and Gerall (1978) suggest a consideration critical for correct interpretation of such lesion studies. Male rats with MPOA lesions before puberty that had social interactions with peers exhibited normal adult male sexual behavior patterns. Thus, it appears that enhanced social experience can offset the effects of MPOA lesions. In addition, a critical factor could be the age when such lesions are made.

FUNCTIONAL NEUROANATOMY

The study of **functional neuroanatomy** (the size, structure, and arrangement of cells within the nervous system, particularly the brain) has resulted in several important findings concerning the brain's role in behavior.

Working primarily with cats and monkeys, Hubel and Wiesel (1962, 1977, 1979) explored how light stimuli are processed by the brain, specifically, how visual information is encoded in the cortex. They investigated how the receptive fields of retinal ganglion cells are translated into cortical projections. Hubel and Wiesel discovered several projection levels of increasing complexity within the cortex that appear to correspond to a layered arrangement of the retinal cells. There is also a correspondence between the receptive fields of the retinal ganglion cells and the responses of particular cells in the visual projection areas of the brain. In some way, at some levels in the visual cortex, responses are transformed so that cells respond to oriented line segments rather than to spots of light. Hubel and Wiesel

shared the 1981 Nobel Prize with Sperry for their fine work on functional architecture of the visual system. Significant progress is currently being made toward untangling the relationships between incoming stimuli that influence behavior and the exquisitely organized neuroanatomy of the brain.

Let's consider another example of the importance of understanding the connections between brain organization and behavior, the songs of canaries (*Serinus canarius*) (Nottebohm 1975, 1980, 1981; DeVoogd and Nottebohm 1981; Nottebohm and Nottebohm 1976; Nottebohm, Nottebohm, and Crane 1986; Paton and Nottebohm 1984). Canaries produce sounds in the syrinx during expiration as air is forced out of the lungs and air sacs. Male birds sing; females generally do not. When a young canary matures at one year of age, it learns a song repertoire. Each succeeding year, the bird learns a new repertoire. The muscles of the syrinx that control the frequency and amplitude of sound are innervated by the hypoglossal nerve. Canary songs are composed of phrases (figure 6–18) made up of syllables, each containing one or more elements. Nottebohm has shown that the left hypoglossal nerve and left cerebral hemisphere exert dominant control over the song, though shifts in hemispheric dominance are possible. Thus, much of their research has involved the dominant side of the brain.

Nottebohm and colleagues have uncovered some fascinating aspects of the canary song control system:

1. The neurons in one of the cortical nuclei implicated in song control show an increase in dendritic growth when the birds are treated with testosterone.
2. In males this nucleus is 76 percent larger in the spring (when they sing) than in the fall (when they do not sing).

FIGURE 6-18   **Fragment of a song from an adult male canary**

Canary song includes 20–30 syllable types, each repeated several times forming a phrase. Figure 6–18 shows three phrases. Each syllable in the first and second phrase is composed of two elements, whereas syllables in the third phrase consist of a single element. The vertical axis represents sound frequency (in kHz), and the horizontal axis is in seconds.

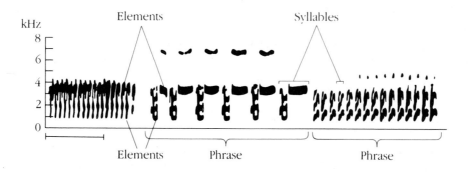

3. Female birds treated with testosterone show anatomical changes similar to those observed in males, and they start to sing.
4. Experiments with adult canaries using radioactively labeled thymidine (a marker for DNA synthesis) indicate that new neurons are being formed in one of several brain nuclei involved in song control. Electrodes detected impulses resulting from auditory stimuli, and confirmed that these new neurons are parts of functional circuits.

Thus, neuroanatomical changes occur on a seasonal basis that corresponds to the annual behavioral cycle of the birds. As a new song repertoire is learned each spring, new connections, possibly involving newly generated neurons, are formed in brain areas associated with song control. At the end of the breeding season, there is a reduction in size of the nuclei responsible for controlling the canary's song (Nottebohm, Nottebohm, and Crane 1986) that might correspond to the birds' forgetting the songs of that year. These data also reveal a plasticity in brain structure allowing the addition of new cells and connections. This could greatly facilitate our understanding of other brain-behavior interrelationships.

PSYCHOPHARMACOLOGY AND NEUROTRANSMITTERS

In recent years, the identification of various brain neurotransmitters (e.g., acetylcholine, serotonin), and the development of synthetic drug compounds chemically related to these neurotransmitters, have permitted investigators to explore relationships between the nervous system and behavior. This new area of brain research, called **psychopharmacology,** incorporates chemistry, physiology, and psychology.

Consider the following examples. Lorden and Oltmans (1978) studied learned taste aversion in rats. Rats that became ill after eating a particular food ate less of that food or avoided it altogether on subsequent presentations. Lorden and Oltmans conditioned rats to avoid saccharin water by making them ill with lithium chloride injections. Rats with brain lesions in a particular nucleus still learned the aversion. However, lesioned rats that were given an injection of a serotonin precursor before conditioning failed to learn the aversion. The serotonin precursor may have acted within the brain to decrease sensitivity to the lithium chloride injections; this change in sensitivity may have involved serotonin as a neurotransmitter.

The copulatory behavior of male rats involves a series of mounts with intromissions leading to ejaculation. This is followed by a refractory period when the males are not affected by external stimuli (e.g., presentation of a receptive female). In a series of three experiments McIntosh and Barfield (1984a,b,c) investigated the role of neurotransmitters in the duration of the postejaculatory refractory period in male rats. The first test involved serotonin. Rats were given a pretreatment sexual behavior test, followed by an injection of a chemical that blocks brain serotonin synthesis. When posttreatment sexual behavior was compared to pretreatment behavior, the length of the refractory period was significantly reduced from an average of 458 seconds before treatment to 272 seconds after treatment.

In a similar procedure, McIntosh and Barfield tested the importance of dopamine's effect on the refractory period length. They used a chemical that blocks dopamine receptor sites on the postsynaptic membrane. The effect was a significant lengthening of the average postejaculatory refractory period, from 326 seconds to 546 seconds. A third test examined the role of norepinephrine on the refractory period. Intraperitoneal administration of a norepinephrine synthesis inhibitor resulted in significant lengthening of the average refractory period, from 343 seconds to 526 seconds.

We can derive several conclusions from the McIntosh and Barfield investigations. First under normal conditions, a serotonergic system inhibits the resumption of mating after ejaculation. Second, dopamine and norepinephrine pathways are involved in the arousal of copulatory behavior and in the maintenance of the normal postejaculatory refractory period.

Other investigators have explored the role(s) of various drugs on the normal processes of synaptic transmission within the central nervous system. For example, chlorpromazine, a tranquilizer, acts by blocking the postsynaptic membrane receptors for the neurotransmitters acetylcholine and epinephrine. Nicotine mimics the effects of acetylcholine and facilitates synaptic transmission. Sedatives such as the barbituates, secobarbital and pentobarbital, act by interfering with the synthesis and release of serotonin and norepinephrine.

CANNULATION

Hollow electrodes or fine tubes (cannulas) or needles can be inserted into specific areas of the brain in order to introduce substances. Davidson (1966) used the **cannulation** technique to test the relationship between androgens and male rat sexual behavior: Does sexual behavior result from direct action of hormones on specific brain regions? Davidson castrated sexually experienced male rats. In various areas of the brains of rats in the treatment group he implanted crystalline testosterone propionate (TP, a synthetic androgen) and implanted cholesterol in various areas of the brains of rats in the control group. Implants of TP in the hypothalamus and medial preoptic region brought about renewed sexual activity; implants of TP in other areas of the brain had lesser effects or no effect on sexual behavior; and implants of cholesterol had no effect at all on sexual behavior. Davidson concluded that sexual behavior is dependent on testosterone activation in the hypothalamic-medial preoptic region of the brain.

TRANSPLANTATION

Another technique that we should mention is the **transplantation** of neural tissue from one organism to another, or occasionally from one location to another within the same organism. Transplantation of preoptic tissue from neonatal male rats into the preoptic area of female littermates resulted in changes in behavior for females of the species (Arendash and Gorski 1982). They exhibited increased levels of both masculine and feminine sexual behavior as adults. The female's ongoing nervous system development was affected by the presence of male brain tissue. The transplantation technique has also been used recently to explore biological clocks. An extended example of the use of the technique involving activity rhythms in cockroaches is presented in chapter 8.

METABOLIC ACTIVITY OF NEURONS

One of the most recent advances in neurobiology involves the use of autoradiography to measure the rate of metabolic activity of localized regions of the nervous system, particularly the central nervous system (Sokoloff et al. 1972; Gochee, Rasband, and Sokoloff 1980). The technique is based on the fact that neurons utilize glucose to meet their metabolic needs and couple energy usage with functional activity, as do all cells (Gallistel et al. 1982). Deoxyglucose (2-DG) labeled with ^{14}C was injected into the test animal. This compound is taken up by cells differentially, depending upon their rate of activity and energy utilization. Once inside the cell, the 2-DG is converted to 2-deoxyglucose-6-phosphate (2-DG6P). The 2-DG6P is not metabolized further and is trapped within the cell.

The test animal was then sacrificed, its brain tissue sectioned, and the sections placed directly on X-ray film (autoradiography). When the film was later developed, the remaining images showed varying intensities of gray that depend upon rate of uptake and utilization of 2-DG. Neuronal activity can thus be measured using the labeled compound and autoradiography. A further development of this procedure involves the use of a computer-assisted scanning technique to assess the varying intensities of the gray in the autoradiographs (Gallistel et al. 1982). This technique has been used to study the olfactory bulb of rats (Sharp, Kauer, and Shepherd 1975), the effects of imprinting on specific areas of the chick brain (Kohsaka et al. 1979), and the changes in metabolic activity in specific regions of the frog's brain during presentation of visual stimuli (Finkenstadt et al. 1985).

POTENTIAL PROBLEMS

We should pause briefly to consider some of the potential problems associated with the experimental techniques just described. A problem common to most of these techniques is how to determine which nerves were cut, which were used for recording impulses, or which cells were stimulated or lesioned. There are two partial solutions. First, we can use X-rays to verify the locations of electrodes in a neuron or in the brain. However, ascertaining the exact position of an electrode is sometimes difficult. A second more common solution is to use histological techniques to map the positions of electrodes or lesions. After completing an experiment, we make thin sections of brain or other neural tissue, stain the section, examine mounted slides under a microscope, and pinpoint the areas where recordings or lesions were made. Although the electrode is not left in place for this procedure, electrode paths through neural tissue remain visible after the electrodes have been withdrawn.

Two other problems are associated with lesion studies, and in some instances, with electrode implants for recording or stimulation. It is difficult to perform a lesion on an animal in such a way that only the desired neural tissue is destroyed. Slides made with histological techniques often reveal that either less or more tissue has been destroyed than was intended. During electrode implantation, the passage of the electrode through

neural tissue may destroy some cells or impair their capacity to function normally. The possible side effects of this tissue destruction may not be accounted for in the design of an experiment, in the hypothesis, or in the explanation of the results.

Lesioning an area to destroy tissue, or unintentionally destroying nerve cells when implanting an electrode, can lead to yet another problem. Once tissue damage interrupts neural pathways, **degeneration** (the death or functional impairment of cells connected to destroyed or severely damaged neurons) often begins. Degeneration of additional tissue, beyond that of the initial lesion or nerve transection, may have effects on behavior unpredicted by the experimental hypothesis. It may be difficult to assess completely and accurately the influence of these side effects on behavior, or to separate these side effects from the intended behavioral effects.

Many nervous system investigations use anesthetized or restrained animal subjects. It is impossible to be certain that the responses and neural discharge patterns of an anesthetized or restrained animal are identical to those of a normal, awake, free-moving animal. Some techniques, like radiostimulation, have provided a partial solution to this problem.

Finally, if a particular area of the brain is lesioned because we hypothesize that this area is involved in certain specific behavior patterns, we must also ask whether lesions in other areas of the brain might produce a similar behavioral result: Are the lesion effects general or specific? To test this possibility, we should make control lesions in regions of the brain not thought to be associated with the behavior.

Because of the problems associated with the techniques used to study the neural regulation of behavior, investigators must proceed with caution and consider alternative explanations when drawing conclusions from experimental data that take into account the possible side effects of treatments. As animal behaviorists, we must remain constantly aware of the limitations and consequences of the methods and techniques we employ in our research investigations.

EXTENDED EXAMPLES OF NEUROBIOLOGY AND BEHAVIOR

To augment the foregoing material about the relationships between the nervous system and behavior, let us consider two examples. Over several decades, both cricket song and auditory communication in frogs have been explored using a variety of research techniques.

CRICKET SONG

Males of most species of crickets (*Teleogryllus* spp., *Gryllus* spp.) emit species-specific sounds via **stridulation,** a process that involves rubbing together a portion of the cuticle of one wing (the scraper) against a series of ridges on the other wing (Bentley and Hoy 1974). The songs serve to attract females; the orientation of the females to the calling males is termed *phonotaxis.* The species-specificity of the calls is important in maintaining species isolation; otherwise, confusion would result when crickets of different species in the same geographic area are maturing and seeking mates at the same time. The processes involved in sound production and reception, particularly the detection of the males' calls by females, have been studied in some detail (Bentley and Hoy 1974; Huber 1983a).

From early work, we know that the central nervous system of some adult crickets (e.g. *Teleogryllus commodus*) produces stereotyped motor output patterns that develop during the nine to eleven molts after hatching. The neurons involved in sound production are present at hatching, but they undergo changes in size, and new synaptic connections are made during the molts that lead to the adult stage. By examining specific nymphal crickets' neurons known to be involved in production of calls in adults during various instars (stages between two molts), Bentley and Hoy (1970) demonstrated that the neural processes underlying the production of sound develop progressively with each molt. Only at the last instar is the neural machinery complete for calling.

The cricket song is also under a rather high degree of genetic control (Hoy and Paul 1973; Bentley and Hoy 1974). Males of two species of Australian field crickets (*Teleogryllus commodus* and *T. oceanicus*) emit species-specific sound patterns to attract females. Female crickets were tested with songs from the males played over high-fidelity speakers. Test females were placed on the Y-maze globe (figure 6–19). They could indicate their preference for one of the two songs by "walking" on the globe which rotated beneath and entering one arm of the Y-maze. Females of each species preferred the calls from conspecific males. When hybrids produced by mating males and females of the two species were tested, the females exhibited a clear preference for the calls from hybrid males over the calls of the separate species. Thus, there appears to be a genetic basis for call reception in these crickets, just as there is genetic basis for the sound patterns generated.

FIGURE 6-19   Cricket sound preference
When crickets are suspended in the air via a small rod attached to the junction between the head and thorax, they will grab onto and walk along a Y-maze sphere such as that shown here. When sounds are played from two different directions corresponding to the choices of the Y-maze, the cricket will walk along one arm or the other, indicating a preference.
Source: Photo by Charles Walcott.

The neuronal events involved in sound production have been investigated using a series of recording and stimulating electrodes placed at designated locations in the cricket nervous system (figure 6–20). A nerve impulse begins when the command interneurons connecting the head region and thoracic ganglion are stimulated (Bentley and Hoy 1974). This leads to a muscle contraction, followed by changes in wing position that result in stridulation, producing the sounds we hear from crickets.

FIGURE 6-20   Production of cricket song

A series of microelectrodes can be used to study the neural bases for production of cricket song. The electrodes can be used either to track the impulses moving through the system or to stimulate particular structures in the system. The song of the cricket is elicited when the command interneurons, located in the group of neurons between the head region and the thoracic ganglion, stimulate thoracic interneurons and eventually the motor neurons leading to the muscles involved in song production. The vertical muscle fibers are involved in wing closure for sound production, and the horizontal muscle fibers open the wings during sound production.

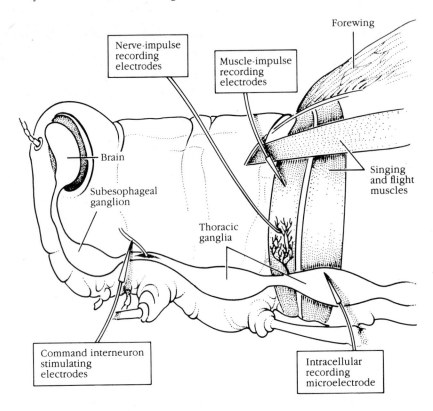

The processes involved in reception of the sounds by female crickets have also been studied in some detail (Pollack and Hoy 1979; Hoy, Pollack, and Moiseff 1982). How is it that the female crickets can recognize the songs of conspecific males? For one species of Australian field cricket (*T. oceanicus*), the song pattern comprises three classes of interpulse intervals; these are arranged into a stereotyped sequence (figure 6–21). The responses of female *T. oceanicus* females to male songs and synthetic calls were tested in two ways. In one test, crickets were placed in an arena with two gauze-covered speakers, one on each side of the arena. When cricket songs were played from the speakers, they measured the response of the female by her movement toward and onto the gauze. In the other test, they observed tethered females in a windstream to determine the response by the direction of the cricket's turn. The females were tested with three types of songs (figure 6–21). Results from both behavioral assays indicated that the females could successfully discriminate between their species' song and that of the related *T. commodus*. When the females were presented with their own species song and the shuffled song (III in figure 6–21), they failed to discriminate between them. Thus, we might conclude that the species-specific song and the shuffled song are equivalent from the perspective of the female *T. oceanicus*. This finding was substantiated by the fact that female *T.*

FIGURE 6-21

The upper line (I) depicts an oscillograph picture of *Teleogryllus oceanicus* song. The horizontal bar beneath the tracing denotes 1 second. Note, in particular, the three types of intervals within the song, the four intrachirp intervals, and nine pairs of alternating inter- and intratrill intervals. The lower line (II-IV) contains three types of electronically synthesized calls; (II) *T. oceanicus* song, (III) *shuffled song*, and (IV) *T. commodus* song. The time bars beneath each call denote 1 second.

FIGURE 6-22   Representative frog calls

Oscillograph tracings of representative mating calls of *Hyla versiclor* (upper tracing) and *Hyla chrysoscelis* (lower tracing). The small bar represents 100 milliseconds. Notice that the temporal form along the axis as the song proceeds, and the emitted energy of the call (deviations from the central line) differ dramatically for these two species of tree frogs.

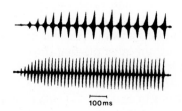

100 ms

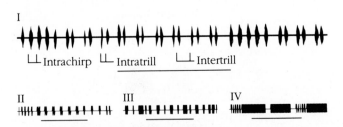

I

LL Intrachirp   LL Intratrill   LL Intertrill

II   III   IV

oceanicus, given a choice between shuffled species-specific song and *T. commodus* song, selected the shuffled song. These results are interpreted to mean that the female *T. oceanicus* recognize their own species' song by the dosages of the three interval types within the song; the sequential ordering of the intervals within a song is not necessary, only the presentation of sufficient numbers of intervals in species-specific proportions (Hoy, Pollack, and Moiseff 1982).

Considerable work has also been done on the neurophysiology of sound reception in female crickets (see Huber 1978, 1983a, 1983b). The ears of crickets are located on the tibiae (on the forelegs) near the animal's body. Intracellular electrodes in a group of about sixty receptor cells at each ear revealed that some neurons are tuned specifically to the frequency of conspecific males' courtship songs. Other neurons appear to be tuned to frequencies of various intraspecific calls and to sounds from potential predators. As sound intensity increases, the response latency of these neurons decreases; small intensity differences between the ears, which lead to slight variations in latency, could be used by central ganglia to obtain direction information (Elsner and Popov 1978; Kleindienst 1980). The neurons originating at the auditory receptors project to specific locations in the prethoracic ganglion of the cricket. Details of neural impulse processing in the central nervous system of the cricket are still the subject of intensive study (Boyan 1981; Huber 1983a).

Thus, the cricket provides an excellent example of the way studies of the nervous system can be integrated with observed behavior patterns. In a system such as

that involving cricket song, it is also possible to hypothesize about the relationships between neuroanatomy and physiology, and the probable evolutionary selection pressures that have shaped these structures and processes.

SOUND COMMUNICATION IN FROGS

As a second example of the integration of knowledge about the relationship between the nervous system and behavior, consider the calling behavior of many frogs. Through the investigative efforts of many, we have a growing knowledge of the anuran's nervous system and behavior that relates to how they detect and hear the calls, and why these animals differentiate inter- and intraspecific calls.

Much of the research on anuran calls has been done on the American toad (*Bufo americanus*) and on a group of small tree frogs (*Hyla* spp.). From early spring until well into the summer months, frogs and toads use this strategy to find mates: males vocalize (figure 6–22) in the evening hours to attract females (Blair 1958). The pair form a bond, termed *amplexus*; and as the female lays a cluster of eggs, the male fertilizes them. In many locales, there are a variety of frogs and toads of different species and genera calling on the same evening. As it turns out, female frogs are capable of making species-specific discriminations and can selectively find mates of their own kind from the cacaphony of noise by phonotaxis (e.g., Gerhardt 1974a,b; Oldham and Gerhardt 1975). The basis for these discriminations lies in part with the nature of the species-specific calls made by the male anurans (Gerhardt 1975). As we shall explore momentarily, there is a close relationship between the characteristics of the calls and the sensory and central nervous system processes in the females.

First, however, what are the advantages of this species-specific discrimination? The most obvious advantage is that males and females of the same species will find each other and mate; the calls serve as a premating isolation mechanism. Without such discrimination, much energy would probably be expended on costly reproduction attempts that do not lead to successful egg fertilization. Furthermore, in some anuran species (e.g., *Hyla* spp., *Rana catesbiana*), it may be possible for females to make distinctions among conspecific males (Howard 1978; Gerhardt 1982). That is, female anurans may exhibit some degree of mate choice based upon the calls' features. This could lead to differential reproductive success for some females, and, for those species where this type of discrimination occurs, could be a selective force in the evolution of calling behavior and mate selection processes. Lastly, the calls may be how the females localize males and move to their locations to mate (Rheinlander et al. 1979).

How then do females make their selective inter- and intraspecific discriminations? In studies of phonotaxis, it appears that the females' frequency preference match the sensitivity measurements of their auditory system (Lombard and Straughan 1974; Capranica and Moffat 1983; Gerhardt 1987). Two types of mechanisms can account for female's discrimination of calls (Capranica and Moffat 1983). She could process the incoming information as waveform over time; by possessing a temporal template encoded in the brain, she could compare it with the incoming messages and determine which signal(s) match the template (the species-specific signal). Alternatively, she could process the frequency of the incoming calls. Because incoming calls are mixed with other sounds with many frequencies the anuran will need to employ a matched filter system (figure 6-23) to detect and sort out the male conspecific call. Though our knowledge about these two signal-processing mechanisms is still limited, from what we do know about the anuran peripheral and central nervous system, it appears that the females use a matched filter system to assess incoming auditory stimuli (Capranica and Moffat 1983).

FIGURE 6-23   Frog matched filter system . . .
Depiction of how a matched filter, such as that of the green tree frog, may work. The emitted signal has a particular spectral pattern, in this instance bimodal (a). However, in nature, the call would normally be embedded in noise (b). If the receiver has a matched filter system, then we would expect the frequency response in the sensory input system of that receiver (c) to be very similar to the pattern emitted by the sender.

MATCHED FILTER DETECTION OF KNOWN SIGNAL

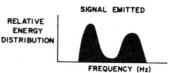

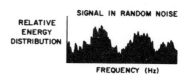

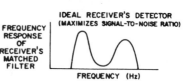

What within the nervous system accounts for this filtering system? The anuran's peripheral auditory system contains two organs: an amphibian papilla and a basilar papilla. The amphibian papilla is more sensitive to low to mid-range sound frequencies, and the basilar papilla is more sensitive to higher frequencies (Feng, Narins, and Capranica 1975; Capranica and Moffat 1977). Impulses from these sense organs travel via the olfactory nerve to the brain and project to several locations. By recording from neurons in the olfactory nerve of the green tree frog (*Hyla cinerea*) while sounds of various frequencies are played (see figure 6–4 regarding tuning curves) we learn that the females' highest sensitivity is in two ranges: a low range at

600–1200 Hz, corresponding to fibers from the amphibian papilla and a high range at 3000–3800 Hz, corresponding to fibers from the basilar papilla. Male frog calls for this species have a bimodal frequency spectrum with peaks at 1000–1100 Hz and 3000–3800 Hz. Thus, the neurophysiological apparatus for a matched filter system is operating in green tree frogs. The final step in this matched filter system must be in the central nervous system where input from both amphibian (low) and basilar (high) papillae converge on the same location (Mudry 1978; Capranica and Moffat 1983). Ideally, the maximum female behavioral response would occur only when the two channels in this system were excited simultaneously. Further studies on behavioral phonotaxis and the neurophysiology and anatomy of the anuran auditory system will be necessary to verify the more intricate aspects of this mechanism for matching the characteristics of male anuran calls to the reception system of the females.

SUMMARY

The *neuron,* which consists of a *cell body,* an *axon,* and *dendrites,* is the basic structural unit of the nervous system, and conducts the electrical impulses that result from changes in membrane permeability and movement of Na^+ and K^+ ions. Transmission of impulses across *synaptic junctions* between neurons occurs through a chemical process involving *neurotransmitters.*

Organisms possess mechanisms for transducing energy from different environmental stimuli into neural impulses—a process defined as *sensation.* The interpretation (*perception*) of these sensory neural impulses takes place in aggregations of cells called *ganglia,* or in the *central nervous system.* The sensory and perceptual worlds of animals differ from our own in three general ways. Some organisms (e.g., electric fish) have totally different modes of sensation. Some organisms (e.g., silkworm moths) may share the same mode as humans, but the range over which they can transduce stimuli from the environment may differ. Finally, some organisms (e.g., octopodes) may transmit the same sensory information to the central nervous system, but may differ in their perceptual interpretation of this information. Organisms have *sensory-filtering mechanisms* that help select only pertinent stimuli to receive, process, and respond to.

Several key developments characterize the evolution of the nervous system. These range from the phenomenon called *irritability* in unicellular organisms, to *nerve nets,* to the *radially symmetrical nervous system,* to the *bilaterally symmetrical nervous system,* and finally, to the *generalized vertebrate nervous system.* They include changes in the numbers and types of sensory receptors; formation of specific neural pathways through aggregations of individual neurons into fiber tracts; and *encephalization,* the concentration of neural material near the organism's anterior end. These developments result in flexibility and complexity changes of behavior.

Investigators can use several techniques to study the functional processes within different portions of the nervous system. They can make *direct recordings* of neural impulses through implanted electrodes. They can *transect* nerves and observe the resulting effects on behavior. They can use *electrode stimulation* (sending of an electrical current to an electrode implanted in the nervous system) to trace neural pathways associated with specific behavior patterns. Two variations of this latter method are *telestimulation,* attachment of a radio receiver to the subject's head, and *self-stimulation,* depression of a lever by an animal to receive stimulation.

Investigators use chemicals or electrical currents to destroy specific areas of the brain and thus assess these areas' role in affecting behavior patterns. They can inject neurotransmitters or drugs that affect the neurotransmitters. Finally, researchers can introduce chemicals via a cannula or needle into specific areas of the brain to assess effects of chemical substances on behavior.

These various techniques have permitted us to formulate some theories of how the brain mediates between incoming stimuli and observed behavior. We know that sensory systems like the eye receive and encode visual stimuli in highly specific ways—for example, *ganglion cells* in the retina are characterized by an *on-center off-surround receptive field* arrangement. Information is carried to the brain along discrete pathways. The structure and arrangement of cells within the brain is highly ordered, with stimuli from each modality and region of the body projecting to specific cortical regions. Some functions appear to be predominant in one hemisphere of the vertebrate brain, while other functions are represented primarily in the other hemisphere.

Further studies relating behavior and brain structure have produced information on the functional flexibility of canary brains. Seasonal changes in behavior such as singing appear to be related to structural changes within specific areas of the brain. We are rapidly expanding our knowledge of the ways specific neurotransmitters affect behavior, and of the close relationships between hormonal and neural control of behavior. Throughout our investigation of the neural control of behavior, we must constantly bear in mind that our conclusions should reflect an understanding of the limitations and problems associated with the research techniques employed.

Studies of cricket song and sound communication in frogs provide detailed insights about the complex variety of roles the nervous system plays in mediating the responses of organisms to their surroundings and in their interactions with one another.

Discussion Questions

1. Peripheral and central sensory filters are the mechanisms by which an organism selects among incoming stimuli. You are attending the evening program of thoroughbred horse racing at a local track and have $200 to spend on bets, food, drink, and so on. Describe the peripheral and central filters that are operative on you in this situation.

2. The sensory and perceptual systems of various organisms may differ in at least three ways: (a) organisms may have different sensory equipment and may be capable of receiving different types of stimuli; (b) organisms may have the same sensory equipment but different thresholds, so that they receive a particular stimulus over different ranges; or (c) organisms may interpret the same stimulus in different ways. Can you provide several specific examples of these differences in the various sensory modalities?

3. As we noted in the text, a number of changes—including changes in numbers and types of sensory receptors, the aggregation of neurons into fiber tracts, and encephalization—characterize the evolution of nervous systems. Changes in behavioral capacities are associated with these evolutionary changes. In terms of flexibility and adaptation, what advantages and disadvantages do the following animals have: (a) unicellular organisms, (b) worms, (c) insects, (d) amphibians, (e) mammals.

4. Permanent electrode implants in regions of the brain have been used to elicit particular behavior patterns in animals. Passing an electrical current through an electrode implanted in one region of a fish brain produces attack behavior. Below are the results of a hypothetical study of electrical stimulation and attack behavior in fish. Explain why each treatment group is necessary to the overall design. The results are shown as the number of attack responses elicited from ten fish of each treatment type, each of which was given ten electrical impulses in the presence of another conspecific male.

Electrode location	Number of attack responses (max. = 100)
No electrodes (normal fish)	77
Electrodes in three brain regions, but no electrical stimulation	73
Electrodes in brain region *A* stimulated	11
Electrodes in brain region *B* stimulated	81
Electrodes in brain region *C* stimulated	96

Should any additional treatment groups be included? Discuss the possible interpretations of these data.

5. We have all enjoyed watching the flight of a beautiful butterfly, be it a monarch, tiger swallowtail, or other species. Describe the series of neural events that are occurring in the butterfly as it flies a short distance, alights briefly on a flower to obtain some nourishment and then flies onward to another stopping place. What experiments would you perform to gain a better understanding of the events taking place inside the nervous system of the butterfly during this flight sequence?

Suggested Readings

Atrens, D., and I. Curthoys. 1982. *The Neurosciences and Behavior.* Sydney: Academic Press Australia.
An introductory text—short, but packed with good information. Somewhat elementary, but well-written and with excellent drawings.

Bentley, D., and M. Konishi. 1978. Neural control of behavior. *Ann. Rev. Neurosci.* 1:35–60.
Review of the field, with emphasis on insects and birds. Both authors have done extensive research at the interface between studies of the nervous system and investigations of functional aspects of observed behavior patterns.

Camhi, J. 1984. *Neuroethology.* Sunderland, MA: Sinauer.
An up-to-date treatment of this subject with a solid emphasis on the behavioral aspects of the relationship between events in animal nervous systems and the observed actions. Excellent for the novice as well as for the advanced student.

Ewert, J. P. 1985. Concepts in vertebrate neuroethology. *Anim. Behav.* 33:1–29.
A nice summary of the goals and progress being made in the area at the interface between the neurosciences and animal behavior. Provides an introduction to the wide variety of techniques being used in this rapidly growing subdiscipline.

Ewert, J. P., R. R. Capranica, and D. J. Ingle. 1983. *Advances in Vertebrate Neuroethology.* New York: Plenum.
A hefty volume, but one which contains many excellent articles on almost all aspects of the rapidly developing field of neuroethology. Topic areas include sensory systems, communication and motivation.

Guthrie, D. M., ed. 1987. *Aims and Methods in Neuroethology.* Manchester, Eng.: University of Manchester Press.
A small volume of eight studies of neural processes and behavior. Very current and with excellent invertebrate examples. The book also has a strong flavor of European ethology, giving students a different perspective on a current hot topic.

Northcutt, R. G. 1981. Evolution of the telencephalon in nonmammals. *Ann. Rev. Neurosci.* 4:301–350.
Review article covers evolution of the forebrain in fish, reptiles, amphibians, and birds. Draws connections between evolutionary changes in overall brain structure/function and correlative changes in behavior. Heavy on terminology, but packed with information.

7

HORMONES AND BEHAVIOR

Hormones are chemical substances produced either by specialized ductless glands in various parts of the body, or by neurons called neurosecretory cells within the nervous system, in which case they are called **neurosecretions.** Hormones are carried by the circulatory system, and neurosecretions travel along nerve axons or in the blood. Both are messengers to various target organs, and they influence such processes as growth, metabolism, water balance, and reproduction. In this chapter we consider how the endocrine system acts as a behavior-regulating mechanism through the

production of hormones in both invertebrates and vertebrates, and how behavior can influence the endocrine system.

A male rat placed with a sexually mature female rat will mount the female within a matter of seconds and begin a sequence of behaviors that ends with copulation (figure 7–1). A castrated male rat takes much longer to initiate mounting (figure 7–2). How might we explain this difference in behavior? The testes are a source of androgens, hormones that affect reproductive behavior. Injections of synthetic androgens can replace the

FIGURE 7-1   Male rat mounting female rat
The male may perform only a mount or a mount with intromission. The sexual performance and timing of sexual behavior patterns in male rats is influenced by the presence of testicular hormones.
Source: Photo by Ronald J. Barfield.

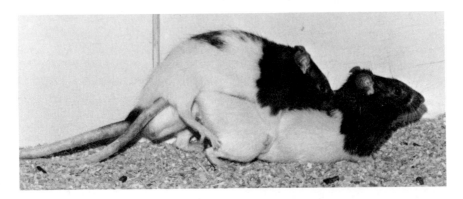

hormone normally produced by the testes and restore mount latencies to near normal levels in castrated male rats.

Many insects, like grasshoppers (*Melanoplus* spp.), hatch from eggs and reach adult form after going through a series of nymphal stages (figure 7–3). Between the stages, the insect molts, shedding its exoskeleton and undergoing other external and internal changes. What internal processes regulate this sequence? If, in an early nymphal stage, we remove the prothoracic gland, which secretes **ecdysone** (the hormone that controls molting), the normal developmental sequence stops. If, however, a prothoracic gland from another grasshopper is transplanted, or if ecdysone is injected, the sequence of nymphal stages resumes. The hormonal changes are also correlated with certain behavior transitions relating to diet, activity rhythms, and choice of habitat, for example. Both the dispersal of grasshoppers—individually or in large numbers—and the mating behavior of grasshoppers are partially under hormonal control.

We take a general look at the endocrine systems of invertebrates and vertebrates in the first part of this chapter. The major portion of the chapter explores the ways that hormones influence behavior; by examining various studies we gain an appreciation of the effects of

hormones on behavior and look closely at the techniques that have been used to explore these relationships. Our primary focus is on the mechanisms of the hormonal system, but we will also note some functional aspects of the hormone-behavior relationship. Later, in chapter 14, we will discuss more thoroughly the ecological and evolutionary significance of endocrine systems.

COMPARISONS OF THE ENDOCRINE AND NERVOUS SYSTEMS

It is important to understand that the nervous system and the endocrine system are feedback systems, that both systems are key parts of the body's mechanisms for interfacing with the environment, and that both systems are critical for adaptation. In general, the nervous system provides faster and more specific responses to external and internal events; the endocrine system provides slower responses and more general effects. The nervous system regulates the fine-tuning responses to specific external conditions or stimuli. The endocrine system regulates more generalized responses to external and internal conditions.

FIGURE 7-2 Sexual behavior in castrated rats
The average latency to the first mount varies in intact rats, castrated rats, and rats given daily injections of doses of synthetic androgen.

Source: F. A. Beach and A. M. Holz-Tucker, "Effects of Different Concentrations of Androgen upon Sexual Behavior in Castrated Rats," *Journal of Comparative Physiology and Psychology* 42:433–453. Copyright © 1949 American Psychological Association.

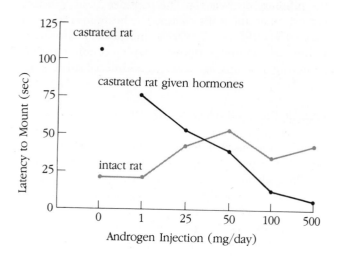

FIGURE 7-3 Grasshopper development
Hoppers, young grasshoppers that hatch from eggs, proceed through a series of five to seven molts, changing in body size, and in internal and external characteristics with each molt, until they attain the adult stage in several months. Molting and other changes in physiology and behavior are controlled largely by the neuroendocrine secretions produced by the brain and various specialized glands within the insect.

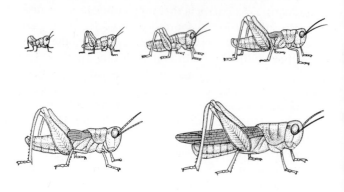

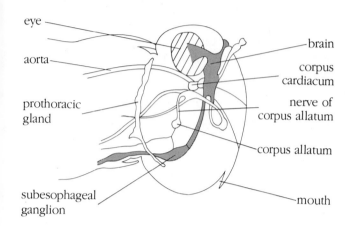

FIGURE 7-4   Brain neurosecretory cells and glands of the head region in the grasshopper

eye

aorta

prothoracic gland

subesophageal ganglion

brain

corpus cardiacum

nerve of corpus allatum

corpus allatum

mouth

INVERTEBRATE ENDOCRINE SYSTEMS

Different types of invertebrates—for example, echinoderms (e.g., starfish), segmented worms (e.g., leeches), mollusks (e.g., clams) and crustaceans (e.g., lobsters)—possess different endocrine systems. Perhaps the role of hormones in affecting behavior is more clearly understood for insects. Let's examine one such system closer, that of the grasshopper. Neurosecretions or hormones that affect behavior are released from five major locations: (1) the **neurosecretory cells** in the brain (figure 7-4); (2) the paired **corpora cardiaca** just behind the brain, near the aorta; (3) the paired **corpora allata** alongside the esophagus; (4) the elongated **prothoracic gland** at the rear of the head, near the neck membrane; and (5) the male and female **gonads** in posterior body regions (not shown). Most insect species have similar structures, with some variations and exceptions.

One important characteristic of invertebrate neuroendocrine structures is that they either involve neurons directly, as do the neurosecretory cells, or they are closely tied to the nervous system by nerves that connect them with the brain and with one another. The glands are also interconnected by the circulatory system. This dual relationship of the neuroendocrine glands and the nervous system is an important feature of hormonal systems in both vertebrates and invertebrates. Thus, the mechanisms that exert controlling influences on behavior have apparently evolved in close harmony.

In invertebrates, a neurosecretion or a hormone does not necessarily exert a direct influence on behavior. Instead, the neurosecretion or hormone may have another endocrine gland as its target. These secretions are termed **trophic neurosecretions** or **trophic hormones.** For example, neurosecretions from the brain of many insects exert a trophic effect on the prothoracic glands, which in turn produce and secrete the hormone ecdysone that controls molting, the developmental sequence of insects (see figure 7-3). Other trophic hormones influence dispersal and mating behavior.

VERTEBRATE ENDOCRINE SYSTEMS

What are the characteristics of vertebrate endocrine systems and how do they differ from those of invertebrates? During the course of evolution, many changes have taken place in animal endocrine systems. In particular, the endocrine system of vertebrates has evolved into two major components. (1) The **pituitary gland** located near the hypothalamus on the ventral side of the brain (it had its developmental origins as an outpocketing from the roof of the mouth) has close connections to several central nervous system structures (figure 7-5). The pituitary gland and the hypothalamus are closely connected, and together they form an important bridge between the nervous and the endocrine systems. This bridge between the two systems is the key to integration of the two control systems. Two connections exist: the posterior pituitary is mainly composed of neurons that originate in the hypothalamus; the anterior pituitary is linked to the hypothalamus by the hypothalamic-pituitary portal system. The pituitary produces trophic hormones, which affect other endocrine glands, and direct-acting hormones. (2) A series of endocrine glands, including the **thyroid, pineal gland, adrenals, pancreas,** and **gonads,** are located throughout the body. The trophic hormones released by the various parts of the pituitary are *peptides*. The peptides' basic structure consists of amino acids chains. The hormones from the adrenal glands, testes, ovaries, and placenta are *steroid* hormones, organic compounds with a common carbon atom ring-structure and varying additional side chains.

ENDOCRINE GLAND SECRETIONS

What are the major endocrine glands of vertebrates, and what behavioral processes do they influence? Several pituitary gland secretions exert controlling effects on behavior and related physiological mechanisms in vertebrates (see table 7-1). Oxytocin and vasopressin are produced by hypothalamic neurons and are stored by nerve terminals in the posterior pituitary (*pars nervosa*);

FIGURE 7-5   Hypothalamus and pituitary of a generalized vertebrate neuroendocrine system

The products of various portions of the pituitary gland regulate other endocrine glands and secrete hormones that directly influence physiology and behavior. Posterior pituitary cells are extensions of nerve cells in the hypothalamus; these cells store and release neurosecretions. Other neurosecretions released in the hypothalamus are carried by blood vessels to the anterior pituitary, where they directly influence hormone production and release.

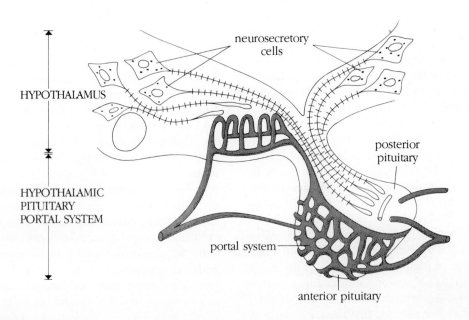

TABLE 7-1   Sources of hormones that affect behavior, hormones produced, and regulatory and physiological effects

Source	Hormone	Regulatory and physiological effect
pineal gland	melatonin	annual reproductive cycle
posterior pituitary gland	oxytocin, vasopressin	milk ejection; parturition; water balance
anterior pituitary gland	luteinizing hormone (LH)	corpora lutea formation; progesterone secretion; Leydig cells stimulation (androgen secretion)
	follicle-stimulating hormone (FSH)	follicle development (♀); ovulation (with LH and estrogen); spermatogenesis
	prolactin	milk secretion; parental behavior in birds
	adrenocorticotrophic hormone (ACTH)	adrenocorticosteroid secretion
intermediate pituitary gland	melanophore-stimulating hormone (MSH)	color change
adrenal cortex	steroids	water balance; metabolism; electrolyte balance
adrenal medulla	adrenaline noradrenaline	blood sugar level; stress reaction
testis	androgens	testis development; spermatogenesis, secondary sex characteristics
ovary and placenta	estrogens	uterine growth; mammary gland development
	progestogens	gestation

they are then released as neurosecretions into the bloodstream. Oxytocin acts to stimulate uterine contractions that may facilitate sperm movements in the female genital tract after copulation and that also help expel the fetus during parturition. Oxytocin also stimulates the ejection of milk from the mammary glands. Vasopressin influences the physiology of the kidneys; it alters urine concentration and thus helps regulate water balance. For example, excretion of highly concentrated urine and retention of body water are physiological and behavior adaptations of many desert mammals, such as camels, kangaroo rats, and gerbils.

The intermediate pituitary (*pars intermedia*) secretes melanophore-stimulating hormone (MSH), which affects the concentration or dispersion of pigment granules in chromatophores, or color cells, found in many vertebrates, particularly fish, reptiles, and amphibians. In the absence of MSH, pigment granules remain clumped; MSH stimulation leads to dispersion of the granules and a color change. For example, adult male three-spined stickleback fish (*Gasterosteus aculeatus*) normally have pale-colored sides. But when two males engage in a territorial boundary display, release of MSH triggers dispersion of pigment granules, and the fish's sides take on a bright blue color. Color changes in vertebrates may serve as communication signals, or they may produce coloration patterns that render an animal inconspicuous against a particular background (camouflage).

Four hormones that indirectly affect behavior are secreted by the anterior pituitary (*pars distalis*); three of these are trophic hormones that exert their effects on other endocrine glands. In females, follicle-stimulating hormone (FSH) and luteinizing hormone (LH) affect the cycle of egg maturation in the ovaries, sexual receptivity, conception, and pregnancy. In males, FSH and LH control sperm production and secretion of male hormones, or androgens. A third trophic secretion, adrenocorticotrophic hormone (ACTH), affects the production and secretion of adrenal cortex steroid hormones. Prolactin, a pituitary hormone present in birds and mammals, is important for maternal behavior and influences milk production in mammals and crop milk accumulation in some birds. Prolactin is also found in amphibians; in certain species, its role may be the stimulation of migration to water to breed.

The pineal gland within the brain secretes several hormones that are indoleamines, proteins, and polypeptides. In the study of behavior, the most important of these is probably melatonin. Melatonin functions in the modulation of reproduction and of annual breeding activity patterns in mammals (Reiter 1980), among other

things. We will postpone further discussion of the role of the pineal gland and melatonin until chapter 8, where we discuss biological rhythms.

The gonads are stimulated by trophic secretions from the pituitary. The testes produce androgens, and the ovaries produce estrogens and progestogens. During pregnancy, progestogens secreted by the placenta play a key role in maintaining gestation. These hormones affect reproduction, maternal behavior, and aggression, as well as some secondary sex characteristics that may be important communication signals. Adrenal hormones are involved in maintaining the body's water balance, metabolism, and electrolyte balance. Adrenaline and noradrenaline, from the adrenal medulla, play important roles in emergency stress reactions.

The specificity of action of all hormones is dependent upon the specificity of receptor sites at the target tissues. This is true for peptide and steroid hormones, whether the targets are other endocrine glands or nonendocrine tissues. At any particular time, the bloodstream may be circulating many "messages" in a variety of hormones, but a hormonal effect on physiology and behavior occurs only when the hormone makes contact with the appropriate receptor sites. Thus, the testes secrete androgens into the bloodstream, and the hormones are carried throughout the body. The androgens affect seminal vesicle growth, changes in secondary sex characteristics, and behavior through target tissues in the brain. In each instance there are appropriate receptors for the androgens located on cells of the target tissues.

ENDOCRINE GLAND INTERACTIONS

FEEDBACK LOOPS. How do endocrine glands interact with each other and with target tissues? Many endocrine gland secretions are characterized by **feedback loops**; an example is diagrammed in figure 7–6. Pituitary secretion of FSH and LH is controlled by **releasing factors** that travel via the hypothalamic-pituitary portal system from the hypothalamus. The trophic hormones FSH and LH are then carried in the blood to the testes, where they stimulate the gonadal processes of spermatogenesis in the seminiferous tubules and the production and release of testosterone from the interstitial cells. Testosterone in turn is carried in the bloodstream to other locations, including the sex accessory glands (e.g., the seminal vesicles, where semen production is stimulated) and the hypothalamus. Specialized hypothalamic sensory cells are part of the body's

FIGURE 7-6 Feedback loops in endocrine physiology of a male mammal or bird
Feedback loops are a key feature of hormone-behavior relationships. Some of the pathways involve direct effects (e.g., FSH and LH act to stimulate the testes). Other pathways involve negative feedback effects. (e.g., testosterone in the bloodstream passing through the hypothalamus is monitored by specific receptors. High levels of testosterone may result in reduced output of the releasing factors, which affect the anterior pituitary.)

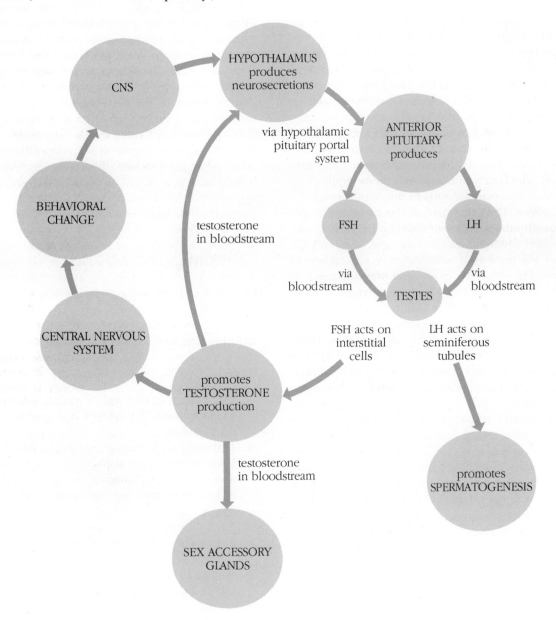

homeostatic machinery and continuously monitor blood levels of various chemicals, including testosterone and its metabolites. Thus when we castrate an animal, the amount of testosterone present decreases, but at the same time, the levels of FSH rise. To what should we then attribute any observed changes in behavior—to the lower levels of testosterone or to the higher levels of FSH? This last point illustrates one of the difficulties encountered when conducting research on behavior-affecting hormones.

The levels of circulating testosterone influence the secretion of hypothalamic releasing factors to the pituitary in a negative feedback relationship. As the testosterone level in the blood increases, the secretion of hypothalamic releasing factors decreases. Conversely, as the testosterone level in the blood decreases, more releasing factors are secreted into the hypothalamic-pituitary portal system, travel to the anterior pituitary, and cause an increase in output of FSH and LH.

There are also feedback loops involving FSH, LH, and the female gonadal hormones, estrogen and progesterone. The feedback effects may be exemplified by tracing the sequence of events in the estrous cycle of a female mammal. At the start of the cycle, the hypothalamus stimulates release of FSH and LH from the pituitary; FSH predominates at this time. FSH and LH stimulate the growth of follicles (or of one follicle in some species) within the ovary, and the production of estrogen from the follicles. When the estrogen level in the blood reaches a peak indicating the follicles are mature, the estrogen has a negative feedback effect on the pituitary, and it decreases the release of FSH. However, estrogen has a positive feedback effect on LH; the pituitary releases more LH when the estrogen level rises, and LH becomes the predominant pituitary hormone. In those mammals that ovulate spontaneously, the follicles rupture and the ova are released at about the time of the transition to LH dominance. The ruptured follicles secrete estrogen and progesterone under the continuing influence of LH and prolactin. However, rising blood levels of progesterone exert a negative feedback influence on the pituitary that causes it to diminish its release of LH.

Estrogen and progesterone are also responsible for changes in behavior. Species vary in their reproductive physiology and endocrinology, but for mammals in general, the estrous phase of the cycle, when the females are receptive to males for courtship and mating, is influenced by the two ovarian steroid hormones. The

complex physiological and behavioral changes are the product of several endocrine pathways that involve positive and negative feedback systems.

Similar feedback loops exist for ACTH and adrenal hormones. The feedback principle is important for a complete understanding of hormone interactions and their effects on behavior. Feedback loops can also be influenced by factors in the environment (e.g., day length) to alter or set hormone levels.

SYNERGISM AND ANTAGONISM. Two key endocrine interactions involve synergism and antagonism. Female estrogens and progestogens, which together affect sexual behavior in vertebrates, are often in circulation simultaneously. These effects are termed **synergism.** Sexual receptivity in female rats depends on the synergistic action of the two hormones (Beach 1976). In contrast, some hormones when they occur in circulation together have **antagonistic** actions; the effects of the two hormones appear to be in opposition to one another. Male pigeons initially act aggressively toward a female that is a potential mate, yet eventually their aggression is inhibited as a result of the antagonistic interactions of testosterone and progestogens.

EXPERIMENTAL METHODS

What types of techniques have investigators developed to explore the relationships between hormones and behavior? In the sections that follow, we will explore each technique in more detail as we examine the various ways that hormones can affect behavior. These techniques include:

1. Extirpation, or removal, of a particular endocrine gland to assess the absence of a specific hormone on behavior
2. Hormone replacement therapy, in which the investigator injects a specific hormone into an animal, or transplants a gland from another animal to replace one previously removed by surgery
3. Excess hormone provision to augment the animals' own hormone supply with exogenous hormone
4. Antagonist or competitive chemical provision, which involves giving the animal a chemical that antagonizes a particular hormone or a chemical that competes with the hormone and blocks its receptor sites, interfering with the effectiveness of the hormone

5. Blood transfusions to transfer the "hormonal state" of one animal to another in order to observe possible behavioral effects
6. Bioassays to assess circulating hormone levels indirectly by measuring a secondary characteristic that is dependent on a particular hormone such as skin gland size
7. Radioimmunoassay to measure circulating levels of a hormone directly
8. Autoradiography to localize the sites where hormone uptake occurs

ACTIVATIONAL EFFECTS

We can divide hormonal influences on behavior roughly into two categories: activational and organizational effects. In activational effects, hormones act as triggering influences on the expression and performance of behavior patterns. Direct activational effects are occurring when hormone secretion or inhibition of secretion leads to a relatively rapid response. Indirect activational effects require more complex stimulation and hormone secretion sequences. The organizational effects of hormones are manifested during an organism's development. For example, sex differentiation and patterns of growth for body tissues are partly under hormonal control. Let us first consider some examples of activational effects on behavior and related processes, and then some examples of organizational effects.

SECONDARY SEX CHARACTERISTICS

In addition to their direct activational effects on behavior, hormones can affect secondary sexual characteristics. The characteristic cock's comb of the male chicken decreases greatly in castrated roosters. Male cats, which spray urine probably as a marking behavior, often cease to spray after their testes are removed. In these two examples, a change in a secondary sex characteristic (cock's comb) and in a sex-related behavior (urine spraying) both affect behavioral processes of communication.

AGGRESSION AND SEXUAL BEHAVIOR PATTERNS

When male ring doves (*Streptopelia risoria*) are castrated, they exhibit decreased levels of aggression, courtship, and copulation behavior. When castrated birds are treated with implants of crystalline testosterone propionate in specific sites in the hypothalamus, the normal levels of these behavior patterns are restored (Barfield 1971; Hutchinson 1969, 1971, 1978). Such experiments clearly demonstrate the activational effects of testosterone on sexual and aggressive behavior and suggest that specific brain sites may be stimulated by testosterone—sites that influence sexual behavior.

The results of other research indicate that the presence or absence of testosterone influences aggressive behavior in birds and mammals, such as ring doves, roosters, mice, rats, and domestic cats (Tollman and King 1956; Barfield et al. 1972; Leshner 1975; Guhl 1961; Bennett 1940). Intact male birds and mammals show more aggression, have shorter latencies to initiate fighting behavior, and fight more frequently than do castrated subjects of the same species. Castration is used with some domestic livestock, such as swine and cattle, to alter behavior patterns, to reduce aggression, and thus to make the animals easier to handle and less likely to injure one another.

Both male and female Syrian golden hamsters (*Mesocricetus auratus*) have paired sebaceous flank glands that they use to mark objects in their environment by depositing sebum from the gland (figure 7–7). The flank glands are androgen dependent; measurement of their size and pigmentation can be combined in an index that is a bioassay measurement of relative levels of circulating androgens (Vandenbergh 1971, 1973; Drickamer and Vandenbergh 1973; Drickamer, Vandenbergh, and Colby 1973). When groups of four intact male hamsters that had been isolated since weaning and that were equal in body weight were allowed free social interaction in a large pen, there emerged a significant positive correlation ($r = .77$) between the outcomes of encounters and the index of gland size and pigmentation. The key feature of this experiment was that the investigators measured the glands before they placed the hamsters together and could predict the outcomes of social encounters on the basis of the bioassay (gland size) of relative levels of circulating androgens. In a related experiment, the investigators gave injections of testosterone propionate to castrated male hamsters in four groups. Each hamster in a group was given a different dose, and when the four were allowed to interact, the investigators could again predict the outcomes of social encounters ($r = .81$) on the basis of dose level and corresponding gland index. Interestingly, investigators could also predict the outcomes of social interactions among female hamsters on the basis of the measurement and pigmentation of their flank glands (Drickamer and Vandenbergh 1973).

FIGURE 7-7   Hamster flank glands
Hamsters have paired sebaceous flank glands that are used to mark objects in the environment with sebum. The size and pigmentation of the glands are androgen dependent. A normal, intact male (left), and (right) a castrated male.
Source: Photo from John G. Vandenbergh.

 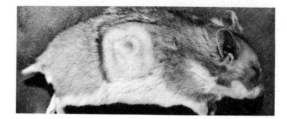

SEXUAL ATTRACTION

In certain species of cockroaches (e.g., *Periplaneta*) and moths (e.g., *Bombyx*), females release pheromones, which act as sex attractants for males. A **pheromone** is a chemical produced by an animal and released in the external environment where conspecifics' behavior or physiology may be affected by it. If a female cockroach's corpora allata are surgically removed after she has molted to the adult state, she does not produce pheromones when she becomes reproductively active (Barth 1965). Females treated in this way are incapable of attracting males and do not mate. However, if a corpus allatum from another adult female is transplanted into the test subject, she is again capable of pheromone production and release. Apparently a hormone that is produced by the corpus allatum acts on peripheral glandular tissues to stimulate production of the sex-attractant pheromone. Interestingly, if a female is gonadectomized as an adult, the processes of pheromone release, mate attraction, and mating are not affected.

Why has this system of mate attraction evolved in some moths, cockroaches, and other insects? There are several possible answers to this question. First, emission of a species-specific sex-attractant chemical may be a species-isolating mechanism. If several closely related species are reproducing at the same time, it would be advantageous—to avoid gamete waste and to allow insects to locate potential mates of their own species. Thus when each species has a different sex-attractant pheromone, individuals can easily locate and mate with partners of their own species. Second, some pheromones may serve to synchronize the reproductive activities of the two sexes of a given species so that males and females are ready to mate at the same time. Third,

chemical attractants could serve a sexual excitatory function by bringing both members of the pair to peak reproductive activity simultaneously, thereby ensuring greater reproductive success. Fourth, these pheromones may set the stage for internal hormone and behavioral output changes that are reflected in the nest-building and parenting behavior occurring later when eggs are laid and young are born. Finally, of course, the chemical attractant ensures that a male will find a female and mate with her. One practical application of the species-specific quality of pheromones is that we can use them to attract and capture designated target animals. Thus, for example, traps that contain the scent from female Japanese beetles can be used to attract males as a pest control procedure.

ECLOSION

The process whereby the adult form of an insect emerges from the pupa after metamorphosis is called **eclosion** and is another activational effect controlled largely by hormones. Many moth species eclose at a species-specific time of day. The eclosion hormone, produced by neurosecretory cells in the brain, plays a critical role in this process (Truman 1971; Truman and Riddiford 1970). If the eclosion hormone is injected into pupae that are near the end of metamorphosis, eclosion behavior, such as abdomen movements and wing spreading after emergence, can be activated at any time of day. Moths that have had their brains removed usually emerge successfully; therefore the presence of the eclosion hormone is not an absolute requirement for eclosion to take place. The process, however, is not as coordinated in brainless subjects, and some activities (e.g., wing spreading) are usually absent. Thus, although the hormone may not be necessary for eclosion, it does appear to be necessary for proper coordination of the sequence.

LIFE STAGES

Adult male desert locusts (*Schistocera gregoria*) fail to exhibit sexual behavior when the corpora allata had been removed. When corpora allata from other adult males are transplanted into allatectomized males, sexual behavior is restored (Lohrer 1961; Pener 1965). However, similar investigations have revealed that the corpora allata are not needed for sexual behavior in certain grasshoppers (Barth 1968).

Several locust species exhibit both solitary and gregarious phases; young hoppers (see figure 7–3) reared in isolation exhibit moderate activity levels and do not engage in sustained flights; hoppers reared under crowded conditions do engage in long flights (Johnson 1969). Some evidence supports the conclusion that this difference has a hormonal basis. First, solitary hoppers have larger prothoracic glands than do gregarious forms (Carlisle and Ellis 1959). Second, adult locusts that develop from solitary hoppers retain the prothoracic glands, but the glands are absent in adults that develop from gregarious hoppers. Third, if homogenates of prothoracic glands are introduced into gregarious hoppers, general activity is reduced (Carlisle and Ellis 1959), and if prothoracic glands from solitary adults are transplanted into gregarious adults, sustained flight activity is diminished (Michel 1972).

MOLTING

Studies of hormones and neurosecretions in invertebrates other than insects have revealed common patterns of effects. We know that crustaceans, and some worms, and mollusks, have one or more endocrine glands, the secretions of which affect sexual differentiation, gamete maturation, and reproduction stimulation (Charniaux-Cotton and Kleinholz 1964; Wells and Wells 1959; Golding 1972). In some crustaceans, several behavior-related phenomena are under partial endocrine control. Many crustaceans (as well as other arthropods) **molt** (shed their exoskeletons) periodically as they grow. Removal of both eyestalks in these animals shortens the interval between molts. If the crustaceans are given extracts of a particular neurosecretory gland,

molting is prevented. Clearly the gland produces a molt-inhibiting factor. It is also possible that some cells in the eyestalks are involved in the production and secretion of a substance that shortens the interval between molts (Kleinholz 1970).

COLOR CHANGE

As we noted earlier, a melanophore-stimulating hormone (MSH) affects color changes in fish. We can also see MSH action in short-tailed weasels (*Mustela erminea*), which undergo seasonal changes in **pelage,** or coat color, during spring and fall molts (Rust 1965; Rust and Meyer 1969). In the spring, MSH secretions increase, and new brown hairs replace the white winter coat. During the fall months, MSH release is inhibited by the action of melatonin, another hormone that is secreted by the pineal gland. The developing hairs at this time of year are not pigmented, and the weasel's coat returns to its white winter color. These seasonal shifts in coat color may be both behaviorally and functionally significant because the result matches the general background color of the weasel's environment. Camouflage coloration serves a possible dual function because it enables the weasel to hunt inconspicuously and to be "hidden" from potential predators.

MSH has also been implicated in the color changes that occur in fish, amphibians, and reptiles. For example, the color of two fish competing at a territory boundary may change or deepen during the course of the interaction. One interesting feature of the way MSH affects color changes in fish, amphibians, and reptiles is the rapidity of the physiological changes—some take only seconds.

ORGANIZATIONAL EFFECTS

Studies of quail, zebra finches, rats, guinea pigs, mice, and rhesus monkeys, and other animals have provided clear evidence that certain hormones exert critical effects on sex differentiation during early development (see reviews by Toran-Allerand 1978; Dörner and Kawakami 1978; Adkins-Regan 1987). Most of the investigations of mammals have concentrated on the effects of gonadal hormones on organizational processes as they affect later adult sexual and aggressive behavior (Young et al. 1964; Harris and Levine 1965; Barraclough 1967; Davidson 1966a; Luttge and Whalen 1970; see also review by Feder 1981). Research on birds has been involved primarily with the organizational effects of gonadal hormones on later sexual behavior (Adkins-Regan and Ascenzi 1987; Schumacher, Hendrick, and Balthazart 1989). Let's consider some specific examples of the ways hormones exert organizational effects on behavior and related processes.

SEXUAL AND AGGRESSIVE BEHAVIOR

If we castrate a male rat within the first four or five days of his birth, he will not show normal sexual behavior as an adult. If we give a neonatally castrated male rat estrogen and progesterone as an adult, he will exhibit female sexual responses, such as the lordosis posture that a receptive female adopts to permit mounting and intromission by a male (figure 7–1); lordosis is characterized by shifting the tail to one side, raising the hindquarters, and lowering the abdomen. If we give a male rat that was castrated as an adult estrogen and progesterone, he will not show female sexual behavior.

In an experiment, neonatal male rat pups were injected with estrogen. Histological examination after sacrifice as adults revealed some degeneration of the seminiferous tubules where sperm are produced. Although these males showed mounting behavior when near a receptive female, the behavior was irregular: the mounts were often incorrectly oriented, and no ejaculation occurred.

Female rats treated within the first four or five days after birth with either artificial androgen (TP) or artificial estrogen (EB) did not show normal estrous cyclicity as adults and did not respond with lordotic behavior when injected, as adults, with estrogen and progesterone. When these neonatally injected females were given injections of TP as adults, they exhibited malelike sexual behavior. In fact, the neonatal injection of androgen alone will produce malelike sexual behavior in adults of some inbred strains of mice (Manning and McGill 1974). Females given estrogen injections neonatally sometimes had irregular estrous cycles; females injected with TP neonatally, exhibited permanent vaginal cornification. (Vaginal cornification, the sloughing off of the epithelium lining, is produced by the action of estrogens on the lining of the vaginal tract and is a sign of estrus). Masculinized females also exhibited higher propensities and shorter latencies to fight, compared to those of adult males (Edwards 1968).

Similar studies of the organizational effects of hormones on female behavior have been conducted on guinea pigs (Phoenix et al. 1959) and rhesus monkeys (Goy 1970; Goy, Bercovitch, and McBrair 1988). Female progeny of females treated with androgens during pregnancy have external genitalia that are masculinized (smaller vaginal opening and hypertrophied clitoris), and they exhibit malelike sexual behavior.

These data, in combination with the previously cited studies of rats, point up an important aspect of early hormone injections. There is a critical period during which hormone injections must occur for certain behavior patterns to be affected. In rats and mice, this critical period occurs during the initial four or five days after birth; in guinea pigs and rhesus monkeys the critical time period occurs prenatally. Neonatal injections do not have the same effects on guinea pigs and rhesus monkeys that they do on rats and mice. Rhesus monkeys and guinea pigs prenatally attain the stage of development that rats and mice attain postnatally. The effects of the injected hormones occur at about the same relative stage of development in these different species of animals. There are also critical periods for the organizational effects of hormones on the physiology and behavior of quail (Schumacher, Hendrick, and Balthazart 1989) and zebra finches (Adkins-Regan and Ascenzi 1987). For quail, the critical time period for organizational effects of estradiol occurs during the nestling phase, in the first two weeks after hatching (figure 7–8).

One interesting aspect of the organizational effects has emerged in recent years: Not all aspects of physiology and behavior are influenced in the same manner. Thus, for example, different groups of quail were treated with three different doses of estradiol (0, 5, or 25 μg) on either the 9th or 14th day of embryonic development (Schumacher, Hendrick, and Balthazart 1989). The birds had been castrated at day 4 after hatching to avoid the confounding effects of post-hatching hormone release, and they had then been given exogenous testosterone as adults. Male birds treated at day 9, but not on day 14, exhibited demasculinized sexual behavior and increased cloacal gland growth. However, males treated at either day 9 or day 14 exhibited diminished crowing activity and lowered levels of plasma LH levels. Thus, the process of demasculinization may not be restricted entirely to a specific critical period; different behavior patterns may be affected at different times during development.

In rhesus monkeys (Goy, Bercovitch, and McBrair 1988), androgenization can take place early (days 40–64 of gestation) or late (days 115–139 of gestation). Early,

FIGURE 7-8   Organizational effects of estrogen treatment on male and female quail
Estradiol benzoate was injected into the eggs on the 10th day of incubation. As adults, the birds were held in short day-length conditions to produce regressed gonads. They were then injected with testosterone and tested for three behavior patterns when placed with normal female partners.

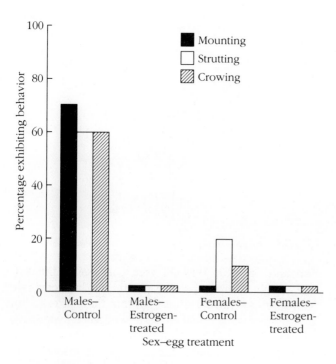

but not late, androgen treatment results in masculinization of the female genitalia. In a series of behavioral test measures that included mounting, initiation of play, rough play, and grooming, androgen treatment at neither time resulted in masculinization of all four types of behavior. Early androgen treatment produced females that exhibited more mounting behavior (with mothers and peers) and less grooming, whereas females treated with androgen later in gestation exhibited more rough play and more grooming. Thus, it appears that in those animals where the phenomenon has been tested, organizational effects may occur over longer periods than were previously thought; the timing of the effects and the sensitivity of various systems to the organizing hormones varies.

Recent research on reptiles from three different orders indicates that the organization of behavior via hormonal effects occurs in alligators (*Alligator mississippiensis*), turtles (*Trionyx spiniferus*), and geckos (*Eublepharis macularius*) (see Adkins-Regan 1981). From each

species, eggs were obtained and were injected at an appropriate age before hatching with estradiol. The artificial estrogen caused female development in all of the treated animals. Thus, estradiol apparently feminizes the gonads of a variety of reptiles.

OTHER BEHAVIOR PATTERNS AND INTRAUTERINE POSITION

In addition to organizational effects on male and female sexual behavior and aggression, we know there are organizational effects on other behavior patterns. Meany and colleagues (1982) investigated the effects of glucocorticoids on play-fighting behavior of Norway rat pups. Glucocorticoids administered prenatally to male rat pups suppressed the normally high levels of play-fighting found in males. Similar treatments given to female rat pups did not influence the level of play-fighting behavior. Thus there appears to be a sex-dependent, organizational effect of glucocorticoids on the development of play-fighting in rats.

As an interesting adjunct to the foregoing discussion, organizational effects of hormones on sexual behavior have been suggested by some data for humans (Money and Ehrhardt 1972). Two such effects involve masculinization of genetically female fetuses. Mothers who had been given progestogens to help maintain pregnancy often gave birth to females whose appearance was masculinized. (This form of treatment has since been discontinued for pregnant women.) Among nonhuman mammals, some newborn females have a hereditary disease that results in high levels of adrenal gland androgen output—this disease in a pregnant woman could masculinize a female fetus. Genetic males may have target tissues that cannot respond to the androgens secreted from their own testes, and this results in a more feminine appearance. It is important to note that in humans, social experiences often predominate in shaping behavior; individuals usually respond to the sex of rearing regardless of the genetic sex.

The female of a pair of heterosexual twins in cattle is called a freemartin, and is unable to produce her own offspring in adulthood. In the early part of this century, scientists recognized that hormone action was probably involved; presumably at some time during early embryonic life, a circulatory link connected the blood flow of the two fetuses of opposite sexes. This process could result in the transfer of some androgens to the heifer, resulting in a sterile individual (Cole 1916; Lillie 1916). We now know that this hypothesis is substantially correct.

We know more from recent investigators who have tested the effects of postnatal androgen treatment on behavior in heifers (Bouissou 1978; Bouissou and Gaudioso 1982). Heifers were given 100 days of treatment with testosterone propionate beginning before puberty when they were three months old. When these heifers were tested three to twelve months after cessation of treatment, they tended to be more dominant than untreated control animals. The treated heifers were not, however, more aggressive than controls, nor were there any differences between treated and untreated heifers with respect to physical appearance or normal sexual cycles. The investigators concluded that there were some persistent effects of the early hormone administration: the treatment resulted in heifers that exhibited less fear (and presumably more boldness) in encounters with other cows. Similar effects of androgen treatment resulting in higher social rank have also been demonstrated in mares (Jussiaux and Trillaud 1979) and hinds of the red deer (Fletcher 1975).

The intrauterine position of a female fetus in rats and mice can influence her genital morphology and sexual behavior (Clemens 1974; vom Saal and Bronson 1978; vom Saal 1989). Rats and mice release multiple eggs at each cycle and bear litters of sizes ranging from one to twenty pups. Both species have bicornate uteri; fetuses are arranged sequentially in each arm of the uterus (figure 7–9). Thus, fetuses of one sex can be positioned between two other fetuses of the same sex, between two of the opposite sex, or between two fetuses of opposite sexes. In rats and mice, there is a brief period late in gestation near the time of parturition when testicular androgens are produced and released. Clemens (1974) proposed that females in utero could be masculinized by exposure to testosterone from neighboring littermates. It turns out that anogenital distance, a measurement of the distance between the anal opening and the genital opening, is a reliable measure of androgen exposure. For rats and mice, anogenital distances for genetic females positioned between males in utero are larger than for females positioned between two females in utero. In addition, certain behavioral and physiological traits are affected. In rats, females that have been masculinized due to in utero position exhibit more mounting behavior (Clemens, Gladue, and Coniglio 1978), while in mice the estrous cycles are longer, levels of aggressive behavior are increased, senescence occurs at earlier ages, daily activity levels are lower, and the tendency to exhibit malelike sexual behavior is increased in females between two males compared with females between two females or between a male and a female (Gandelman, vom Saal, and Reinisch 1977; vom

FIGURE 7-9   Rat uterus
Both rats and mice have bicornate uteri. The often numerous fetuses of a pregnancy are arranged sequentially in each of the two horns of the uterus as shown here for the rat. Pups of one sex may be positioned between two pups of their own sex, between two of the opposite sex, or between fetuses of opposite sexes. The location of a female next to a male fetus results in masculinization of the genetic female.

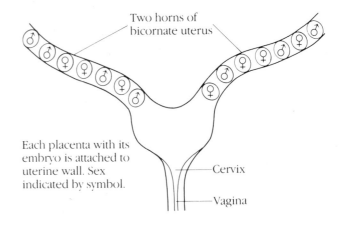

Two horns of bicornate uterus

Each placenta with its embryo is attached to uterine wall. Sex indicated by symbol.

Cervix

Vagina

Saal and Bronson 1980a,b; vom Saal 1989). Interestingly, male fetuses that are positioned between two female fetuses have higher concentrations of estradiol and show organizational effects as well; such males have altered genital and brain morphology, higher levels of activity, less sexual behavior, and less responsiveness to doses of testosterone used to induce intermale aggression than do males positioned between two other males in utero (vom Saal 1989).

From all of these studies we can reach some tentative conclusions about the effects of gonadal hormones on the organization of behavior in animals. Gonadal hormones (either estrogens or androgens) secreted at the appropriate critical time during early development affect the undifferentiated brain by their production of a malelike adult organism, or one exhibiting malelike behavior, regardless of the actual genetic sex. Early gonadal hormone secretion apparently sensitizes the brain to hormones that circulate in the blood later. In the absence of gonadal hormone secretion, a typical cyclical female pattern develops (see review by Arnold and Gorski 1984). Again, the absence of hormones establishes certain patterns of sensitivity in critical areas of the brain, such as the hypothalamus, so that later hormone circulation produces particular physiological and behavioral responses. The general sensitivity of particular areas of the brain, and these areas' sensitivity to different chemicals' temporal patterns or blood levels of different chemicals are affected (Dörner and Kawakami 1978).

THYROID AND ADRENAL GLANDS

Are there other endocrine glands whose products can have organizational effects on behavior? Both the thyroid and adrenal gland hormones have been implicated in organizational processes. Thyroidectomized rats exhibit characteristics of cretinism, slower body growth, delayed sexual maturation, and retarded development of the nervous system. Their actions are slower, their learning is slower and occurs with great difficulty.

Young rats that are given a few minutes of handling each day during early infancy exhibit less extreme responses to stressful situations as adults (Levine 1967, 1968). When adult rats that were handled when pups are placed in a stressful situation, they secrete less adrenal steroid hormones and show less fear response to presentations of novel stimuli. When they were pups they secreted more adrenal steroid hormones in response to shock or ACTH injections. Early in development, these higher levels of early adrenal steroid hormones may have affected the brain mechanisms related to stress responses, and thereby had established permanent changes that continued into adulthood. These findings exemplify another principle regarding organizational effects on behavior: Some of the effects organize behavior in a manner that does not require hormones for activation in adulthood, whereas other effects do require hormones later in life for activation of the behavior.

INVERTEBRATES

All of the preceding material has dealt with vertebrates: Are there organizational effects of hormones in invertebrates? There are, in fact, several processes in invertebrates that involve organization effects of hormones on behavior and related physiological processes (Truman and Dominick 1983), though additional studies in this area are needed. The processes involved in insect metamorphosis from pupa to adult, and in the molting that takes place periodically as the insect forms a new exoskeleton and sheds its old one, involve the organizational effects of specific hormones. The last stage of metamorphosis is called eclosion, and the shedding of the old skin is called ecdysis. There are apparently some close parallels involved in the hormonal control of these processes. For the molting sequence, there are three behavior patterns associated with ecdysis: the finding of a suitable perch, the stereotyped movements involved in removing the old exoskeleton, and the expansion of the new cuticle to fit the body of the animal. Many of the steps in this sequence depend upon the level of eclosion hormone present in the insect. Changes in hormone titers and the timing of these changes are activational hormone effects that influence the behavior sequence. However, it also appears that for some insects, e.g., the tobacco hornworm (*Manduca sexta*), certain hormone changes in the sequence prime or organize the system for later effects. Thus, the decline of eclosion hormone level at one stage early in the sequence is necessary to trigger certain preparatory behavior, e.g., perch finding, and also to make the system responsive to increases in the eclosion hormone titer at a later stage, inducing the next step in the sequence. If an exogenous hormone that mimics the actions of eclosion hormone is given to the hornworms at the time that they would be experiencing a decline in the natural levels of the hormone, then the proper preparatory stages are delayed. This is a dose-dependent effect, with larger doses of the exogenous hormone leading to longer delays before the next step in the sequence occurs.

Two invertebrate hormones, MH (molting hormone, also called ecdysone, or ectysteroid) and JH (juvenile hormone), interact in the control of growth and metamorphosis in insects. When levels of JH in the blood are high, and MH is also present, the insect continues to grow and differentiate, but will not molt to the adult stage. However, if MH acts alone, molting is induced and the insect will metamorphose and differentiate into the adult form. There is additional evidence from studies of some insect species that the differentiation and maturation of sexual organs depends partly on gonadal hormones (Fraenkel 1975; Gilbert 1974; de-Wilde 1975).

ENDOCRINE-ENVIRONMENT-BEHAVIOR INTERACTIONS

Some activational effects of hormones involve complex interactions between behavior, hormones, and environmental stimuli. We shall now discuss in some detail four examples that illustrate these interrelationships: reproductive sequence in ring doves, parturition and maternal behavior in rats, reproduction in lizards, and reproduction in house sparrows.

FIGURE 7-10   Reproductive behavior cycle of the ring dove
This cycle provides an example of indirect environmental determinants of behavior. The sequence involves (1) courtship and copulation, (2) nest building, (3) egg laying, (4) incubation, and (5) feeding crop milk to the young squabs after they hatch. The cycle then repeats.

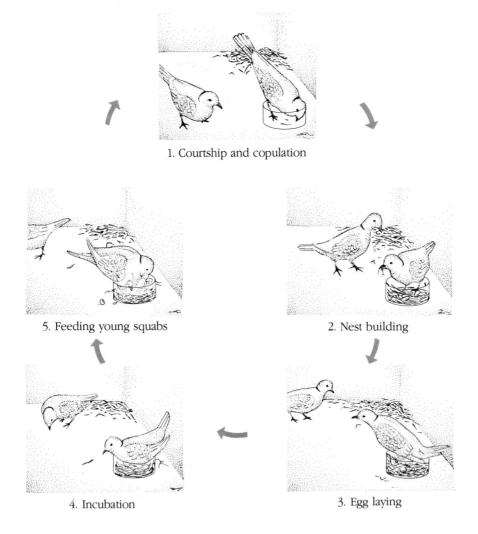

1. Courtship and copulation

5. Feeding young squabs

2. Nest building

4. Incubation

3. Egg laying

REPRODUCTIVE SEQUENCE IN RING DOVES

Figure 7–10 illustrates the reproductive sequence in ring doves. A male begins courtship display shortly after being placed with a female. The failure of castrated males to court females indicates the importance of a continuous supply of androgens for the initiation of the cycle. Male courtship stimulates pituitary release of FSH in the female dove; FSH in turn stimulates follicle development in the ovaries. The follicles secrete estrogen, which affects uterine growth and development. Within a day to two, the birds begin nest construction; during this phase of the cycle, they copulate and continually add to their nest. The presence of a nest stimulates the production and secretion of progesterone in females. Progesterone in both sexes promotes incubation behavior after eggs are laid. Egg laying is activated partially by secretion of LH by the female's pituitary.

Incubation, maintained by progesterone secretion, lasts fourteen days; the male and female take turns on the nest. Under the influence of the presence of eggs in the nest and as a result of stimulation from incubation behavior, both the male's and the female's pituitary glands secrete prolactin. Prolactin acts to inhibit FSH and LH secretion, and all sex behavior ceases. Prolactin also stimulates crop development and the production of crop milk (a nutrient-rich fluid secreted in the gullets of males and females) and may also help to maintain incubation behavior. When young squabs hatch after two weeks, the parents immediately feed them with crop milk. During the next ten to twelve days, both parents continue to feed the young with crop milk; however, feeding behavior wanes toward the end of this period, in part because the secretion of prolactin decreases. As prolactin decreases, the pituitary secretes FSH and LH, the same pair of doves resumes courtship, and the sequence begins again.

At each stage of this sequence, the internal state of each bird interacts with external variables to produce the observed behavior patterns. The variables consist of (1) the hormonal state of both the male and female dove, including the feedback loops; (2) the behavior of each member of the pair that stimulates changes in the hormonal levels and behavior of its mate; and (3) environmental cues, such as nests and eggs, that influence hormonal and behavioral changes in both (figure 7–11).

Daniel Lehrman and his colleagues conducted many ingenious experiments to clarify the interactions and changes that together comprise the reproductive sequence in ring doves. A description of a few of their experiments gives us a glimpse of the logic and experimental method they employed (see Lehrman 1955, 1958a,b, 1959, 1961, 1964, 1965; Lehrman, Brody, and Wortis 1961; Erickson and Lehrman 1964; Cheng 1979).

To determine whether the presence of a mate or of nesting material affects incubation behavior of female ring doves, they used three experimental groups: (1) control females housed alone, (2) females housed with a male only, and (3) females housed with a male and nesting material. They assessed the results of these pairings in terms of the percentages of test females in each group that exhibited incubation behavior when presented with a nest containing eggs (figure 7–12). Control females never incubated eggs. By days 6, 7, and 8 after pairing, increasing percentages of females in the second group (those housed with a male only) incubated eggs; by day 8, 100 percent of females caged with a male and nesting material incubated test eggs. Thus, we can conclude that the presence of both the male and

FIGURE 7-11   Interrelationship between hormones and behavior in ring doves
Hormones and behavior patterns of each individual, interactions between individuals, and external cues affect the synchrony of reproductive behavior. Bidirectional arrows indicate feedback relationships, and unidirectional arrows indicate direct effects.

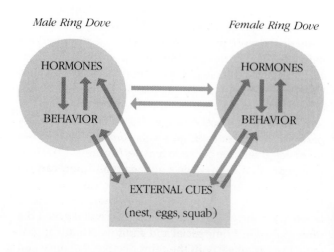

FIGURE 7-12   Incubation behavior in ring doves
Development of incubation behavior in female ring doves is effected by association with a mate or with a mate plus nesting material. Each point on this graph is derived from tests of twenty different birds.

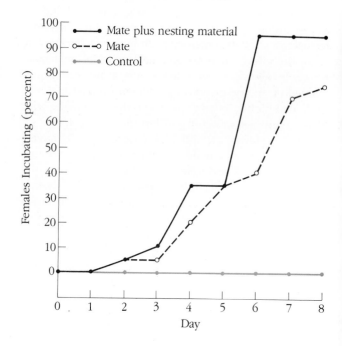

the nesting material is necessary for complete incubation behavior in the female ring dove. Similar research with the male ring dove shows that the presence of both the female and the nesting material are necessary for males to show complete incubation behavior.

A combination of several techniques has been used by Silver and her colleagues to explore the stimuli that affect males and possible corresponding hormonal changes during the early portion of the ring dove reproductive cycle (Silver 1978; O'Connell et al. 1981a,b). Male doves were presented with varying stimuli, their behavior was observed, and radioimmunoassay measurements were made of plasma hormone levels. Males presented with intact courting females had higher levels of testosterone than males paired with ovariectomized females. Males exposed to females caged behind glass partitions had testosterone levels comparable to those for males given free access to females (figure 7-13). Thus, it appears that the female's gonadal condition can influence the male's response. When surgically deafened males are exposed to intact females, their testosterone levels were lower than those of normal males given the same exposure. Some social and auditory stimuli appear to be important early in the reproductive sequence and result in high male testosterone levels and stronger behavioral responses to females.

By about the eighth day after pairing, when incubation begins, the male's testosterone levels have declined to precourtship baseline level. Courting males were exposed to either an incubating female or another courting female. Those in the first group exhibited reduced testosterone levels within a few days, whereas testosterone levels remained high in the second group. Also, the effects of nests and nest material on testosterone levels were tested. Males whose nests were destroyed each day, or who were given no nest material at all, had higher testosterone levels than did males who were given nest material and left undisturbed.

Both external stimuli and female dove behavior appear to influence male behavior and male hormone levels at several early stages in the sequence. In the male's behavior, transitions appear to be strongly mediated by context stimuli; whereas for females transitions appear to be influenced more by hormonal changes, as we noted previously.

In another study, male ring doves were allowed to court and mate with females and to participate in nest construction. After the eggs had been laid and incubation had begun, each male was placed behind a partition so that he could see his mate sitting on the eggs but he could not incubate the eggs. These males, stimulated only by the visual cues from an incubating mate,

FIGURE 7-13   Ring doves courting
Female ring doves separated from their male partners by a glass partition can still induce both behavioral and hormonal changes in the males. Males in this situation begin to court the female and exhibit increased levels of plasma testosterone.
Source: Photo from Rae Silver.

underwent normal crop development, and when permitted, they fed the young squabs normally after hatching. Males that were not permitted to watch their mates incubate eggs failed to develop crops or to feed the young. Although participation in nest building is necessary for crop development, direct involvement in incubation is not.

Why has such a finely tuned system of complex interactions between behavior, hormones, and external cues evolved in ring doves? The almost lockstep system that characterizes the reproductive cycle of ring doves—where each stage in the sequence is dependent on certain interactions and cues from preceding stages—ensures that as the cycle proceeds, both members of the pair are in synchrony and both are in the proper "frame of mind" to perform the required behavior as needed. Since the system involves the participation of two birds and necessitates that their activities be coordinated throughout, the reproductive success of the pair is guaranteed only if both perform certain actions in synchrony. For example, females that laid eggs before a nest was completed would contribute little to future generations. If the male developed a crop or began to produce crop milk just as the female laid the eggs, long before any squabs had hatched, and ceased to produce crop milk during the two weeks they are normally fed by both parents, only the female would be capable of supplying the squabs with nourishment. Her food supply might be insufficient, and both of their reproductive energies would have been wasted because some or all of the squabs would die before fledging.

PARTURITION AND MATERNAL BEHAVIOR IN RATS

In some ways, maternal behavior in the laboratory rat is similar to the cyclical reproductive behavior of ring doves: The internal state of the female rat is partly a response to external stimulation of the presence of a nest and pups. Three major events of the maternal cycle will illustrate the interaction between internal state and external stimulation: (1) parturition and the events surrounding the birth of a new litter, (2) nest building, which occurs both before and after parturition; and (3) the period of lactation (Lehrman 1961; Zarrow 1961; Rosenblatt and Lehrman 1963; Richards 1967; Lott and Rosenblatt 1969; Rosenblatt 1970; Lubin et al. 1972; Rosenblatt, Siegel, and Mayer 1979).

First let's examine events at the time of parturition. Progesterone, sometimes called the pregnancy hormone, helps maintain proper internal conditions prior to parturition. Internal changes in hormone secretions and shifting external cues occur in the days just before birth and at the time of birth. The previously low level of estrogen begins to increase, and progesterone level may decrease; the overall effect is a shift in the estrogen/progesterone ratio. These hormonal changes help trigger parturition, and after birth, to trigger the retrieving and nursing behavior that characterize a mother's treatment of her newborn pups. Estrogen acts synergistically with oxytocin released by the pituitary to promote the secretion of milk from the mammary glands. Prolactin from the pituitary acts to promote milk production (Masson 1948; Lott 1962; Grota and Eil-Nes 1967).

To demonstrate that hormonal changes occurring around the time of birth are at least partially responsible for producing changes in maternal behavior, Terkel and Rosenblatt (1968, 1972) used a system of chronically implanted heart catheters in rats. They mounted catheters on the rats' necks and connected them to a swivel pump that permitted the shunting (transfusion) of blood from one rat to another. When blood was shunted from newly parturient females to virgin females provided with young pups, the virgin females exhibited maternal behavior, such as retrieving pups and crouching over the pups to nurse, after fourteen to fifteen hours of exposure to the pups.

FIGURE 7-14   Female rat with pups in litter nest
Prior to parturition, the female rat constructs a prepartum nest, a flat mat of material. After giving birth, she builds a better nest with higher sides, partially due to stimulation provided by the presence of her pups.
Source: Photo by Ronald J. Barfield.

Some of the events that occur during parturition do not have to take place for the female to exhibit maternal behavior, for example, the birth process itself and the consumption of birth fluids or the placenta. If a female is prevented from experiencing the events associated with normal birth through the removal of the fetuses by Caesarian section and then is shortly thereafter presented with newborn pups, she still exhibits characteristic maternal behavior (Moltz, Robbins, and Parks 1966).

What about the hormonal relationships during the period of nest building? Observations of preparturient female rats tell us that nest-building behavior begins to increase in intensity four to six days before parturition. The construction of the first nest, called a prepartum nest, is apparently a behavior that is partially under hormonal regulation, although investigators have obtained conflicting results. Several experiments have demonstrated that the change in the estrogen/progesterone ratio is an important nest-building trigger. When pups are present, a female builds a litter nest with higher sides (figure 7–14); the trigger for this behavior seems to be the pups' presence.

Finally, consider the events that occur during lactation. Maternal behavior during lactation and up to the time of weaning (three to four weeks of age) depends on close behavioral synchrony between mother and pups. Most of the interactions between them operate through nonhormonal channels. For example, the retrieving responses and the nursing crouch have been induced in virgin females and males. The test subjects exhibited maternal behavior when they were presented with stimulus pups each day for up to several weeks; they did not exhibit maternal behavior at the first presentation of pups. In addition, a female that has been presented with a replacement litter at any time up to the midpoint of her lactation period will often continue to lactate (Nicoll and Meites 1959). As the development of a normal litter proceeds, pup behavior interferes with maternal behavior—that is, the pups make it difficult for the female to perform maternal behavior. Maternal rats retrieve pups frequently up to the second week of lactation, and then this behavior declines. At first, pups are nursed only in the nest and with the mother's assistance; but later, nursing can be initiated by pups and can take place away from the nest, wherever pups and mother encounter each other. After three to four weeks of nursing, FSH increases again in the female, and the estrous cycle resumes (Bruce 1961; Rosenblatt 1967; Terkel and Rosenblatt 1971).

REPRODUCTION IN LIZARDS

Using primarily the green anole lizard (*Anolis carolinensis*), Crews (1975, 1977, 1979, 1980, 1983; Crews and Greenberg 1981) and his colleagues conducted field and laboratory investigations of the behavioral endocrinology of reproduction and the significance of various endocrine-behavior events in environment adaptation. To understand these hormone-behavior interactions better, we first need to make note of the annual sequence of events in these lizards. Louisiana populations of the green anole, which investigators have studied most extensively, exhibit a four-part annual cycle:

1. From late September to late January, the lizards are quiescent and live under tree bark and rocks.
2. In February, the males emerge from dormancy and establish breeding territories.
3. In March, the females become active; mating begins in late April, and by May, they are laying one egg every ten to fourteen days, a pattern that they continue for several months.
4. In August, both sexes enter a refractory period for about one month; during this time, the same environmental and social cues that in the spring bring about gonadal recrudescence are no longer effective.

Combined laboratory and field techniques have led to significant conclusions about reproduction in the green anole lizard. Let's examine a sequence of observations and see what each can tell us about hormones

FIGURE 7-15   Dewlap display of male *anolis* lizard
The male *Anolis* lizard extends the dewlap toward the female as part of his courtship display. The dewlap display patterns are important for species-specific mate selection. This behavior pattern also appears to be critical for stimulating ovarian activity in the female.
Source: Photo from David Crews.

and behavior. The seasonal rise in temperature in the spring (Licht 1973) and the male's courtship behavior (Crews, Rosenblatt, and Lehrman 1974) affect ovarian development and egg laying in the female. The extension of the male's dewlap during displays is the key to selection of males as mates by the females and to the promotion of ovarian activity (figure 7-15). Females are receptive to advances by males only when conception is most likely to occur. This receptivity is partially regulated by the estrogen secreted by the developing follicle. Mating activity inhibits further receptivity by the female—the inhibition begins within twenty-four hours after mating, and lasts for the duration of that cycle.

Several points regarding the adaptive significance of this reproductive pattern emerge from the data. The mating inhibition in females is adaptive, because copulation in this species of lizard is prolonged and usually takes place in the open—for example, on tree limbs—where the mating pair is vulnerable to predation. Field data support the contention that more lizards are captured when they are mating than when they are engaged in other activities. The restriction of receptivity and mating to the time when fertilization is most likely to occur is thus evolutionarily adaptive (Valenstein and Crews 1977). Cessation of reproduction during the refractory period that precedes the winter dormancy period is adaptive in at least two ways. First, it ensures that young do not hatch at a time of diminished food resources and poorer environmental conditions. Second, the female is able to build up fat reserves for the winter rather than devote energy toward reproduction.

Sexual behavior in male and female lizards depends on gonadal hormones, a pattern that we have noted in many other vertebrates. Investigators found that when a female lizard was given an injection of progesterone and twenty-four hours later was given an injection of estrogen, a synergistic effect—induction of female receptivity—resulted. We noted a similar synergistic effect on receptivity in rats with these same two hormones (see Beach 1976, for review).

Data obtained by autoradiography on the uptake of sex steroids by specific areas of the brain augment these parallel findings in lizards and rats. Experimenters inject the animal with radioactively labeled estrogen; several hours later they sacrifice the animal; section the brain with a freezing microtome; and place the sections on special glass plates that have been treated with a photographic emulsion. They leave the plates in the dark for periods ranging from days to months depending upon the exact procedure used; and then they develop the plates to find the specific sites where uptake occurred. The results from rats and lizards are strikingly similar. Concentrations occur in the septum and preoptic regions of both species (McEwen et al. 1979; Morrell, Kelley, and Pfaff 1975). These and related findings of comparative investigations point toward an exciting avenue for future research on common patterns of endocrinological events and interactions between hormones and behavior in different classes of vertebrates.

Further investigations of the reciprocal relationships between males and females in lizards have been conducted with the whiptail lizard (*Cnemidophorus inornatus*). Like the anoles, these lizards exhibit highly seasonal patterns of reproduction, and the gonads regress at other times of the year (Lindzey and Crews 1988). The presence of a male whiptail facilitates ovarian recrudescence in females. Conversely, the presence of a reproductively active female will stimulate testicular development in males. This is one of the few species in which either sex can directly affect the coordination of the reproductive condition of the opposite sex. Two additional findings augment our understanding of the degree of reciprocation that occurs in this species. Female whiptail lizards can stimulate some, but not all, of the male courtship behavior sequence in castrated lizards; exogenous androgens are necessary for reappearance of the entire courtship sequence. Female whiptail lizards are responsive to male courtship only

at a specific time in the process of egg formation; they will actively reject males that approach too early or after ovulation has occurred. The high degree of coordination is extremely important for successful reproduction in a seasonally reproducing species, but with few exceptions, such as these studies on whiptail lizards, the complete nature of the male-female interaction has not been demonstrated.

REPRODUCTION IN HOUSE SPARROWS

For our final example, consider another example of the manner in which hormones and behavior interact during reproduction in the house sparrow (*Passer domesticus*). In a pair of studies Hegner and Wingfield (1987a,b) demonstrated the interactions of hormones and environmental cues that affect reproduction in free-living house sparrows. The first study involved manipulation of male sparrows' testosterone levels to determine testosterone's effect on the amount of parental investment exhibited by the sparrows or on their breeding success (number of nestlings fledged) (Hegner and Wingfield 1987a). The investigators hypothesized that testosterone levels are somewhat elevated early in the breeding season as males compete for establishment and maintenance of territories. However, for males to exhibit normal parental behavior, including assistance with nestling feeding, the levels of testosterone must drop as the reproductive sequence proceeds. Males with high levels of testosterone when they would normally be collecting food for the nestlings might spend too much time defending the territory, and thus would not provide sufficient food for the nestlings. As the nestling period ends, testosterone levels again rise as they prepare for another reproductive sequence.

To test the foregoing hypothesis, Hegner and Wingfield captured breeding male house sparrows in the spring and implanted a small capsule beneath the skin. The capsule contained one of three substances: (a) testosterone, (b) flutamide, an androgen antagonist that inhibits testosterone uptake and binding at receptor sites, or (c) nothing, as a control procedure. They then monitored the behavior of the birds during reproduction. Blood samples were taken from the birds periodically to assess levels of circulating hormones. The changes in circulating levels of testosterone in the birds are shown in figure 7–16. Overall, there was a significant decrease in reproductive success in the testosterone-treated birds (mean of 2.6 young fledged per nest) compared to the birds that were treated with flutamide (3.8 young fledged) or those receiving an empty capsule

FIGURE 7-16   Hormone levels in sparrows
The diagram depicts the relationship between the age of the oldest nestling and circulating testosterone level in male house sparrows under three different treatment conditions. The top line represents males given implants of testosterone, the middle line consists of data from males given an empty capsule implant as a control, and the bottom line represents males given flutamide, a substance that blocks the actions of testosterone. The vertical bars indicate ± 1 standard error of the mean.

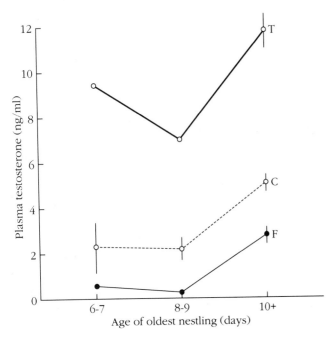

and endocrinology. They manipulated the number of nestlings present in house sparrow nests by (a) adding two nestlings to some nests, (b) removing two nestlings from some nests, or (c) leaving nests alone so that they contained the normal numbers of nestlings, ranging from three to six birds. They then recorded the number of feeding visits by the parents, the number of young fledged, and by obtaining blood samples late in the nestling phase of the cycle, the levels of androgens, corticosterone, and luteinizing hormone. House sparrows were able to rear the larger clutches successfully; those clutches where nestlings were added resulted in a greater number of fledged young than the other treatments, though the average weight per fledgling was less in the larger clutches. Also, parent birds spend much more time feeding larger clutches than smaller clutches. Those birds that reared larger clutches had longer intervals until the next brood, and the next brood was reduced in size. These latter findings may be an outcome of the extra energy cost of rearing the larger brood, particularly for the female, and the extra time needed to rear a larger brood of nestlings to the stage where they become independent. Male parents with larger broods had higher levels of dihydrotestosterone circulating in their blood, but all other hormone measurements were similar across the three treatments. The results indicate that there is not any stress for the parents related to rearing the larger broods, though there are consequences for the reproductive output of the next brood. When the number of nestlings is greater, parent birds of this species apparently invest more in the current brood and less in the subsequent brood.

While hormone levels play a significant role in affecting reproductive success within a brood as demonstrated by the first study, it is not clear that hormone levels have any important influence on reproductive success between broods as shown by the second study. However, additional hormone measurements taken from blood samples obtained at various times during both the first and second clutches in the second experiment would be needed to clarify this conclusion further.

(4.2 young fledged). One major contributing factor to this difference was that males treated with testosterone made only about half as many feeding visits to the nest during the nestling phase as birds given flutamide or an empty capsule, a significant difference. Indeed, males given testosterone were active defending their nests two to six times more than the other birds, which detracted from the time they could have been providing food.

In a companion investigation, these same investigators (Hegner and Wingfield 1987b) studied the relationships between the number of nestlings present in the nest and parental investment, reproductive success,

HORMONE-BRAIN RELATIONSHIPS

In the previous chapter and in this one we have mentioned the ongoing research concerning the mechanisms by which the endocrine and nervous systems interact with one another. A few additional comments will provide an appropriate finale to our examination of these behavior control systems.

Caribou migration (*Rangifer tarandus*)
Caribou migrate seasonally (chapter 15) to ensure an
adequate food supply (chapter 16) and to provide adequate
conditions for rearing their young (chapter 14).
Source: Photo by Daniel D. Roby.

PLATE 4

White-footed mouse (*Peromyscus leucopus*)
These ubiquitous rodents have been used to study olfactory
communication (chapter 11), aggression (chapter 12), and
habitat selection (chapter 16).
Source: Photo by Sharon Cummings.

PLATE 5

We now know there are regions of the brain with cells that possess specific receptor sites for many of the hormones that influence behavior. The experimental procedures involve a combination of autoradiography to pinpoint the receptor locations precisely, and electrodes (to measure electrophysiological activity from the same receptor sites) (Morrell and Pfaff 1981; Pfaff 1981). Thus, for example, when radioactively labeled sex steroids are given to rats (Pfaff and Keiner 1973), frogs (Kelley, Morrell, and Pfaff 1975), or chaffinches (Zigmond, Nottebohm, and Pfaff 1973), the label becomes concentrated in similar brain areas in all three animals. The preoptic area and portions of the limbic system and hypothalamus are among the sites that commonly have the highest amounts of the labeled steroid.

When electrode recordings are made from cells in these same preoptic and limbic locations, some important relationships between hormones, behavior, and neural activity emerge. Firing rates for neurons in the highly labeled portions of the limbic system and hypothalamus vary during the course of the rat estrous cycle; generally, they are highest shortly before ovulation (Kawakami, Terasawa, and Ibuki 1970; Terasawa and Sawyer 1969). When ovariectomized female rats are given injections of estrogen, firing rates in these same sites are elevated (Cross and Dyer 1972).

As we have noted earlier, progesterone facilitates sexual behavior in female rats. In a recent study, Pleim and Barfield (1988) demonstrated that intracranial administration of progesterone, but not estradiol facilitates sexual receptivity. Small cannulas were placed into either the midbrain or ventromedial hypothalamus of adult female rats. Progesterone, estradiol, cholesterol, or nothing was given to the female via the cannula; the latter two treatments served as control procedures. When females were placed with sexually active adult male rats at one and at four hours after treatment, only the females that had been treated with progesterone in the ventromedial hypothalamus, and that were tested four hours after hormone administration, exhibited sexual receptivity. Studies such as this further our understanding of the location of key hormonal receptor areas that trigger behavioral output.

In the lobster, various hormones and neurosecretions act upon the nervous system to influence aggressive behavior (Kravitz 1988). In particular, various neurotransmitter substances (e.g., serotonin, octopamine) are found throughout the nervous system. These amines apparently act to regulate the sensitivity of tissues within the nervous system and in effector organs

FIGURE 7-17   Hormone sensitization
Schematic diagram of the manner in which hormonal sensitization of various tissues can occur. The responses of certain neurons can be primed by doses of hormones or neurosecretions that precede the input from particular environmental stimuli.

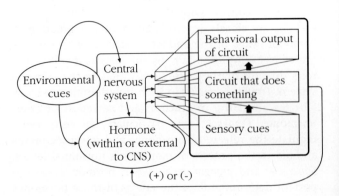

to environmental stimuli (figure 7–17). In lobsters, injections of serotonin resulted in stereotypical postures seen in dominant animals, and injections of octopamine produced postures characteristic of subordinate animals. Further investigations led to the conclusion that a key effect of the aminergic neurotransmitter substance is the priming of receptors in exoskeletal muscles to respond in particular ways when appropriate stimuli occurred, e.g., another lobster. Serotonin directed the muscle program toward flexion of the postural muscles involved in more threatening, dominant postures; whereas octopamine produced the opposite effect, inhibiting flexion and producing postures characteristic of a subordinate animal. Through this regulatory process, the input to the nervous system, the processing of incoming information, and the behavioral output are biased toward certain stereotyped responses.

Thus, there appear to be some clear connections between hormonal and neural activity. We are now nearing the time when we will understand mechanisms that relate hormones, nervous system, and behavior at the level of the individual cells and the biochemistry within and between those cells (Komisaruk 1978).

SUMMARY

Hormones are chemical substances produced by specialized *ductless glands* or by *neurosecretory cells.* They influence the processes of growth, metabolism, water balance, and reproduction. The endocrine system has evolved in concert with the nervous system; the two systems are closely interrelated in their regulation of behavior. Most of the endocrine structures in invertebrates, as exemplified here by the insects, involve neurons directly or they are closely tied to the nervous system. *Trophic hormones* in both invertebrates and vertebrates affect the production and secretion of other hormones. *Feedback loops,* particularly in vertebrates, regulate the release of hormones and neurosecretions and aid in the process of monitoring hormonal levels. *Synergism* and *antagonism* between hormones are important determinants of the effects of those hormones on behavior.

Three major regions of the vertebrate pituitary gland produce hormones that affect behavior. The *posterior pituitary* secretes *oxytocin,* which affects uterine muscles and milk ejection, and *vasopressin,* which plays a key role in regulating water balance. The *central* portion of the pituitary secretes *MSH,* a hormone that affects pigments, and thus skin and hair color. The *anterior pituitary* secretes three trophic hormones—*FSH* and *LH,* which affect the gonads and reproduction, and *ACTH,* which influences production and release of hormones from the adrenal cortex. The anterior pituitary also produces *prolactin,* a hormone that affects maternal behavior, milk production, and related processes in birds and mammals. The *gonads* and *adrenals* are endocrine glands secreting hormones affecting behavior.

Some hormones exert *activational,* or *triggering,* effects on behavior. Examples of activational effects include coat color changes in the weasel, the release of pheromones and the process of eclosion in insects, molting in crustaceans, sexual and courtship behavior in rats and ring doves. Other hormones exert *organizational effects* because these influences are manifested during development. In particular, sexual differentiation is an organizational effect in some mammals, birds, and reptiles because it is under some degree of hormonal regulation. Some hormones can exert both activational and organizational effects.

There are complex *environment-hormone-behavior interactions,* as exemplified here by the reproductive sequences in ring doves, maternal behavior in rats, seasonal reproductive patterns of lizards, and reproduction in house sparrows. The hormonal effects within individual animals, the effects of behavior on hormones, and the interactions between the hormonal/physiological state of the animal and features of its environment combine to produce observed behavior patterns. The hormonal state of the animal can affect nestling production in house sparrows and the number of nestlings can influence levels of circulating hormones.

The research techniques investigators use to explore hormone-behavior relationships include the removal of endocrine glands, hormone replacement therapy, blood transfusions, administration of hormone substitutes or use of chemicals that inhibit particular hormone effects, autoradiography, and assays for hormone levels. Using several of these techniques, investigators have explored the mechanisms by which the endocrine and nervous systems are interrelated in regulating behavior.

Discussion Questions

1. Both male and female golden hamsters possess sebaceous flank glands that produce sebum, a substance that is used to mark objects in their environment. Suggest three or four experimental questions (hypotheses) you would want to ask in studying the hormonal regulation and control of this scent-marking behavior. For each question, identify the treatment groups you would use for testing your hypotheses and justify the need for each.

2. The following summary table is from a review of early organizational effects of hormones on behavior by Adkins-Regan (1981). A number of reptiles have been tested using various techniques to manipulate the early hormone environment. What general conclusions can you make about organizational effects on sex structures in these reptiles? How do these results compare with those discussed in the chapter on mammals?

Summary of experiments on sex differentiation in Reptilia

Species	Age at Treatment	Treatment	Effect on[a]		
			Gonads	Gonaducts	Other Sex Structures
Several	Embryos	Gonadectomy		M F	
Several	Embryos	Estrogens	m f		
Lacerta vivipara	Embryos	Gonadectomy		M F	F M
	Embryos	Parabiosis			M F
	Embryos	Androgens			f m
	Embryos	Estrogens	m f	M F	M F
Anguis fragilis	Embryos	Testosterone		F M	
Anolis carolinensis	Juveniles	Estrogen			M F
Sceloporus sp.	Juveniles	Estrogen			M F
Alligator mississippiensis	Juveniles	Testosterone		m f	f m
Crocodylus niloticus	Juveniles	Testosterone	f m	m f	
Thamnophis sirtalis	Embryos	Testosterone		m f	f m
	Juveniles	Estradiol	m f	m f	
Chrysemys marginata	Embryos	Testosterone	f m		
Emys orbicularis	Embryos	Estradiol or Testosterone	m f		
Emys leprosa	Recently hatched	Androgens		m f	f m
	Recently hatched	Estrogen	m f	m f	
Testudo graeca	Embryos	Estradiol	m f	m f	

From E. Adkins-Regan, "Early Organizational Effects of Hormones," in *Neuroendocrinology* edited by N. Adler, p. 188. Copyright © 1981 Plenum Press, New York.

[a]FM indicates extensive or complete masculinization of the character; M→F indicates extensive or complete feminization. f→m indicates slight or partial masculinization; m→f indicates slight or partial feminization.

3. The reproductive sequences of ring doves and lizards and the maternal behavior of rats illustrate feedback loops involved in internal hormonal conditions, behavior, and environmental cues. Can you describe or explain other behavior sequences that have similar complex feedback interactions? How would you use available techniques to investigate the various steps in the hormones-environment-behavior sequence?

4. The testes of twenty-five males of a bird species were weighed in each of four seasons. To organize the data, the birds were divided into groups of five on the basis of their social dominance interactions; the five top-ranking birds were placed in category I, the next five in category II, and so forth. The following table shows average testes weights in milligrams for the birds in each category and in each of the four seasons. What general conclusions can you draw from these data? What conditions of interpretation would you place on these conclusions?

Average weight of bird testes (milligrams)

	Male dominance category				
	I	II	III	IV	V
Spring	7.6	7.0	6.8	6.4	5.8
Summer	8.0	7.7	7.2	6.9	6.2
Fall	4.1	3.9	3.8	3.9	3.8
Winter	2.6	2.5	2.6	2.4	2.5

5. Hormonal mediation of environmental stimuli and the role of the endocrine system in regulating behavior in response to external stimuli may affect reproduction and the reproductive success of organisms. Using the work of Wingfield and his colleagues described in this chapter as a model system, can you devise an experimental scheme to explore the relationship between hormone levels and reproductive success in (a) red-winged blackbirds, (b) garter snakes, and (c) woodchucks?

Suggested Readings

Adkins-Regan, E. 1981. Early organizational effects of hormones: An evolutionary perspective. In *Neuroendocrinology of Reproduction,* ed. N. Adler. New York: Plenum.
The most comprehensive summary to date of the organizational effects of hormones on behavior. Covers the entire taxonomic spectrum from coelenterates to mammals. Excellent references.

Beach, F. A. 1965. *Sex and Behavior.* New York: John Wiley and Sons.
An edited volume of contributions from two conferences held in the early 1960s. Though some of the research has been replaced by newer findings, the basic concepts and processes outlined by the authors of the various chapters remain as classic paradigms for the exploration of relationships between hormones and behavior.

BioScience. 1983. Vol. 33 (October Issue).
This particular issue of the journal BioScience contains a series of summary articles on the relationships between hormones and behavior in invertebrates and vertebrates.

Crews, D. 1980. Interrelationships among ecological, behavioral, and neuroendocrinological processes in the reproductive cycle of *Anolis carolinensis* and other reptiles. *Adv. Stud. Behav.* 11:1–74.
Crew's investigations on lizards parallel those of Lehrman and Hinde on avian systems. Strong ecological and evolutionary flavor. Provides integrated approach of the animal in its natural habitat and laboratory studies at the chemical level.

Hutchinson, J. B. 1976. Hypothalamic mechanisms of sexual behavior, with special reference to birds. *Adv. Stud. Behav.* 6:159–200.
Comments on research conducted at the interface between the hormonal and nervous systems. Combines a review approach with presentation of new research findings.

Truman, J. W., and L. M. Riddiford. 1974. Hormonal mechanisms underlying insect behavior. *Adv. Insct. Physiol.* 10:297–352.
Good summary of relationship between hormones and behavior for invertebrate organisms. Good bibliography. Solid starting point for techniques, and methods of studying insect behavior from neural and hormonal perspectives.

8

BIOLOGICAL TIMEKEEPING

Many birds are heard producing their melodious songs in the early morning hours just before and just after sunrise. Moths and butterflies generally emerge from their cocoons at dawn when the still, somewhat moist air provides the best conditions for the slow drying of their unfolding wings. Thus, we notice that many animal behavior patterns occur daily. Many songbirds of the north temperate deciduous forest fly south during autumn, overwinter in tropical or subtropical zones, and begin their return flight north as spring approaches. Ground squirrels emerge from their winter dens in spring, remain active for four to six months, and return to hibernate in autumn. During their months above ground, they mate, produce offspring, rear the young, and prepare for another winter. Thus, we note further that certain animal activities occur on an annual cycle. We can observe other animal behavior patterns that occur with frequencies that approximate cyclic features in the environment, e.g., tides, lunar phases. **Biological rhythms** are occurring when animal activities and behavior patterns can be directly related to distinct environmental features that occur with regular frequencies. Biological rhythms are the external manifestations of biological clocks and are regulated by them. **Biological clocks** are internal timing mechanisms that involve both a *self-sustaining physiological pacemaker and an environmental cyclic synchronizer (zeitgeber)*.

In this chapter we first examine the characteristics of biological rhythms. Next we look at pacemakers and their physiology: the mechanistic, or proximate, questions. We also ask questions about the functional significance of biological rhythms within ecological contexts: the ultimate questions. We would expect natural selection to favor individuals whose peak activity periods of feeding, for example, coincide with peak activity periods of their prey. Finally, we cover the significance of biological clocks.

BIOLOGICAL RHYTHMS

Let us first consider the terminology and concepts that have been developed for describing and characterizing biological rhythms. Each biological rhythm is composed of repeating units called **cycles.** The length of time required to complete an entire cycle is the rhythm's **period;** twenty-four hours is the period in the example shown in figure 8–1. The magnitude of the change in activity rate during a cycle, the difference between peaks and troughs, is the **amplitude.** Any specified recognizable part of a cycle is called a **phase;** the designated portion of the cycle shown in figure 8–1 would be an active feeding phase.

FIGURE 8-1 Biological rhythm

The characteristics of a biological rhythm include period, phase, and amplitude. Each of these characteristics can vary over time for the same animal, between animals of the same species, or between animals of different species.

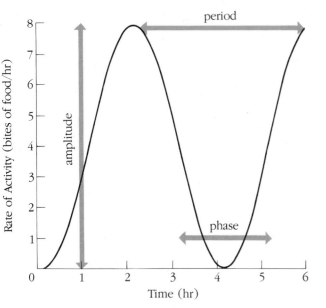

Biological rhythms may be characterized by several properties. First, whereas temperature changes alter the rate of most chemical reactions and cellular processes, biological rhythms are **temperature-compensated.** Generally, the rate of a chemical reaction doubles for each 10° C increase in temperature. However, biological rhythms are relatively insensitive to change in temperature (Sweeney and Hastings 1960; Rawson 1960). This fact is significant, for if biological rhythms were sped up or slowed down by ambient temperature changes, they would not help the organism keep accurate time.

Second, biological clocks are generally unaffected by metabolic poisons or inhibitors that block biochemical pathways within cells. We might expect that application of a metabolic poison such as sodium cyanide would alter the period of a biological rhythm, yet this does not happen.

A third property is that periods of biological rhythms occur with approximately the same frequency as one or more environmental features. The Latin word *circa* (about or around) usually is part of the name of this kind of rhythm, as we will see. Fourth, biological

rhythms are self-sustaining, maintaining approximately their normal cyclicity even in the absence of environmental cues. Fifth, and last, biological rhythms can be entrained by environmental cues. The self-sustaining pacemaker mechanism(s) may be set and adjusted according to input from the external environment.

TYPES OF RHYTHMS

EPICYCLES OR ULTRADIAN RHYTHMS. Different organisms exhibit a variety of biological activities with varying frequencies and periods (table 8–1). Some of these cycles are of short duration and are termed **epicycles** or **ultradian rhythms.** Lugworms (*Arenicola marina*) that live in burrows on sand flats in the intertidal zones feed every six to eight minutes. Some small mammals like meadow voles (*Microtus pennsylvanicus*) that are active primarily during daylight hours show short cycles of activity bursts followed by periods of rest that vary from twelve minutes to two hours (Madison 1985). These epicycles do not generally fit the definition provided for biological rhythms, and we will not consider them here except to note that they are cyclic, and as the two examples indicate, they are important activity rhythms in the lives of some organisms.

TIDAL RHYTHMS. A primary environmental feature of seacoasts is the ebb and flow of the tides. Tidal rhythms affect activity periods in many organisms that inhabit these zones. Tides, the result of unequal gravitational forces of the sun and the moon, exhibit about a 12.4 hour period from one low-tide phase to the next (though there are some variations in time period from one location to another). Many species of small crabs (e.g., *Uca minax*, *Uca crenulata*) inhabit these seashore regions; they time their activity cycles so that they are feeding on the sandy shore when the tides are out, but are returning to burrows when the tide flow returns (figure 8–2). In this way they avoid being stranded above the tide line where they might desiccate or be swept out to sea by the inrushing tide (Barnwell 1966; Honegger 1976).

LUNAR RHYTHMS. Based on the approximately twenty-eight-day cycle of the moon, lunar rhythms are clearly related to the tidal rhythm. Some marine insects, like the midge (*Clunio marinus*), coordinate eclosion, mating, and egg-laying activities with the lunar cycle. They lay

TABLE 8-1   Summary of biological rhythms

Type of cycle	Organism	Behavior
Ultradian (variable)	lugworm	feeding (every 6–8 min.)
	meadow vole	feeding/resting (every 15–120 min. during daylight)
Tidal (12.4 hours)	oyster	opening of shell valves
	fiddler crab	locomotion/feeding
Lunar (28 days)	midge (marine insect)	mating/egg laying
	grunion (marine fish)	egg laying
Circadian (24 hours)	deermouse	drinking/general activity
	fruit fly	emergence of adults from pupa
Circannual (12 months)	woodchuck	hibernation
	chickadee	reproduction
	robin	migration/reproduction
Intermittent (variable—several days to several years)	desert insect	reproduction (triggered by rain)
	lion	feeding (triggered by hunger)
	shiner (river fish)	reproduction (triggered by flooding)

FIGURE 8-2   Tidal rhythms affect fiddler crabs

Fiddler crabs inhabit shorelines where they forage along the beaches. They must do their foraging between the tides and return to their burrows with each new incoming tide. Without a biological clock to warn them, they would either be swept out to sea by the water, or be left high on the beach away from their burrows, where they would desiccate in the sun.

Source: Photo by Robert and Linda Mitchell.

their eggs at very low tide, thereby ensuring that the larvae will hatch in the proper marine environment (Neumann 1966). Grunion (*Leuresthes* spp.), a marine fish, spawns during the spring tides on the sandy beaches of California and uses the moon as a timing cue for these activities (Walker 1949).

CIRCADIAN RHYTHMS. Many organisms exhibit biological rhythms of about twenty-four hours' duration that are governed by self-sustaining internal pacemakers. These we call **circadian rhythms.** In their daily cycle, some animals exhibit peak activity during the daylight hours (**diurnal**); some are active primarily at night (**nocturnal**); and still others exhibit peak activity around dusk and/or dawn (**crepuscular**). Activity periods may shift seasonally. Many bird species that are year-round residents of northern temperate zones are primarily crepuscular through late spring and summer, but shift to a more diurnal pattern during the cold winter months; thus they avoid the very cold temperatures of early winter mornings. Circadian activity periods may also show age-dependent shifts. Young woodchucks restrict most of their activity to early evening hours, whereas adult woodchucks are more diurnal in their pattern. Similarly, young dragonflies fly during the hours just after dawn, but adults of the species fly mostly in the middle of the day (Corbet 1957, 1960).

CIRCANNUAL RHYTHMS. As the term implies, **circannual rhythms** are behavioral and physiological patterns, which are governed by self-sustaining internal pacemakers, that occur within a period of about one year (Gwinner 1986). Some mammals enter **hibernation**, a condition of deep sleep and reduced metabolic activity, during the winter months. By doing so they avoid the harsh conditions of winter. Many bird species escape the rigors of winter in northern and temperate climates by migrating to southern latitudes. The annual life cycle of many insects that live where there are seasonal climatic shifts incorporate a **diapause phase,** a period of dormancy, during the more rigorous portions of the climatic cycle. For example, silkworm moths (*Bombyx* spp.) and mosquitos (*Aedes* spp.) lay eggs that are dormant during the winter (Krogure 1933; Vinogradova 1965); nymphs of the dragonfly (*Tetragneura cynosura*) overwinter as larval forms and complete development the following spring (Lutz and Jenner 1964); both the parasitic wasp (*Nasonia vitripennis*) and its host, the flesh fly (*Sacrophaga argyrostoma*) enter diapause as larvae (Saunders 1978); still other insects go through diapause in the pupal stage or as adults.

ENDOGENOUS PACEMAKER

One of the critical characteristics of biological rhythms is the existence of an internal self-sustaining pacemaker. What is the evidence for the existence of **endogenous** clock mechanisms? As we examine the following sequence of information in support of the existence of internal chronometers, bear in mind that what we are measuring are generally the overt, observable manifestations of the clock—the periodic changes in physiology and behavior—not the mechanism itself.

FREE-RUNNING RHYTHMS. First, let us consider what happens when we place an animal in constant environmental conditions; for example, in constant darkness. When this is done, many organisms exhibit activity rhythms, called **free-running rhythms,** with a period different from that of any known cyclic environmental variable. This provides indirect evidence for an endogenous pacemaker. The pattern of activity of a flying squirrel (*Glaucomys volans*) housed in constant darkness for several weeks is shown in figure 8–3 (DeCoursey 1960, 1961). Since the animal's activity has a clearly rhythmic pattern in the absence of any obvious cyclic cue, evidence is in favor of an endogenously based system of timekeeping.

For most species studied, the free-running periods follow what has become known as **Aschoff's Rule** (Aschoff 1960, 1979). When animals are kept in constant darkness, their activity rhythm continues with a period of nearly twenty-four hours, but it **drifts** slightly, becoming somewhat shorter (as in the flying squirrel) or somewhat longer each day. Aschoff's Rule states that the direction and rate of this drift away from the twenty-four hour period are a function of light intensity and

FIGURE 8-3   Free-running periods and entertainment
When a flying squirrel is placed in constant darkness, it adopts a free-running rhythm with a period different from any known cyclic environmental variable. When the squirrel later is placed in an environment with a varying cue, for example, a light-dark cycle, the squirrel's activity onset becomes locked onto the lights-off signal.
Source: Photo from Patricia DeCoursey.

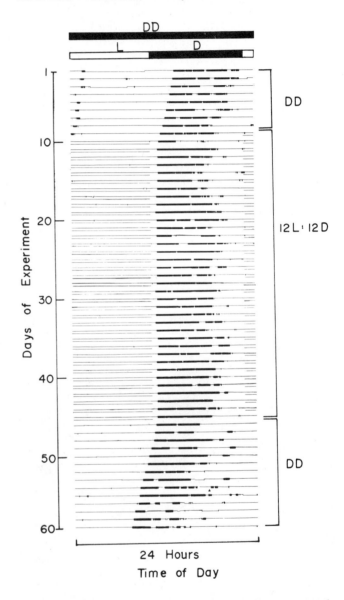

of whether the animal is normally diurnal or nocturnal. For nocturnal animals like the flying squirrel, housing under constant dark conditions results in a free-running rhythm period shorter than twenty-four hours; the activity begins slightly earlier each day. Conversely, for a diurnal animal housed in the dark, the free-running period is slightly longer each day, and the activity begins slightly later each day. There are some exceptions to this rule; but in general, the data show that a variety of animal species conform to the pattern.

ISOLATION. What about "learning" or other similar influences as a mechanism for biological rhythms? Birds and some reptiles that hatch from eggs can be kept under constant conditions in an incubator before they hatch until after they hatch. If the newly hatched organisms exhibit circadian rhythms, a major component of biological rhythms would appear to be inherited and endogenous. Hoffman (1959) maintained lizard eggs under one of three conditions: (1) eighteen-hour days consisting of nine light hours and nine dark hours; (2) twenty-four-hour days, with twelve light hours and twelve dark hours; and (3) thirty-six-hour days with eighteen light hours and eighteen dark hours. Lizards from all three groups, hatched and maintained under constant conditions, exhibited free-running activity periods of 23.4 to 23.9 hours. We can thus conclude that a component of the biological clock mechanism in these lizards is inherited, is unaffected by various rearing regimes, and is thus endogenous.

GENETICS. Another piece of evidence for an endogenous pacemaker comes from reports of mutations in the gene(s) that regulate the basic program of biological clocks in various invertebrates, e.g., *Drosophila* spp. (Konopka 1979). A mutant golden hamster (*Mesocricetus auratus*) with a free-running period of 22.0 hours, considerably shorter than the species' norm of 24.1 hours, was recently found (Ralph and Menaker 1988). Further work revealed that the mutant gene in the hamster was at a single locus. In hamsters with the heterozygous condition, such as the animal in which the phenomenon was first discovered, the mutant gene produced the 22.0 hours free-running period. In hamsters with the homozygous condition, the free-running period decreased to about 20 hours. Animals with mutant alleles exhibited abnormal patterns of entrainment to a twenty-four-hour period, or would not entrain. Mutations such as these provide direct evidence for the inherited endogenous machinery for circadian rhythms and for the basic underlying program of activity.

TRANSLOCATION. Additional support for the hypothesis of endogenous control of biological rhythms comes from **translocation** experiments. Honeybees (*Apis* spp.) visit particular feeding sites at the same time each day.

This behavior is functionally significant because many flowers that provide bees with food are open only a specific time of day. Renner (1957, 1959, 1960), working in a specially designed room with constant conditions, trained bees to leave the hive to forage at a specific time each day. He trained the bees in Paris and then flew them to New York during the night, where he placed them in a similar enclosed room with constant conditions. On their first day in New York, the bees began to forage at the same time as that that would have been expected had they remained in Paris—twenty-four hours after their last foraging. In related studies, bees that inhabited outdoor hives were translocated from Long Island to California. Although the bees initially foraged at the new site at the same time as that that they foraged at the original site, they gradually adjusted to local time by foraging later and later each day (figure 8–4).

A key aspect of current biological clock models illustrated by these experiments is that while the clocks are endogenous, external conditions can cause the biological rhythms to be reset. Similar relocation studies of circadian skin color changes in fiddler crabs and shell-valve-opening time in oysters reveal that these animals also react initially as if they were still in the original location, but gradually their clocks become reset to local time. The jet lag experienced by humans making long-distance airline flights is another example of this phenomenon.

VARIATION IN PERIOD. A final source of evidence for the endogenous nature of biological rhythms comes from the variation that exists in the natural activity periods of most organisms. For example, if we measure the activity periods of two rattlesnakes (*Crotalus* spp.) under constant conditions, we find circadian rhythms of twenty-three hours, ten minutes and twenty-four hours, thirty-three minutes. These rhythms do not match the period of any known environmental variable. The animals must have internal chronometers that, due to individual differences, lead to deviations from a twenty-four-hour rhythm.

ZEITGEBERS

Many organisms exhibit circadian and circannual periodicities that appear to be closely adjusted to patterns of daylength, temperature fluctuation, or other environmental features. The mechanism whereby the period

FIGURE 8-4   Visiting frequency of bees

Bees were trained to forage at a particular time period (1:00 P.M. to 2:30 P.M. EST) in New York and were then flown to California at night. During the first two days at the new site, the bees gradually shifted the time of their feeding. These data support the notion of an endogenous biological rhythm that shifts in response to local conditions.

Source: Data from M. Renner, "The Contribution of the Honey Bees to the Study of Time-Sense and Astronomical Orientation," *Cold Spring Harbor Symposia on Quantitative Biology* 25:361-367, 1960.

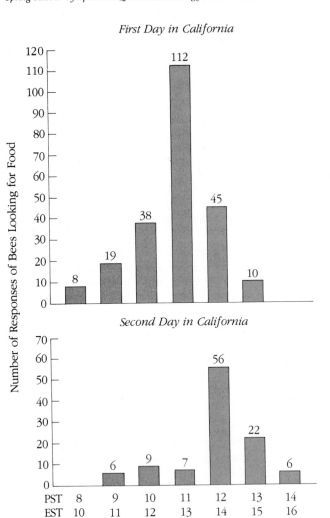

as those cyclic environmental cues that can entrain free-running endogenous pacemakers. We know that zeitgebers can influence rhythms by affecting both the phase and the frequency.

Return for a moment to the example in figure 8–3. When the flying squirrel is provided with a daily cycle of light and dark, its activity rhythm generally conforms to the light cycle, and the onset of activity occurs at about the same time each evening (DeCoursey 1961). The squirrel is thus said to be entrained to the daily light-dark cycle. The light-dark cycle is the critical cue for entrainment in most endothermic vertebrates tested so far. Further, in most terrestrial organisms, the daily light-dark cycle is the zeitgeber. This is not surprising, since the daily light cycle is usually the most consistent environmental cue in the terrestrial habitat. What may be surprising is that the animal does not need to be exposed to the entire light-dark regime to have its clock set or adjusted. In fact, fifteen-minute pulses of light given to flying squirrels as they awaken from their sleeping through the lighted portion of a twelve-hour light–twelve-hour dark regime will induce a phase shift in their activity pattern, causing a nap followed by a delay in starting the activity phase (DeCoursey 1986).

What other cues can serve as zeitgebers? For animals in the intertidal zone, the ebb and flow of the tides may be the prime zeitgeber. Ectotherms such as lizards and insects that cannot fully control their own body temperatures may use temperature or light cues as zeitgebers. When entrained, these ectotherms exhibit increased activity under warmer conditions and decreased activity during the cooler portions of the daily cycle. To measure this effect in tsetse flies (*Glossina* spp.), the number of animals that are caught during each hour of the day were sampled with nets (Crump and Brady 1979; Brady 1988). During cooler seasons, they were active only at midday when ambient temperatures reach 20° C; they tended to be active only when conditions were above about 18° C. During hotter seasons, they were active during the early morning and evening hours when the temperature drops to about 20° C and they avoided the peak heat (40° C) during the midday hours.

There is at least one question that arises from the foregoing information on daylength as a zeitgeber: Why should photoperiod be the primary cue that sets and alters biological clocks? Because, for most animals, the cue with the greatest degree of predictability is photoperiod. This is true whether we are looking at circadian or circannual rhythms; day in and day out, year in and year out, the daily and seasonal patterns of changes in photoperiod will be extremely consistent. Thus, during the course of evolution, animals that developed biological timekeeping systems that used photoperiod

of a rhythm occurs repetitively and coincides approximately with the presence of some external stimulus is called *entrainment*. As we noted in the introduction, cues that provide information to animals about periodicity of environmental variables are called zeitgebers ("time givers"). **Zeitgebers** are the entraining agents defined

to set and reset their clocks would be most likely to have "the correct time." As we shall see in the paragraphs that follow, other cues may interact with or modulate the major reliance on photoperiod as a zeitgeber. The exception to the photoperiod rule is the group of organisms that exhibit tidal rhythms. The most predictable and most important cue in these organisms' environment is the ebb and flow of the tides. We would expect natural selection to favor development of timekeeping processes that are tuned-in to the tides, rather than to the photoperiod.

Other environmental cues that may in some organisms entrain biological rhythms are summarized in Moore-Ede, Sulzman, and Fuller (1982). Cycles of food availability, but not water availability, can entrain activity rhythms in several species of small mammals (Richter 1922; Moore 1980). In humans, social cues may be involved in the entrainment of biological rhythms. Two groups, each with four volunteer subjects, were isolated in separate identical chambers (Vernikos-Danellis and Winget 1979). Each individual's activity rhythm was synchronous with those of the others in that room. However, the average free-running periods for the two rooms differed; 24.4 hours for one room and 24.1 hours for the other. To avoid any confounding effect from self-selected light-dark cycles, both groups were constantly illuminated. The possible role of social cues in synchronization of biological clocks within the rooms, but not between the rooms, was tested further by moving one subject from one group to the other. That subject soon went through a phase shift and became synchronized with the activity rhythm of the new group. The exact nature of the social cues involved in the entrainment of circadian rhythms in humans is not yet known.

Social cues may also act in concert with other environmental cues to entrain biological rhythms. When individuals who have just flown by jet across six time zones are required to remain in their hotel rooms, they entrain more slowly to the new local time than do individuals who are permitted to leave their rooms and who thus obtain more social-environmental cues from their new surroundings. One method for overcoming jet lag is to spend extra time in the daylight (sunshine) in the new location (Daan and Lewy 1984).

Resetting of the human circadian pacemaker using exposure to bright light has recently been reported (Czeisler et al. 1989). Subjects were assessed to determine the normal pattern of circadian rhythmicity. They were then subjected to five-hour periods of bright light. The phase shift effects produced by the bright light were dependent upon when during the daily cycle the treatment was provided. Preliminary evidence from medical studies indicates that melatonin may also help to alleviate jet lag and related problems (Arendt, Aldhous, and Marks 1986). In a test, eight subjects were given melatonin for three days before and four days after a San Francisco–London flight, and nine subjects were given a placebo before the flight. None of the subjects who had taken melatonin had jet lag effects, whereas six of the nine who had been given the placebo did have appreciable jet lag effects.

Daylength and social cues also act together to entrain certain aspects of the annual rhythms in various animals, as for example in the house sparrow (*Passer domesticus*). In this species, both hormone levels and behavior are influenced by a combination of photoperiod duration and conspecific competition (Hegner and Wingfield 1986).

If we bring an animal such as a skunk (*Mephitis mephitis*) into the laboratory, and reverse the light and dark portions of its natural cycle with artifical lighting, the animal's activity phase will shift to the light portion of the cycle within a few days. The ability to manipulate the activity phase of animals' daily rhythms has been used by zoological parks to display some nocturnally active animals. By having special buildings that are kept darkened when the zoo is open to visitors and illuminated when the zoo is closed, the public is able to watch them be active, instead of watching them sleep. As investigators, we can manipulate the rhythms of nocturnal animals, and under darkened conditions, conduct behavioral observations during our own diurnal activity phase.

MODEL

Using information obtained on biological rhythms, we can depict what we know in a model (figure 8–5). The model system we have drawn accounts for the two major elements known to be important for biological timekeeping: an endogenous self-sustaining pacemaker and a system for entrainment to environmental zeitgebers. In the example illustrated here, we note that some form of environmental input, e.g., light, is picked up from the environment by the appropriate receptor system and transmitted to the circadian oscillator, e.g., the suprachiasmatic nucleus in the brain. From there a message, which may be neural, hormonal, or both, is sent to various tissues and organs, and results in behavioral output (e.g., feeding), and thus produces the rhythm we observe and record. The same model may be used with other types of rhythms involving entrainment to tidal, annual, or other cues.

FIGURE 8-5   Model of a mammalian pacemaker system portraying the various component parts and how they may function together, resulting in the overt behavioral and physiological rhythms which can be observed

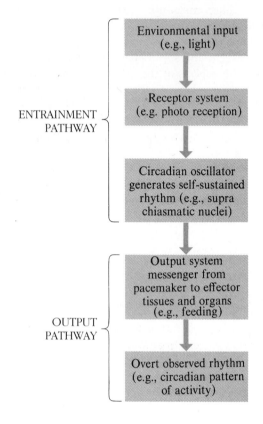

LOCATION AND PHYSIOLOGY OF THE PACEMAKER

Having described biological rhythms and having examined how they appear to be controlled, we might now wish to ask: Where is the pacemaker device responsible for doing the timekeeping? During the past several decades, great progress has been made on locating the pacemakers and studying their physiology. Much of this work has been summarized by Brady (1982, 1988); Menaker, Takahashi, and Zatz (1982); Aschoff (1981); and DeCoursey (1983). The work has been conducted using cockroaches, sea hares, birds, and rodents.

COCKROACHES. Are clock mechanisms primarily neural or hormonal? Harker's work (1960, 1964) suggested that hormonal and neurosecretory products play key roles in the cockroach's biological clock mechanism (figure

8-6). Harker made one cockroach (*Leuccophaea moderae*) arrhythmic by keeping it in continuous light. She then interconnected its blood system with that of a cockroach kept in a normal light-dark cycle; that animal showed regular circadian periodicity. The arrhythmic recipient cockroach adopted the regular circadian periodicity of the donor. However, other studies (Brady 1967a,b; Roberts 1966) have failed to replicate Harker's work and instead have shown that cockroaches kept in constant conditions can maintain regular circadian periodicities; the recipient cockroach in Harker's studies may simply have been reverting to the free-running rhythm.

Decapitation of a cockroach results in arrhythmic activity; the surgery apparently removes the source of the key oscillator, but not the ability to exhibit rhythmic locomotor patterns. Experimenters have used these headless organisms in implantation and transection experiments to test various organs and tissues as candidates for endocrine or neural clocks. Brains, corpora allata, and corpora cardiaca transplanted from hosts exhibiting rhythmic activity to headless cockroaches have all failed to produce regular rhythmic activity in the recipients. The relatively unimportant role of brain neuroendocrine secretions in cockroach clock mechanisms has been ascertained by surgical transection operations shown at locations B, E, and F in figure 8-6.

Brady (1969), in further transection studies, found that the optic lobes play a key role in cockroach circadian rhythms. Transection at A does not affect periodicity, but transection at B, which destroys the influence of the optic lobes, eliminates rhythmic activity. The pacemaker thus appears to involve the optic lobes and possibly the lateral brain neurosecretory cells.

Information about periodicity is hypothetically communicated to effector organs in the body through three possible pathways; (1) hormones in the blood, (2) hormones in the nerves, and (3) electrical impulses in the nerves. Cutting the NCA II pathway (figure 8-6) and removing the corpora cardiaca without affecting rhythmicity would indicate that (1) is unlikely, since lateral brain neurosecretory cells send axons only to the corpora cardiaca, where their secretions must either be released or travel down the NCA II pathway to the subesophageal ganglion. Since surgery on the NCA II pathway has no effect on rhythmicity, the only remaining pathways involve the circumesophageal connections (figure 8-6). However, no neurosecretory activities have been demonstrated for these connections, and thus we may also rule out (2).

FIGURE 8-6   Parts of the neural and hormonal systems of the cockroach

This schematic diagram illustrates a compilation of studies of biological clock phenomena. The connected spheres represent the ganglia of the central nervous system. Endocrine tissues that can be removed without affecting biological rhythms are shown as dotted boxes. Arrows designate organs that have been transplanted into arrhythmic, headless recipients from rhythmic donors. 0 indicates a host with implant that shows no detectable rhythm. Cuts in nerve trends are shown as heavy broken lines. Cuts that are made at B, E, and F or by splitting the protocerebral lobes (part of the interior of the brain) bilaterally appear to stop the rhythm, but cuts at A, D, C, or by splitting the pars interocerebralis (part of the interior of the brain) do not stop the rhythm.

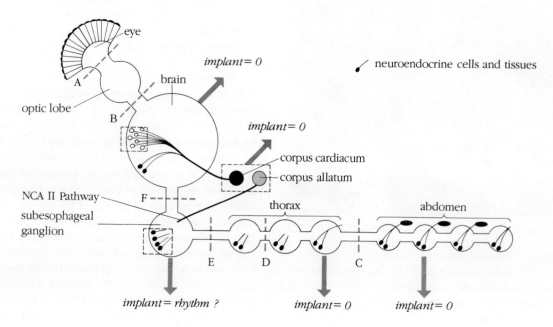

These steps bring us to a third experiment: disconnecting the brain from the remainder of the nerve cord. If rhythmicity is lost when we separate the brain and nerve cord, the third alternative would appear to be correct. Severing the nerve cord posterior to the thorax at C does not affect the rhythm. However, cutting the nerve cord between the thoracic ganglia at D reveals a graded effect: the more anterior the cut, the higher the proportion of cockroaches that exhibit disrupted activity rhythms. Thus, it appears that a pacemaker located in the optic lobes communicates with effector tissues (muscles and organs) through electrical impulses that travel to the thoracic region of the ventral nerve cord.

Further studies of the optic lobes of the cockroach lead us to the tentative conclusion that the critical area for the pacemaker lies primarily in cell bodies in the inner region of the optic lobes (Roberts 1974; Sokolove 1975). Another series of experiments (Page, Caldarola, and Pittendrigh 1977) revealed that separate pacemakers in each optic lobe interact with one another. In addition, the presence of either compound eye alone can serve to entrain the pacemakers in both optic lobes.

Page (1982) extended our understanding of cockroach circadian pacemakers one step further. The optic lobes of a cockroach were surgically removed and replaced by the optic lobes of a second cockroach with a different circadian rhythm. After a four-to-eight-week lapse, the recipient cockroach again exhibited a circadian activity pattern—and the restored rhythm's free-running period matched that of the donor cockroach. These findings support the hypothesis that the optic lobes contain a circadian pacemaker, and that after a lapse of some weeks, neural connections between the optic lobes and the brain regenerate in the recipient cockroach—resulting in the reestablishment of a recognizable pattern of circadian activity.

FIGURE 8-7   Sea hare

The sea hare's eyes are embedded in the integument of its head. When we remove the eyes and place them in sea water, they continue to exhibit a circadian pattern of optic nerve impulses.

Source: Photo by Runk/Schoenberger from Grant Heilman.

SEA HARES. Experiments on the sea hare (*Aplysia*) have provided additional evidence of the relationships between neural structures, optic system, and biological rhythms (Jacklet 1969, 1973; Jacklet and Geronimo 1971). When sea hares' eyes are isolated and kept in total darkness in sea water, their optic nerve impulses exhibit a circadian rhythm (figure 8-7). Eyes taken from sea hares that have been under the twelve hour light/twelve hour dark regimen before eye excision show a peak firing frequency at "dawn" on the day after removal. Eyes from sea hares kept in constant light before surgery exhibit regular optic nerve impulse rhythms. Thus some type of oscillator, possibly related to neurosecretory processes, must be present in the eye. The phase of the eye rhythm is readjustable at each dawn, and the oscillator in the eye may control other sea hare rhythms. A clock mechanism that uses a pacemaker in the eye to adjust to day-to-day changes in daylength could be adaptive for an organism that lives in a variable environment.

The circadian firing frequency in the sea hare's optic nerve apparently results from a coupled, interacting population of 950 retinal cells. Two findings support this conclusion: First, cells, down to a critical level of about 20 percent of the original total, can be removed from the population without altering the basic period and

amplitude of the rhythm (Jacklet 1973, 1976). Second, below that level, removal of additional cells results in a shortening of the rhythm's period and a dampening of the amplitude. When the number of cells has been reduced to twenty or fewer, the rhythm becomes ultradian—shorter than a normal daily pattern. Apparently, a population of noncircadian oscillators located in the retina produces a circadian rhythm in the optic nerve when they are coupled to each other. Circadian rhythms in the body's effector systems may be partially regulated by this optic system oscillator. The cockroach appears to have two coupled pacemakers in each optic lobe; sea hares have pacemakers in their eyes (Hudson and Lickey 1977).

Further work on *Aplysia* has resulted in an increased understanding of its circadian oscillator's biochemical events (Eskin and Takahashi 1983). Using isolated eye preparations from sea hares, these investigators determined that the amount of cyclic adenosine 3′, 5′-monophosphate (cAMP) formed in homogenates of the eye was affected by the amount of (exogenous) serotonin, but not other neurotransmitters, present. The action of serotonin is mediated by the activity of adenylate cyclase, an enzyme critical for synthesis of cAMP. The investigators also demonstrated that forskolin, a specific activator of adenylate cyclase activity, produced both advance and delay phase shifts in the timing of the circadian rhythm of the isolated *Aplysia* eye. From these findings we can conclude that cAMP is important in mediating phase shifts in the pacemaker located in the sea hare eye.

VERTEBRATES. What about pacemaker mechanisms in vertebrates? Investigations of biological clock mechanisms in rats (*Rattus norvegicus*) have led to the finding that a direct neural connection exists between the retina and the hypothalamus, a pathway that terminates in the suprachiasmatic nuclei (SCN), a region of the hypothalamus (Hendrickson, Wagoner, and Cowan 1972; Moore and Lenn 1972). When investigators made lesions to cut connections to these nuclei, the rats lost circadian rhythms of drinking behavior and wheel-running activity (Moore and Eichler 1972; Stephan and Zucker 1972). Additional studies on rats reveal that other circadian rhythms are stopped or radically altered by lesions of the SCN (see Rusak and Zucker 1979; Moore 1979).

Studies on the SCN lead some investigators to hypothesize that the SCN is the pacemaker. There is now ample evidence to indicate this is not the case; at least the SCN is not the sole pacemaker. For example, bilateral SCN lesions in monkeys, which result in a circadian arrhythmicity of the animal's activity-rest cycle, do not eliminate the ongoing circadian rhythm for body temperature (Fuller et al. 1981). Therefore, there must be at least one other pacemaker located somewhere else in the animal. Several bits of evidence suggest that another pacemaker may be located in the ventromedial nucleus of the hypothalamus (VMH). There may be some form of mutual coupling between the various pacemakers—neural pathways connecting the SCN and VMH. Also, it is possible to entrain the second pacemaker by food; food entrainment persists after removal of the SCN (Stephan, Swann, and Sisk 1979a,b; Boulos, Rosenwasser, and Terman 1980). Further, Krieger (1980) has shown that circadian rhythms retained after destruction of the SCN are lost when the VMH is destroyed.

Investigations of rhythms in birds and mammals have implicated the pineal gland as a probable receptor of light stimuli that entrain and affect circadian patterns (Gaston 1971; Menaker and Zimmerman 1976). Existing evidence supports the hypothesis that both neural processes (Block and Page 1978) and hormone-neuroendocrine products (Zucker, Rusak, and King 1976) are involved in the mechanisms underlying biological rhythms in vertebrates.

The pineal gland in mammals lies within the brain, near the midline; but in some earlier terrestrial vertebrates, this structure was positioned on the top surface of the brain and served as a third or median eye. (We have all seen the mythical movie monsters with the third eye on the forehead!) In today's reptiles, birds, and amphibians, the pineal gland is located just under the skull—indeed, it is still sensitive to light in many of these organisms. We should not find it surprising that the pineal gland plays a key role in regulating certain rhythms based on photoperiod.

The most important of the variety of endocrine products from the pineal may be melatonin, an indoleamine (Reiter 1980). Daily subcutaneous injections of melatonin given to pinealectomized male hamsters induce regression of the gonads, mimicking the effect of shortened photoperiods in these animals (Tamarkin, Brown, and Goldman 1975; Reiter 1974a). A series of investigations using hamsters, rats, mice, and other mammals have repeatedly demonstrated the antigonadotrophic effects of melatonin. In these mammals, the pineal normally receives photoperiod information via neural circuitry from the eyes. Reiter (1974b) proposed that one key function for the pineal may be the partial control of the annual rhythm of reproduction. Thus, the seasonal changes in photoperiod may be translated into physiological effects by the pineal and its endocrine products. Using radioactively labeled melatonin and autoradiography, Reppert et al. (1988) determined that for humans, specific sites in the hypothalamus differentially bind the melatonin. In particular, the site where the most binding of the labeled melatonin occurs is the suprachiasmatic nucleus.

Additional information of the control mechanisms of circannual rhythms, primarily in vertebrates, has also accumulated. The data collected so far largely concern correlated responses that are probably several steps removed from the actual clock mechanism. In thirteen-lined ground squirrels (*Spermophilus tridecemlineatus*), selected mixtures of dialysates (materials that pass through the membrane in dialysis) and their residues obtained from the blood of other ground squirrels or from woodchucks (*Marmota monax*) accelerate or impede the induction of hibernation (Dawe and Spurrier 1972). The brown fat found in some animals, which was thought for many years to contain chemicals responsible for inducing hibernation (Johansson 1959) has been shown instead to contain a substance that produces arousal (Smith and Hock 1963). When turtles (*Testudo hermanni*) are housed outdoors, the levels of two compounds from their pineal gland—serotonin and melatonin—exhibit both circannual and circadian rhythms (Vivien-Roels, Arendt, and Bradtke 1979). The turtles synthesize serotonin during the day, and melatonin at night; this pattern of synthesis disappears entirely during hibernation. During the breeding season, concentrations of both chemicals and the amplitude of circadian fluctuations increase. Additional investigations are needed to determine the relationship between the concentrations of these chemicals and the observed circannual and circadian rhythms.

SIGNIFICANCE OF BIOLOGICAL TIMEKEEPING

ECOLOGICAL ADAPTATIONS

Pittendrigh (1960), Enright (1970), and Daan and Aschoff (1982), have summarized the ecological and evolutionary significance of biological rhythms. Physical factors of the environment, such as light, temperature, and humidity, can be critical for some organisms: those whose integument loses water to the atmosphere, those who cannot fully regulate their own body temperature, and those who live in temperate, arctic, or desert environments where physical factors undergo dramatic seasonal or daily variation.

Many insect species live in environments where the photoperiod (daylength) differs from the periods of physical factors in the environment that may have either positive or deleterious effects, such as temperature, moisture, and chemical substances. Photoperiod exerts no direct beneficial or harmful influence, but may serve as an adaptive timing cue, or a predictor of environmental conditions critical for survival (Beck 1968).

Field observations and laboratory experiments indicate that woodlice exhibit a circadian rhythm of activity that is regulated by photoperiod: they are active at night and quiescent during the day. At night, temperatures are lower and humidity is higher than during the day. Cloudsley-Thompson (1952, 1960) found that during the day, woodlice respond negatively to light and positively to moisture; thus during the daytime they tend to remain in dark, damp areas of the test chamber. At night woodlice are photonegative, but their positive reactions to humidity are not as pronounced. Under conditions of extremely low relative humidity, the woodlice become weakly photopositive.

Together these behavioral reactions can be seen as adaptive traits related directly to water conservation and the nocturnal habits. Since response to moisture is weaker at night, woodlice can move into and through dry places they would never go during the day. The woodlouse's stronger photonegative response at night ensures that when daylight comes, the insect will go into hiding. Finally, weakly photopositive responses when humidity is very low enable the woodlouse to move out of its resting place to seek a moister environment, even during daylight hours when prevailing conditions dictate such a move to ensure survival; for example, when remaining in a dry location could lead to death by desiccation.

DIURNALITY

For animals that have an impervious integument—birds, mammals, and organisms that live in environments where conditions remain relatively constant, like the ocean—a number of biotic factors, including reproduction, feeding habits, interspecific competition, and the possibility of predation, help to determine the timing of daily activities. Whether the animal is nocturnal, diurnal, or crepuscular appears to depend on its responses to various biotic factors. It is exceedingly difficult and complex for us to make determinations of the ecological significance of circadian rhythms in birds and mammals, and few concrete answers have emerged.

For example, let us consider the evolution of primates from primitive insectivores, an evolutionary sequence that is represented today by some species of primates in Africa and Asia (Simons 1972)—lorises, tarsiers, and some lemurs (figure 8–8). Primates usually exhibit a diurnal activity phase in their circadian rhythms; the shift from nocturnal to diurnal activity during primate evolution can be traced in part to the effects of certain biotic factors.

Rodents probably evolved concurrently with early primates from insectivorelike predecessors and rapidly became dominant in the terrestrial habitat. The resulting strong competition between rodents and early primates for food, nesting sites, and living space may have brought about two adaptive changes that characterized primate evolution: a shift by primates to arboreal habitats and a phase shift in the timing of their activity peak to the daylight hours. In addition, if the first change involved a move to the trees (the evidence supports this contention) then there may have been additional selection pressures for diurnal activity, because arboreal locomotion is safer in daylight.

An alternative explanation of the evolution of diurnality in primates is that the insectivorelike stem group, from which the primates evolved, was originally composed of small, diurnal, arboreal animals. Predation pressure might have led to natural selection for a nocturnal activity phase, since adequate cover during the daylight hours exists only in the forest canopy. Later, as the true primates evolved, the food habits of some species shifted to a diet containing more fruits and leaves, and body size increased. With increased body size, the primates were better able to defend themselves or escape from predators, and some species adopted a terrestrial lifestyle. Having at least two equally plausible explanations (there may be others) for the diurnal activity of primates illustrates for us the problem of the use of post-hoc reasoning to account for the evolution of observed present-day behavior.

HIBERNATION

Circannual rhythms in invertebrates and vertebrates have clear ecological significance. Some mammals, such as ground squirrels (*Spermophilus* spp.), chipmunks (*Striatus* spp.), and jumping mice (*Zapus hudsonicus* and *Napeozapus insignis*), exhibit annual patterns of hibernation (Kayser 1965; Pengelley and Asmundson 1974). From mid-fall until mid-spring, these mammals enter a physiological state in which their normal endothermic

FIGURE 8-8   Loris and hamadryas baboon
Today's primates have evolved from insectivorelike
predecessors. The loris (left) is a descendant of one of the
earliest forms to evolve; note the large eyes for night vision.
The hamadryas baboon (right) has returned to a largely
terrestrial existence, possibly aided by its large size and
social defense mechanisms against predation.

Source: Photo by Ron Garrison. Copyright © by Zoological Society of
San Diego.

body temperature (ca. 37°C) falls to within a few degrees of the environmental temperature. Hibernators
begin preparation for the winter season well in advance of the arrival of colder conditions. Some hibernating mammals, such as woodchucks, alter their diet
by late summer and add a layer of body fat, which provides extra insulation and a food supply for the winter.
Other animals, such as chipmunks and hamsters, which
inhabit shallow burrow systems during the spring and
summer, extend their underground system deeper or
dig special burrows with a hibernation chamber. These
winter sleeping sites are always located deep beneath
the ground surface, usually below the frost line, so there
is no danger of being frozen during the coldest periods
of winter.

Some of the above observations about hibernation
must be reexamined based upon the finding that body
temperatures in hibernating arctic ground squirrels
(*Spermophilus parryii;* figure 8–9) are as low as −2.9° C

(Barnes 1989). How can the animal survive such conditions? Two ways that this might occur involve the
ground squirrel's having sufficient solute materials in
its vascular system to depress the freezing point, or its
having antifreeze molecules in its system. Studies indicate that neither of these occurs for arctic ground
squirrels. In this instance, a phenomenon known as supercooling is involved; the temperature of the liquid
(blood and other body fluids) can be cooled below the
freezing point without crystallization. The ability of
arctic ground squirrels to hibernate with core temperatures below 0° C may be related to the prolonged hibernation period (up to 8–10 months) and the extreme
ambient conditions of their habitat in the far northern
latitudes. The physiology of hibernation and some
closely related behavioral phenomena have been reviewed by Lyman et al. (1982).

FIGURE 8-9
The arctic ground squirrel lives in a harsh environment where conditions are suitable for above-ground activity only 2–4 months out of each year. The squirrels spend the remainder of the year in hibernation, with body temperatures as low as −2.9°C. These animals are able to tolerate temperatures below freezing because of supercooling of the blood and other body fluids without crystallization.
Source: Photo by Leonard Lee Rule/Photo Researchers, Inc.

The most obvious adaptive value of a circannual rhythm of hibernation is that hibernating creatures avoid many problems of survival: they do not have to find and possibly compete for limited food supplies, and they do not have to maintain a warm and suitable nest site during adverse winter conditions. However, other mammals in the same regions do not hibernate but have adopted other strategies for survival. Some mice, for example, produce great numbers of progeny during the summer months; a portion of these will survive the rigors of winter to breed the next spring. Other mice live in communal nests during the winter and gain warmth from each other. Some mammals, like raccoons, do not actually hibernate, but during the coldest periods of winter, they may remain in their dens for several days in a state of quiescence or torpor.

A very important aspect of the circannual rhythms in mammals that do hibernate is that they are able to anticipate the coming of winter. During late summer and fall their behavior changes; they shift their dietary habits and prepare hibernation burrow systems. If these animals waited until the food supply disappeared or the ground was frozen or covered with snow, it would be too late.

Hibernating mammals must mate soon after emerging from the winter den so that their progeny can develop before the onset of the next winter season only a few months away. How is this possible? When the animals emerge, both males and females are in nearly prime reproductive condition; some of their physiological systems have "awakened" before the end of hibernation. For example, thirteen-lined ground squirrels (*Spermophilus tridecemlineatus*) are in estrus within one to two days after emergence from the winter burrows (Landau and Holmes 1988), and they mate successfully the first week out of hibernation. The internal changes accompanying the observed behavioral phenomena are apparently triggered by an internal clock—either an endogenous mechanism or a clock that is set exogenously when the animal goes into hibernation in the fall.

MIGRATION

The annual rhythm of migratory behavior exhibited by birds that breed in north temperate and polar regions, but winter in more southern latitudes, offers interesting parallels to hibernation. By flying south for the winter, these birds escape severe winter conditions and likely food shortages; the lack of sufficient physiological adaptation generally precludes their remaining in northern climates for the entire year. When they return to the north in the spring, these birds, like the hibernating mammals, must reproduce soon after their arrival at summer breeding locations to ensure that their progeny are fully developed by the time of their return flight south in the fall. As with hibernation, we might ask: How is this possible? Possession of a circannual clock mechanism permits these birds to anticipate the coming of fall in time to prepare physiologically for and to initiate migration. At the birds' wintering grounds to the south, where fewer dramatic seasonal changes in the environment occur, the same clock mechanism "tells" the birds when it is time to start their northbound flight in order to arrive back at the breeding area at the best time for reproducing. Bird species are entrained by light, and possibly in some cases by temperature zeitgebers. Furthermore, as we shall see in chapter 15, biological clocks also play a role in the orientation and navigation of many birds.

Some marine organisms exhibit circannual rhythms. Longhurst (1976) studied vertical distribution and daily vertical migration in plankton, and found that daily vertical movements are true circadian rhythms controlled by an internal clock mechanism, but that the nature and timing of the vertical migration varies with species, sex, age, and even location. Light is the principal, but not the only, cue that can affect these rhythms. Three ecological and evolutionary explanations of vertical migration have been proposed (McLaren 1963; Longhurst 1976). First, migrating upward at night (a common, but not universal, feature of the vertical movement of plankton) may be a means of avoiding predation. Marine fish that are potential predators are active in the upper, better-lighted zone of the ocean during the day and move to a lower zone at night. Thus, the plankton are in the upper zone at night when the fish that would most likely consume them are not. However, many of these plankton species are found at a depth during the day where enough light still penetrates to permit some predators to be active. Moreover, some of the plankton that rise to the upper level or to the surface at night are bioluminescent and thus announce their presence. Second, by migrating vertically between two levels of water, plankton may use the current that is generated as a means of dispersion. Third, the plankton may derive bioenergetic or nutritional benefits from these daily vertical movements.

SUMMARY

Animals exhibit patterns of activity and inactivity that recur at regular intervals. The timing of these behavior patterns is apparently controlled by *biological pacemakers*, regulated by *circadian, lunar,* or *circannual rhythms.* Biological rhythms involve two elements: an endogenous self-sustaining pacemaker and external cues which can entrain the pacemaker. Evidence for the existence of an internal chronometer comes from several sources. When an animal is placed in an environment with constant conditions, it exhibits an activity rhythm with a period different from that of any known environmental variable; these are called *free-running rhythms.* Animals hatched and reared in isolation and under constant conditions exhibit *circadian* (daily) free-running rhythms. Animals translocated from one time zone to another will, initially, exhibit an activity phase matching that of their original environment. After some period of time in the new location they exhibit a *phase shift* and adjust the time of their activity to the new local time. Last, the activity rhythms for several animals of a species differ from one another; there are individual differences in the chronometers.

The activity rhythms of animals can be *entrained* to a variety of environmental cues called *zeitgebers.* The most prominent and consistent environmental variable is the *light-dark cycle.* It is therefore not surprising that many daily and annual rhythms are entrained to the light-dark cycle in a wide variety of animals. Other rhythms may be entrained to tidal or lunar cues, which are prominent in some environments. The availability of food and social cues have also been demonstrated as probable zeitgebers for rhythms in some organisms.

Research on cockroaches, sea hares, birds, and rodents has begun to provide evidence on the location of the pacemakers and to elaborate how they function. Most biological clocks are relatively unaffected by either temperature changes or metabolic inhibitors. Evidence to date implicates neural, neuroendocrine, and hormonal events in the pacemaker systems of different organisms. For both cockroaches and sea hares the experimental results indicate that there are pacemakers in the optic lobes and eyes. In mammals the evidence suggests the existence of at least two pacemaker systems, one in the suprachiasmatic nuclei and another in the ventromedial hypothalamus. The pineal gland, and in particular, the melatonin produced there, have been implicated in the regulation of annual cycles of reproduction in some vertebrates.

The ecological and evolutionary significance of biological timekeeping can be assessed best with respect to physical variables such as daylight, temperature, and moisture and to related biotic factors, like competition for limited resouces, feeding habits, and predation. For example, the evolution of diurnality among primates may have resulted from their competition with nocturnal rodents. An alternative theory is that as primates evolved larger body sizes, they may have been better able to cope with potential predators, and thus could adopt a diurnal activity phase that enhanced feeding opportunities. Hibernation by some mammals and migration by some birds in temperate and polar climates are evolutionary adaptations to avoid the rigors of the winters in those regions.

Discussion Questions

1. Inventive research techniques, such as translocation, the use of magnets to assess the role of geomagnetic forces, and microsurgery to explore clock mechanisms, have been employed in the study of circadian biological rhythms. If you were about to embark on the study of the clock mechanisms of circadian rhythms and the factors that influence these rhythms, what new techniques might you employ?

2. Let us suppose that a new species of small mammal, the yellow-striped scamp, has just been discovered in the woodlands of central Ontario, Canada. Scamps are both arboreal and terrestrial; they are about 10 cm long, including the 4 cm tail; and they eat seeds and insects. What types of laboratory and field observations and experiments would you conduct to describe the biological rhythms of this species? In your answer, consider behavior patterns and underlying physiological mechanisms and the adaptive significance of the observed rhythms.

3. Certain behavior patterns (e.g., feeding) are regulated by circadian clocks, while other behavior patterns (e.g., hibernation) may be regulated by circannual clocks. What other behavior patterns that may be regulated by circadian or circannual rhythms can you name? What is the adaptive significance of each of the behavior patterns you have named and of their being regulated, at least in part, by a biological clock mechanism?

4. The chart below represents the singing behavior of a wood thrush over a fourteen-day period. For the first seven days, the bird was housed in constant darkness. For the second seven days the bird was subjected to a daily light cycle with the lights on at 0600 hours and off at 1800 hours. What is the period of the activity rhythm in constant darkness and in the light-darkness cycle? What changes has the light-dark cycle brought about in the frequency and phase of the activity period?

5. As was noted in the chapter, birds that inhabit the north temperate zones migrate to warmer climates for the winter months and return the following spring to breed. In recent decades, some birds of certain species, e.g., American robins, have started to remain throughout the year in the more northern zone where they spend the summer. How can this sudden change in behavior be explained? What are the consequences for the birds? Is it possible that some form of natural selection will change the overall annual behavior pattern of these birds?

Actogram for singing of a thrush; note the animal has two active phases.
Shaded area indicates darkness; black oval denotes singing.

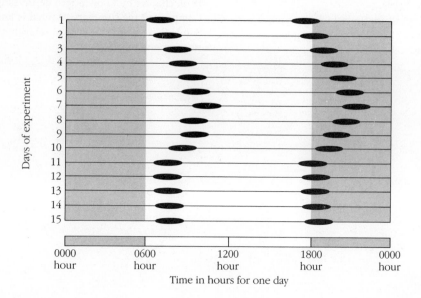

Suggested Readings

Aschoff, J., ed. 1981. *Handbook of Behavioral Neurobiology*. Vol. 4, *Biological Rhythms*. New York: Plenum.
A thorough review summarizing our current knowledge of various biological clock phenomena. Chapters cover all aspects of biological rhythms and the methods used to investigate these phenomena, including models.

Block, G. D., and T. L. Page. 1978. Circadian pacemakers in the nervous system. *Ann. Rev. Neurosci.* 1:19–34.
Review of knowledge on pacemaker locations and related nervous system physiology. Some historical perspective and good bibliography.

Brady, J. 1982. *Biological Timekeeping*. Cambridge: Cambridge University Press.
A basic textbook on biological rhythms. Includes solid elementary explanations of clock processes and good references to current material.

DeCoursey, P. J. 1983. Biological timing. *Biology of the Crustacea*, vol. 7, 107–162. New York: Academic Press.
An excellent review article in two parts. The first part gives an up-to-date review of the various theories and explanations for biological clock phenomena. The second part deals with the application of these ideas in crustaceans.

Gwinner, E. 1986. *Circannual Rhythms*. Berlin: Springer-Verlag.
A short but packed book on seasonal changes in behavior and physiology with an emphasis on endogenous clocks. Excellent bibliography.

Saunders, D. S. 1974. Circadian rhythms and photoperiodism in insects. In *Physiology of Insects*, vol. 2, ed. M. Rockstein. New York: Academic Press.
Book chapter by knowledgeable investigator. Comprehensive summary of what is known on the topic. Some discussion of ecological relevance of biological clock phenomena in insects.

PART THREE

DEVELOPMENT AND LEARNING

9

DEVELOPMENT OF BEHAVIOR

Chapters 5 through 8 have covered the causal mechanisms and processes that control and regulate animal behavior. Now, in part 3 we consider those processes, development and learning, that interact with the genotype of each organism to shape the observed behavior of the phenotype. We begin the chapter by presenting some basic methods considerations that affect much of the research on behavior development. We then examine the key events and factors that affect the development of behavior in chronological sequence, starting with the embryology of behavior and following with prenatal events, juvenile events, play behavior, and some comments on continued development into adult life. The chapter concludes with an examination of the nature-nurture issue and the modern epigenetic theory of behavior development.

METHODS IN BEHAVIOR DEVELOPMENT

Many studies of the development of behavior are designed to investigate the mechanisms that underlie behavior patterns observed in adult animals. Such a study may be relatively direct (for example, determining what factors affect nest-site selection by a bird species), or may be complex (for example, establishing the developmental determinants of feeding strategies in omnivorous rodents). Some studies apply conditions of either deprivation or enrichment; others alter the quality of

specific types of stimulation provided during ontogeny, rather than vary the overall quantity of stimulation. What types of methods are employed to investigate the various questions pertaining to behavior development?

For many investigations of behavior development, we use direct observational techniques. (Some of the basic principles discussed in chapter 3 are pertinent here, and a brief review of that material may be beneficial.) In some studies, particularly in those conducted under field conditions, we may employ only observation, with little or no manipulation. This type of investigation provides useful information on the development of traits within the natural ecological and social context. Other studies of behavior development utilize a variety of manipulations to test the effects of various treatments. Examples of these approaches are presented when we examine the chronology of development. First, a bit of elaboration is needed on some aspects of the experimental design and testing procedures used by many investigators when studying the development of animal behavior.

ASPECTS OF EXPERIMENTAL DESIGN

King (1957) set forth seven parameters relevant to the study of early experience. His scheme, or portions of it, can rather broadly be used to design many of the experiments in behavior development, particularly those that attempt to assess the effects of early experience on behavior.

Parameters of the early experience treatment:

1. Age of the animal at the time the early experience treatment is given
2. Type or quality of the early experience treatment
3. Duration or quantity of the early experience treatment

Parameters of the tests administered to determine the effects of early experience treatment:

4. Age of the animal at the time of testing
5. Type of test used to assess the effects of early experience treatment
6. Testing for the persistence of the early experience treatment

The parameter of the genetics of the species being investigated:

7. Testing different strains or species to determine the effects of early experience on different animals

TESTING PROCEDURES

We can test the long-term influence or persistence of early experience treatments by two types of experimental design, as outlined in King (1969). One provides an animal with early experience treatment, and tests the effects of that experience at various ages using the same animal for each test. Appropriate repeated-measures statistical tests are required to analyze the results. This type is called **longitudinal design.** Alternatively, the animal can be provided with an early experience treatment, and at a later age, tested only once, because in the same animal's second test, the observed effect might result from either the treatment, or the experience of the first test. Therefore, reuse of the test subject is avoided, even though some information about the persistence of early experience treatment on that subject may be lost. This second type of design is called **cross-sectional design.**

In some studies of early experience, investigators subject animals to stimulus deprivation or enrichment by manipulating either the quality or quantity of one (or a combination) of three types of stimulation: (1) sensory, (2) motor, or (3) social. In all cases, the test performances of animals reared under deprived or enriched conditions should be compared with the test performances of control subjects reared under "normal" conditions. "Normal rearing" in this context usually refers to the common laboratory environment: a cage that restricts activity amount and type, and provides contrived social contacts, a prepared diet, and constant ambient conditions. These conditions are extremely simplified when compared with those of a natural environment. Thus we should keep in mind that the control conditions are often stimulus-poor and not at all comparable with the subject's natural surroundings.

In the chronological sequence that follows, we provide some general remarks and several examples of investigations of behavior development at each stage or period.

EMBRYOLOGY OF BEHAVIOR

When does behavior development begin? Some integral processes important to behavior development take place before hatching or birth. Investigators have documented a variety of pre- and postnatal influences on sensory and motor development, and more general effects on subsequent behavior. Developmental biologists are accumulating considerable information about chemical and physical processes involved in the epigenesis of the tissues and organs of the body, including the nervous system. **Epigenesis** can be defined broadly as the interaction of genetic inheritance and environment during the course of development of an organism. The integration of information from studies of embryology and of behavior development is leading to a new synthesis of what takes place during development, and to new frontiers in research.

NERVOUS SYSTEM DEVELOPMENT

During development, continual changes take place in the nervous system, including cell proliferation, migration of neurons, and differentiation of cell types (see Jacobson 1978 for a full review of this topic). We are concerned here specifically with how these embryological events are related to the ontogeny of discrete, observable behavioral events. According to Wolff (1981), the neural developmental process can be divided into two major categories of events; (1) the production and distribution of neurons, and (2) the establishment of appropriate synapses between neurons. In the critical developmental process, not only neuron proliferation is important, but also their selective death. Cell death is partially a consequence of another process characteristic of nervous system development: overproduction of neurons. This overproduction leads to competition for appropriate connections, and ultimately to either death or survival of neurons, depending on the interactions between the neurons and their targets.

Oppenheim and his colleagues explored the development of motor neurons that innervate the limbs of embryonic chicks (Oppenheim, Chu-Wang, and Maderut 1978). When chick embryos are immobilized with neuromuscular blocking agents for varying periods between days 4.5 and 9 of incubation, the chicks have more motor neurons in the spinal cord motor columns than untreated chicks (Pittman and Oppenheim 1979). Apparently the number of motor neurons undergoing cell death during this period is related to the embryo's muscular activity. Treatment started after day 12 of incubation did not influence the rate of cell death. Also, if the immobilization treatment was stopped on about day 10, the cell death rate in treated chicks was merely delayed; they had the same total number of cells as control chicks by days 16 through 18. Apparently, some functional interactions that determine cell death or survival occur at the developing neuromuscular junctions.

In another study, the developmental effects of removing the limb bud on one side were compared with the same processes in the intact limb bud, which served as a control. The nerve cell death rate was greatly enhanced on the side where the limb bud was ablated. However, the characteristics of the developing motor neurons that grow outward from the central nervous system were no different than those of normal motor neurons; the neurons exhibited normal cell morphology, axon growth, and synaptic enzyme systems. Thus it appears that the motor neurons of the chick spinal cord motor column have the capacity to initiate differentiation. Neurons deprived of their target, the limb bud, are no different from those not deprived.

In a study of chick development, Bekoff (1978) used electrodes to record embryo muscle movements. A video camera focused on the chick through a window in the shell was used to record movements. Simultaneously, recordings called electromyograms (EMGs) were made from designated muscles (figure 9–1). Combining these techniques is an excellent way to study the development of coordinated actions. Patterned output from the

---

FIGURE 9-1   Ontogeny of coordinated behavior
Schematic diagram of the experimental setup for simultaneously recording behavioral and EMG data from a spontaneously behaving chick embryo. A hole is made in the shell over the embryo and the egg is placed in a heated, humidified chamber. Extracellular suction electrodes are placed on individual, identified muscles; the recorded EMGs are amplified 1000 times, viewed on the oscilloscope screen, and stored on magnetic tape for later filming and analysis. A stimulator is used to produce a time mark on one channel of the EMG record at a frequency of 2.5 Hz. This same signal drives a counter in the field of view of the videotape camera. The videotape records of behavior can then be synchronized on a frame-by-frame basis with the simultaneously recorded EMG records.

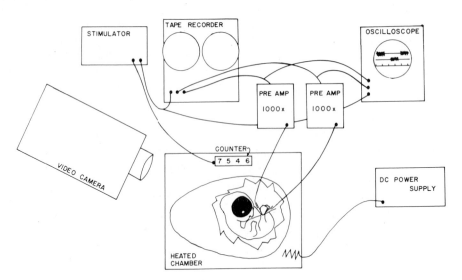

EMGs can be related to behavioral observations of wings, feet, and head. From these data, we also recognize that the neural circuitry underlying specific coordinated behavior is often developed and operative some time before a recognizable pattern emerges (Bekoff 1976; Bekoff, Stein, and Hamburger 1975). Bekoff hypothesizes that complex behavior in adult animals may arise from the coordination of simpler actions occurring during the course of development.

SENSORY AND MOTOR STIMULATION

Can sensory and motor responses be evoked and measured prenatally? The sensory and neural systems of an embryo at various developmental stages can be tested to see if they are mature enough to respond to appropriate stimuli. For example, we can measure an embryo's reflex responses to tactile stimulation. Hamburger (1963, 1973), and others studying various avian species, produced evidence that both general and specific embryonic movements can be induced by stimulating an embryo with a fine brush or blunt probe. The embryos of some bird species begin to respond to stimulation as early as the first week of development. Mammalian embryos, which are much more difficult to study because they are isolated in the uterus, respond with reflex-type movements to stimulation of different body zones (see summary in Barron 1941). Touching the snout of a sixteen-day-old rat fetus produces head movements; touching the side of the face of an eight-week-old human fetus produces general movements of the body and limbs.

In studies of the auditory mode, Gottlieb (1968) has shown that several days before hatching, Peking duck embryos (*Anas platyrhynchos*) respond selectively to sounds of its species' maternal call. Recordings of other species' maternal calls produce a quickening of the heart rate, which is a general activation response; but only the species-specific maternal call induces a heightened frequency of bill-clapping responses.

A slightly different, but related, set of experiments revealed that vocalizations made by prehatching birds may play an important role in synchronizing the hatching time of the birds in a clutch for some avian species (Vince 1964, 1966, 1969, 1973). Vince first demonstrated that developmentally advanced bobwhite quail (*Colinus virginianus*) embryos accelerated the development of less-advanced embryos in a clutch. Early experiments suggested, and later work confirmed, that a "clicking" vocalization made by the embryos is important for hatching stimulation and synchronization

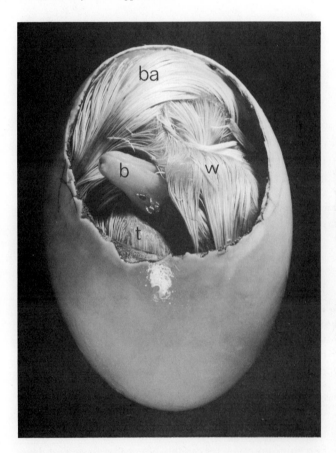

FIGURE 9-2 Bird embryo after removal of portions of eggshell
By using a delicate technique, we can observe development in avian embryos without disturbing the normal ontogenetic processes. (ba = back, b = bill, w = wing, t = thigh)
Source: Photo by R. W. Oppenheim.

in bobwhite quail. Vince has also provided evidence that in some instances, slower-developing embryos may retard those that are more advanced. Thus sensory-motor systems that operate prenatally in certain animal species may have important functional significance.

The motor development of prenatal organisms has been investigated descriptively and experimentally. In a study of duck and chick embryos, Oppenheim (1970, 1972) recorded general activity levels and body movement sequences from the embryos' first weeks of incubation to its hatching. His method (see Kuo 1967; Gottlieb 1968) was to remove a portion of the eggshell, exposing the developing embryo (figure 9–2), which remained viable and continued to develop normally.

FIGURE 9-3   Measurement of activity in duck embryos
The mean duration in seconds of activity (dotted line) and inactivity (solid line) varies for each day of incubation in duck embryos.

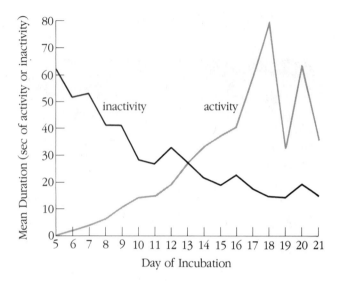

Figure 9–3 illustrates activity and inactivity of embryos at different stages of development. Oppenheim also analyzed the individual movements of the head and various parts of the body, and related the motor development sequences to pipping of the shell and hatching of the duckling or chick. His work demonstrated the functional relationships between embryonic motor movements and successful hatching.

Prenatal development of the motor system is not limited to general body, limb, or head movements. Duck and chick embryos are capable of vocalizing several days before hatching (Gottlieb and Vandenbergh 1968). Three types of calls—distress, contentment, and brooding—can be elicited by prodding the embryonic bird's oral region. All of these calls resemble vocalizations made by birds shortly after hatching and are utilized at that time for communication with the mother.

Other investigators have reported generalized prenatal stimulation affecting later behavior. Subjecting pregnant female rats (*Rattus norvegicus*) to stress can affect the activity of the young at postnatal ages of 30 to 140 days (Thompson 1957; Ader and Conklin 1963). Offspring of females that were stressed during pregnancy were less active than those of unstressed mothers.

Studies of human mothers and neonates (Ferreira 1965; Sontag and Wallace 1935) have revealed differences between infants born to mothers who experienced stress during pregnancy and infants born to mothers who did not experience stress. During the weeks immediately after birth, neonates of stressed mothers were more active, cried more, and gained weight slower than did neonates of unstressed mothers.

MATERIAL EXPERIENCE AFFECTS OFFSPRING BEHAVIOR

Knowing that an organism's sensory and motor systems are functionally capable of some responses before birth or hatching, we might then want to know, What are the effects of maternal experience on later behavior of progeny? Denenberg and Whimbey (1963) conducted an experiment in which young female rat pups were handled by the experimenters during their first twenty days after birth. The pups were removed from the dam by hand, placed on clean wood shavings in a small can for three minutes and then returned to the mother by hand. At maturity, a group of these handled females, and another group of nonhandled females, were bred with colony males. The litters from these dams were all weaned at age 21 days and housed with like-sexed littermates until they were 50 days of age. Starting at that age, the young rats of both sexes were tested for activity rates and defecation scores in a 45-in² open-field arena that consisted of a circular or square pen with lines marked off on the floor. Activity was measured by the number of lines crossed and/or the number of different squares entered by the test subject in a standard time period. Defecation scores were the number of fecal boluses dropped in the same time period. (Open-field tests are designed to measure emotionality; the more emotional animals cross fewer lines, enter fewer areas of the pen, and defecate more often. For a review of this method, see Archer 1973.)

When results for rats born to handled mothers were compared with results for rats born to nonhandled mothers, the former exhibited significantly lower activity levels and more defecation. Thus, treatments administered to female rats before they were bred had a strong influence on their pups' behavior. These effects may be mediated both by the prenatal mother-fetus relationship and by the interaction between mother and pup after birth. There could also be some effects attributable to interactions between littermates.

In a related investigation, Denenberg and Rosenberg (1967) tested whether these effects could be extended a second generation to the grandpups of the original handled females. The same general procedures were employed as in the previous experiment, except that the grandpups were tested in the open field at 21 days of age. Female grandpups of handled females were significantly less active than female pups descended from nonhandled mothers, but male grandpups were only slightly affected. Thus, the effects of handling female rats in infancy may have effects on behavior two generations later. It should be noted that in this second experiment, the mother rats were also provided with a modified housing experience involving objects present on the sawdust-covered floor of the cage for the period from 21 to 50 days of age; this alternative caging may have enhanced the carryover effects to a second generation. The observed effects, particularly on second-generation female rats, could result from a variety of mechanisms, including physiological influences on the fetus, alterations of the milk provided during lactation, or extrachromosomal inheritance.

EARLY POSTNATAL EVENTS

What do we know about events that occur immediately after hatching or birth? The young of some species are born relatively helpless, often with little or no fur or feathers, and are generally incapable of locomotion or ingestion of solid foods; these we call **altricial young.** The young of other species are born in a more advanced stage, are capable of locomotion and other behavior patterns, and are often able to consume at least some solid foods; these we call **precocial young.** A variety of events at or near the time of birth have important consequences for behavioral processes immediately and later in life. Among these are social attachments and feeding behavior. We now examine each of these in more detail using examples of precocial and altricial young.

IMPRINTING

Imprinting occurs when an animal learns to make a particular response *only* to one type of animal or object. There are two kinds of imprinting, filial imprinting and sexual imprinting.

FILIAL IMPRINTING.   **Filial imprinting** is the process by which animals develop a social attachment for a particular object. The phenomenon of imprinting has been observed for centuries. One of the earliest scientific descriptions of imprinting is that of Spalding (1873), and the first thorough investigations of the process were carried out by Lorenz (1935). Originally this effect was thought to be both instantaneous and irreversible, but we now know that although the process occurs in many animals, it is not necessarily instantaneous, and it can be reversed or otherwise altered (Hinde 1970; Salzen and Meyer 1967; Salzen and Sluckin 1959).

The basic experimental method we use to study imprinting in precocial birds is to expose them to an imprinting stimulus soon after hatching, and then to test them later (usually several days later) using the same stimulus and a different stimulus. Measures of following and approaching behavior are used to assess the success of imprinting. However, to test the possibility that birds may have been predisposed to prefer the imprinting stimulus, a second test group is necessary: one of birds that are initially exposed to the second stimulus. If the majority of the responses are again directed at the initial stimulus, then imprinting is judged to have been successful, and both objects are usable imprinting stimuli. For imprinting stimuli, we can use live or stuffed birds of the same species, flashing lights, colored spheres or cubes, and stationary colored lights—many animate and inanimate objects work. The same objects are used to test the effectiveness of imprinting. A common apparatus used for imprinting and testing birds is a circular arena with a rotating boom from which the stimulus object can be suspended (figure 9–4).

For clarity, we will define the time the animal can develop an attachment as the **sensitive period,** and the part of the sensitive period that the attachment response performance and reinforcement is greatest, the **critical period.** In mallard ducks (*Anas platyrhynchos*) (Hess 1959), the sensitive period occurs from about five to twenty-four hours after hatching; the critical period when imprinting is most successful encompasses the interval from thirteen to sixteen hours after hatching (figure 9–5).

FIGURE 9-4   Circular arena apparatus used for filial imprinting and testing the following response of young precocial birds

Note the stuffed bird on the boom, the young duckling, and the lines on the arena floor that can be used to measure the movement of the test bird as well as the proximity of the duckling to the moving object. Here a mallard duckling is following a mallard maternal call in preference to following a moving visual replica of a mallard hen.

Source: Photo from the laboratory of Gilbert Gottlieb.

FIGURE 9-5   Period of sensitivity for imprinting in mallard ducklings

At each four-hour age interval the percentage of ducklings that show successful imprinting is plotted. The peak percentage occurs at thirteen to sixteen hours after hatching; we call this the critical period.

Source: From E. H. Hess, "Two Conditions Limiting Control Age for Imprinting," in *Journal of Comparative & Physiological Psychology,* 52:516, 1959. Copyright © 1959 by the American Psychological Association.

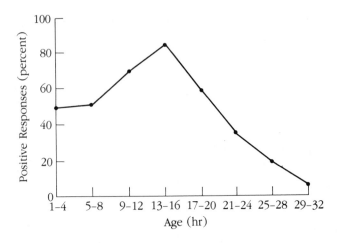

FIGURE 9-6   Locomotion/fear dichotomy
As a young bird develops a fear of novel objects (left axis), it also improves in locomotive ability (right axis). In the middle region of the graph, where the fear response is still low but locomotive ability is relatively high, imprinting is most likely to occur and to have the highest degree of success.

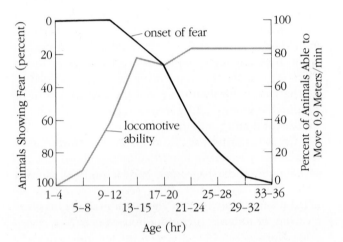

One characteristic of imprinting, most often referred to as the **locomotion-fear dichotomy,** may be closely related to the timing of the critical period (figure 9-6). Immediately after hatching, a young duck or chick shows little fear of strange objects, as measured by its avoidance and vocalizations, but it is also not yet very mobile. The bird gradually develops locomotive abilities, but at the same time, its fear of strange objects increases. Thus, there is a period when locomotion is high and fear response is low; and the interval between thirteen and sixteen hours after hatching is the time that imprinting in ducklings is maximally successful.

A final aspect of filial imprinting concerns the importance of different sensory modalities. In his original investigations of imprinting, Lorenz (1935) found that ducklings responded to both auditory and visual stimuli, but the strongest responses were to the mother's call. The work of Gottlieb (1971) and his colleagues with precocial birds has revealed that either visual or auditory stimuli used alone are effective in producing successful imprinting in some species. However, when both auditory and visual cues are provided, the degree of successful imprinting exceeds that achieved with either used alone. Stimuli of several sensory modes presented simultaneously may represent natural conditions more than do some single-mode experimental regimes.

What about imprinting phenomena in mammals? In goats (*Oreamnos* spp.) the offspring are imprinted on the adult female parent. This imprinting is based primarily upon olfactory and possibly visual cues (Klopfer, Adams, and Klopfer 1964). The sensitive period for formation of the attachment between the doe and kid may last for several hours after the kid's birth, but the critical time period for formation of the bond is during the initial hour after parturition. During the first postpartum hour, a foster kid can be substituted for the neonate and the doe accepts the alien kid as her own (Klopfer and Klopfer 1968). A similar imprinting phenomenon occurs in sheep (*Ovis aries*). This has been used by Price et al. (1984) and Martin et al. (1987) to study cross-fostering in sheep; an application of animal behavior that has consequences for sheep ranchers in many western states. In one experiment (Price et al. 1984), alien sheep were placed into stockinettes (a body sock made of synthetic material that fits over the head and limbs of the lamb) that had previously been on lambs belonging to a particular ewe. The alien lambs were then placed with the ewe; 84 percent of the alien lambs were adopted by the ewes. Alien lambs placed with ewes while retaining their own stockinettes served as controls; all were rejected by the foster ewes. Thus, the odor cues that impregnated the stockinettes served as a critical cue for this fostering process. In a follow-up experiment, Martin et al. (1987) demonstrated that the stockinette technique could be used to transfer a second lamb to ewes that already had single lambs.

What is the functional significance of filial imprinting? The imprinting or following response is important in some species, particularly in ducks, which may build nests located at some distance from water, the eventual home of the birds. Within several weeks of hatching, the mother leads the ducklings to the water. Imprinting with visual and/or auditory species-specific cues may enable young birds to follow the mother successfully through dense vegetation or other obstacles between the nest and the water. Gottlieb (1963,1971) has shown the importance of auditory communication in wood ducks (*Aix sponsa*). Shortly after hatching when the young are ready to leave the nest, which is in a tree hole several feet above the water or other suitable site, the mother descends to the water and calls to the ducklings. In response to the call, the young jump up to the nest opening, and drop down into the water to follow the mother.

Miller and Gottlieb (1978) have shown that incubating female mallard ducks emit species-specific vocalizations. These vocalizations are similar to those of the mother when she calls her young from the nest. In this instance, the imprinting process actually begins before hatching.

Both visual and auditory imprinting may play an important role in predator avoidance. The ability to follow the mother's lead quickly to safety in the seconds between the sighting of a predator and its attack may mean the difference between life and death. Alternatively, young birds may respond to the maternal alarm call by freezing and remaining motionless until danger has passed (Miller 1982; Blaich and Miller 1986; Miller and Blaich 1987). In addition, the recognition of conspecifics that may result from imprinting may be significant in the young birds' socialization process, and in conspecific cooperation in a social organization. Associating with conspecifics may be important for surviving, for locating food and shelter, and for migrating.

SEXUAL IMPRINTING.   A different type of imprinting, in which individuals learn selectively to direct their sexual behavior at some stimulus objects, but not at others, is termed **sexual imprinting.** Sexual imprinting may serve as a species-identifying and species-isolating mechanism. By appropriate exposure to an imprinting stimulus and later testing, the sexual preferences of birds have been shown to be imprinted to the stimulus they were exposed to. Sexual imprinting generally involves longer periods of exposure to the stimulus than filial imprinting. Both precocial birds like turkeys (*Meleagris gallapvao*) (Schein and Hale 1959) and ducks (Schultz 1965), and altricial species like zebra finches (*Taeniopygia guttata*) (Immelmann 1965, 1972) and doves (*Columba* spp.) (Craig 1914), have exhibited mate preferences based on early rearing experiences. Most of these studies have involved cross-fostering young birds to another species or rearing them with models and postpubertal testing for mate selection preference.

It is interesting to note that male and female sexual imprinting differs in two species: mallard ducks and zebra finches. Both species are sexually dimorphic: males have morphological (color, plumage) and behavioral (display) characteristics that trigger sexual responses; females lack these traits. When male mallards are crossfostered with other ducks, and zebra finches are crossfostered with other finches, the resulting males court females of the foster species and not of their own; however, mate selection by female mallards and zebra finches is not affected by the rearing experience. In a review of this subject, Bateson (1978) presented a model

for testing sexual imprinting that comprises four parameters: (1) the age of the animal, (2) the length of exposure to the foster species, (3) the actual imprinting stimulus value of the foster species, and (4) the exposure to other species before and after exposure to the foster species.

We should note that most birds in nature have limited contact with members of other species during the first few days or even weeks after hatching. This ensures that young birds will imprint on members of their own species and that they will court only conspecifics when they engage in reproductive activities as adults. Thus we can see that imprinting as a species-isolating mechanism helps guarantee that the investment of reproductive energy and gametes is not wasted on unsuccessful mating activities.

Cross-fostering has also been used to study the effects of early exposure to particular stimuli on rodents' sexual preferences (Lagerspetz and Heino 1970; Murphy 1980; Huck and Banks 1980; see also summary by D'Udine and Alleva 1983). These studies have involved a variety of rodents, including house mice (*Mus domesticus*), Norway rats (*Rattus norvegicus*), lemmings (*Lemmus trimucronatus* and *Dicrostonyx groelandicus*), and hamsters (*Mesocricetus auratus* and *Mesocricetus brandti*). In all of these studies, we find that beginning shortly after birth, a period of residence with a dam from another species resulted in the adult foster pup's reduced preference for conspecifics of the opposite sex and enhanced preference for the foster species. Clearly, some type of imprinting process or similar early experience phenomenon is occurring that strongly affects sexual preferences and potential reproduction. These early established preferences may serve as species-isolating mechanisms and/or may play critical roles in mate-seeking and mate-selection behavior.

DEVELOPMENT OF FEEDING AND FOOD PREFERENCES

EFFECTS OF LACTATING MOTHER'S DIET AND CONSPECIFIC ODORS ON OFFSPRING FOOD PREFERENCES.   Is it possible that the food ingested by a lactating female influences the food preferences of her progeny? To test this question, Galef and Henderson (1972) experimented with thirty-two pups born to four female rats (*Rattus norvegicus*). After the birth of the pups, the investigators removed the lactating mothers from their home cages and placed them in separate compartments during three one-hour feeding periods each day. They gave two of the females food prepared by Purina, and

FIGURE 9-7   Feeding preferences in rat pups

The mean amount of Purina chow eaten as a percentage of total food intake by pups reared by mothers fed either a Purina or a Turtox diet varied. The numbers above each bar indicate the sample size for that day.

Source: Data from B. G. Galef and P. W. Henderson, "Mother's Milk: A Determinant of the Feeding Preferences of Weaning Rat Pups," in *Journal of Comparative and Physiological Psychology*, 78:213–219, 1972. Copyright © 1972 by the American Psychological Association, Arlington, VA.

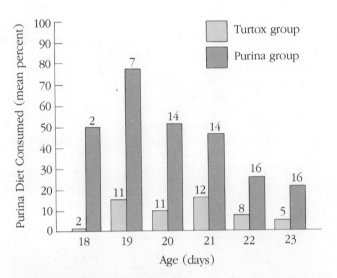

two food made by Turtox. When untested rats are given a choice test, they normally prefer the Turtox diet; the foods differ in taste, texture, and color. Galef and Henderson measured the food preferences of the pups for seven days, beginning at seventeen days of age. Figure 9-7 shows that the young rats reared by Purina-fed lactating mothers ate proportionately more Purina chow than Turtox; and pups reared by lactating mothers fed Turtox preferred Turtox chow.

A rat pup may acquire information about the food its mother is consuming in at least three ways. Young rats may consume feces dropped by the mother; they may ingest or smell particles of food adhering to the mother's fur or oral region; or the flavor or odor of the mother's food may be transmitted in her milk. Further experiments conducted by Galef and Henderson confirmed that information about diet is transmitted through the mother's milk. The results of these experiments have some interesting implications for our understanding of the development of food habits in young rodents as they leave the natal homesite to fend for themselves. Prior to this work, it was assumed that young rats develop their own preferences of solid foods without any assistance from adult rats (Barnett 1956).

However, the results outlined here and in other work by Galef and his colleagues confirm that young rodents may obtain some of this information from their mothers.

At the age of weaning, the feeding site selection of young rats is influenced by interactions with adults, and in particular, by olfactory cues associated with conspecifics (Galef 1977, 1981; Galef and Clark 1971; Galef and Heiber 1976). When presented with a choice, rat pups select a feeding site associated with either conspecifics or their excreta in preference to a clean site. If pups are reared without contact with conspecifics, they select a feeding site regardless of whether it is clean or has conspecific cues. When pups are reared away from conspecifics, five days of conspecific exposure before testing, is sufficient for them to prefer feeding sites associated with conspecific stimuli.

TURTLE FOOD PREFERENCES.   Can early experience with foods affect the dietary habits in turtles? Burghardt (1967) investigated the feeding preferences of snapping turtles (*Chelydra serpentia*). He fed separate groups of newly hatched turtles either horsemeat or worms for one day. One week later, each turtle was given its choice between the two types of meat. The results revealed that (table 9–1) the turtles exhibited a clear preference for the food they had eaten immediately after hatching. We conclude from these results that the initial feeding experience may be a critical factor in later diet preferences in snapping turtles. Further testing would be needed to confirm and extend these conclusions to the retention into adulthood of the initial feeding experience effect, but some type of food imprinting may be occurring. The turtle's food preference may have become fixed, even though the turtle will still consume other foods.

FEEDING BEHAVIOR OF GULL CHICKS.   The feeding behavior of laughing gulls (*Larus atricilla*) has been studied in both field and laboratory settings (Tinbergen and Perdeck, 1950; Hailman 1967,1969). In the field, Tinbergen and Perdeck made observations and found that hungry chicks use their bills in a pecking and stroking pattern directed at the bill of a parent, which induces the adult bird to regurgitate food for the chick (figure 9–8).

Hailman examined how this food-begging behavior in gull chicks developed. His results indicate that the behavior resulted from an interaction of genes and

TABLE 9-1   Food preferences of snapping turtles

		First meal		Second meal		
Group	N	Food	Pieces eaten	Food	Pieces eaten	Number choosing first-fed food
1	12	Horsemeat	26	Worm meat	19	12
2	13	Worm meat	28	Horsemeat	34	8
Totals	25		54		53	20

Source: Data from Burghardt (1967).

FIGURE 9-8   Normal feeding behavior of laughing gull chick

Chicks engage in two separate types of pecking. Here the chick aims an accurately coordinated peck at the beak of a parent, which prompts the parent to regurgitate food. In other instances the chick may peck at food (e.g., fish) the parent has regurgitated.

FIGURE 9-9   Accuracy of chick pecking

Gull chicks' improving accuracy in pecking ability is illustrated here. By the time the chicks have been in the nest two to four days, they exhibit 75 to 90 percent accuracy in directing their pecks at the bill portion of a painted card representation of the parent's head.

Source: Data from J. P. Hailman, "How an Instinct is Learned", in *Scientific American* 221(1969):100. Copyright © 1969 Scientific American, Inc., New York, NY.

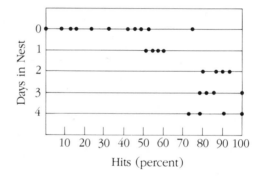

environment. The genome of the young gull provides the information necessary for the correct maturation of the bird's sensory and motor systems. As it matures, the bird learns to peck at the parent's red bill, which contrasts with its black head. By using cardboard models of a gull's head to test pecking behavior, Hailman found that it took the chicks several days to attain 75 to 90 percent accuracy in hits (figure 9–9). Thus, we can conclude that full development of the begging behavior requires genetically programmed development of sensory-neural systems to receive and interpret the stimuli, motor capacities to respond to the stimuli, and experience to learn the pecking behavior.

JUVENILE EVENTS: BIRDS AND MAMMALS

The period that lasts roughly from fledging (birds) or weaning (mammals) animal's full independence on the way to maturity can loosely be termed the **juvenile stage.** Development is a continuous process that is usually marked by various identifiable events, such as birth or fledging. For convenience, we as investigators often divide the continuum into stages or periods. Because animals of different species—and indeed, animals of the same species—often proceed through development at varying rates, the timing and length of these periods will vary. Several pivotal events usually occur during the juvenile period, as defined here, including natal site dispersal, puberty, appropriate communication signal learning, various experiences that influence later behavior patterns, and for birds, the first migration toward the equator. Play behavior, which for some species constitutes an important activity during the juvenile period, will be discussed in more detail later in the chapter.

DEPRIVATION EXPERIMENTS

One method we can use to assess the possible significance of a particular experience or stimulation during development involves depriving the organism of one or more of these experiences or stimuli and measuring the effects on subsequent behavior. Some of these treatments may begin before the juvenile stage; others start during this stage. The effects may be noticeable and measureable immediately or later in life. Deprivation may be social, sensory, motor, or some combination of these.

SOCIAL DEPRIVATION IN DOGS. Fuller (1967) investigated the effects of experiential deprivation on the social behavior of dogs (*Canis familiaris*). Beagle and terrier puppies were placed in isolation for varying lengths of time. Periodic contact with a human handler or another dog was included at prescribed intervals in some of the isolation regimes. The dogs were given an open-field test and several learning tasks, and were observed during the periods they were allowed to make contact with a towel, a ball, or another puppy. The results were similar to the findings for monkeys reported by Harlow and his coworkers (see next column): The more severe the deprivation, the more pronounced the dogs' behavioral deficits and abnormalities. The dogs reared under the most restricted regimes sometimes failed to leave the starting area of an open-field test, were generally less active than dogs from other treatments, and lost most frequently in competition with other puppies.

Before Fuller's work, two explanations of the observed behavioral differences between deprivation-reared and normal animals had been proposed. Some investigators ascribed the differences to the deterioration of previously organized neural patterns; this theory is sometimes referred to as the **disuse hypothesis.** Other investigators postulated that the behavioral deficits resulted from a lack of critical information input necessary for proper development of neural patterns and normal behavioral responses; this concept is often called the **loss-of-information hypothesis.**

Fuller proposed a third explanation, which he termed the **stress-of-emergence hypothesis.** Animals that have been in isolation, and that are suddenly thrust into a test situation loaded with many novel stimuli, may suffer from stimulus overload—that is, they are faced with a multitude of competing emotional responses. To test this hypothesis, Fuller suggested three treatments: (1) provide the isolate-reared dogs several brief pretest exposures to the testing arena to habituate them to the situation; (2) give the dogs a tranquilizer like chlorpromazine before testing to reduce the effects of stimulus bombardment; or (3) let the dogs have several brief periods of handling by the experimenter before testing. These methods have been tried (Fuller 1967), and with varying degrees of success, have produced some reductions in the stress of emergence. Fuller's hypothesis appears to be valid in some experimental deprivation studies, but we must consider the two earlier hypotheses as primary explanations for the behavioral deficits that result from isolation treatments.

SOCIAL DEPRIVATION IN MONKEYS. From the 1950s onward, Harlow and his colleagues conducted a series of studies on the effects of social deprivation on the behavior of young, adolescent, and adult rhesus monkeys (*Macaca mulatta*) (Harlow 1962; Harlow and Harlow 1962a, 1962b, 1969; Harlow and Zimmerman 1959; Arling and Harlow 1967; Mason 1960, 1961a, 1961b; Mitchell 1970). Similar studies have also been conducted by Hinde and his associate in Great Britain (Hinde and Spencer-Booth 1967, 1970, 1971; Hinde and Davies 1972; Hinde and McGinnis 1977; Hinde, Leighton-Shapiro, and McGinnis 1978). Throughout these investigations, young rhesus monkeys were reared under various treatment conditions for periods of time during the first two years of life. The conditions included:

1. Rearing in total isolation in chambers that remove the infant from all social contacts and most external stimulation
2. Isolate rearing, but with a cloth or wire surrogate mother (figure 9–10)
3. Peer-group rearing with other monkeys of similar age
4. Rearing with the mother only
5. Rearing with the mother, but with varying periods of separation at specified age intervals
6. Rearing in small social groups in the laboratory (often used as a control condition)

FIGURE 9-10   Young rhesus monkey with surrogate mother

When given a choice, most infant rhesus monkeys preferred a cloth surrogate mother to a wire surrogate mother, even when the nursing bottles were attached to the wire mother.

Source: Photo by Harlow, Primate Laboratory, University of Wisconsin.

FIGURE 9-11   Socially deprived rhesus monkey mother
Rearing young rhesus monkeys under varying conditions of social deprivation produces several types of deviant and abnormal behaviors. When mature, the socially deprived female may be a very poor mother, at least with her first infant.

Source: Photo by Harlow, Primate Laboratory, University of Wisconsin.

Both the age of the monkey at treatment initiation and the duration of treatment are critical parameters of social deprivation. This list of rearing treatments is not intended to be exhaustive; researchers have employed others as well as gradations or combinations of these.

As we might expect, the severity of the observed effects due to deprivation rearing varies with the degree of social isolation and treatment duration. The first five treatment conditions are listed in the order of approximate degree of deprivation. Behavioral deficits (e.g., failure to perform complete behavior patterns, performing abnormal behavior patterns, lack of responsiveness to conspecifics) are produced in varying intensities, depending upon the severity of the isolation. These effects include withdrawal, engagement in fewer social interactions, deficiency in understanding communication, and inability to perform normal sexual behavior. Among the better-known effects of social deprivation are rocking and swaying, self-clasping and other self-directed actions, huddling behavior (exhibited by monkeys reared in peer groups), and poor maternal behavior by surrogate-reared females, who are

abusive toward their infants (figure 9–11); hence the name "motherless mother." Results of another type of effect investigated—the reestablishment of the bond between the mother and infant after deprivation or periods of separation—also depend upon the severity of the separation treatment, but indicate there is a great deal of individual variation.

Studies performed by Harlow, Hinde, and their associates have taught us a great deal about the nature and development of bonds between mothers and their infants and between maternal members of conspecific groupings of primates. These studies show that several qualities of the rhesus mother, such as contact (even that provided by cloth surrogates), warmth, and some types of movement, are important for normal infant development and development of affection. Young rhesus monkeys spent more time clinging to cloth surrogate mothers than wire mothers, even when the source of milk was associated with the wire surrogates. The effects from experiments using brief separations can last up to two years. These investigations have provided information helpful for bettering primate rearing conditions in zoos and laboratories, and insight regarding the importance of parent-offspring contact during the human rearing process.

We should note two additional points in connection with these studies of social deprivation. First, some of the behavior that Harlow and others have recorded as "abnormal" occur occasionally in rhesus monkeys reared under "normal" conditions (Erwin, Mitchell, and Maple 1973); self-directed aggression has been observed on a number of occasions in nonisolate-reared subjects. Another study reported that in free-ranging rhesus monkeys living on an island off the coast of Puerto Rico, only 50 percent of the young born to primiparous females (those giving birth for the first time) survive to the age of twelve months (Drickamer 1974b). Harlow and his coworkers have also noted that "motherless mothers" exhibit much better maternal behavior with their second infant. Part of the deficiency in maternal care recorded in "motherless mothers" may result from their ignorance of how to be good mothers, and not strictly to the emotional abnormalities associated with isolate rearing.

Secondly, and quite significantly, Harlow and his associates (Harlow and Suomi 1971; Suomi 1973; Suomi, Harlow, and Novak 1974) have succeeded in socially rehabilitating isolate-reared monkeys. They accomplished this by exposing six-month-old social isolates to three-month-old normal monkeys, called "therapy monkeys," for two hours per day, three days per week, for one month. The effects produced by some types of early social isolation are not irreversible, as was once thought, though even rehabilitated monkeys continue to exhibit some behavior deficits.

BIRD SONG

We have already explored control of bird song in chapter 6, and some of its functional significance will be explored in chapter 11. What about the development of song in birds? In the past several decades, an increasing amount of animal behavior investigators' attention has been given to bird song pattern—their control, development, and evolutionary significance. There is a diversity of developmental strategies that led to the variety of songs and calls we are all familiar with (Marler and Mundinger 1971; Kroodsma 1978,1981; Irwin 1988; Slater 1989). Comprehensive analyses of many bird species show that there are at least two major strategies for song development: (1) imitation of the songs of others, particularly of adult conspecifics; and (2) invention or improvisation. Underlying both strategies are the questions: What kind of templates exist to provide some genetic basis for the song learning process, and to help shape or guide it? and Does a sensitive

phase for development of the song repertoire exist? Let's consider several examples that should help explain song development.

The song learning of the marsh wren (*Cistothorus palustris*) has been studied extensively. In nature, male marsh wrens sing over one hundred types of songs; neighboring males generally sing identical song types; they often interact by countersinging with one another using the same song type. To test their song learning development, males were reared in special housing conditions where the songs they heard could be completely controlled. Males were played specially prepared tutor tapes containing nine different songs each. Males were exposed to one tape for age 15 to 65 days, a second tape for age 65 to 115 days, and a third tape the following spring. By imitating them, the males learned the nine songs on the tape they were exposed to before 65 days of age, but did not learn the songs on the subsequent two tapes. The males never sang any invented songs. Further investigation revealed a more refined estimate of the peak sensitive period, which falls between about age 35 and 55 days (Kroodsma 1978). In addition, males learned song types played for either three days or nine days. Some males improvised some songs in these additional tests, presumably from elements of the songs on the tutor tapes.

The learning situation used for these marsh wrens involved only tutor tapes played over loudspeakers. In another test, young wrens were exposed to a number of song types on tutor tapes and then were given a period of social interaction with adult males with varied song repertoires. The period of social interaction occured in the fall of the first year for some birds, and not until the following spring for others. Data on song repertoires for these birds indicated that song learning can occur early from the tapes, or at either of the periods of exposure to adult males. Hence, song learning is somewhat flexible, and social interaction may be an important feature of the process. These findings also make it clear that we should be careful when using only artificial stimuli like loudspeakers (see Kroodsma and Pickert 1984). Social interaction has also been shown to be critical for language acquisition in human children (Jerison 1973; Freedle and Lewis 1977).

A longitudinal study of song development has been done on male swamp sparrows (*Melospiza georgiana*) that were hatched in the wild and brought into the laboratory. Beginning at age 16 to 26 days, the birds were given song training with species-specific songs twice daily for forty days (Marler and Peters 1977; Peters, Searcy, and Marler 1980). Weekly recordings of each bird's songs began shortly after training when they were

FIGURE 9-12   Song development in a swamp sparrow
Samples of each of the seven stages of developing song in a male swamp sparrow. The stage is indicated at the left and major stages above. Imitations of the six training syllables portrayed at the top of the figure are identified by number. Training syllable 6 is a song sparrow syllable; the remainder are from swamp sparrows. Syllables were presented in one-part songs, except for syllables 5 and 6, which appeared together in the same "hybrid" song. In nature, swamp sparrows typically have a repertoire of three to four song types, and, although there is some syllable-sharing in local populations, most individual repertoires are unique. The experimental male shown here had a final repertoire of two song types, shown in stages I and II. This crystallized song is virtually identical to the song of the local wild male that served as the original source of the training syllable.

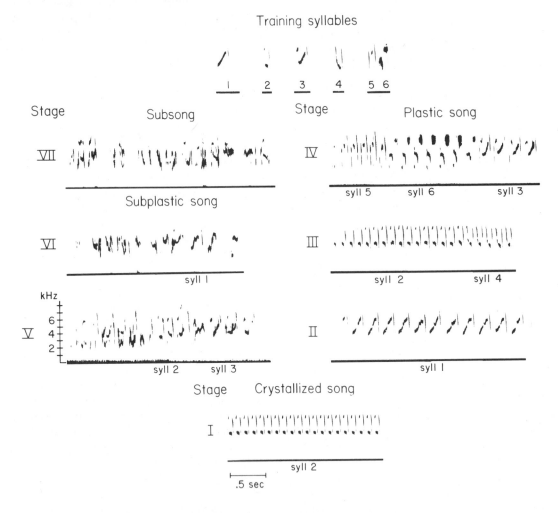

just over three months old, and continued until they were older than one year. When these recordings were analyzed, a seven-stage sequence of song development was discernible (figure 9-12). The syllables used in training are shown at the top of the figure. Young birds began by singing what Marler and Peters call subsong (stage VII) at an average age of 272 days. They progressed through subplastic song (stages V and VI) and began to sing the plastic song (stages II to IV) at an average age of 299 days.

Crystallized song (stage I) began at an average age of 334 days. During the course of this developmental sequence, the duration of the song decreased. Syllabic structures began to emerge during subplastic song. Syllable analysis revealed that about 30 percent were imitations of the training songs and 70 percent were inventions or improvisations. By the time the crystalized song emerged, the number of syllables sung was only 23 percent of the potential repertoire; both imitated and improvised syllables were included in the crystallized song of most birds (Marler and Peters 1981, 1982). During development it was not unusual to record songs characteristic of more than one stage on the same day. However, once the crystallized song singing began, the birds rarely reverted to an earlier stage. Marler and Peters (1982) hypothesize that the pattern of song development noted for the swamp sparrow may be characteristic for many songbird species.

What internal events can be related to the song development process? The neural bases for song development have been explored in a series of studies using white-crowned sparrows (*Zonotrichia leucophrys*). Birds of this species learn their songs from hearing adult birds, and have a sensitive period from about age 50 to 100 days. Birds that are isolated during this period develop abnormal song (Marler 1970). Further, white-crowned sparrows that live in different regions exhibit dialects in the basic song pattern (Marler and Timura 1962). A series of neurons that are critical for song production have been located in a brain region called the pars caudale in the hyperstriatum ventrale (termed the HVc) (Margoliash 1983, 1986, 1987). These neurons respond selectively when the bird's own song, when songs from within the same dialect, or when sequences of synthetic song with specific properties are played to the bird. This responsiveness may be important during song development—modification of these neurons may occur during the sensitive period. Song recognition may be important for white-crowned sparrows' territory definition and mating partner recognition and could thus contribute to greater reproductive success.

The relationship of ontogeny and phylogeny to bird song has also been examined recently (Irwin 1988). Using a group of sparrows as a test group, Irwin notes that the ontogeny of song parallels phylogeny: song development proceeds from general stages to more specialized stages. At the earliest ontogenetic stage, song is continuous, the repertoire size is undefined, syllable structure is minimal, and the number of syllable types per song is not defined. By the middle stages of ontogeny, the song is still continuous and the song repertoire remains undefined, but the syllable structure is now stereotyped and each song contains many syllable types. Still later in development, the song has become discrete and the repertoire contains many songs. In the adult stage, there are species variations. These variations differentiate this stage from the previous stage where patterns were generally common for all species in a group. For example, within the three species of sparrows studied, the song sparrow (*Melospiza melodia*), the swamp sparrow (*Melospiza georgiana*) and white-crowned sparrow (*Zonotrichia leucophrys*), birds as adults

all have discrete songs and stereotyped syllable structure. However, the song and swamp sparrows have many songs in their repertoires; the white-crowned sparrow has only one. The song and white-crowned sparrows have many syllable types in each song, but the swamp sparrow has only one. Thus, general patterns in song development follow phylogenetic lines that lead to differentiation in adult song types.

In a recent review of bird song learning, Slater (1989) has made two key points that provide a good summary of this topic. The first is that bird song has different functions in different species. The second is that bird song probably has more than a single function in many species. Taken together, these two summary statements provide a fine backdrop for understanding what has been discovered to date about the manner in which birds learn their songs and a stimulus for further exploration of the topic.

PUBERTY IN FEMALE HOUSE MICE

Puberty is a critical event in the lives of most organisms. For many, sexual maturation marks the onset of reproductive behavior and the production of progeny. Puberty is also important in the population biology of many species, as we note further in chapter 18 (see also Drickamer 1986; Vandenbergh and Coppola 1986). House mice's (*Mus domesticus*) deme structure generally involves one to several adult males, three to seven females, and their young offspring. Some juvenile females may disperse from the natal site, whereas others may remain within the deme, but virtually all juvenile males disperse (Bailey 1966; Delong 1967; Crowcroft and Rowe 1957; Crowcroft 1973). The timing of puberty in female mice can be affected by various social and environmental factors within this social context. The age of puberty is measured by the occurrence of first vaginal estrus; the heightened levels of estrogen associated with ovulation result in cornification of the cells that line the vagina. This phenomenon can be detected by microscopic examination of a vaginal smear.

The presence of a mature male mouse or daily exposure to urine from mature males accelerates the onset of puberty in young female mice (table 9–2), as does daily exposure to urine from lactating females and urine from individually-caged females in estrus. Daily exposure to group-caged females or their urine, regardless of the age of the grouped females, results in delays in the onset of puberty. Female mice exposed daily to clean bedding or to urine from singly caged diestrous females reach puberty at ages that are intermediate between the

TABLE 9-2   Mean ages and ranges for sexual maturation of female house mice (*Mus domesticus*) given various treatments with urinary chemosignals
Each young test female was treated starting at 21 days of age, and all test females were housed under the same conditons of photoperiod, food and water availability, and individual caging.

Treatment	Mean age at puberty (days)	Age range (days)
Control females—exposed to water treatment daily	35	29–42
Females caged with an adult male	27	24–31
Females exposed daily to urine from adult males	31	27–36
Females exposed daily to urine from grouped females	41	36–45
Females exposed daily to urine from estrous females	30	26–36
Females exposed daily to urine from diestrous females	36	30–41
Females exposed daily to urine from lactating females	30	27–35
Females exposed daily to urine from pregnant females	31	26–36

acceleration and delay effects produced by exposure to urine from the other sources (table 9-2). These effects have been demonstrated in both laboratory and wild stocks of house mice (Vandenbergh 1967,1969; Drickamer 1974a, 1979, 1982a, 1983; Vandenbergh, Drickamer, and Colby 1972; Colby and Vandenbergh 1974; Drickamer and Hoover 1979; Drickamer and Murphy 1978), and several of the effects have been replicated in stocks of wild *Mus domesticus* maintained in the cloverleaf islands of superhighways (Massey and Vandenbergh 1980, 1981; Coppola and Vandenbergh 1987).

When mice are exposed simultaneously to urine from two or three sources, the outcome depends on which sources were used and the relative proportions of urine from each source (Drickamer, 1988). If treatment involves equal proportions of urine from different sources and any exposure to urine from grouped females, puberty is delayed in young test females—regardless of what other urine sources are used. When the proportion of grouped female urine mixed with acceleratory chemosignal increases to a ratio of about 10:1 relative to urine that delays puberty, then the acceleration effect overides the delay. If all donor sources involve urine that accelerate puberty, then the combination treatments result in earlier maturation, but not any earlier than using one of the sources alone (Drickamer 1982b). Diet and daylength have also been shown to affect the timing of puberty in female house mice (Vandenbergh, Drickamer, and Colby 1972; Drickamer 1975).

Hence, a variety of contextual cues from the developing mouse's environment influence its internal physiological events. These effects, in turn, have important consequences for the onset of reproductive behavior in the mouse. The chemosignal effects will also influence its **generation time,** the length of the interval between birth and onset of reproduction. Generation time is a key element in determining the rate of growth or decline in the numbers of young mice entering the population over a given length of time.

JUVENILE EVENTS: INSECTS AND FISH

Much of the research on the ontogeny of behavior in animals has concentrated on birds and mammals as exemplified by the material we have just covered. What about other types of animals? A number of excellent studies of behavior development have been conducted using insects, fish, and other animals. Let's consider a few examples.

INSECTS

DROSOPHILA. Fruit flies (genus *Drosophila*) have been used in a wide variety of investigations of the development of behavior; studies utilizing field and laboratory conditions have been done both on larvae and on adults. The primary activity of insect larvae is feeding. As a larva moves across the food surface, it probes with its mouthparts and ingests food with each cycle of extension and retraction of its body. The larvae of *D. melanogaster* go through three molts, or **instars,** and then pupate before emerging as adult flies. The rate of feeding reaches a peak early in the third instar and then declines during that instar (Burnet and Connolly 1974). Before pupation, the larva may migrate away from the food source to a pupation site, or it may remain at the food source during pupation (de Souza, da Cunha, and

FIGURE 9-13   Behavior of *Drosophila*
Frequencies of eight behaviors in *Drosophila grimshawi* males
at ages of 1, 8, 15, 22, and 29 days, post-eclosion

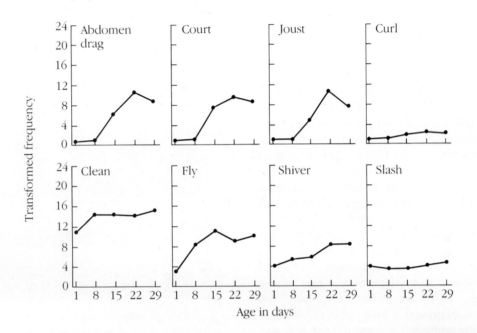

Age in days

dos Santos 1970); the pattern varies for different species and sometimes even within a species. Some species may burrow into the soil to pupate, probably to decrease the risk of predation.

The adults of many *Drosophila* species, particularly females, have been studied more extensively than have larvae, because adults perform a considerably greater number of behavior patterns. Two major behavior patterns related to reproduction, sexual receptivity and oviposition, undergo developmental changes in females. Manning (1966, 1967) studied sexual receptivity in *D. melanogaster*. As the corpora allata, which release juvenile hormone (JH), and ovaries grow larger, the female becomes receptive. This generally occurs about forty-eight hours after the eclosion of the adult fly from the pupa. Immature females reject courting males that attempt copulation. One hypothesis that we can suggest relates the physiological and behavioral events: JH may affect the brain directly or indirectly, lowering the threshold for sexual receptivity (Schneiderman 1972). If actively functioning corpora allata are implanted in females at the pupal stage, the emergent flies exhibit earlier sexual receptivity and have larger ovaries than nonimplanted controls.

After mating, females become unreceptive again and soon oviposition behavior increases. The turning off of receptivity appears to be due to at least three factors: the presence of sperm in the females' receptacles, the act of copulation itself, and a secretion from the paragonial gland of males (Burnet et al. 1973; Manning 1967). The increase in rates of oviposition is probably due to the increased size of the ovaries, which provides information to the nervous system via stretch receptors, and to the presence of male paragonial gland secretion (Grossfield and Sakri 1972; Merle 1969).

Ringo (1978) discusses the development and maturation processes in females and males of a group of *Drosophila* species inhabiting the Hawaiian Islands. Development takes a longer time in these species, continuing for days or weeks. For males of *D. grimshawi* a series of eight behavior patterns were observed and recorded at four ages during a one-month period following eclosion (figure 9-13). In general, the diversity of behavior observed increased with age, and the relative frequency of each behavior increased with age. Males of this species form leks, in which groups of males display communally and mate with females attracted to the lek. For days 15 and 22 there were pronounced increases in sexual and agonistic behaviors—courting, jousting, and abdomen dragging (see Spieth 1966, and Ringo 1976, for detailed descriptions of the behavior). The increases in behavior correspond to the time when the males are most likely to be competing for copulations.

BEES.   Many bee species live in large colonies in which there are **castes**—physiologically, behaviorally, and often morphologically different forms occurring together (Michener 1974; Wilson 1971). For example, in the honey bee (*Apis mellifera*) and other related species of social bees, there are at least three castes: queens, drones, and workers. Queens are larger, they mate, lay eggs, eat proteinaceous food, and often do not forage or defend the colony. In contrast, workers generally are smaller, do not mate or lay eggs, and actively engage in foraging, defense of the colony and nest-building. The eggs of the queen(s) hatch into larvae, and after these pupate, they emerge as adults.

One key question concerning bee development is: How and when does the determination of caste occur? Several factors appear to be important in this process (Wille and Orozco 1970). Which factors are important varies between major groups of bees, but in general they include: size of the brood cell, amount of food mass provided to the developing larva, quality of the food supplied to the larva, and possibly some chemical cues transmitted with the food. For most species studied, these effects begin immediately upon hatching or very early in larval life (Ribbands 1953; Jung-Hoffman 1966; Darchen and Delage 1970; Weaver 1966). Conditions both within the hive (e.g., loss of a queen) and outside the hive (e.g., changing seasons) can influence the numbers of workers and queens produced.

Most worker bees of these social species undergo changes in behavior as they develop which correspond to changes in their functional roles within the colony or hive (Free 1965). The activity changes for a worker honey bee are shown in the histograms in figure 9-14. Resting and patrolling occurred throughout the twenty-four days surveyed. Many other activities occur in a sequential pattern, beginning with high levels of cell clearing in the early days after emergence. This is followed by activities related to the comb and tending the brood. For the last portion of their lifespan, the workers become foragers; the average worker bee living in a temperate zone climate survives to the age of six weeks. For many of the activities, there is some overlap: Bees shift among behavior patterns of all types on a given day, up to the start of foraging activity. In the last phase, they concentrate primarily on foraging for food for the hive.

FIGURE 9-14   Time occupied by various activities during the first 24 days of life of a single marked worker honeybee living in a colony
The columns of figures at the left and right represent percentages of its time during which the bee engaged in each activity, out of the two to ten hours of observation daily.

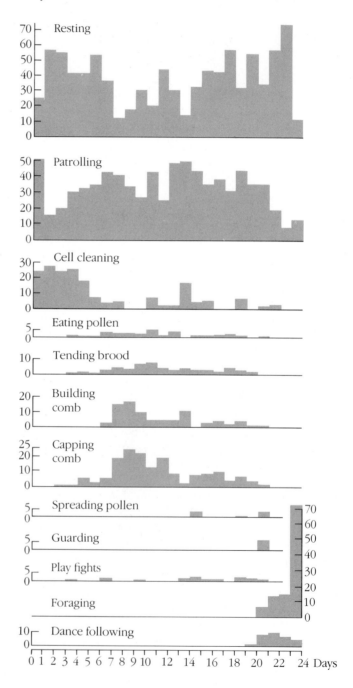

Several physiological changes are correlated with the shifting patterns of functional roles that we have just noted. Young bees have enlarged hypopharyngeal glands. These produce a major component of the bee milk, part of the larvae's diet. Levels of secretion for invertase, an enzyme involved in the conversion of nectar to honey, are highest during the middle portion of the lifespan (Simpson, Riedel, and Wilding 1968). Wax glands are small in the youngest workers, but there is a gradual increase to maximum size by about days 16 to 18, and then the wax glands decline rapidly (Snodgrass 1956). These changes correspond with the higher levels of comb-related activities during this particular age range. Many other traits follow similar patterns of correlation between developmental events throughout the life of the bee and its changing functional activities in the hive.

Most social bees actively guard and defend their colonies. How do they recognize nestmates and discriminate them from other conspecifics? Work by Breed (1983) indicates that bees will use environmental odor sources, or if the environmental cues are controlled, they will use genetically based cues to discriminate nestmates from non-nestmates (see also review by Hölldobler and Michener 1980). Investigators have shown that the cues necessary for making these discriminations are acquired prior to emergence as adults.

FISH

While considerably fewer studies of the ontogeny of fish behavior have been conducted than on some other vertebrate groups, recently there has been some progress (Noakes 1978; Huntingford 1986). Perhaps the most extensive investigations were those of Ward and his colleagues (Wyman and Ward 1973; Cole and Ward 1970; Quartermus and Ward 1969). After observations of the orange chromide (*Etroplus maculatus*), a theoretical model for the ontogeny of numerous behavior patterns was proposed (figure 9–15). All subsequent behavior patterns of this species are developed from two initial movements: glancing and micronipping. Data from a number of other cichlid species, and from some salmonid species, seem to fit this model, though extensive additional testing is needed.

Other research on salmonids (*Salmo* spp. and *Oncorhynchus* spp.) reveal some general patterns of behavior development in these genera, though there is considerable variation at each stage (Abu-Gideiri 1966;

Jacobssen and Jarvik 1976; Dill 1977; see Huntingford 1986 for a summary of fish development). The eggs begin their development buried in the gravel of a stream several months after fertilization. The first signs of life are contractions of the heart muscle. Soon after, the body muscles begin contracting. Initially, these contractions do not occur in coordinated or regular patterns, but they develop into characteristic undulations that look like swimming movements. Shortly before hatching, the jaws and fins also show coordinated movements. Hatching results from swimming motions that are rigorous enough for larvae to break out of the egg. Soon after hatching, the yolk sacs are fully absorbed, and larvae rest in an upright position on the gravel bottom of the stream. At this early stage, fish are photonegative, and they tend to swim into the current. Making movements to flex the tail and push off the substrate, the fish succeed in leaving the gravel and entering the stream; at about this time they switch to being photopositive. Beginning before emergence from the egg and increasing in frequency after emergence, movements, such as darting toward and biting small objects in front of them begin to occur; the fish often direct some of these same biting actions toward conspecifics. Young fish of most salmonid species exhibit one other general trait during development: they form a school in the presence of a potential predator, and if attacked, they flee and then freeze, remaining motionless on the substrate, or occasionally near the surface of the stream. Salmonids in later stages of ontogeny show greater variability across the various species, though apparently some species temporarily maintain a form of territoriality before leaving the natal stream for larger bodies of fresh or salt water (Kennleyside and Yamamoto 1962).

PLAY BEHAVIOR

Many mammals, some birds, and possibly a few other animals exhibit play behavior during their course of development. What is play behavior and how does it function in the developmental process? Play behavior has been defined and characterized in many different ways. Fagen defines play as "an inexact term used to denote certain locomotor, manipulative and social behavior characteristic of young (and some adult) mammals and birds under certain conditions in certain

FIGURE 9-15   A model representing the ontogeny of behavior in *Etroplus maculatus*

The arrows marked "+" indicate facilitation; the solid arrows marked "−" indicate inhibition. The dotted arrows, for diagrammatic simplicity, represent environmental feedback to the organism. The environmental factors implied by the dotted arrows are those indicated feeding into the model from the right. The dotted arrows marked "+" are proposed routes of facilitation between behavioral units.

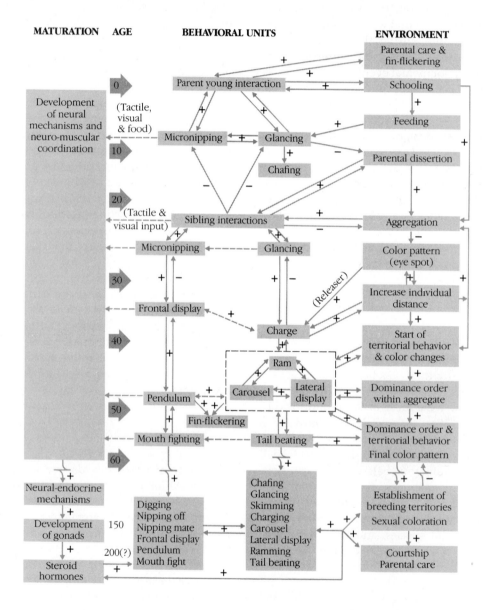

FIGURE 9-16   Canid play behavior
Source: Photo by Heather Parr-Fentress.

environments" (Fagen 1981, 21). We can characterize play in various animal groups according to the actions of the animals and the contexts in which the play behavior is observed. Most investigators recognize at least three types of play that have some overlap. The first type is *social play*, exemplified by wrestling, chasing, and tumbling activities of the young of many species (Bekoff 1978) (figure 9–16). A second type provides *exercise* for developing muscles, locomotor patterns, and other movements. The third type is often labeled *diversive exploration*. This type of play generally involves sensory inspection of an object followed by extensive repeated manipulation of the object.

By reviewing the available literature, we find that one or more of these types of play behavior have been recorded in mammals ranging from rodents and bats, to bears, cats, elephants, and whales. Play has also been recorded in a large number of avian species, including raptors, passerines, aquatic or oceanic birds, and parrots. To date, only a few anecdotal bits of evidence exist regarding play behavior in other vertebrate or invertebrate groups. However, we should gather additional observations before reaching any firm conclusions that limit play to birds and mammals.

FUNCTIONS.   Investigators have attributed a wide range of functions to play behavior (Loizos 1967; Fagen 1981; Bekoff 1974b; Bekoff and Byers 1981). In the most general sense, play functions as practice for adult activities. In the course of play, animals perform many actions that contain elements of behavior seen in later adult life. Perhaps aggressive behavior is the best example. As young cats or primates engage in various forms of social play, their mock attacks, chases, and mild, noninjurious bites are practice for "real life"—they will use these same patterns a few months or years later. A second function often ascribed to play behavior is to aid in the process of maturation—growth and development. As young foals or lambs cavort about alone or in small groups, they use their muscles and develop coordinated movements. A third function of play may be to gain information about the environment; this would be particularly true of diversive play. By exploring and manipulating objects found in their environment, young animals accumulate information that may prove useful later in life. Finally, a function of play can be to establish social relationships with peers and adults. Some of these may be pure affiliations, whereas others could be related to the establishment of dominant-subordinant relationships. The aggressive play observed in young juveniles of many primates and some canids gradually becomes more intense and adultlike. The patterns of dominance established in play encounters as juveniles can be retained into adult life. Let us consider several examples of the nature of play behavior and its functional significance during development, for better understanding.

CANIDS

One animal group in which play behavior has been studied extensively is the canids (Bekoff 1974a, 1974b, 1974c; Scott and Fuller 1965; Zimen 1972; Moehlman 1979). Bekoff (1974a) studied social play and play soliciting in coyotes (*Canis latrans*), wolves (*C. lupus*), and beagles (*C. familiaris*) (figure 9–16). His observations indicated some species differences and some age-specific trends for several aspects of play behavior; for example, play soliciting and agonistic behavior (figure 9–17). Bekoff notes that the onset of a play soliciting in the beagles was early, and that the beagles exhibited very little agonistic behavior during play—no fighting occurred at all, only mild threats. Wolves showed moderate levels of play soliciting, which increased during the last age interval recorded in the sample. Like beagles, wolves displayed low levels of agonistic behavior that consisted primarily of threats. In contrast, coyotes exhibited high levels of agonistic behavior throughout the observation period, and a correspondingly low rate of play-soliciting actions. Coyotes generally establish dominance relationships through fights at an early age (Fox and Clark 1971); this may account for the differences between these animals and the other two species.

FIGURE 9-17   Play-soliciting and agonistic behavior in canids
The median frequency of occurrence (percent) of action patterns observed during both play-soliciting and agonistic interactions in relation to the total number of actions performed during the stated time periods.

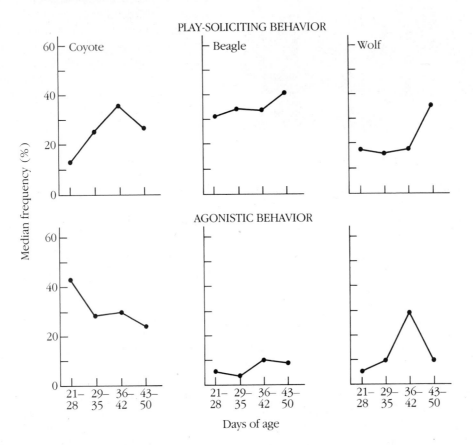

When all play behavior was summarized, the beagles were seven times more playful than were the coyotes, and three times more playful than were the wolves. It is interesting and worthwhile to speculate on the possible correlation between differences in observed play behavior and differences in social structures of the species tested. Wolves are generally social, group-living animals, though some individuals may lead a solitary existence (Mech 1970). Beagles are somewhat social animals, though we rarely observe domestic canids in seminatural or natural situations where their feral social structure can be fully recorded. Coyotes, on the other hand, are generally much more solitary in their social organization, with each animal roaming a large home range.

In a recent review, Price (1984) suggested that neoteny explains why some domesticated breeds of animals are more playful than their wild ancestors were. Man has selectively bred his domestic animals so they would retain their juvenile characteristics in adulthood. Thus, the beagles in the foregoing example may be more playful than the wolves because of the effects of man's selection.

KEAS

Keas (*Nestor notabilis*) are large parrots that inhabit parts of New Zealand. A variety of play behavior has been observed and reported for this species (Derscheid 1947; Jackson 1963; Keller 1975, 1976). The most extensive observations were those of Keller on captive keas in zoos. The young keas perform a variety of acrobatic maneuvers, that included somersaulting, sliding on snow-covered slopes, hanging by their bills from tree branches, and hanging by their feet, upside down. Social play in groups is also quite common and involves wrestling and activities that anthropomorphically resemble "hide-and-go-seek" and "king of the hill." In social situations, the solicitation to play may involve tactics like assuming a "defensive" posture—head lowered and one foot raised, or walking stiff-legged, as is commonly seen in some mammals and a few other birds. Lastly, keas engage in a great deal of object play. They manipulate objects of all sorts using feet and bills, they toss things into the air and fly at them, and they play with snow when it is present.

HUMAN CHILDREN

Numerous studies have been conducted both by developmental psychologists and by ethologists on play behavior in young *Homo sapiens* (reviewed by Bruner, Jolly, and Sylva 1976; Fagen 1981). Hutt and Bhanvani (1972) conducted a follow-up study of earlier work by Hutt (1966, 1967a, 1967b, 1970a, 1970b) to predict differences in play behavior based on longitudinal sampling. A young child confronted with a new toy will first investigate and inspect it (specific exploration) and then play with the toy (diversive exploration). Three- to five-year-old nursery school children can be classified into three mutually exclusive categories: (1) *non-explorers* who may visually inspect a new toy, but do not handle or play with it; (2) *explorers* who thoroughly investigate a new toy, but fail to do more than that with it; and (3) *inventive explorers* who investigate a new toy and then play with it in a variety of innovative ways.

Do these individual differences in exploratory behavior provide any insights into predicting traits of the children at a later age? To test this question Hutt and Bhanvani (1972) obtained data for about fifty children from the original sample of one hundred used to generate the three categories above. The children were seven to ten years of age at the time of the second sampling procedures. Each child was given a series of tests to measure creativity and a personality questionnaire.

Each child was rated by parents and teachers on behavior, adjustment, and development. The results provided support for several suggestive conclusions (which were really more like hypotheses for further testing):

1. Lack of exploration was related to later lack of curiosity and adventure in young boys, and to diffficulties in personality and social adjustment in young girls.
2. Children who had been more creative and imaginative in their early play behavior were more likely to be creative in the later ages; this was particularly true for boys.

Hutt and Bhanvani note that some of these observed effects may be attributable to early childhood differences between the sexes; boys are more exploratory in their play activity for a longer period of their life, and girls are socially and linguistically more advanced than boys at nursery school age. We might also note that longitudinal follow-up assessment of longer term effects of early differences would be both appropriate and necessary to strengthen, extend, or refute these conclusions.

DEVELOPMENT INTO ADULT LIFE

The development of behavior, as we noted earlier, is a continuum. Many key events take place early in an organism's life. However, this does not mean that developmental processes cease when the organism reaches some particular chronological age. Rather, depending upon the species and other conditions, continuing developmental changes in some traits occur throughout most or all of the organism's life. We have already discussed in some detail the worker honeybee's chronological sequence of changes in physiology and behavior that continues throughout life. In the preceding section, we noted that both the young and the adults of many species engage in play behavior. Some plasticity for particular traits may be important to the ecology and ultimately to an organism's reproductive success, as the following example illustrates.

White-footed mice (*Peromyscus leucopus noveboracensis*) primarily occupy a woodland habitat, but they are ubiquitous in a wide range of habitats within their geographical range (midwestern and eastern United States). Although they are most abundant in a variety of types of forests, they also live in fields, in croplands, in marshes, and often in human dwellings. Prairie deermice (*P. maniculatus bairdi*) live in grasslands—usually open fields or croplands—and are almost never captured in wooded areas, brushland, marshes, manmade structures, and so on.

Adult mice of these two species were caught and brought into the laboratory, where they were presented with a cafeteria arrangement of seeds and grains from the various habitats where the mice had been caught (Drickamer 1970). Mice from both species ate large amounts of corn, but there were species differences in preferences to the other foods. The *P. leucopus* ate more elm and maple seeds than *P. maniculatus*; the latter species ate more bush clover seeds and wheat. Could there be a relationship between the mice's occupancy of particular habitats and their food preferences or feeding strategies?

In a further test, Drickamer (1972) explored the effects of dietary experience on subsequent food choices both for young mice (21 to 90 days of age) and for adult mice (over 90 days of age) of these two species. To circumvent the problems of inadequate nutrition that occurred when the mice were given only diets of seeds, laboratory mouse chow was used—but three different flavors were generated by placing an odor source (a drop of an essential oil such as anise or pine on a small piece of cotton) beneath the food in each dish. Mice were provided with a two-week training experience either as young or as adults, and were tested either immediately after the training or one month later. The mice were

tested two ways to determine whether the training experience influenced their choice of food-odor combination. The results of these tests are summarized in table 9–3. Preferences for particular food-odor combinations were influenced by prior training experience for young *P. maniculatus* and for both young and adult *P. leucopus*. Adult *P. maniculatus* were not affected by the training experience. Further tests assessed the patterns of visitation to various food sources by young and adult mice of both species. Adult *P. leucopus* changed feeding sites more frequently than young *P. leucopus*, and more frequently than either young or adult *P. maniculatus*. If the feeding site arrangements are shifted about, the young and adult *P. maniculatus* demonstrated a strong position preference, whereas *P. leucopus* of both ages either followed the foods to new locations, or changed to another food.

How may these observed differences in feeding habits be related to the ecology of these species? Clearly, the adult *P. leucopus* are more flexible in shifting their feeding preferences and, apparently, in using a more diversified strategy, in comparison with the adult *P. maniculatus*. This suggests that *P. leucopus* may successfully occupy and utilize a wider variety of habitats because of more flexible feeding habits as adults. In contrast, *P. maniculatus* with age became more rigid in their food habits and feeding strategy. This correlates with their occupancy of a much more limited range of habitat types. The young of both species exhibit some

TABLE 9-3 Effect of food-odor-training on mice
Results from tests on two species of *Peromyscus* to determine the effects of training with various food-odor combinations on the selection of diet from a series of food-odor combinations presented cafeteria style or on the chewing of balsa wood pegs to obtain a preferred food-odor combination. '+' indicates that the testing resulted in significant preferences for the food-odor combination with which the mice were trained. '—' indicates no significant effects of the training experience on food-odor combination preference.

	Appetitive	Consummatory
Peromyscus leucopus		
Young mice—tested immediately	+	+
—tested one month after training	+	—
Adult mice—tested immediately	+	+
—tested one month after training	+	—
Peromyscus maniculatus		
Young mice—tested immediately	+	+
—tested one month after training	+	+
Adult mice—tested immediately	—	—
—tested one month after training	—	—

flexibility in feeding, but by the time the animals are about 90 days of age there has been a change for *P. maniculatus*. Thus, we can see that developmental processes can continue into adulthood, and that the results of such processes may have important consequences for the life history patterns of various species.

NATURE/NURTURE/EPIGENESIS

What about the relative roles of genetics and development experiences in shaping behavior? What lines of reasoning have been applied to development by animal behaviorists to explain the appearance and performance of behavior patterns? For a number of decades, a controversy raged within the animal behaviorist camp, over which forces shaped and determined behavior, particularly during development. Some investigators operated with the hypothesis that observed behavior was largely under genetic control (nature), whereas others operated with the hypothesis that observed behavior was primarily a function of developmental experiences and environmental influences (nurture). We often associate the notion of instinct with the ethologists in Europe, and the hypothesis of environmental influences with the zoologists and comparative psychologists in North America; however, this distinction is too stereotyped. In fact, although the basic concepts for these opposing viewpoints have received more emphasis from one group or the other, research in this century has been conducted using both viewpoints on both sides of the Atlantic Ocean. Both viewpoints have portions of their historical roots in Europe *and* North America (see chapter 2).

Today, most animal behaviorists subscribe to the concept of epigenesis, the integrated process of behavior development involving the interaction of the genome and experience. In order to fully understand and appreciate these two viewpoints we now examine the nature and nurture hypotheses and conclude the chapter with a discussion of epigenesis.

GENETIC INFLUENCES

Ethologists, the most important proponents of the theory of genetically preprogrammed (innate) behavior, developed a special terminology for describing the manner in which behavior was controlled:

- **Sign stimulus**: an external signal that elicits specific responses from conspecifics.

- **Innate releasing mechanism (IRM)**: a neural process, triggered by the sign stimulus, that preprograms an animal for receiving the sign stimulus and mediates a specific behavioral response.

- **Fixed action pattern (FAP)**: an innate behavior pattern that is stereotyped, spontaneous, and independent of immediate control, genetically encoded, and independent of individual learning (Tinbergen 1951).

According to the theory of genetic or innate control of behavior, we analyze animal actions as a sequence of events: sign stimulus, innate releasing mechanism (IRM), and then fixed action pattern (FAP). As an example of the use of this terminology applied to behavior, consider the adult male three-spined stickleback fish (*Gasterosteus aculeatus*). These fish establish territories and engage in aggressive displays and fights at territorial boundaries (Tinbergen 1948, 1951). One way to discover which characteristics (sign stimuli) elicit aggressive responses is to present fish models (figure 9–18) to a male stickleback and see which one he attacks. The males display various standard, easily recognizable threat and aggressive attack postures (FAPs) in response to sign stimuli (figure 9–19). Crude models that have a red belly are attacked more often than are normal-looking stickleback models that lack the red underside. The aggressive posture exhibited by male sticklebacks appears rigidly stereotyped. This similarity of display patterns by all male sticklebacks may have some strong evolutionary advantages. If each male performed the threat behavior differently, the male being threatened might not be certain of the meaning of the posture. Confusion over an actual threat could lead to misinterpretation, and in turn, to an attack involving physical damage to one or both males.

The hypothesis of genetic control of behavior and the experimental methods used to test various aspects of this hypothesis have been criticized on several grounds (Lehrman 1953,1970; Moltz 1965). The term *innate* can have two different meanings to many animal behaviorists. First, it may refer to variations in a trait among individuals in a population. For example, human eye colors are blue, brown, green, and mixtures of these colors, with the color genetically based. In this instance, we might say that differences in a trait are inherited or innate, but that external influences during ontogeny may still affect the development of that trait. Second, some ethologists have used the term *innate* to refer to the notion of fixed development of a specific behavior pattern—that is, the organism exhibits behavior that is

FIGURE 9-18   Fish models used to test aggression in male sticklebacks
The first model of fish resembles the normal three-spined stickleback except that it lacks the red belly characteristics of males of this species; the other four crude models all have red bellies. When these models are presented to a male, he attacks the last four models more than the first.

FIGURE 9-19   Aggressive threat displays of male sticklebacks
These postures of the adult male three-spined stickleback fish are stereotypical, are performed in a similar manner by all males of this species, and are called fixed action patterns (FAPs).

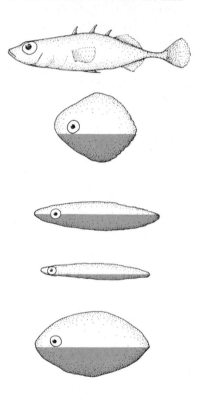

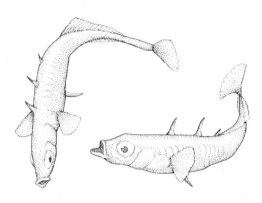

If we define **experience** to include the effects of all interactions between an organism and its environment that influence behavior, then we must consider when a developing animal is capable of receiving internal or external stimuli that affect the organization of brain and the body structures involved in behavior. Lorenz (1937) argued that the neural structures, as encoded in the genes, appear first and control both the reception of external and internal stimulation, and the behavioral responses of an animal. In support of his ideas, reports on neurobiological development (see review by Jacobson 1978) have indicated that experience probably does not affect basic neural structure in some organisms. But recall also the work of Oppenheim, Chu-wang, and Maderut (1978) from earlier in this chapter and the studies of Nottebohm and his colleagues reported in chapter 6 on canary song, in which evidence is provided that neural structures can be altered by ongoing experiences. Also, Moltz (1965) and Lehrman (1970) presented arguments that the developmental processes involved in the establishment of the neural circuitry for receiving stimuli and producing responses are a function of both the genome and the experiences of an organism, beginning at conception.

preprogrammed in the genes, as is the case in the fixed action patterns of sticklebacks. Unfortunately these different meanings create a great deal of confusion, and some animal behaviorists use the term indiscriminately. Many psychologists, zoologists, and ethologists prefer to use *innate* to refer only to differences between individuals or populations.

We know that genes carry information that codes for proteins and not directly for behavior patterns, morphological structures, or physiology. Biochemists and geneticists have discovered that the sequences of molecules in DNA, the chemical that carries genetic information, are codes for the production of specific protein molecules (see chapter 5) that are involved in the structures and processes within the cells and tissues of the body. To the extent that the genome provides the basic framework for ontogeny, genes have a critical role in determining the eventual behavior patterns exhibited by an organism.

One common method for demonstrating that a behavior pattern is performed without any prior learning experience is the **isolation experiment.** For example, male stickleback fish reared away from all conspecifics will perform the species-typical zigzag courtship dance correctly the first time they are introduced to a gravid

FIGURE 9-20   Inter-gobble intervals of turkeys
The inter-gobble intervals (IGIs) for 35 samples of 100 gobbles each are plotted as the cumulative distribution in logarithmic probability coordinates. Data were obtained from 9 different male turkeys. Data such as these that show variation in a particular parameter of an FAP are useful in assessing the fixity of such fixed action patterns.

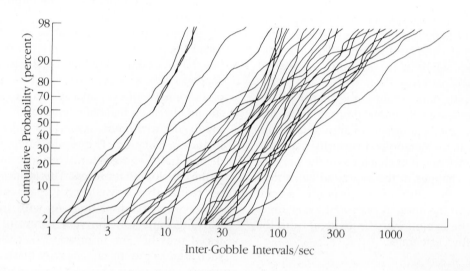

female or model of a gravid female (Cullen 1960). Isolation experiments may involve a range of treatments—from housing in the absence of conspecifics to prevention of motor movements. When considering an isolation experiment, we find it difficult to know exactly which stimuli an animal is isolated from. Rearing in the absence of any conspecifics may not involve solely social deprivation; certain forms of sensory and motor stimulation necessary for normal development may require social contacts. Moreover, rearing under conditions of restricted motor or sensory stimulation may also deprive an animal of normal social contacts. We should thus be quite cautious when interpreting the results of isolation experiments.

Those who favor instinct theory have postulated that genetic encoding accounts for similarities in FAPs among the members of a species. Moltz (1965) has argued in opposition that the stereotyping of a particular behavior pattern may result from environmental consistencies. Thus, for example, the rhythmical activity patterns controlled by biological clocks (see chapter 8) are, at least in part, functions of external cues from the environment. Thorpe and Jones (1937) experimented with insect species that always lay their eggs

on other specific insect hosts. If the eggs of one generation are reared on a different insect host, the next generation of adults may show a preference for laying their eggs on the new host. Other behavior patterns, similarly influenced or channeled by the environment, may also seem to be stereotyped (e.g., see King 1968).

Schleidt (1974), who carefully examined aspects of the stereotyping of FAPs, concluded that more quantitative studies of stereotypy are needed. He utilized data from the measurement of gobbling calls of male turkeys to illustrate the variability that exists in an FAP (figure 9-20). The fixity of an FAP is a relative judgment that depends on the population of animals being sampled, the method of assessing the variation, and the limits of our own perception.

In response to these criticisms of the concept of FAPs, Barlow (1968) developed the concept of **modal action patterns (MAPs)**. Barlow defined the MAP as a spatiotemporal behavior pattern that is common to members of a species; different individuals tend to perform the pattern in a recognizably similar or modal fashion. This sounds very much like the species-typical behavior pattern concept proposed by Beach (1960). The

MAP concept takes into account the possibility of variation around the modal pattern, the necessity of possessing some flexibility in behavior for individual adaptation, and the possibility that environmental input or the sign stimulus can vary the action pattern.

We should also make note of the fact that in a general sense, different animals apparently possess varying degrees of genetic preprogramming of behavior. Invertebrates appear to have more of their behavior under a higher degree of genetic control: they possess more stereotyped responses and less flexibility. This is in contrast to vertebrates, where more flexibility is apparent in many behavior patterns. Even within the vertebrates, some investigators would argue that there are differences in the degree of genetic programming; fish show more stereotyped patterns than amphibians; amphibians exhibit more stereotyped patterns than reptiles; etc. Thus, as we noted in chapter 6 on the nervous system, there are changes in the degree of behavioral programming and flexibility of the responses accompanying changes in the nervous system; these changes are reflected in the evolutionary sequences, from the invertebrates to the vertebrates.

ENVIRONMENTAL INFLUENCES

What about the "opposite" hypothesis that ongoing events during the course of development of the animal could influence its behavior patterns? Proponents of the hypothesis that the environmental influences and experience are of critical importance in behavior development have supported their viewpoint with evidence from several types of investigations. First, the emphasis, particularly in the first half of this century, of some investigators on learning behavior led to the conclusion that behavior in many animals was largely acquired through experience and practice. Among these investigators E. L. Thorndike, R. M. Yerkes, and J. B. Watson emphasized the study of animal intelligence and learning processes. Watson (1930) developed the approach called **behaviorism.** This theoretical and experimental approach operates with the premise that much of behavior results from previous experience. Later, B. F. Skinner (1938, 1953) espoused and supported with experiments the idea that most animal actions can be analyzed functionally in terms of combinations of stimulus and response. According to Skinner, these S-R combinations are learned by the organism. Since learning, by

its very definition, implies an important role for experience and practice (see chapter 10), it is likely that we have historically ascribed a heavy environmental emphasis to animal psychologists that investigate learning behavior, when many of them actually did not hold such an extreme position (see Dewsbury 1984). Watson, for example, was very interested in the development of instincts in young animals and the improvement of such behavior with practice.

Second, many comparative psychologists have tested the effects of early experience on later behavior (we have examined some of these studies in this chapter). The sensory, motor, and social aspects of experience have all been manipulated to ascertain the resultant effects on behavior. Some of these investigations involved enrichment, and others utilized deprivation, including isolation. The results often, but not always, demonstrate key roles for experience and environment in the development process. This is not surprising when we realize that it was exactly that which the investigators were manipulating. As with investigations of learning behavior, the very nature of what is being tested provides the impression that those conducting such studies are, in fact, heavily biased toward an environmental point of view. Many studies of experiential effects also involve genetics, in the form of testing different species or strains.

One criticism of studies of learning and experience as environmental influences on the behavior development is that too often, the work has been conducted only in laboratory settings, and that too often, the laboratory rat has been used as a test subject. Beach (1950) presented a numerical analysis of research published in the *Journal of Comparative and Physiological Psychology* during the period from 1930 to 1948. He found that between 60 and 70 percent of the reported studies had used laboratory rats as subjects, and less than 10 percent of the published studies had used invertebrates or nonmammalian vertebrates as test subjects. Hodos and Campbell (1969) have also decried the lack of a truly comparative psychology.

Isolation and deprivation experiments may be criticized because we cannot be certain what we are isolating the subject animal from, and therefore, we have difficulty interpreting the results. In a similar vein, experimental regimes that provide an animal an enriched environment in the laboratory may involve a control situation that consists of a bare cage—hardly a natural setting.

EPIGENESIS

What has been done to reconcile these two seemingly opposed perspectives of the processes of behavior development and behavior control? The modern theory of behavior development is called **epigenesis,** the integrated process of behavior development that involves both the genome and the environmental influences. According to the epigenetic approach, the expression of the genetic material, which leads to the synthesis of tissues, organs, and thus to behavior patterns, is dependent on the environmental context. Thus in different environmental conditions, the same genes may be expressed differently. For example, individuals of a species of fruit fly (*Drosophila*) reared at different temperatures, develop wings capable of normal flight, irregular weak flight, or no flight at all (Harnly 1941). We have already noted in this chapter several ways the environmental regime can influence the behavior and sexual selection of various animals.

The general structure of the organism, which develops through the processes of maturation, is dictated in large measure by its genome, but the development of various structures and behavior is influenced by experience. Genes set the limits, and through interaction with the environment, the final product or phenotype is determined. What an animal inherits—that is, what is dictated by the genome—consists of a range of possible expressions of each measureable physical, physiological, and behavioral trait; genes set the limits on the phenotypic expression of traits. For some morphological, physiological, or behavioral traits, the prescribed limits of expression may be quite flexible, as

dictated by the genetic makeup; whereas for other traits the limits of potential expression may be quite narrow. Also, as noted previously, within the animal kingdom, there are some distinct patterns of the degree of genetic control of behavior; invertebrates are less flexible than vertebrates.

The theory of epigenesis involves elements of both genetic and environmental viewpoints (see Miller 1988). Today, most scientists interested in the development of behavior are exploring questions about the mechanisms that underlie particular behavior patterns. This often involves working closely with embryologists investigating gene expression and with physiologists and anatomists investigating relationships between structure and function in embryos and neonates. Others interested in the development of behavior are concerned with questions about life history strategies, the evolution of rates of development, and the relationships between patterns of development and the ecological habitats of various species.

Modern behavior development can be viewed in terms of the four P's: **Preprogramming** involves the genetic make-up of the organism and its predisposition to receive particular stimuli and output particular actions. **Practice** involves experience and other environmental inputs that affect the development of behavior throughout the life of the organism. **Potentiation** involves the particular contextual situation at a given point in the life of the animal, and the nature and sources of stimuli that it is receiving. **Performance** is the resulting series of behavioral actions and associated internal physiological changes in the animal.

SUMMARY

Seven parameters are important in the design of experiments to investigate the effects of early experience on later behavior. These include the age of the animal at the time of early experience treatment, the type and duration of treatment, the age of the animal at the time of testing, the type of test used, the persistence of effects, and the use of different strains or species to assess relative differences in the effects of experience treatments. Both *longitudinal* and *cross-sectional* designs can be used to test the persistence of treatment effects.

The embryology of behavior encompasses investigations involving events that are integral to the process of behavior development and occur prior to birth or hatching. Both sensory and motor responses can be

evoked prenatally. Recent studies have linked aspects of development within the nervous system and the development of motor actions by the organism. For some mammals, experiences of the mother (e.g., stress) during pregnancy can affect the later behavior of progeny.

A variety of events that occur just after birth have important consequences for social behavior processes, both immediately and later in life. *Filial imprinting* involves the development of a social attachment for a particular object. Imprinting, studied primarily in birds, has been investigated with regard to *critical* and *sensitive periods*, a variety of stimulus objects, the *locomotion-fear dichotomy* and the importance of both the auditory and visual sensory modes. Filial imprinting is significant with respect to the young following the parent(s), predator avoidance, and recognition of conspecifics.

Sexual imprinting involves learning to direct sexual behavior at particular stimulus objects and may serve as a species-identifying mechanism, important for species isolation.

The diet consumed by a female rat may affect the food preferences of her offspring. Also, young rats may be influenced in their selection of feeding sites during weaning by social cues from interactions with conspecifics. Young snapping turtles are strongly influenced in their selection of diet by their first feeding experiences after hatching. Young gulls learn to peck at the bills of their parents to obtain food through a combination of genetic predisposition and practice over time.

A variety of events occur during the juvenile period of development, lasting from fledging or weaning until full maturity and independence are achieved. Investigators have studied the effects of *deprivation* (social, sensory, motor) and *enrichment* during the juvenile period on subsequent behavior. Deficits in behavior due to deprivation may be attributable to loss of information, disuse, or stress of emergence. The severity of the behavioral deficits occasioned by deprivation treatments in dogs and monkeys are related to the degree of deprivation.

Birds learn to sing based on both a genetic template that varies with the species and learning from conspecifics which may take the form of imitation or improvisation. Longitudinal studies of song development indicate that some birds progress through a series of stages in song acquisition: *subsong, subplastic song, plastic song,* and, finally, *crystallized song.*

Puberty in house mice is influenced by a variety of social and environmental cues. Some social cues (e.g., urinary chemosignals) accelerate sexual development, whereas others retard the process. Both diet and daylength also influence puberty. The timing of puberty is important both for the onset of reproductive behavior and for the population biology of the mice.

Behavior development has been explored in a variety of insects and fish. For fruit flies there are clear relationships between internal physiological events and changes in sexual receptivity after eclosion, and later, the tendency to oviposit. After honey bees emerge from pupae they progress through a series of functional roles in the colony— again, there are clear developmental correlations between the roles and physiological and morphological changes in the bees. One model for the development of behavior in fish demonstrates how the complex and varied behavior patterns of the young adult derive from a few simple patterns in the newly hatched animal.

Play behavior is an important component of the developmental sequence in many mammals and birds. Studies on canids, keas, and children illustrate the varied types of play: *social, exercise/maturation,* and *exploration.* The functions of play behavior include practice for adult activities, aiding the processes of growth and development, learning about the environment, and establishing social relationships. Development does not end when the young animal becomes independent of its parent(s), but rather continues into adult life. Possessing some capacity to remain *flexible* in certain behavioral traits may have important consequences for the organism. For example, differences in habitat occupancy of two species of deermouse may be partly a function of the differential dietary flexibility of the adults of these two species.

A critical problem in animal behavior has been the *nature/nurture issue*—that is, the relative importance of genetic inheritance and of experience for the expression of behavior patterns. Instinct theorists have proposed that *fixed actions patterns* (*FAPs*), like morphological traits, are inherited. FAPs result from preprogrammed neural circuitry that predisposes an organism to make stereotyped responses when it receives specific environmental stimuli. FAPs can be characterized as *stereotyped, spontaneous,* and *independent* of immediate external control; they are also genetically encoded and independent of individual learning. The theory that behavior is inherited has been criticized on the following grounds: (1) genes code for proteins, not for behavior; (2) isolation experiments are difficult to interpret; and (3) FAPs are not as fixed as was once assumed. The recently developed concept of *modal action patterns* (*MAPs*), or species-typical behavior, accounts for the close similarities between the behavior patterns of members of a species and also for individual variation.

Proponents of the viewpoint that experience and environmental input are central to the processes of behavior development have supported their viewpoint by manipulating experiential conditions and recording the effects on behavior development, and by studying learning behavior. That this viewpoint has historically been labeled as emphasizing environmental influences may be largely because the very processes being investigated are, by definition, environmental. Many investigators exploring environmental influences on behavior development have also, for example, manipulated genetics using different species. Studies by these investigators are subject to criticisms about the overemphasized use of the white rat and use of laboratory environmental conditions that bear little resemblance to the natural habitat of the animal.

Animal behaviorists today study behavior development in terms of *epigenesis*— the interaction of the genome and environmental stimuli present during all phases of ontogeny. Genetic inheritance programs the basic characteristics and overall range of flexibility for trait expression in the organism, and environmental influences determine the final nature of behavior exhibited by the animal. The mechanisms of these interactions are the focus of much current research. Modern behavior development may be viewed in terms of *preprogramming, practice, potentiation,* and *performance.*

Discussion Questions

1. Careful definition of terms is important in the study of animal behavior. Write out your own definitions of the eight terms important in behavior development listed below. Locate the definitions of these terms in this book and in other books on behavior. How do the definitions differ? (a) epigenesis, (b) experience, (c) play behavior, (d) critical period, (e) ontogeny, (f) filial imprinting, (g) stress-of-emergence hypothesis, and (h) species-isolating mechanism.

2. Suppose we are interested in comparing the development of behavior in two species of snakes. Specifically, we are interested in their habitat preferences. What methods would you employ in such an investigation? Assume you will conduct your studies in both field and laboratory settings. What advantages and disadvantages can you cite for each of the methods or approaches you have suggested?

3. Many animals we use for studies of behavior are from domesticated laboratory stocks or are derived from wild animals that have been kept for varying periods of time in the laboratory. What differences in developmental processes might you expect to find between these laboratory-reared animals and their counterparts in the wild?

4. Pratt and Sackett (1967) raised three groups of young rhesus monkeys with different degrees of contact with their peers. They allowed one group no contact; the second, only visual and auditory contact; and the third, full normal contact with peers. Next they allowed animals of the three groups to interact socially; then they gave each individual a three-choice preference test where the alternative choices were conspecifics— one from each of the three different treatment conditions. The data in the table represent the preferences of the test subjects. What conclusions can you draw from these data?

Rearing condition of test animal	Rearing condition of stimulus animal		
	No contact	*Visual and auditory contact*	*Normal contact*
No contact	156	35	29
Visual and auditory contact	104	214	103
Normal contact	94	114	260

Source: Data from Pratt and Sackett (1967).

The numbers in the table are the mean number of seconds spent by a test monkey with each of the three possible partners in a choice test.

5. Behavior development actually is a continuing process that occurs throughout the life of an organism. Birds of a number of species migrate to the equatorial regions for the winter months and return to the temperate zones to breed in spring and summer. What types of ongoing developmental processes may be taking place in these birds that survive and make this annual journey two, three, or more times? What types of experimental tests can you propose to explore these lifelong developmental processes?

6. Insect behavior development is an area that is likely to receive considerably more research attention in the next several decades than it has to date. Given what you know about physiological processes in insects from earlier chapters in this textbook, what types of experiments would you propose concerning behavior development in each of the three species listed next? (a) cockroaches, (b) blowflies, (c) grasshoppers.

Suggested Readings

Bateson, P. P. G. 1978. How does behavior develop? In *Perspectives in Ethology*, Vol. 3, ed. P. P. G. Bateson and P. H. Klopfer, 55–66.
Theoretical, thought-provoking presentation. Bateson is best known for his work on imprinting, but this treatment is broader and more general.

Burghardt, G. M., and M. Bekoff. 1978. *Development of Behavior*. New York: Garland STPM Press.
A book based on a symposium of the Animal Behavior Society. The wide range of topics covered and the short, informative articles provide an excellent state-of-the-art summary of this subfield of animal behavior.

Fagen, R. 1981. *Animal Play Behavior.* New York: Oxford University Press.

A thorough synthesis of all aspects of play behavior. The book is quite well-written, and presents detailed examples and comprehensive reference listings. The illustrations are an added bonus.

Halliday, T. R. and P. J. B. Slater, eds. 1983. *Animal Behaviour, 3, Genes, Development and Learning.* San Francisco: W. H. Freeman.

Chapters 1–4 of this small paperback book provide an excellent summary of the theory and examples for development from genes to social relationships. Tightly written and therefore probably better suited for the somewhat advanced student.

Hess, E. H. 1973. *Imprinting.* New York: Van Nostrand Reinhold.

Both theory and experimentation of imprinting, primarily on birds, are treated by one of the leading investigators of this phenomenon. Bits and pieces of ecological and evolutionary significance of imprinting processes.

Immelmann, K., G. W. Barlow, L. Petrinovich, and M. Main. 1981. *Behavioral Development.* New York: Cambridge University Press.

This volume resulted from a series of interdisciplinary seminars held in West Germany. Each article is a new contribution; some are presentations of new data, while others are theoretical or review papers. Together they represent one of the most comprehensive collections on the topic of behavior development. There are ample references with each of the twenty-eight separate contributions.

Oppenheim, R. W. 1982. Preformation and epigenesis in the origins of the nervous system and behavior: Issues, concepts, and their history. In *Perspectives in Ethology,* ed. P. P. G. Bateson and P. H. Klopfer, Vol. 5, 1–99. New York: Plenum.

An excellent chapter that covers both the history of many aspects of the nature-nurture controversy and provides insights into modern theories concerning behavior development. Oppenheim has been a pioneer in regard to exploring neural aspects of ontogeny.

10

LEARNING AND MOTIVATION

In this chapter we explore learning and motivation, the processes underlying them, and their functional significance. We first briefly examine several types of learning and their characteristics. We then survey learning in a variety of organisms. This is followed by several in-depth examples of the interaction between learning and ecology in animals' lives. We next explore the constraints that affect studies of learning behavior. After a brief review of current knowledge about memory, we conclude the chapter with a discussion of motivation and the use of this concept to explain internal processes for which we do not yet have biological explanations.

Learning is best defined as a relatively permanent modification of behavior that occurs through practice or experience. **Motivation** refers to internal processes that arouse and direct behavior. We cannot measure motivation directly, thus we must make inferences from observed behavior. We find that motivation is manifested as **drives,** which correspond to various specific **needs**—the basic requirements for an animal to maintain **homeostasis** (relatively constant internal body conditions necessary for life). As we shall see, both learning and motivation are critical processes that affect the behavior of animals of all ages.

TYPES OF LEARNING

Our knowledge of learning in animals has accumulated from the work of psychologists and ethologists. A thorough historical review is beyond the scope of this text (see Thorpe 1956; Hilgard and Bower 1975). Two major types of association learning are currently recognized by animal behaviorists: classical conditioning and operant conditioning. There are many similarities and some differences in the key characteristics of these two major types of learning. We also recognize habituation as a distinct type of learning.

HABITUATION

We define **habituation** as the relatively persistent waning of a response that results from repeated stimulus presentations not followed by any form of reinforcement. Habituation is specific to the particular stimuli involved and can be distinguished from fatigue and from sensory adaptation by its relative persistence. **Fatigue** involves a loss of efficiency in the performance of a motor act when that act is repeated in rapid succession. **Sensory adaptation** generally occurs at the level of the peripheral sensory receptors and consists of a slowing down or cessation of nerve impulses transmitted to the central nervous system. Both fatigue and sensory adaptation last a relatively short time, but habituation is a persistent central nervous system process involving changes in the brain or spinal cord.

Consider, for example, a group of students listening to a lecture. If a classroom radiator clanks loudly, everyone will be momentarily startled. If after a short spell the radiator again makes its offending noise, the degree of response will be considerably decreased. After several additional clanks and bangs, almost no one will be startled by each new episode of noise and the class will go on as usual.

Habituation can be a functionally important aspect of an animal's behavior in its natural surroundings. Young ducklings scurry for cover when any shadow passes overhead, an adaptive response to avoid predators. Gradually the ducks learn, partly through habituation, which types of shadows signal potential danger and which are harmless. Ground squirrels, fiddler crabs, and marine worms all live in burrows. When danger threatens, these animals dash for their burrows for protection. They habituate to specific nonharmful stimuli in their environment, but retain escape reactions to threatening or unusual stimuli.

CLASSICAL AND OPERANT CONDITIONING

Classical conditioning starts with a stimulus, the **unconditioned stimulus (UCS)**, that elicits a specific response, the **unconditioned response (UCR)**. At approximately the same time as the UCS, a second, neutral stimulus is presented that does not customarily elicit the UCR. When the neutral stimulus and UCS are paired for a number of trials, the response will eventually be elicited by the neutral stimulus. At this point, the neutral stimulus (in figure 10–1, the bell) is termed the **conditioned stimulus (CS)** and the response is the **conditioned response (CR)**. Classical conditioning works most effectively if the CS precedes the UCS by a brief time interval for each trial; in fact, little or no conditioned learning may take place if the CS follows the UCS.

The best known example of classical conditioning is that of Pavlov and the salivary responses of dogs (figure 10–1). During his investigations of digestion and related physiological processes, Pavlov described the paradigm illustrating classical conditioning. The UCS is meat powder, which elicits salivation when presented (UCR). Presentation of a bell tone (potential UCS) alone

FIGURE 10-1 Pavlov's testing apparatus
Ivan Pavlov discovered classical conditioning through his work on the salivary reflex in dogs. The dog in the restraining apparatus is ready to be tested using the classical conditioning paradigm.

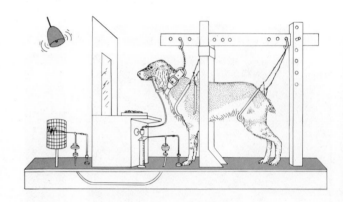

does not elicit the salivary reflex. Now, on a number of trials, the bell is sounded a short time (one to three seconds) before the meat powder is presented. Soon presentation of the bell alone is sufficient to elicit the salivation (now a CR). Thus, classical conditioning results from the association of two stimuli in the environment. In general, it is also true in this type of learning that the sequence of events occurs regardless of what the test subject does.

Operant conditioning, or as it is sometimes called, **instrumental learning,** involves the animal's learning to associate its behavior with the consequences of that behavior; the sequence of events is dependent upon the behavior of the animal. Usually some type of reward or punishment is involved. The task performed by the animal may be relatively simple, as in a rat trained to press a lever in a Skinner box (figure 10–2). Pressing the bar is rewarded with food or water. Another example is an animal learning to run a maze to a goal box to receive the reward. In nature, a weasel may learn to associate the odor of mice with locating and catching a meal.

CLASSICAL AND OPERANT CONDITIONING COMPARED

Many who study learning phenomena today postulate that the processes underlying classical and operant conditioning are similar processes, while others argue that they are different. We should thus examine some similarities and differences.

FIGURE 10-2   Skinner box
The interior of the box contains a lever, a light, a food bin, and a grid floor. Additional apparatus for automation and for monitoring the rat's behavior is housed behind the back panel and on the left side of the cage.
Source: Photo by Will Rapport from the office of B. F. Skinner.

FIGURE 10-3   Characteristic forms of learning curves
(a) The percentage of CRs and response magnitude increase with practice. (b) Latency measures and other time measures decrease, as do errors.

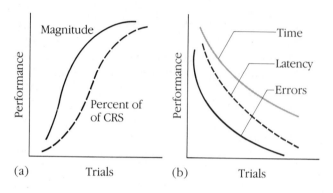

ACQUISITION.   The act of first developing a response is termed **acquisition.** We find that several different response measures are used, some common to both types of learning. A graphic presentation of the response measures is a **learning curve** (figure 10–3). A sequence of trials given over time often serves as the scale for the horizontal axis. The vertical axis is usually some measure of performance or strength of CR. Among the common dependent measures used are the percentage of conditioned or correct responses, the magnitude of the response, the time required to complete a response, the error rate, or the latency (time interval) between some signal (e.g., the CS) and the response. For several of these (figure 10–3a) the curve increases over the trials, whereas for other measures the curve decreases (figure 10–3b). We must emphasize that these are merely measures of performance. It would be incorrect to draw conclusions from these measures regarding the underlying mechanism(s) of learning.

SCHEDULES OF REINFORCEMENT. **Reinforcement** in classical conditioning consists of presenting the UCS after the CS. In operant conditioning, the reinforcer is presented when the subject gives the appropriate response. Reinforcers in operant conditioning may be positive (such as food, water, or an opportunity to manipulate some object) or they may be negative (such as electric shock or other punishment). The most basic schedule provides reinforcement for each trial or response. Other schedules provide partial or intermittent reinforcement: **Fixed ratio** schedules require the test subject to make a fixed number of responses to receive reinforcement. **Fixed interval** schedules provide reinforcement for the first response after the passage of a prescribed time interval following the last reinforced response. Similarly, we can use **variable ratio** and **variable interval** schedules to program reinforcement to occur after a variable number of responses, or for the first response after a variable time interval, respectively.

EXTINCTION. **Extinction** refers to the decrease of response rate or magnitude with lack of reinforcement. If we eliminate the UCS for a period of time in classical conditioning or eliminate the reward or punishment in operant conditioning, the response is extinguished. The number of responses that must occur without reinforcement before extinction occurs is a function of several variables. In general, the longer the conditioning procedure has been (the more well learned the response), the harder it is to extinguish that response. Interestingly, responses that have received partial reinforcement during acquisition are the most difficult to extinguish. Animals trained on partial reinforcement schedules resist extinction for several reasons: they learned to persist in responding when faced with some nonreinforced responses during training, and there is less difference between partial reinforcement and no reinforcement than between continuous reinforcement and no reinforcement. An excellent example of a behavior in humans that is affected by partial reinforcement schedules and that is resistant to extinction is gambling.

SPONTANEOUS RECOVERY. Depending upon the species and experimental conditions, if a conditioned response has been extinguished and is then followed by a rest interval of several minutes to a day or more, the animal may exhibit **spontaneous recovery** upon reintroduction to the test situation. For example, when a rat trained to press a bar for food reinforcement is no longer given the reward, the response will be extinguished. After an hour's rest, the rat will exhibit spontaneous recovery, pressing the bar many times without reward until the response is again extinguished.

GENERALIZATION. We find that if an animal has been conditioned to respond to a certain stimulus, the response usually will also occur to stimuli similar to that used in the original acquisition trials. In classical conditioning, dogs will respond by salivating to tones similar to that of the original bell, but not to tones with quite different pitches. As an example of operant conditioning in humans, consider the learned association between the type of music that leads up to various action or scene in a television program and the action or scene itself. We learn to associate particular cadences and musical motifs with particular effects, such as danger, sadness, or mystery.

DIFFERENCES. Several key differences exist between classical and operant conditioning. First, the basic paradigms used to demonstrate the two types of learning differ. In classical conditioning, the situation is thought by many to involve a stimulus-stimulus pairing; the CS and UCS are paired. In contrast, operant conditioning involves pairing the stimulus and response. A second difference has already been noted: in classical conditioning, the subject does not control the sequence of events; whereas in operant conditioning, the sequence

of events is contingent upon the responses of the test animal. In other words, in classical conditioning the responses are elicited, but in operant conditioning the responses are emitted by the subject. A third difference involves the **shaping** process characteristic of operant conditioning. The experimenter can reinforce some of the subject's responses and not others, enhancing the frequency of some actions and extinguishing other actions. For example, while training a rat to press a lever for food reinforcement, at first the experimenter may control the process by rewarding the rat for being in close proximity to the lever, then for exploring the lever, and then for touching the lever, until the rat begins to press the bar to obtain the food.

OTHER ASPECTS OF LEARNING

In addition to habituation and the two major types of association conditioning, several additional types of learning should be noted; they are generally considered categories of operant conditioning (instrumental learning).

AVOIDANCE LEARNING. **Avoidance learning** is sometimes called **aversive conditioning** and is really a type of operant conditioning. Many predatory animals learn to avoid distasteful prey by associating sensory cues from these prey with negative aftereffects, such as upset stomach. Birds learn to avoid eating monarch butterflies (*Danaus plexippus*) because these insects contain poisons that are emetics and make the birds ill (Brower 1958; Brower and Brower 1964).

INSIGHT LEARNING. Using **insight learning**, the animal makes new associations between previously learned tasks in order to solve a new problem. This, too, is really a form of operant conditioning. An early example of insight learning is the work of Köhler (1925) on chimpanzees (see also Birch 1945). Chimps learned to connect a series of small poles into one longer pole to obtain bananas suspended above the cage floor. In a similar way, they learned to stack boxes and climb the stack of boxes to reach the bananas. When the chimps were given a new problem, with the bananas suspended higher above the cage floor, they learned to put the poles together, stack the boxes, and climb the stack of boxes with pole in hand, to knock down the bananas (figure 10–4).

FIGURE 10-4   Insight learning in chimpanzees
The chimp has previously learned to connect the poles or to stack the boxes in order to obtain a reward of bananas. When the bananas are out of reach even if the chimp uses either the poles or the boxes, the chimp must use insight learning to deduce that a combination both of stacking the boxes and then climbing up them with the pole will provide access to the bananas.

LATENT LEARNING. An association made with neither immediate reinforcement nor particular behavior evident at the time of learning, has sometimes been labeled **latent learning.** The processes involved are not readily elucidated, but the phenomenon appears to be real, as the following example illustrates.

The predatory digger wasp (*Philanthus triangulum*) inhabits burrows in sandy soil. The way these insects relocate their nest burrows after flying some distance to capture prey was the subject of studies by Tinbergen and Kruyt (1938; see also Tinbergen 1958). Tinbergen provided landmarks around the wasp's nest holes; upon emergence, the wasp surveyed the area and then flew off. If Tinbergen removed or rearranged some landmarks, the returning wasps became disoriented to

varying degrees. Additional studies support the conclusion that the returning wasp uses the entire configuration of landmarks as a guide to its burrow's location.

OBSERVATIONAL LEARNING. The tendency to perform an appropriate action or response as the result of having observed another animal's performance in the same situation may involve either classical conditioning or operant conditioning, and has been called **observational learning.** For example, Klopfer (1957) demonstrated that ducks (*Anas platyrhynchos*) can learn a discrimination task by observation. He conditioned a group of ducks to feed from one of two dishes by placing the incorrect dish on a wired shock grid. During the conditioning he restrained observer ducks nearby, where they could see the subject ducks learning the discrimination but could not participate. He then released the observer ducks and placed them in the test situation with both food dishes present. The observer ducks avoided the incorrect dish with its wired shock grid.

Imitation occurs when an animal immediately copies the actions of another while they are both in each other's immediate presence (Thorpe 1963). The act performed is one that would not normally be expected in the species' behavioral repertoire. Consider, for example, the way a group of Japanese macaques (*Macaca fuscata*) of Koshima Island learned some of their food habits. Imo, an inventive young female monkey in the group, introduced two new techniques that were imitated by other group members. One new technique was taking the sweet potatoes they were fed to the water and washing them before eating (figure 10–5). Washing removed the gritty sand that adhered to the potato skins. The salt water may also have added some flavor.

Imo also took handfuls of wheat, another food given to the provisioned monkeys, to the water and allowed a little at a time to fall into the water; the sand that was mixed with the wheat grains sank, and the wheat floated on the surface of the water. Imo then picked up and ate the individual wheat grains (Itani 1958; Kawai 1965).

IMPRINTING. We have already discussed imprinting as a process of attachment formation and following behavior in chapter 9. This is sometimes classified as a separate, special type of learning.

LEARNING SETS. One general phenomenon important for an understanding of learning was first elucidated by Harlow and his associates (Harlow 1949,1951; see also

FIGURE 10-5   Japanese macaques washing sweet potatoes

A young female macaque initiated the practice of washing potatoes in water before consuming them—a process that was learned by other macaques through imitation.

Source: Photo by M. Kawai, courtesy of Kyoto University Primate Research Institute.

Warren and Barron 1956; and Shell and Riopelle 1957). **Learning sets** are defined as the acquisition of a learning strategy by the animal. We find that given a series of problems, an animal will transfer some of what it has learned about solving the first problem to solving subsequent problems. The learning curves of monkeys given blocks of test problems (figure 10–6) demonstrate this effect. As the series proceeds, by the second or third trial on each new problem, the animal scores an increasingly high percentage of correct responses. In effect, the formation of learning sets occurs as the animal "learns how to learn." Learning sets may also restrict the behavior of an animal: Negative consequences may result when an animal forms a learning set when solving one problem and cannot shift strategies as readily as an animal without the learning set.

FIGURE 10-6   Discrimination learning curves

Each dot on the graph represents the average percentage of correct responses on a particular trial for the 8 problems or 100 problems given to a rhesus monkey. Note the improvement that occurs with successive trials and particularly the large improvement by the second trial in the 100-problem set compared to the small improvement in the set of 8 preliminary problems.

Source: Data from H. Harlow, "The Formation of Learning Sets," in *Psychological Review* 56:51–56. Copyright © 1949 by the American Psychological Association.

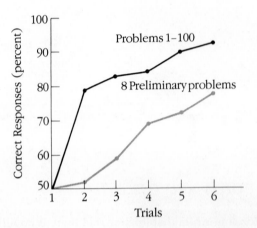

PHYLOGENY OF LEARNING

We cannot retrace the steps in the evolution of learning, but a brief survey of the learning capacities and capabilities of organisms from various animal phyla should provide an insight into the generality of learning principles and a sampling of the differences in learning abilities. Before proceeding, we must note four cautions. First, across the various phyla, learning has been examined disproportionately. More studies have been conducted on vertebrates than on invertebrates; and within the vertebrates, much of the attention has been focused on mammals. Second, scientists disagree about what constitute the proper criteria for demonstrating various types of learning. Third, exploring the physiological and behavioral aspects of learning requires more refined techniques and objective, bias-free methods. Fourth, performance, and not genetically determined ability or mechanism, must be measured. Many factors (notably, test design) can influence performance.

FIGURE 10-7   Interaction between starfish and sea anemone

When a starfish (*Dermasterias*) makes contact with the sea anemone *Stomphia cocinea*, the anemone will release from its attachment, "swim" free for a time, and eventually reattach itself to the substrate.

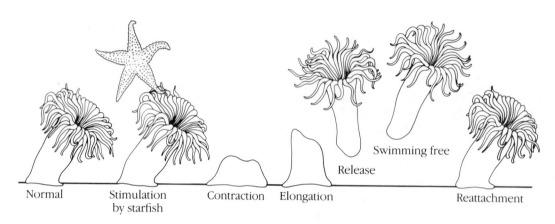

Normal   Stimulation   Contraction   Elongation   Release   Swimming free   Reattachment
                  by starfish

PROTOZOA AND COELENTERATA

Protozoans exhibit habituation, but there is no fully accepted evidence for association learning (Corning and von Burg 1973). For example, Patterson (1973) demonstrated that the contraction response of *Vorticella convallaria* to mechanical and electrical stimuli habituates with repeated stimulus presentations. Gelber (1965) claimed that *Paramecium aurelia* learned to associate presentations of a fine wire in their environment with food (bacteria). After numerous presentations of a wire coated with bacteria, elicited an approach response to the wire alone more paramecia "trained" in this way than did untrained paramecia or populations trained with an uncoated wire. An alternative interpretation of these findings was provided by Jensen (1965). He tested to see if the unicellular paramecia were merely responding to a food-rich zone created by the presence of the bacteria-coated wire. The debate over these interpretations has not been resolved.

There is also evidence of the habituation response in coelenterates, but little evidence of association learning. In the early years of this century, investigators generally assumed that coelenterates exhibited only certain involuntary stereotyped reflexes. We now know that many coelenterates have endogenous neural rhythms, and that rather than being totally "passive," many coelenterates are capable of spontaneous, active responses when interacting with their environment (Rushforth 1973). One demonstration of a form of association learning (Ross 1965) is based on the fact that sea anemones (genus *Stomphia*) respond to chemostimulation from certain starfish (e.g., *Dermasterias imbicata*) by stretching their bodies, detaching from the substrate, and "swimming" away (figure 10-7). They soon land and eventually reattach to the substrate. If chemostimulation is paired with gentle pressure applied near the base of the anemone in repeated trials, the application of pressure alone leads to a reduction in the swimming response.

PLATYHELMINTHES AND ANNELIDA

Research has been conducted mostly on planaria, and little is known about the other flatworm groups, some of which (tapeworms, flukes) are internal parasites. Conclusive evidence exists for habituation and several types of association learning in planarians, which have relatively simple bilateral nervous systems. Thompson and McConnell (1955) first reported classical conditioning in these organisms, but it was not until the report by Block and McConnell (1967) that satisfactory control procedures were employed. Among the issues was whether there was classical conditioning or whether the observed effects were due instead to either sensitization or pseudoconditioning (Dyal and Corning 1973; Corning and Kelly 1973). **Sensitization** is an increase in the strength of a conditioned response (CR), originally evoked by a CS, as a result of pairing the CR with a UCS. **Pseudoconditioning** is an increase in the strength of a response to a previously neutral stimulus as a result of repeated elicitation of that response with a different stimulus, with no pairing of the two stimuli's presentation. In general, researchers today agree that planaria are capable of exhibiting some form of conditioned learning.

Between then and now, considerable research on learning has been conducted in planaria, and has resulted in some startling conclusions. For a time, there was even a journal, *Worm Runner's Digest*, which included both serious and humorous articles on learning behavior in worms. Two of the most striking conclusions that arose from these experiments were: (1) Learning, or some component of learning was retained or passed on to the two regenerated planarians that resulted from severing a pretrained worm in half. The two regenerated planaria learned a T-maze choice task faster than naive worms or naive regenerates (McConnell, Jacobson, and Kimble 1959). (2) Worms that were permitted to cannibalize worms that had previously been taught a task learned that task much faster than planaria that consumed naive conspecifics (Corning and Kelly 1973). One potentially serious problem with some planarian learning studies is that they leave a slime trail as they move along. It is thus possible that using the same apparatus for repeated trials without cleaning would confound the results.

Annelids have a segmented nervous system with ganglia in each body segment and some concentration and coalescing of ganglia in the anterior body segments. Habituation to air puffs has been demonstrated for earthworms (*Lumbricus terrestris*); the backward movements that occur in response to puffs of air eventually decrease (Ratner and Gilpin 1974). Test paradigms used with worms utilize either aversive conditioning or some form of punishment training such as mild shock. Classical conditioning has been shown by pairing a light stimulus and mild vibrations (Ratner and Miller 1959). Operant conditioning has been demonstrated for annelids in a variety of experiments using a two-choice T-maze apparatus and mild shock for incorrect choices. Investigations of annelids, like those of flatworms, have the potential problem of persisting mucus trails that contain chemical cues.

MOLLUSCA

The learning capabilities of two major groups of mollusks have been investigated: gastropods (snails and slugs) and cephalopods (squids, octopodes). These organisms have one great advantage: the nervous systems of many species are readily accessible and contain large neurons. Thus, it is possible to explore some of the neurophysiological processes that accompany learning.

Habituation has been studied extensively in snails of the genus *Lymnaea*. Both visual and mechanical stimuli result in a withdrawal movement: the snail's shell is drawn downward and forward (Cook 1971). The response habituates with repeated stimulation. Habituation of the gill withdrawal reflex in *Aplysia* occurs with repeated stimulation from a jet of water (Pinsker et al. 1970). Kandel and coinvestigators (see Hawkins et al. 1983) have demonstrated differential facilitation of excitatory postsynaptic potentials in the neuronal circuit for this withdrawal reflex. Activity at the synapses between the sensory and motor neurons involved in this reflex arc is facilitated more by tail shock (UCS) if the if the shock is preceded by spike activity in the sensory neuron, than when the spike activity and shock occur in an unpaired pattern, or with the shock treatment alone.

Association learning has been reported in various species of both gastropod and cephalopod mollusks. Much of what we know about this group comes from studies conducted with octopodes (reviewed by Sanders 1973). Recall the material in chapter 3 regarding tactile discrimination in the octopus; members of this group can learn to discriminate cylinders based on the amount of grooved surface, but they fail to learn to discriminate different patterns when the amount of grooved area is roughly the same. Through convergent evolution, the peripheral aspects of the octopus's visual system are quite similar to the visual system found in vertebrates. However, the results of visual discrimination tests (figure 10–8) reveal that the octopus's interpretations of the stimuli must be quite different than that of most vertebrates: octopodes fail to make successful choices when the paired items are mirror images or when there are strong similarities between the arrangements of horizontal and vertical lines of the two stimuli. Finally, using conditioned learning, we can train octopodes to discriminate between various foods (Boycott 1965). If the octopus is presented with a fish and a geometric figure such as a disk, and upon approaching the fish the octopus is given a mild shock, the animal learns to avoid the fish. When the disk and fish are presented together, the octopus will retreat. If the fish alone is presented

FIGURE 10-8   Octopus performance over sixty trials on the discrimination of pairs of shapes differing in orientation only

When an octopus is given discrimination trials with pairs of outline shapes, it can successfully discriminate those that differ in orientation, but not those with the same or nearly the same horizontal and vertical extents.

Problem	Discriminanda	Horizontal extent	Vertical extent	N	Percent correct responses
1				6	81*
2				8	71*
3				8	65.5*
4				6	50
5				6	59*
6				7	56*

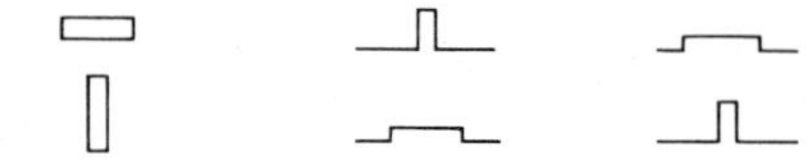

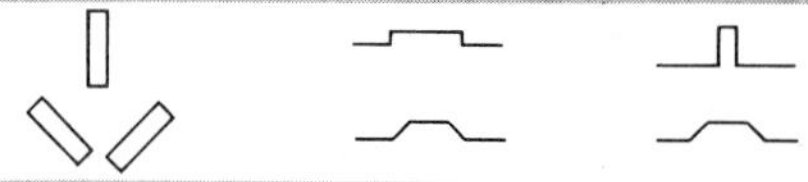

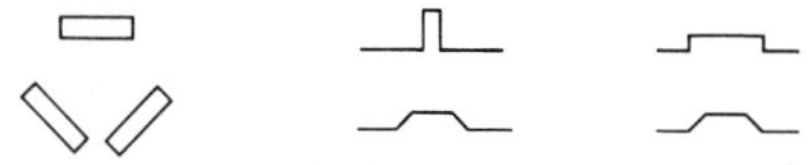

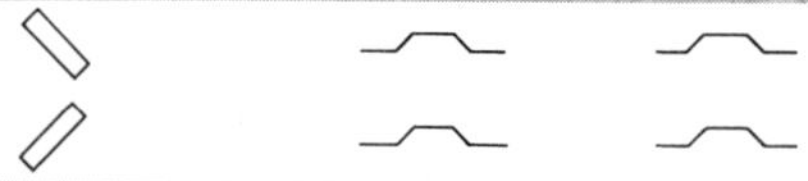

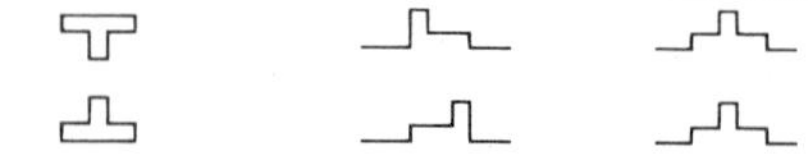

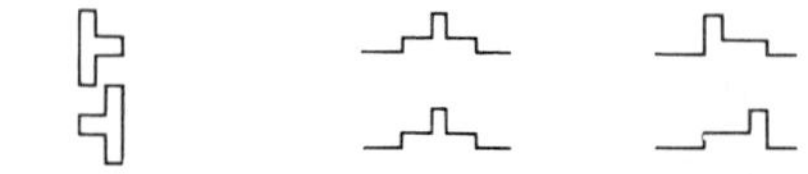

*Better than chance level of performance $P < 0.05$.

the octopus will attack. Such techniques can be used to train octopodes to approach one type of food (e.g., crabs), while avoiding fish, or vice versa.

Alkon (1980, 1983) demonstrated conditioning in the marine snail (*Hermissenda crassicornis*). These snails are normally positively phototactic during the daylight portion of their daily cycle. Snails were trained in glass tubes that were filled with sea water and mounted radially on a turntable. Prior to training, the snails' movement velocities toward the lighted center portion of the turntable were measured. Then the turntable was rotated, producing centrifugal force for the snails, and possibly simulating the water turbulence encountered near the ocean's surface during stormy weather. Thus, the light source and rotation were paired. After training,

the velocities were again recorded; trained snails moved toward the center of the turntable with about one-third of the pretraining velocity. Care was taken to perform a number of control procedures to ensure that the observed effect was an actual form of association learning by the snails. These control procedures included exposure to light alone, rotation alone, alternation of light and rotation, and presention of light and rotation at random. By mapping the neural pathways for reception of information from the light source and from the rotational forces, Alkon (1983) produced a wiring diagram for the pertinent portions of the nervous system of *Hermissenda*. He also showed that one of the consequences of training is that particular receptors and neurons become either more excited or inhibited, depending upon the conditioning.

ARTHROPODA

The arthropods are a large and diverse group comprising 80 percent of all animal species. Only vertebrate learning behavior has been studied more. The arthropods have evolved a wide variety of types of sensory receptors and have a nervous system characterized by an aggregation of ganglia—a "brain"—in the anterior segment of the body, and a pair of ventral nerve cords that pass to the thorax and abdomen. Learning has been studied most in three groups of arthropods: the subphylum Chelicerata (e.g., *Limulus*, scorpions, spiders), and the classes Crustacea and Insecta. Habituation, classical conditioning, and operant conditioning have been demonstrated in a variety of species in all these groups. Consider bees, for example, which can learn to discriminate colors by associating different hues with either sugar solutions or with plain water (von Frisch 1967; Wells 1973). Spiders can be trained to associate flies coated with either sugar or quinine with sounds of particular pitches (Walcott 1969). When glass beads are substituted for the flies, the spiders either discard or bite the beads depending upon the pitch of the sound presented with the bead. Various species of crabs and crayfish have been successfully trained in T-mazes (Gilhousen 1927; Datta, Milstein, and Bitterman 1960). Krasne (1973) reported that for crabs there may be some transfer of the knowledge gained from learning a simple maze to the learning of more complex mazes.

VERTEBRATA

We know considerably more about the complex processes which underlie learning phenomena for this phylum, particularly the mammals (reviewed by Masterton et al. 1976) than for most invertebrates. Rats and rhesus monkeys have been used in many of these studies. We will consider an example of both to provide some flavor of the research and how it is conducted.

Rats can be trained in an operant conditioning apparatus to press a lever for food. They can also be trained in a two-lever apparatus (L for left lever and R for right) to press them alternately, LR or RL, for a food reward. Can rats be conditioned to press the levers in a LLRR or RRLL double-alternation sequence? Investigators tested this question by conditioning rats first on the single-lever task, either R or L, followed by the single alternation task, RL or LR (Travis-Niedeffer, Niedeffer, and Davis 1982). They then rewarded only double-alternation performances. The rats learned the double-alternation task at a level exceeding chance expectations, and their performance improved over days. It was necessary in some instances, however, to let the rats give

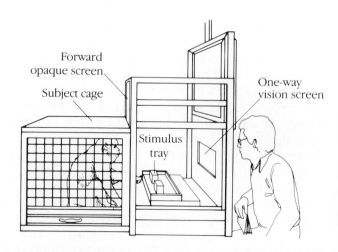

FIGURE 10-9   Wisconsin General Test Apparatus
An early version of the Wisconsin General Test Apparatus (WGTA) used to test aspects of learning in primates and, with some modifications of the apparatus, cats and raccoons.

extra responses to the first lever before pressing the second lever. Thus, LLLRR was rewarded in the same way as LLRR. When the investigators attempted to condition the rats to a sequence of LLRRLLRR or RRLLRRLL (double alternation with a fixed ratio schedule of two repetitions), no rat could successfully perform at better than a chance level.

One particular apparatus, the **Wisconsin General Test Apparatus (WGTA)** (figure 10–9) has been used in many rhesus monkey learning behavior studies (see Meyer, Treichler, and Meyer 1965, for the history of this and related techniques). One type of learning problem studied with this apparatus is the **delayed response problem** (Fletcher 1965). The basic procedure starts with the test tray out of reach but in full view of the test animal. Food was placed in one food well, and then two identical objects were placed over the wells. After a prescribed delay, the test tray was moved closer to the monkey, which then responded by lifting one object or the other. A trial ended when it picked up one of the objects, uncovering either the food well containing a reward or the empty well. A number of variables can be investigated with this procedure, and other animals such as cats and raccoons have been tested with slight apparatus modifications. Among the variables that have been manipulated are length of the delay phase, nature and size of the reward, nature and the similarity or dissimilarity of the objects covering the food wells, and presence or absence of an opaque screen in front of the tray (whether or not the animal is permitted to watch the test tray during the delay phase) (Meyer and Harlow 1952).

Among the conclusions arrived at were (a) rhesus monkeys are capable of learning basic discriminations in this procedure with delays of up to 30 seconds or more, (b) larger food reward leads to better performance, and (c) imposition of the opaque screen during the delay phase increases error rates by up to 50 percent. One interesting and striking finding in these studies is that the behavior and performances of individual monkeys differed markedly. In general, monkeys that exhibited hyperactivity in the test situation, and those that were more easily distracted during the delay phase, exhibited lower levels of performance. Clearly, in studies of learning behavior we must consider the significance of individual differences in performance, regardless of the species being tested or the task being performed (Warren 1973).

We have not yet discovered any consistent patterns relating phylogeny and learning among vertebrates. In different types of learning tasks, different species may be faster or slower than other species and may perform with a higher or lower error rate. One general finding was the rapid avoidance learning in feeding situations for virtually all vertebrates studied (Domjan 1980). We find that, in many instances, one trial is sufficient to establish an association between a food and a distasteful or noxious effect occurring from immediately after ingestion to up to one-half hour or more after ingestion. Wild rats that have eaten a sublethal dose of poisoned bait usually become bait shy and avoid that food source thereafter, making it difficult to use poisoned food in rodent control (Barnett 1963). While it is highly likely that learned food aversions are important evolutionary adaptations, we should be cautious in applying this logic without first carefully examining the nature of each species, its normal life patterns, and the test situations utilized to investigate the food aversion learning.

COMPARATIVE LEARNING

Over the past several decades, various comparative psychologists have attempted to make comparisons of learning behavior across the various phyla. Most notable of these are the reports by Bitterman (1960, 1965a, 1975) and Dewsbury (1978).

Bitterman's approach has been to look at certain organisms (e.g., fish, rats, turtles, pigeons, and monkeys) and their learning capabilities, and to classify organisms according to whether their learning abilities are like those of his test species. Basically, Bitterman believes different processes occur in different species; different learning phenomena may result from the same process in different organisms; and conversely, what appear to be similar learning phenomena in different organisms may be from different underlying mechanisms. He argues strongly against the principle of **equation,** the attempt to equate situations and procedures for learning in different species. He favors instead the concept of control by **systematic variation,** and investigates qualitative differences in learning of a wide variety of species from diverse taxonomic groups. Bitterman's work is significant because it cautions animal behaviorists who would plunge into comparative learning studies without a close look at the capacities of the various animals under study and the variety of methods used to test learning phenomena.

Dewsbury (1978) classified attempts to investigate comparative learning into two categories; quantitative comparisons and qualitative comparisons. Dewsbury first examined the cross-species comparisons for acquisition of learning or of avoidance learning (see also Brookshire 1970): there appear to be no general patterns across the various phyla for acquisition or avoidance-learning rates. The different species' development of learning sets for particular conditioned learning responses has been widely used to compare learning in vertebrates. Again, the data do not support any clear relationship between phylogeny and the learning set phenomenon. Dewsbury summarizes the problems of making such quantitative comparisons under several headings: (1) individual differences, which make it difficult to utilize mean performance levels as species-typical (see Rumbaugh 1968); (2) motivational differences; for example, in food-deprived animals that are given learning tasks; and (3) species differences, which reflect variations in biological constraints on learning. There is little evidence to suggest the possibility of constructing any sort of "scale of intelligence" for vertebrate organisms, let alone comparisons that attempt to include the invertebrate phyla.

CONSTRAINTS ON LEARNING

Over time we have discovered that two general types of constraints are operative in studies of learning in general and in studies of comparative learning in particular. These are the biological constraints brought to the learning situation by the animal, termed "preparedness" by Seligman (1970), and the methods constraints imposed by the investigator and test situation (see Hinde and Stevenson-Hinde 1973).

PREPAREDNESS

The generality of laws of learning has been questioned by Seligman (1970), who introduced the notion of **preparedness,** the genetically based predisposition to learn that is manifested during development. According to this notion, an animal faced with a particular learning task in nature or in the laboratory, may be prepared, unprepared, or contraprepared to perform that task. The constraints and limitations imposed by inheritance affect an animal's relative preparedness to learn (Hinde and Stevenson-Hinde 1973). In the evolutionary process, each species has been provided with tools and capacities that limit its range of potential responses to specified inputs.

For example, rats (both wild and laboratory strains) are prepared to learn to associate the taste of certain foods with subsequent illness, even when the onset of illness occurs up to several hours after they ingest the food (Barnett 1963; Rozin 1968; Garcia, Hankins, and Rusiniak 1976). This lack of a close timing relationship between the stimulus and response is an interesting and important contradiction of the previously established view that there had to be continuity between the presentation of a stimulus and the positive or negative reinforcement. Rats learn to avoid foods that make them ill after one or a few exposures; the evolutionary advantage of rapidly learning to avoid ingesting potentially harmful or poisonous foods should be obvious. Humans also make rapid associations between illness and a food they have ingested. Even though people have actually contracted an illness not connected with a food (e.g., the flu), they may associate something they ate shortly before becoming sick with the illness and may maintain the negative association for some time.

In other instances, it appears that even though an animal is capable of learning a particular task, we find its system is generally unprepared—that is, the animal requires training to complete the learning task. Rats do not naturally press levers for food, but they can be trained (conditioned) to do so. Chimpanzees normally communicate by vocalizations and gestures (van Lawick-Goodall 1968). They can, however, be trained to use the American Sign Language of the Deaf (Gardner and Gardner 1969; Fouts 1973). Chimps learning sign language require a great deal of time and training to become successful at even rudimentary communication with this system (figure 10–10).

A third possibility is that an animal may be contraprepared to perform a particular task—that is, even with many attempts or a great deal of training, the animal appears to be incapable of performing the task. In an

FIGURE 10-10   Chimpanzee signing
Using sign language is not part of the normal behavioral repertoire of chimpanzees. They can, however, be trained to learn a vocabulary of more than 100 words in sign language. This chimp has learned a variety of signs.
Source: Photo by H. Terrace from Anthro-Photo.

investigation of avoidance or defensive reactions, an animal's responses must be selected from species-specific patterns (Bolles 1970). Thus, pressing a lever to avoid an electric shock is an avoidance task for which the rat is contraprepared, whereas it is prepared to avoid the shock by running away, a more natural tendency. For any particular learning task, the capacities of a specific animal species are somewhere along the continuum formed by preparedness, unpreparedness, and contrapreparedness.

METHODS CONSTRAINTS

As we have seen many times in earlier chapters, restrictions or constraints operate in many laboratory and field situations where learning studies are undertaken (Shettleworth 1972); in fact, the apparatus and the situation being used to test the animal may alter the results. Stimulus cues must be related to the animal's sensory capacities and perceptual world, or *Umwelt*. A decrease in an animal's observed performance may result not from its inability to display a certain action, but rather from its being presented with an inappropriate stimulus—a stimulus that the animal cannot interpret.

We must be certain that the response tasks an animal is asked to perform are part of its potential repertoire. In addition, our experimental design must take into account the possibility of sex differences in learning capacities and the probability that animals of different ages learn at different rates, possess different amounts or types of prior experience, and have different sensory/perceptual capabilities.

If we state that a form of learning behavior is present in most animal types, we assume **commonality;** that is, we infer a phyletic, or evolutionary, relationship from the common exhibition of a type of learning. The assumption of commonality of evolutionary relationships brings with it potential problems and limitations (Hodos and Campbell 1969). Too often we merge data on animals of one phylogenetic lineage (a group of related species) with data gathered on animals of a different lineage. To obtain information about the evolutionary history of a behavior pattern, we should use only animals that share a common lineage. Although we may make inferences about a type of learning across divergent lineages, we do so for comparative purposes only, and do not imply direct-lineage evolutionary connections among diverse types of animals.

We must always be aware that behavior does not fossilize. Our discussion of the phylogeny of learning uses living organisms as examples. We should remember that traits considered rudimentary or general in one lineage may be considered specialized in other lineages.

LEARNING AS ADAPTIVE BEHAVIOR

Up to this point in the chapter, we have explored the various concepts and procedures involved in the study of learning, and we have been on a brief phylogenetic tour exploring learning capacities in various animals. What about the ways in which learning serves as adaptive behavior for animals in everyday life? Some instances of such phenomena have already been presented, but now we need to examine specific examples of learning's role in feeding, predator-prey relations, and reproduction and parenting. Each of these topics is covered in greater detail in the chapters that follow; what we will explore now are direct connections between learning and animals' interactions with their environment. The adaptive value of learning can be evaluated in both field and laboratory settings; the use of both locations for the study of learning may prove to be the best approach (Miller 1985).

FEEDING. Several species of corvids that inhabit the western United States have interesting habitats with respect to seed storage. At times, they consume the seeds immediately, but at other times they place the seeds in caches, which usually are holes or crevices in trees or other vegetation (Balda 1980, 1987; Kamil, Balda, and Grim 1986; Tomback 1980; Vander Wall and Balda 1981). The three species that are most conspicuous in their seed storage behavior differ in their general habitat ecology, and also to the extent that they rely on stored seeds. Clark's nutcrackers (*Nucifraga columbiana*) (figure 10–11) live at higher elevations than their cousins and have fewer alternatives to seeds for food, particularly in the winter months when nearly 100 percent of their consumption is stored pine seeds; each of these birds may cache as many as 33,000 seeds each year. Pinyon jays (*Gymnorhinus cyanocephalus*) live at intermediate elevations, have some alternative foods available, and pine seeds are 70–90 percent of their consumption in winter months; they cache some 20,000 seeds each year. Scrub jays (*Aphelocoma coerulescens*) inhabit much lower elevations, do not face severe problems with food location in the winter, and each bird stores only about 6,000 seeds annually.

Balda and Kamil (1988, 1989; see also Kamil and Balda 1985) have tested the cache recovery capacities of these three species of corvids. Their basic hypotheses were (1) learning and spatial memory are important in cache recovery, and (2) these capacities might differ across the three species of corvids as a result of the differences in habitat ecologies and in the importance of storing and remembering cache locations. The experiments were conducted in a large room using captive adult birds taken from the wild. Individual sessions

FIGURE 10-11 Clark's Nutcracker

Learning and spatial memory play critical roles in the feeding habits of Clark's nutcrackers. These birds cache pine seeds during periods of abundance and then return to relocate and consume them during periods (e.g., winter) when food is hard to find. The caching behavior has been studied under laboratory conditions, as shown here, in a chamber constructed so that birds can cache seeds in holes in the floor.

Source: Photo by Dr. Russel P. Balda

were provided for birds from all three species in which they were permitted to obtain seeds from a central feeder and cache them in a series of holes in a raised floor in the test room. The test room also contained rocks, tree branches, and other landmarks. Each bird was provided with habituation sessions so that it could become accustomed to caching in the holes in the floor, rather than in the more customary crevices and holes in vertical surfaces; most birds learned this process without difficulty, though a few did not and were not tested further. After each caching session, the bird was removed from the room and returned to its home cage.

Seven days later with one day of food deprivation, the birds were permitted back into the room with seeds placed in exactly the same holes where they had been cached by the bird in its previous session. The cache recovery efficiency of each bird was measured during the second test session. Cache recovery rates were higher

than chance for all three species, which indicated that the birds do learn and retain in memory the characteristics of the cache sites where they put their seeds. The recovery percentages were significantly higher for the nutcrackers and the pinyon jays than for the scrub jays. These differences in recovery rates may correlate with the birds' natural history, and with possible differences in spatial memory capabilities.

Another series of studies focused on aspects of foraging behavior, using rats in a laboratory setting that simulated the costs and benefits associated with food finding (Collier and Rovee-Collier 1982; Johnson and Collier 1987, 1989). Foraging costs in this test situation were simulated by bar-pressing requirements when a rat was placed into a series of Skinner-box test chambers (see figure 10–2). Individual rats were given choices between test chambers where the size of the food reward, and the cost in terms of bar presses, could be varied. Interestingly, they tended to take in a relatively constant amount of food each day, regardless of the variations in the test scheme. Rats tended to take more meals in the more profitable patches, either because of larger food pellets for the same effort or because of a lower effort required to obtain the food. These types of studies represent a novel approach to food sampling behavior and foraging patterns in animals.

Social learning may also play a key role in the dietary selection and preferences of animals like rats (*Rattus norvegicus*). Learning about foods starts with young rat pups' acquiring information about diet as we discovered in the previous chapter (Galef and Clark 1971). Additional studies by Galef and his colleagues have revealed that rats can learn to prefer or avoid foods by watching a demonstrator rat (Galef, Kennett, and Stein 1985), can transfer information about a distant food source (Galef, Kennett, and Wigmore 1984), and will follow another rat to a food source more readily if they know that the food source is palatable rather than toxic (Galef, Mischinger, and Malenfant 1987). In a most interesting development, Galef and collaborators (Galef et al. 1988) have determined that a chemical in the breath of rats, carbon disulfide (CS_2), acts as an enhancer during the process of learning about unfamiliar foods. For rats living in the wild, all of these processes could be critical when developing food habits, when acquiring new food habits, and when avoiding potentially poisonous foods. Since rats are opportunistic feeders and have a relatively catholic diet, social mediation of preference and avoidance is likely a significant feature in their overall dietary patterns.

FIGURE 10-12
Young African hunting dogs begin obtaining knowledge about prey species while they are still living at the den. Later, they accompany groups of adults on hunting trips, and eventually they learn to participate in running down and capturing various prey such as the wildebeest shown here.
Source: Photo by Harvey Barad/Photo Researchers, Inc.

PREDATOR-PREY RELATIONS. Both predators and prey may exhibit behavior patterns that illustrate the importance of learning. In nature, animals probably learn a great deal about their surroundings during the course of their daily activities. Some of this information does not seem to have immediate functional value, but may be important for survival later. Let's consider an example that involves an owl as the predator and mice as the prey. Metzgar (1967) demonstrated how this process might work for the white-footed deermouse (*Peromyscus leucopus*). Mice in one test group were each given time to explore and live in a room containing logs and trees. Mice in a second test group were held in laboratory cages without the experience in the test room. Each mouse from each group was then placed in the room with an owl present. The owl caught only two of the twenty mice that had prior experience in the room, whereas it caught eleven of the twenty mice that had no prior experience in the room. Spending time in the artificial habitat provided the mice in the first group with more knowledge of the habitat, and enabled them to avoid predation by the owl.

Wild hunting dogs (*Lycaon pictus*) in Africa live in social units consisting of several adults of each sex, juveniles, and pups (van Lawick-Goodall and van Lawick-Goodall 1970). The dogs of a pack hunt cooperatively (figure 10-12), which enables them to run down and capture larger prey than a single individual could take on its own. Hunting in this manner also enables the pack to provide food for their young and other dogs at the den area during the reproductive season. As young hunting dogs develop, they at first join in by tearing apart and consuming prey animals captured and brought back to the den site by other dogs. As they grow a bit older and stronger, they venture out with the adult dogs and start to participate in the hunting process itself. Initially this participation by younger dogs involves going out from the den area with the adult dogs to locate possible prey. Later, the juveniles start running with the adults for part of the chase, learning techniques such as selecting the prey, cutting a prey animal off from the herd, and running down the quarry by having dogs approach it from several directions. Eventually the younger dogs become full participating members of the hunting process, and they learn to take portions of the kill back to the den or to regurgitate food consumed at the kill site for animals that were not on the hunt. Similar learning sequences may occur in other related species that hunt cooperatively, e.g., wolves (*Canis lupus*) and spotted hyenas (*Crocuta crocuta*), and in some cats, e.g., lions (*Panthera leo*).

REPRODUCTION AND PARENTING. A key process pertaining to reproduction for many animals involves the ability to locate conspecifics of the opposite sex. One way this can happen has already been explored in some detail in chapter 9: the phenomenon of sexual imprinting, a special form of learning. The cross-fostering technique has been used to test whether young rodents of various species learn cues that are important for species recognition at early ages. In many instances, when young rodents are transferred at birth to foster species and are later tested for species preference, they exhibit an enhanced selection of the foster species and a reduced selection of their natural species (reviewed by D'Udine and Alleva 1983). Among the species pairs that have been tested in this manner are southern grasshopper mice (*Onychomys torridus*) and white-footed mice (*Peromyscus leucopus*) (McCarty and Southwick 1977); montane voles (*Microtus montanus*) and gray-tailed voles (*Microtus canicaudus*) (McDonald and Forslund 1978); house mice (*Mus domesticus*) (Quadagno and Banks 1970) and Norway rats (*Rattus norvegicus*) (Lagerspetz and Heino 1970); and pygmy mice (*Baiomys taylori*). Results from these studies are mixed; in some instances the fostered animals prefer their own species when tested as adults, and in others they prefer the foster species. In general there are in all instances some effects of the fostering, but the degree to which preferences are affected varies.

Evidence from primates and birds suggests that being a good parent and increasing the success of rearing offspring to independence can depend in part upon learning experiences. For some birds, e.g., white-fronted bee-eaters in Africa (*Merops bulockoides*) (Emlen 1984; Emlen and Wrege 1988), and Florida scrub jays (*Aphelocoma coerulescens*) (Woolfenden and Fitzpatrick 1984), birds other than the biological parents may participate in the processes associated with rearing of a clutch. The helping behavior may include assisting with territory maintenance, guarding the nest, incubating, brooding, removing fecal sacs, and feeding the young. Providing assistance means that the birds delay their own reproduction by one to several years. This topic is

discussed in more detail in chapter 20, where we treat the evolution of social behavior. At this juncture, the point we are making is that the various learning experiences obtained by the birds providing help at the nest, contribute to their own abilities to be good parents and to a higher rate of reproductive success for themselves when they become parents. The process of providing help may have evolved under a variety of selection pressures, but one outcome is that young birds have the opportunity to practice parenting behavior before attempting to nest on their own.

In a similar way, a number of primate species exhibit behavior patterns in which individuals (usually females, but sometimes males) other than the biological mother provide temporary care for a developing individual (see figure 13–20) (Quiatt 1979; McKenna 1981). This phenomenon has been termed *aunting behavior* or *allomothering*. Through handling and carrying an infant, other monkeys, often juvenile females or females nearing the age of first reproduction, will acquire some of the skills necessary for being a good parent. There is also evidence that even with practice acquired through aunting behavior some primate species' (where data are available) infant mortality is high for first or even second births. For example, first-born rhesus monkey infants have only a 74 percent chance of surviving the first month and a 55 percent chance of surviving six months (Drickamer 1974). These percentages increase only slightly for second births. However, by the time the females deliver and care for their third offspring, the survival rate to six months of age has risen to 78 percent, and by the fourth birth to 91 percent. Thus, learning how to be a good successful parent has a relatively high cost even for a social primate.

The foregoing is but a small selection of many examples where learning plays a critical role in the survival and reproduction of various animal species. In chapter 16 on habitat selection, we will see in more detail how learning influences the types of habitat characteristics with which deermice (*Peromyscus maniculatus bairdi*) (Wecker 1963) and chipping sparrows (*Spizella passerina*) (Klopfer 1963) will associate. As part of their orientation mode during migration (chapter 15), many birds use star patterns in the night sky. Studies using birds reared under star patterns that have been altered; e.g., in a planetarium; will orient differently than conspecifics reared under the natural night sky (Emlen 1970). Lastly, many of the communications postures and signals we will discuss in chapter 11 are learned as the animals develop.

MEMORY

What happens when an animal learns? By definition, learning implies some form of retention of experience. **Memory** refers to the capacity of an organism to form lasting connections based on past experiences. Several theoretical explanations for the processes of memory, information storage, and retrieval have been advanced.

THEORIES OF MEMORY

One theory (see Arbib 1972), often called the **dynamic hypothesis,** states that experiences and input of sensory information set up persistent electrical activity in the central nervous system. These so-called reverberating circuits, or populations of continuously active neurons, are the suggested basis for information storage in some coded form. When the active neural processes cease, forgetting occurs, and that bit of information is lost.

Other investigators (John 1967; Deutsch 1973) support the hypothesis that learning and sensory input produce permanent changes in biochemical processes or structures within cells, and memory thus involves structural changes. These structural changes are initially brought about by neural activity, but they are retained in permanent storage form after the neural activity produced by an experience has ceased. At this time we do not know for certain which of these two major hypotheses is correct; additional studies will be needed to solve the mystery of how information is stored permanently in neural tissue.

STORAGE MECHANISMS

Apparently, two separate types of memory storage mechanisms exist, or one mechanism for memory has two separate stages: a **short-term memory,** sometimes called the labile phase, and a **long-term memory,** or permanent phase (although forgetting may still occur). Immediately after an event, the experience is stored in short-term memory, possibly by neural activity alone. During this labile phase, various types of interference (e.g., a concussion or sudden blow to the head) may cause loss of the information. After a **memory trace** (the physical manifestation of learning or sensory input within the central nervous system) is transferred to long-term memory, the memory becomes relatively permanent by processes that are apparently more chemical and

structural in nature. Research on memory processes is still in its early stages; the complexity and exact nature of information storage mechanisms must still be unraveled. Much of the early work on this subject is presented in Lashley (1929) and Hebb (1958). (For thorough reviews of the history of memory theories with particular emphasis on human memory, see Kety 1982 and Woody 1982.)

Localizing and characterizing the memory trace or **engram** has proven to be a most intriguing but difficult undertaking. As we noted in chapter 6 when we discussed canary songs, correlative changes in dendritic growth and synaptic connections between neurons apparently occur as the birds learn new song repertoires each spring (Paton and Nottebohm 1984). In addition, ample evidence exists for increased RNA and DNA synthesis with learning and the storage of information. In the final analysis, models that are developed for memory must take into account both chemical and structural changes.

MOTIVATION

At the beginning of this chapter we defined motivation in terms of internal processes that are manifested by drives, with the attainment of certain goals or needs as rewards. The concepts of motivation and drive are often invoked as intervening variables to explain not yet fully understood events that occur within an animal. The external evidence from which motivation and drives are inferred consists of the known stimuli and the observed behavioral actions.

Why do we need the concepts of motivation and drive? First, differences in responsiveness to particular stimuli can be measured when the stimulus is presented at different times. If a male European robin (*Turdus migratorius*) is presented with a small cluster of red feathers in midwinter, it will show little interest. But if a few red feathers are presented in the spring, the male robin will exhibit threat displays characteristic of its reaction to another male robin. Second, the strength of stimulus needed to elicit a response may vary over time. A satiated dog may be relatively selective about what items it chooses to eat. The same dog when hungry will accept less palatable food. Third, response frequency or intensity may vary over time. Female rats exhibit differences in general activity levels at different phases of the estrous cycle. Investigators have postulated that motivation and drive can account for these variations in responses. Drives, which are correlated with animal needs such as hunger or thirst, are in effect **homeostatic mechanisms:** satisfying a need helps to maintain the proper internal life-sustaining conditions within the animal. For example, when energy expenditures deplete body resources, an animal needs food. Motivation theory interprets this to mean that the hunger drive is increased, and the animal will begin to search for food. Only when the animal locates and consumes food will it satisfy the specific drive and restore homeostasis.

The more information scientists gather about the genetic, neural, and hormonal bases of behavior control, the less important they are finding the concepts of motivation and drive. They are opening the "black box" that necessitates the postulation of an intervening variable, and they are substituting alternative neurological explanations of the observed behavior.

LORENZ'S MODEL

Several models have been proposed for motivation and drive. Here we shall briefly consider only two: one of historical interest and the other more current. Lorenz's (1950) proposed the psycho-hydraulic system for motivation. The system includes a reservoir that fills with fluid, a faucet through which the fluid flows into the reservoir, and a valve through which fluid flows out of the reservoir (figure 10–13). The degree of valve opening is controlled by a pulley connected to a pan on which weights can be added; the more weights in the pan, the wider the valve opens. Since in Lorenz's conception of the model the weights are analogous to sign stimuli, the valve opening is controlled both by the building up of "action-specific energy" in the reservoir and by the strength of the relevant sign stimuli; the valve is thus the innate releasing mechanism (IRM). The model can account for certain cyclical changes in behavior (e.g., feeding), but it fails to account for the sensory feedback that leads to reduction or cessation of an activity when the goal is achieved (see experiment described by Janowitz and Grossman 1949).

DEUTSCH'S MODEL

Another model (figure 10–14), proposed by Deutsch (1960; see also Toates 1986), postulates a set of components or elements located somewhere in the central nervous system. Deutsch proposes that mechanisms

FIGURE 10-13   One early scheme to explain motivation was developed by Lorenz, often called the psycho-hydraulic model

Action-specific energy accumulates in a reservoir until released by the appropriate stimulus (represented by weights on a pan scale), or until the pressure on the valve causes an action pattern to occur spontaneously. The consummatory response or fixed action pattern(s) released vary depending upon how much action-specific energy is released from the valve.

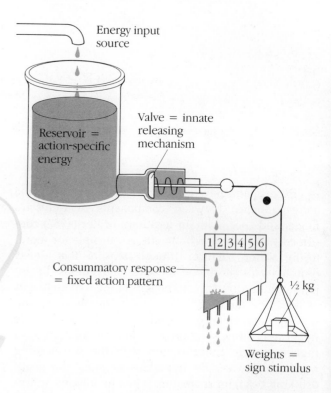

within the nervous system monitor the body for specific needs, such as blood levels of various substances, and water. When conditions in the animal change (when the homeostatic balance shifts) a central link is activated, and the motor system is stimulated. The animal then exhibits a particular behavior pattern, such as eating or drinking, as an interaction with the external environment. Behavior results in receptor discharges within the animal that produce an inhibition of the central link and depress the state that originally stimulated the motor system.

FIGURE 10-14　Deutsch's model for motivation
This model for motivation, proposed by Deutsch in 1960,
involves four separate components and the environment.

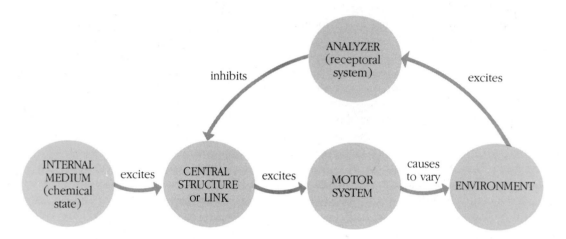

Experiments have provided data that support some correspondence between components of Deutsch's model and specific brain locations. Miller (1957) confirmed that there is a brain site responsible for monitoring water balance through sensors that detect osmolarity of the blood or other body fluids. By injecting fluids directly into regions of the hypothalamus of goats, he identified specific sites involved in controlling water balance. When he injected small drops of hypertonic saline solution (a hypertonic solution has a higher particle concentration than the surrounding medium) into certain hypothalamic sites, the goats' drinking behavior increased. When he injected an isotonic solution (one that has the same particle concentration as the surrounding medium), the drinking behavior was not affected; and when he injected pure water, a hypotonic solution (one that has a lower particle concentration than the surrounding medium), the drinking behavior was reduced, even in an otherwise thirsty animal. Electrical stimulation of particular regions of the hypothalamus of the goat or rat elicits drinking behavior (Anderson and McCann 1955; Greer 1955). In contrast, lesions of the same hypothalamic locus in the dog can greatly reduce water intake (Anderson and McCann 1956).

An experiment on the feeding behavior in dogs by Janowitz and Grossman (1949) illustrates the connection between the analyzer and the central link. They surgically produced an opening in the dog's esophagus so that any food introduced into the mouth exited out the animal's throat and did not enter the stomach. The question was, when the animal is hungry and eats, is feeding inhibited by food in the mouth or in the stomach? When food is passed through the mouth of a test dog, as in the Janowitz and Grossman experiment, further feeding is not inhibited; but when the stomach is filled with food, the animal ceases to feed. Thus the connection between the analyzer and the central link in this system is some type of receptor in the stomach that transmits messages to the central nervous system to halt further eating. The Lorenz model would not correctly predict the outcome of this experiment, but the Deutsch model does.

Halliday (1983) and Colgan (1989) have recently reviewed and revitalized the concept of motivation. Colgan proposes that we examine the idea of motivation from three perspectives, ethology, physiology, and ecology. In this way, a useful picture of animal needs and drives can be pieced together combining the actual behavior observed, the underlying neural and hormonal correlates, and the ecological context in which the animal must meet its needs. In this scheme it is also possible to maintain clarity about causal analysis on the one hand and functional analysis on the other. Halliday provides an integrated view of motivation as a complex of events involving the nervous system, the endocrine system, and the endogenous rhythms, all primarily involved with proximate factors controlling behavior.

The use of a model of motivation like Deutsch's, or like Colgan's three-part approach, can serve as a basis for exploring the operation of homeostatic mechanisms and can stimulate research on location and operation of components of a hypothetical scheme and on the functional aspects of motivation. Eventually as we determine what the actual internal components are, we should no longer need hypothetical constructs like motivation.

SUMMARY

Two important influences on behavior are *learning*, the relatively permanent modification of behavior that occurs through practice or experience, and *motivation*, the internal processes that arouse and direct behavior, but which must be inferred because these processes cannot be measured directly.

Habituation is the relatively persistent waning of a response that results from repeated stimulus presentations not followed by any form of reinforcement. In classical conditioning a neutral stimulus is paired with an *unconditioned stimulus* (UCS) to elicit an *unconditioned response* (UCR). After repeated pairings, the neutral stimulus becomes a *conditioned stimulus* (CS) and elicits the *conditioned response* (CR). In *operant conditioning*, sometimes called *instrumental learning*, the animal learns to associate a behavior (e.g., performing a particular action or task) with the consequences of that behavior, for example, a food reward (*positive reinforcement*) or a mild shock (*negative reinforcement*). Among the many similarities of classical and operant conditioning and of the procedures used to investigate them are those of *acquisition, schedules of reinforcement, extinction, spontaneous recovery,* and *generalization*. They differ in that classical conditioning involves a stimulus-stimulus pairing and the animal does not actually control the sequence of events, whereas for operant conditioning there is a stimulus-response pairing, and the sequence of events is contingent upon the animal's behavior.

Four other types of learning are related to operant conditioning (instrumental learning): *avoidance learning, insight learning, latent learning,* and *observational learning*. When an animal is presented with a series of similar problems and tested sequentially, the animal may develop a *learning set*.

A phylogenetic survey of the capacities of various groups of animals to demonstrate various types of learning reveals some useful patterns. Protozoans and coelenterates exhibit habituation, but not classical or operant conditioning. The flatworms (Platyhelminthes) and segmented worms (Annelida) are capable of habituation responses and also exhibit both classical and operant conditioning in some test situations. Trails of chemical cues left by the organisms may be a problem in experiments on these groups. Mollusks and arthropods are also capable of habituation, as well as classical and operant conditioning. For all of the foregoing groups, very few species have been tested from each phylum, and our knowledge of the overall phylogeny of learning is thus quite limited. Vertebrates have been tested more extensively, but even within this phylum there are large gaps in our knowledge. There are no clear relationships between phylogenetic lineage and learning capacities among the vertebrate classes. Vertebrates, and some invertebrate organisms, do exhibit rapid avoidance learning in feeding situations.

Studies of comparative learning have been both quantitative and qualitative. Quantitative comparisons are impeded by individual differences in learning and motivation and by species differences that are biological constraints on learning. An animal presented with a particular learning situation may be *prepared, unprepared,* or *contraprepared*. *Preparedness* is defined as the genetically based predisposition to exhibit certain behavior. *Methods constraints* in the study of learning make it impossible to equate the test conditions for each species. Also, we must be constantly aware of variations in sensory and perceptual worlds among different species.

Learning serves as the basis for adaptive behavior through which each animal is able to meet the challenges of everyday life and to cope with the problems of survival and reproduction. The types of learning involved in the natural lives of animals are exemplified by birds' caching behavior, rats acquiring food habits, wild hunting dogs' learning to kill prey, mice's learning about refuges to avoid predation, learning species-specific communications signals for mate finding, and practicing to become a good parent.

Memory, the process of information storage, may entail either dynamic reverberating circuitry in populations of neurons, or biochemical or structural changes within cells, or both. Some evidence suggests that *long-term*, or permanent, memory involves chemical-structural changes, and *short-term* memory may consist of neural activity.

Motivation and *drive* are terms used to explain changes in observed behavior due to variations in internal conditions within an animal. As we gather experimental information about the roles of genetics, neurons, and hormones controlling behavior, we will no longer need these terms, nor will we need models explaining motivation.

Discussion Questions

1. Two important aspects of a young animal's life are its diet and its selection of a place to live. For each of the organisms listed below, indicate what type of learning processes are involved in acquisition of food habits and habitat selection: (a) minnow, (b) cardinal, (c) sea anemone, (d) chipmunk, (e) gypsy moth larva. How would you proceed to test your hypotheses with experiments?

2. Drawing upon your knowledge of behavior control, identify situations in which we should be able to propose specific experiments (other than those presented in the chapter) to investigate portions of the Deutsch model for motivation. Where possible, provide suggestions for actual experiments.

3. In this chapter, we have examined how hypotheses about learning can be tested in applied or natural settings. Think of several ways you might test learning behavior in a natural setting using field methods for each of the following: (a) cockroach, (b) mountain goat, (c) crow, and (d) gorilla.

4. Imagine that we have just located a new species of butterfly. We are interested in obtaining an accurate assessment of the sensory/perceptual world of this organism and its learning capacities. Describe the procedures you would use to conduct the experiments necessary to answer this question.

5. One way to look at learning is to view it as a complex of adaptive behaviors that provide an organism with the flexibility and capacity to deal with the problems of survival and reproduction. How would you defend or contradict this statement: Invertebrates have less flexibility and a greater degree of preprogramming in their behavior than do vertebrates, and we would therefore expect invertebrates to exhibit fewer learning capacities than vertebrates.

6. The data in the following histograms are from the work of Galef and Clark (1971) regarding learning of dietary habits in rats. Depicted here are the feeding bouts and approaches by wild rat pups, starting at age 21 days, directed toward two diets, *A* and *B* placed 5 cm apart. The parents of the test animals had been previously poisoned while feeding on diet *B* with the pups present. What conclusions can you draw concerning the dietary preferences of these pups? Of what adaptive value might their responses be in a natural setting where young rats are learning what to eat and what not to eat?

From B. G. Galef and M. M. Clark, "Social Factors in the Poison Avoidance and Feeding Behavior of Wild and Domesticated Rat Pups", *Journal of Comparative Physiology and Psychology*, 75:352.

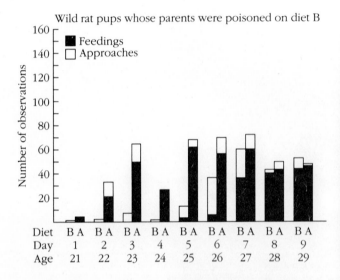

Suggested Readings

Colgan, P. 1989. *Animal Motivation.* New York: Chapman and Hall.
An interesting new monograph that examines animal motivation from several different perspectives. Thorough and well-written with a good bibliography.

Domjan, M. 1980. Ingestational aversion learning: Unique and general processes. *Adv. Stud. Behav.* 11:276–336.
Comprehensive summary of research on this hot topic. Considers theoretical and practical aspects and has a methodological perspective. Good reference source for this topic.

Hinde, R. A., and J. Stevenson-Hinde, eds. 1973. *Constraints on Learning.* New York: Academic Press.
Summary volume on modern learning, largely from ethological and zoological perspectives. Good reading for the scientist or nonscientist. Provides some consideration of ecological and evolutionary issues in the study of learning behavior.

Kamil, A. C., and T. D. Sargent. 1981. *Foraging Behavior.* New York: Garland STPM.
Part III of this book provides extensive treatment of learning behavior and foraging in a variety of animals.

Many of the topics covered in this chapter are given expanded discussion, and many additional examples are presented.

Plotkin, H. C., and F. J. Odling-Smee. 1979. Learning, change and evolution: An enquiry into the teleonomy of learning. *Adv. Stud. Behav.* 10:1–42.
Relates learning theory and evolution. Written for those who have some background in learning theory and modern evolutionary thought.

Staddon, J. E. R. 1983. *Adaptive Behavior and Learning.* New York: Cambridge University Press.
A rather lengthy but well-written volume on all aspects of learning behavior. Deals primarily with vertebrates. A good source book for information and references for those with primarily a psychology background.

Walker, S. 1987. *Animal Learning: An Introduction.* London: Routledge & Kegan Paul.
An up-to-date summary of what is known about classical and operant conditioning. Written by a psychologist, provides a psychologist's perspective on the important aspects of learning behavior.

PART FOUR

BEHAVIORAL ECOLOGY: SOCIAL PROCESSES

11

COMMUNICATION

Animals convey information to members of their own species, and to other species as well, through an incredible diversity of sounds, colors, flashing lights, smells, and postures. The song of a male white-crowned sparrow (*Zonotrichia leucophrys*), for example, is specific not only to the species but to the area and to the individual. By singing, the male may not only be warning neighboring males away, but also be providing potential mates with information about his health, his social status, and even his place of birth.

We begin this chapter by defining communication, by examining how signals are coded to convey information, and by determining how we can measure the information transfer that is taking place. Next we consider the kinds of information conveyed and the sensory channels through which the signals are received, discuss the evolution of conspicuous body structures and the ritualization of behavior, and return to the question of who benefits from communication by considering honesty and deceit in signaling. Finally, we look at two examples of symbolic language in nonhuman animals, one in an insect and another in an ape.

From extensive field observations, ethologists have found that communication behavior occurs in regular patterns that recur nearly unchanged from event to event. These sequences are called *fixed action patterns (FAPs)*. The most striking FAPs are those that occur when animals interact with conspecifics in courtship, territory defense, or dominance encounters. Signaling is the primary function of these behaviors. Julian Huxley (1914) studied great-crested grebes and called their strange postures **displays.** Moynihan (1956) defines *display* as any behavior pattern especially adapted in physical form or frequency to function as a social signal. Ethological interest focuses on the ways that displays have evolved from other noncommunicative behavior patterns.

WHAT IS COMMUNICATION?

Wilson (1975) defined **biological communication** as an action on the part of one organism (or cell) that alters the probability pattern of behavior in another organism (or cell) in a fashion adaptive to either one or both of the participants. The word **adaptive** implies that the signal or response is to some extent genetically controlled and under the influence of natural selection. But this definition presents some difficulty, as Marler (1967) pointed out: What about the mouse that rustles in the grass, making sound that enables the owl to catch it? This case fits Wilson's definition, but would we really say that the mouse is communicating with the owl? One way around this problem might be to add that the sender must *intend* to alter the receiver's behavior; clearly the mouse did not mean to attract the attention of the owl. But now we have another problem: How can we ever tell what an animal intends? Another way to avoid these definitional problems is to require that the sender benefit. For this reason, Slater (1983) defined communication as "the transmission of a signal from one animal to another such that the sender benefits, on average, from the response of the recipient."

Ethologists argued that displays evolve in a way that maximizes the effectiveness of information transfer between sender and receiver to the benefit of both. Emphasis here is on the coevolution of signals, with the implication that natural selection acts at a level above the individual. Although both sender and receiver do frequently benefit, the sender may benefit at the expense of the receiver, as when a parent bird uses a broken-wing distraction display to lead a predator away from her young. The more recent sociobiological view of communication starts with the assumption that natural selection acts primarily at the level of the individual; thus communication is a means by which the sender manipulates others for his or her own benefit, as in advertising. The receiver may benefit or may be harmed. Signals become ritualized so that the sender can control the behavior of the receiver with a minimum of wasted energy (Dawkins and Krebs 1978; Krebs and Dawkins 1984). In this case, the purpose of the display may be not to inform, but to persuade. Exaggeration and redundancy are the rule, and ritualization occurs to increase the persuasive power of the behavior, not to maximize information transfer. Throughout the rest of this chapter, be alert to possible cases of manipulation and deceit.

HOW DO SIGNALS CONVEY INFORMATION?

We saw in chapter 6 that information is coded in the nervous system by frequency modulation of electrical impulses. The information in displays is coded in a much more diverse fashion, as we shall see.

DISCRETE AND GRADED SIGNALS

Some signals are **discrete** (digital) but others are **graded** (analog). For example, equids such as zebras communicate hostility by flattening their ears, and communicate friendliness by raising their ears (discrete signals) (figure 11–1). The intensity of either emotion is indicated by the degree to which the mouth opens (graded signal). The mouth-opening pattern is the same for both hostile behavior and friendly behavior. Graded signals may vary in intensity as a function of the strength of the stimulus.

FIGURE 11-1 Composite facial signals in zebras
Ears convey a discrete signal. They are either laid back as a threat or pointed upward as a greeting. The mouth conveys a graded signal and opens variably to indicate the degree of hostility or friendliness.

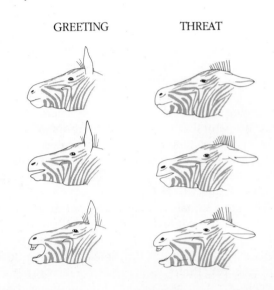

GREETING   THREAT

DISTANCE AND DURATION

Although most species are limited to from twenty to forty different displays (Wilson 1975), animal signals can vary in several other ways that increase information content. Some signals use the modality of smell. In some cases, the amount needed to trigger the receiver's response is no more than a few molecules, as is that of the sex attractants of many insects. Thus the distance a signal travels may vary: A small amount of material produced by a female can be detected by a male several kilometers downwind. At the other extreme, visual displays usually operate over much shorter distances. The duration of a signal may also vary. Alarm signals, such as the chemicals produced by many invertebrates, may have a localized short-term effect, and thus a rapid fade-out time (Wilson 1975). Bright plumage and other male adornments such as antlers may last the entire breeding season. In contrast, although the brightly colored epaulets on the red-winged blackbird are present during the entire breeding season, they are conspicuous only when the male exposes and erects the epaulet feathers during the song spread.

COMPOSITE SIGNALS, SYNTAX, AND CONTEXT

Two or more signals can be combined to form a **composite** signal with a new meaning. In the zebra example, the meaning of the open mouth depends on whether the ears are forward (friendly) or backward (hostile) (see figure 11–1).

Animals can convey additional information with a limited number of displays by changing the **syntax,** or sequence of displays. For example, the two composite signals *A* and *B* would have different meanings depending on whether *A* or *B* came first. There is no evidence of natural syntax use in nonhuman animals; however, language-learning in chimpanzees, in which chimps assemble words in novel ways to communicate with humans, has been accomplished or demonstrated (Rumbaugh and Gill 1976).

The same signals can have different meanings depending on the **context;** that is, depending on what other stimuli are impinging on the receiver. For example, the lion's roar can function as a spacing device for neighboring prides, as an aggressive display in fights between males, or as a means of maintaining contact with pride members. The song-spread display of the male red-winged blackbird serves in courtship with females, as well as in conflict with other males.

METACOMMUNICATION

Increasing the information content of displays by **metacommunication,** or communication about communication, is theoretically possible: one display changes the meaning of those that follow. We can see good examples in play behavior: animals use aggressive, sexual, and other displays in play, but they precede such behavior by an act that communicates the message, "What follows is play, join in" (Bekoff 1977). Canids such as dogs and wolves precede play with the play bow (figure 11–2). Monkeys communicate play behavior through a relaxed, open-mouthed face.

Disagreement exists about the importance of metacommunication (Smith 1984), and some observers have used the word a bit loosely. For example, male rhesus monkeys (*Macaca mulatta*) sometimes communicate dominance by carrying the tail elevated in an S-shape over the back. While this posture conveys status and mood, it does not really change the meaning of behaviors that follow; rather, it does communicate that aggressive behavior is likely to follow.

FIGURE 11-2   Metacommunication in dogs
The play bow performed by the dog on the right communicates that behaviors that follow are play.
Source: Photo by Marc Bekoff.

MEASUREMENT OF COMMUNICATION

Suppose that we wish to learn something about the way crayfish (*Orconectes rusticus*) communicate. If we place crayfish that are strangers together in a tank in the lab, their large claws assume various positions, and a dominance hierarchy related to the size of the animal develops (Bovjberg 1956).

OBSERVATION

Our first step is to identify relatively constant motor patterns by observing the animals and describing the movements they make. We might divide the behavior patterns into the following acts: *retreat*—a rapid, backward swimming motion; *cheliped presentation*—the movement of the large claw from a downward-facing position to one that is horizontal to the substrate (figure 11–3); *cheliped extension*—the rapid movement of the opened claw toward the other crayfish; *forward locomotion*—movement toward the other crayfish; and *fighting*—striking and pinching the other crayfish.

Simple observation might suggest that communication is occurring when certain behaviors occur only in the presence of other animals. Perhaps a behavior performed by one individual (e.g., cheliped extension) is usually followed by a particular behavior performed by the other animal (e.g., retreat). In that case, we suspect that communication has occurred. But the occurrence of any behavior is probabilistic (stochastic). How can we be sure that what one crayfish does affects the behavior of the other?

QUANTIFICATION

OBSERVED AND EXPECTED FREQUENCIES. For a more thorough analysis, we can compare the frequency of each behavioral response against the frequency expected if the response were random and independent of the behavior of the other crayfish. The observed and expected frequencies are shown in table 11–1. A chi-square test is used to see if the responses deviate significantly from random. Note from table 11–2 that some behaviors facilitate responses and others inhibit them. Some, such as forward locomotion, seem to have little effect on other behaviors.

INFORMATION THEORY. Another approach is to quantify the amount of information transmitted from one individual to another, a technique referred to as **information analysis.** Here information is measured as the reduction of the uncertainty of the behavioral response of an individual. The data show that when one crayfish presents its cheliped, the other is less likely to retreat and more likely to fight; there is a reduction in the uncertainty whether fighting or whether retreating will occur. How much information has been transmitted?

FIGURE 11-3 Cheliped (large claw) presentation and extension in the crayfish
In the first phase of cheliped presentation (a), the cheliped tip is lowered and the claw is opened, and in the second phase (b), the cheliped is raised. Part (c) shows cheliped extension. Arrows denote direction of movement. Analysis of these signals (tables 11–1 and 11–2) shows that cheliped presentation by one crayfish increases the chance of a fight with a second crayfish, whereas cheliped extension increases the likelihood that a second crayfish will retreat.

A

B

C

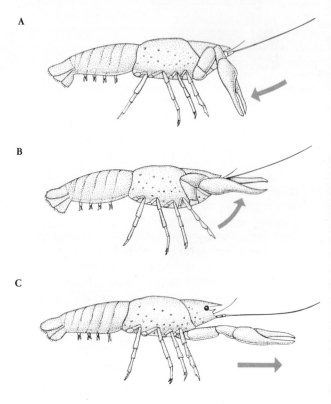

TABLE 11-1   **Frequency distributions of agonistic behavior patterns in crayfish**
Figures in parentheses are the frequencies expected if the behavior were random.

Signaler's signal	Receiver's response						Total observed
	No change	Retreat	Cheliped presentation	Cheliped extension	Forward locomotion	Fighting	
No change	4 (0.9)	1 (5.6)	0 (0.4)	4 (1.2)	0 (0.1)	0 (0.8)	9
Retreat	6 (4.0)	14 (24.3)	0 (1.8)	7 (5.2)	3 (0.3)	9 (2.7)	39
Cheliped presentation	6 (2.7)	6 (16.2)	6 (1.2)	0 (3.5)	0 (0.2)	8 (2.3)	26
Cheliped extension	8 (17.5)	126 (106.1)	3 (7.7)	24 (22.6)	0 (1.3)	9 (14.9)	170
Forward locomotion	17 (8.5)	39 (51.8)	6 (3.7)	15 (11.0)	0 (0.6)	6 (7.3)	83
Fighting	0 (7.4)	63 (44.9)	3 (3.2)	3 (9.6)	0 (0.5)	3 (6.3)	72
Total observed	41	249	18	53	3	35	399

Source: S. de Roth, 1974, "Communication in the crayfish (*Orconectes rusticus*)." Master's thesis, Bowling Green State University, 1974. Reprinted by permission.

TABLE 11-2   **Outcome of statistical analysis of frequency distributions of agonistic behavior patterns in crayfish**

Signal	Receiver's response	
	Facilitates	Inhibits
No change		
Retreat	fighting	retreat
Cheliped presentation	fighting	retreat, cheliped
Cheliped extension	retreat	extension
Forward locomotion		no change, fighting
Fighting	retreat	no change, cheliped extension

Source: S. de Roth, 1974, "Communication in the crayfish (*Orconectes rusticus*)." Master's thesis, Bowling Green State University. Reprinted with permission.

First, we can compute the information contained in the behavioral repertoire of one individual using the method developed independently by communication theorists Shannon and Weiner, usually referred to as the *Shannon-Weiner index* (Shannon and Weaver 1949). The formula is

$$H = -\Sigma p_i \log_2 p_i$$

where H is the information content, i is the occurrence of a particular signal and p_i is the probability of occurrence of that signal ($i/\Sigma i$). Each signal is considered to be discrete: it contains one message, either yes or no. Thus it comprises a binary system. For this reason, we use base two logarithms, and our unit of information is the *bit*. To measure the information in a two-animal system, we measure information at the source (the signaler) and at the receiver. In our example, the signaler information for one crayfish is calculated from data in table 11-1 and appears in table 11-3. The first signal, no change, occurred 9 times out of a total of 399 acts, so p_i = 9/399 or 0.022. $\log_2 p_i = -5.504$ and p_i times $\log_2 p_i = -0.1211$ (table 11-3). The rest of the signals are treated the same way, and then summed to give H. We can calculate the receiver information in the same way. To measure the information actually transmitted, which will be less than either of these values, we must find the conditional probabilities of the signals that evoke each response; Wilson (1975, 194-98) and Attneave (1959) outline the complete procedure, which is not shown here.

The signaler crayfish actually transmitted an average of 1.11 bits of information per behavior pattern, resulting in a 59 percent restriction of the second animal's behavior. One advantage of this analysis method is that it permits comparison between different organisms of the diversity and efficiency of their communication systems. Hazlett and Bossert (1965) compared several species of crustaceans and concluded that the amount of information transmitted per signal remains relatively constant across species. We might expect natural selection to maximize the sender's ability to restrict or control the receiver's behavior if the purpose of displays is the sender's manipulation of the receiver, as the sociobiologists argue.

TABLE 11-3 Computation of signaler information for crayfish data in table 11-1 using Shannon-Weiner formula

Signal (i)	Proportion of total acts (p_i)	$p_i \log_2 p_i$
No change	9/399 = .022	− .1211
Retreat	39/399 = .098	− .3284
Cheliped presentation	26/399 = .065	− .2563
Cheliped extension	170/399 = .426	− .5244
Forward locomotion	83/399 = .209	− .4720
Fighting	72/399 = .180	− .4474
Total	1.000	$\Sigma = -2.1497$
		$= -H$ (signaler information)

Note: $\log_2 x = \dfrac{\log_{10} x}{\log_{10} 2}$

FUNCTIONS OF COMMUNICATION

Signals may also be classified by their function. The ultimate function of any communication is increased fitness; this will be examined closer in the section on the evolution of displays. Although many criteria can be used to classify proximate functions of communication, these classifications tend to be artificial and arbitrary, and are made mainly for researchers' convenience to help keep things organized. The following functions of communication are modified from Wilson (1975) and Smith (1984).

GROUP SPACING AND COORDINATION

Group-living animals use a variety of signals that seem to keep members in touch. The highly arboreal *Cebus* monkeys of the South American rain forest forage in dense vegetation. A group of fifteen may spread out over an area one hundred meters in diameter as they search the treetops for fruits. In addition to the sound of moving branches, an observer hears a continual series of contact calls from the different members of the group. An individual that becomes isolated utters a "lost" call, which is much louder than the contact calls. Marler (1968) suggested that primates use the following types of spacing signals: (1) distance-increasing signals, such as branch shaking, that may result in another group's moving away; (2) distance-maintaining signals, such as the dawn chorus of howler monkeys (*Alouatta* spp.)

that regulates the use of overlapping home ranges; (3) distance-reducing signals, such as the contact or lost calls of *Cebus* monkeys; and (4) proximity-maintaining signals, such as those that occur during social grooming within groups.

Color patterns of some coral reef fish are species-specific; these patterns attract conspecifics and hold them together in schools. Social insects, such as termites, ants, and bees, emit a variety of chemicals that result in assembly of conspecifics (Wilson 1971).

RECOGNITION

SPECIES RECOGNITION. Species recognition before mating is crucial to avoid infertile matings between members of closely related species. For example, note the striking differences in the songs of three species of picture-winged Hawaiian *Drosophila* (figure 11–4). Their sound-producing mechanisms differ radically from mainland populations because they use their abdomens as well as their wings (Hoy et al. 1988). Animals generally communicate messages that are much more individualized than those needed for species recognition alone, however.

DEME RECOGNITION. Local dialects in bird song have been observed in a number of geographically separated populations of white-crowned sparrows. Females from one population in Colorado perform copulation soliciting displays when they hear songs of males from their

**FIGURE 11-4   Courtship songs of three species of
picture-winged Hawaiian *Drosophila*.**
Oscillogram tracings of songs are shown. Note the large
differences among species.

D. fasciculisetae (n = 6 males)

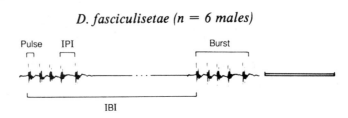

D. cyrtoloma (n = 3 males)

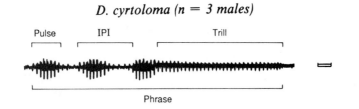

D. silvestris (n = 10 males)

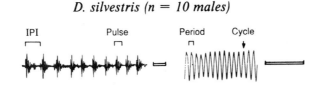

own population, but rarely when they hear songs from another population (Baker 1983) (figure 11–5). This tends to lead to mating with members of locally-adapted demes. The question of why individuals should show a preference for mates from the same deme will be considered in chapter 16.

INDIVIDUAL RECOGNITION. Indigo buntings (*Passerina cyanea*) produce a complex song that is quite variable. Most of the phrases are paired (sweet-sweet, chew-chew, and so forth) (figure 11–6). The playback experiment has been a critical tool in the analysis of the communications used in individual recognition, particularly for bird

song. Researchers record vocalizations with a special directional microphone and a high-quality tape recorder, and loop the tape for repeated playbacks, placing the loudspeaker directly in the field. Observers score the responses of resident birds by the number of times they sing and by the number of times they approach the speaker, before and after playback. Emlen (1972) analyzed tape recordings of the song and spliced pieces of the tapes together in varying orders. When he played them back in the field and observed the responses of territorial males, he discovered the significance of much of the song. Part of the sequence was species-specific, communicating the message, "I am an indigo bunting,"

FIGURE 11-5 The role of early experience in the response of female sparrows to male song.

(a) Sound spectrograms of home-dialect tutor songs and stimulus songs in white-crowned sparrows. The spectrograms lie in the frequency range 3 to 6 kHz (vertical axis) and are about 2 seconds long (horizontal axis). (b) Postures showing the response of each female subject to the two stimulus song dialects. Juveniles are unshaded and adults are shaded. The lordotic posture is the copulation solicitation display. Above each female is indicated the number of displays elicited from the subject during a 21–minute test. Juvenile and adult females responded more to the home dialect than to the alien dialect.

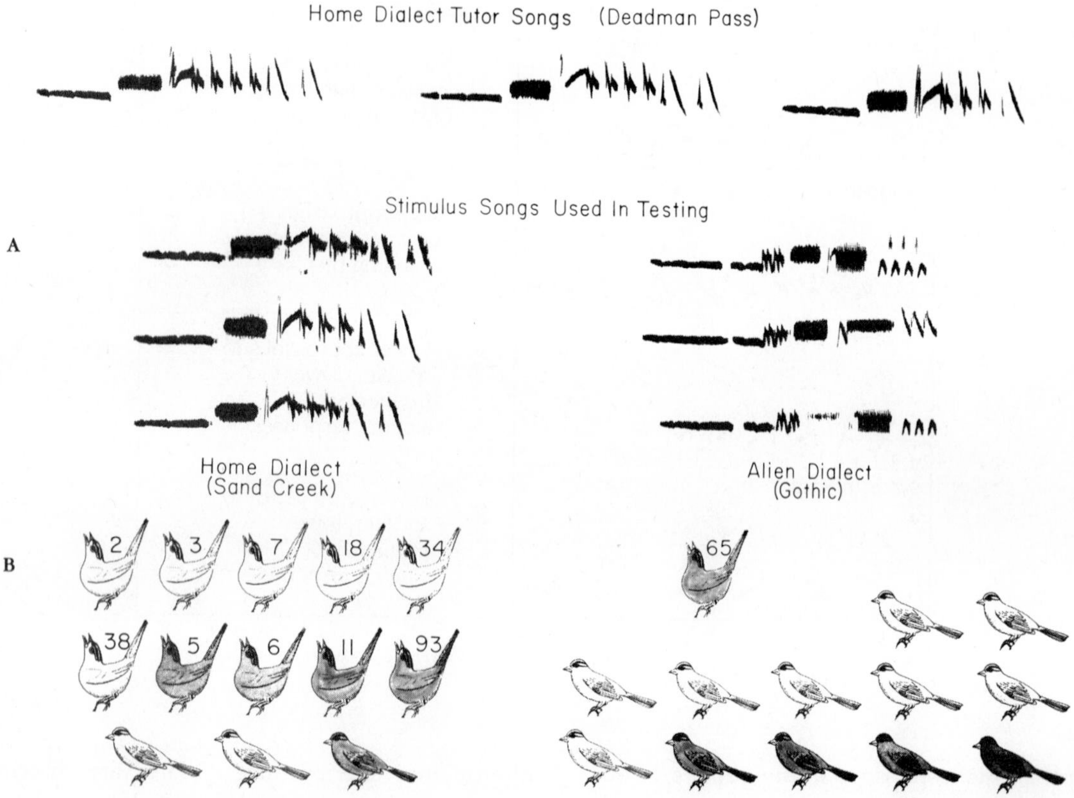

Home Dialect Tutor Songs (Deadman Pass)

Stimulus Songs Used In Testing

A

Home Dialect
(Sand Creek)

Alien Dialect
(Gothic)

B

and part was variable from individual to individual. Changing the ordering of notes (syntax) or eliminating one member of each pair did not greatly affect the response of males, but changing the rhythm or the frequency of the notes themselves reduced the agonistic responses of resident birds, apparently because the sounds were no longer recognized as those of other indigo buntings.

NEIGHBOR RECOGNITION. To test whether white-throated sparrow males recognize neighbors individually or as a class, Falls and Brooks (1975) studied the effect of playback location on the resident male, moving the speaker in and around his territory. When they placed the speaker at the boundary between his territory and his neighbor, his response to a stranger's song

FIGURE 11-6   Audiospectrograms of indigo bunting
Tapes of a normal song were cut and spliced together to test
the function of various attributes of the song. The normal
song and the nonpaired song elicited typical responses, but
temporal changes, such as reducing the time between notes
by one half or increasing it twofold, interfered with song
recognition.

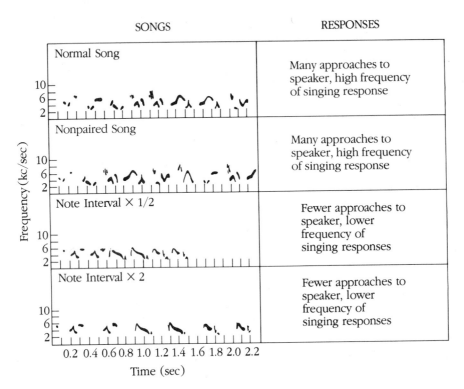

was stronger than his response to his neighbor's, as expected. But when they placed the speaker on the *opposite* boundary, he responded vigorously to the neighbor's song as if it were a stranger's. When they placed the speaker in the center of *his* territory, his response to the neighbor's song was intermediate. These results demonstrate that a territorial male can discriminate between the songs of his neighbors and can also associate each song with its singers' appropriate location.

CLASS RECOGNITION. Class recognition occurs mainly in social insect groups in which castes are treated differentially. The nest queen receives preferential treatment (food and care) from workers because of the

pheromones she produces. Males are discriminated against as a group; they receive less food from the workers and, in times of food scarcity, are even driven from the colony (Wilson 1971).

KIN RECOGNITION. Communication may be involved in the differential responses of many organisms to their close relatives. For example, tadpoles of the American toad (*Bufo americanus*) prefer to associate with siblings over nonsiblings, even after being reared in isolation (Waldman 1982). Waldman hypothesized that some substance, contributed by the mother in the egg jelly, is used as a cue. Other data, from insects, amphibians, birds, and mammals, demonstrate kin recognition even in the absence of interactions with kin early in life (Holmes and Sherman 1983). Chemical, auditory, and visual cues have all been implicated.

FIGURE 11-7   Nest guarding in sweat bees
(a) Percentage of intruders accepted as a function of
relatedness, (b) Female sweat bees were reared in nests of
six bees composed either solely of sisters (colony X and
colony Y) or of three sisters from one nest and three sisters
from another (colony XY). Later, bees from X, Y, and XY
colonies were tested as guards as shown; arrows represent
unfamiliar intruders seeking entrance to the nest.

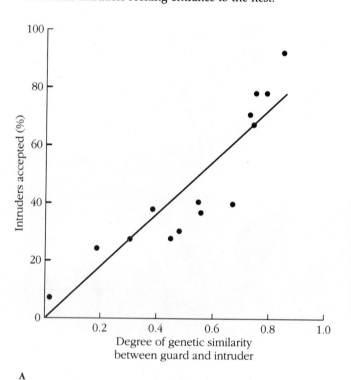

A

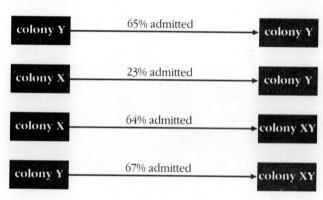

B

How can kin identify each other even if they have
never interacted? One possibility is **phenotype
matching,** where the individual "refers to" the kin
whose phenotypes are learned by association—the referents. The referent is then compared with the stranger
(Holmes 1988; Holmes and Sherman 1983). For example, female sweat bees (*Lasioglossum zephyrum*) guard
the nest entrance and admit only nestmates (usually
their sisters). In a laboratory experiment, Buckle and
Greenberg (1981) reared some guard bees with their
sister bees, and they reared some with nonsisters; these
bees thus became the referents of the second group. In
later tests, the guard bees admitted strangers, but only
if they were sisters of the referents (figure 11-7). A
second possibility for kin recognition—one that does
not require any previous experience—is the existence

of "recognition genes" that enable the bearers to recognize the same genes in others (Hamilton 1964). This
mechanism is called "the green beard effect" since such
a trait could be selected for if the genes controlling it
confer not only the green beard, but the preference for
green beards on others. Although theoretically possible, such a mechanism has not been demonstrated. For
more on mechanisms of kin recognition see chapter 16.

REPRODUCTION

Many of the most striking displays occur during mating.
Receptive females or males may advertise their condition, court a member of the opposite sex, form a bond,

copulate, or perform postcopulatory displays. These behaviors involve species identification, assessment of individual condition, and coordination of the neuroendocrine systems. Identifying the correct species is, of course, critical if viable offspring are to be produced, and many closely related species that have overlapping home ranges also have very distinct courtship patterns, as seen in the Hawaiian *Drosophila* (see figure 11-4).

We have also seen, below the species level, how female white-crowned sparrows preferentially solicit copulations from males of their own population (Baker 1983, see figure 11-5). Another example is the different mating calls of two populations of cricket frogs (*Acris crepitans*) just 65 km apart in central Texas (Ryan and Wilczynski 1988). Females showed a strong preference for the calls of males in their own population; electrophysiological studies also demonstrated that the basilar papillae of the inner ear are most responsive to sounds that match the local dialect. Thus communication between sender and receiver has *coevolved*, and the two populations may be on their way to becoming reproductively isolated.

AGONISM AND SOCIAL STATUS

In social groups, when members of a species are in close proximity, it is sometimes beneficial to the individual for it to compete and fight with others for possession of a resource, be it food, space, or access to another individual. Physical combat is expensive in terms of energy; it increases the risk of death or injury, even for winners. Social species have evolved displays that communicate information about an individual's mood (recall the tail of the rhesus monkey) and the way it is likely to behave in the near future. As a result of previous encounters, the animal may be dominant or submissive, and its behavior may thus be predictable. Presented with a limited resource, the submissive individual will yield to the dominant without an overt fight. Many species rarely compete for resources; thus agonistic interactions make up only a small proportion of their behavioral repertoire. Dominant aggressive displays tend to be opposites of submissive displays, exemplifying the principle of antithesis first mentioned by Darwin (1873) (figure 11-8).

FIGURE 11-8   Threatening and submissive postures in the dog
Note the features that are opposites, such as ear and tail positions, shape of spine, and general posture.
Source: Charles Darwin. On the Expression of the Emotions in Man and Animals. Copyright © 1873. D. Appleton, New York.

ALARM

Animals use vocalizations and chemicals to alert group members to danger. Although male song varies greatly within and among closely related species, sympatric species (those with overlapping ranges) are likely to have simple, hard-to-locate alarm calls that differ little among species. Members of species that live together, and that are endangered by the same predators, benefit mutually by minimizing divergence in alarm vocalizations (Marler 1973). For example, Marler was unable to tell the difference between the "chirp" alarm calls of African blue monkeys (*Cercopithecus mitis*) and red-tailed monkeys (*C. ascanius*), whereas it was easy for him to differentiate the male songs of the two species—they differed greatly. Vervet monkeys (*C. aethiops*) communicate *semantically* by using different signals to warn about different dangers in their environment. Group members climb trees when they hear leopard alarms, they look up when they hear eagle alarms, and they look down when they hear snake alarms (Struhsaker 1967). Young vervets give alarm calls in response to a variety of animals, and their ability to classify predators and give appropriate alarm calls improves with age (Seyfarth et al. 1980).

Both invertebrates and vertebrates produce chemical alarm substances. Sea urchins of the species *Diadema antillarium* move rapidly away from an area containing a crushed member of their own species (Snyder and Snyder 1970); earthworms (*Lumbricus terrestris*) also produce alarm pheromones (Ressler et al. 1968). Mice and rats excrete a substance in their urine when they are given electric shocks, are beaten up by another mouse, or are otherwise stressed. This substance acts as an alarm and may cause others to avoid the area (Rottman and Snowden 1972).

Tracing the evolution of alarm calls presents a challenge to biologists, because such signals seem unlikely to benefit the caller. Sherman (1977) studied individually marked Belding's ground squirrels (*Spermophilus beldingi*) (figure 11–9) and found that whenever a terrestrial predator (such as a weasel or a coyote) was spotted, the calling squirrel stared directly at the predator while sounding the alarm. Sherman suggested many hypotheses to explain this behavior; two of which we will discuss. First, the predator may abandon the hunt once it is spotted by the potential prey (in this hypothesis, the caller is behaving *selfishly*). Second, others in the area may benefit from the warning, even though the caller may be harmed (in this hypothesis, the caller is behaving *altruistically*). Sherman demonstrated that callers attract predators and are more likely to be attacked after calling, and thus they are not behaving selfishly.

Because he kept records on mothers and offspring, Sherman knew that the males leave the area several months after birth, and that the females are sedentary and breed near their birthplaces. He also found that adult and yearling females are much more likely to call than would be expected by chance, and that males are less likely to do so. Furthermore, females with female relatives living in the area (such as mothers or sisters, but not necessarily with offspring), call more frequently in the presence of a predator than those with no female relatives in the area. Sherman concluded that the most likely function of the alarm call is to warn family members. The behavior is phenotypically altruistic (reducing direct fitness) but genotypically selfish (increasing indirect fitness), and evolution by kin selection is indicated (see chapter 4).

FIGURE 11-9 Belding's ground squirrel giving alarm call
This conspicuously calling squirrel, a lactating female, is more likely to be attacked by a predator than is a noncaller. Nearby squirrels benefit since they can remain hidden or take cover. Females with mothers, sisters, or offspring in the vicinity are most likely to call.
Source: Photo by George D. Lepp, courtesy of Paul Sherman.

Models have been powerful tools in the study of the function of alarm calls, as seen in the study of brent geese postures (Inglis and Isaacson 1978). When alarmed by sudden auditory or visual stimuli, geese adopt the extreme head-up posture (figure 11–10). When "flocks" of decoys were placed in grainfields where geese had been causing damage, the real flocks' response depended on the decoys' postures. In general, fields with a high proportion of extreme head-up decoys were avoided by the real flocks, while fields with mostly head-down decoys attracted the real flocks.

HUNTING FOR FOOD

One of the selection pressures in favor of group living is the group's increased food-finding efficiency, which involves both communication about location of food and cooperation in securing it. This is exemplified in the African wild dog (*Lycaon pictus*), a canid distantly related

FIGURE 11-10   Models of brent goose postures
When "flocks" of geese models are placed in grainfields, real geese avoid fields containing models in the extreme head-up (alarm) posture.

Source: Data from I. R. Inglis and A. J. Isaacson, "The Responses of Dark-Bellied Brent Geese to Models of Geese in Various Postures," *Animal Behaviour* 26:953–958, 1978.

Head Down: attractive

Head Up: aversive only in presence of extreme head-up shapes

Extreme Head Up: aversive

to domestic dogs and wolves. Members of the pack engage in a frenzy of nosing, lip-licking, tail-wagging, and circling before they run off to hunt prey that is many times larger than an individual dog (Lawick and Lawick-Goodall 1971). Such behavior seems important in coordinating activities. Social carnivores may even communicate information about what type of prey they are about to hunt. Kruuk (1972) noted that hyenas hunting zebra on some occasions passed by prey they had hunted on other occasions.

Chimpanzees (*Pan troglodytes*) communicate the location of food and may actually lead others to it (Menzel 1971). Menzel removed captive chimps from their home pen. He then showed one chimp where food was hidden in the home pen and allowed that chimp to lead the others to it. No special signals were used; rather, the leader moved toward the hidden food purposively, looking back at the others periodically. The information transferred seemed to be, "Something in those bushes ahead has aroused my expectations of edibles."

In an advancing wave of army ants, movements of the swarm are influenced by tactile stimulation with antennae and by the laying down of a pheromone trail. Having discovered a food source, a scouting forager dashes back and forth between the nearest raiding column and the food; in this way, fifty to a hundred

ants are recruited to the food source within the first minute (Chadab and Rettenmeyer 1975; Franks 1989). For further discussion of feeding behavior, see chapter 17.

GIVING AND SOLICITING CARE

A wide variety of signals is used between parent and offspring, and among other relatives in the begging and offering of food. As Tinbergen (1951) demonstrated, the red spot on the herring gull's lower bill stimulates and directs a pecking response by the chick, and the chick's resultant pecking of the parent's beak stimulates the parent to regurgitate the food. Distress calls by the young are individually recognizable by the parent once the young are capable of leaving the birth site. When they are chilled, baby mice produce high-frequency sounds that are inaudible to humans, but audible to adult mice, who can then assist the young mice.

SOLICITING PLAY

As we discussed earlier, play consists of behavior patterns that may have many different functions in the adult: sex, aggression, exploration, and so forth. The play bow in canids (see figure 11–2) is communication about play and informs others that the motor patterns that follow are not the real thing. The function of play itself is a subject of debate. Observers usually agree that they can recognize play, but that they have had great difficulty ascribing definitions or functions to it. They most often suggest that the function of play is to help develop motor skills and behavior patterns used later in adult life (Fagen 1981) (see also chapter 9).

SYNCHRONIZATION OF HATCHING

Precocial birds, such as pheasants and ducks, lay large clutches of eggs. Synchronous hatching is very important because to feed or to get into the water to escape predation by terrestrial vertebrates, the mother leaves the area with all her young following. Species that nest in tree holes leave the area permanently the day the chicks hatch. Late-hatching chicks are vulnerable to predation. A few days before hatching, the chicks begin to vocalize. This communication contributes to hatching synchrony, since similarly aged eggs incubated separately hatch over a period of several days (Vince 1969).

CHANNELS OF COMMUNICATION

ODOR

From an evolutionary standpoint, the earliest type of communication was chemical, for odor is used throughout the animal kingdom, except by most bird species. Most pheromones are involved in mate identification and attraction, spacing mechanisms, or alarm. The greatest amount of research has been done on insects and mammals. There are several probable reasons why the widespread use of chemical signals has evolved: such signals can transmit information in the dark, they can travel around solid objects, they can last for hours or days, and they are efficient in terms of production cost (Wilson 1975). However, because these pheromones must diffuse through air or water, they are slow to act and have a long fade-out time.

We have a relatively good understanding of insect pheromones. For example, the sex attractant *bombykol*, which is produced by the female silk moth, has been isolated; and we are reasonably familiar with the male silk moth's perceptual system (see figure 6–3). A single molecule of bombykol triggers a nerve impulse in a receptor cell on the male's antenna. About two hundred receptor-cell firings in one second lead to a behavioral response (Schneider 1974). The male responds by flying upwind and by equalizing the pheromone concentration on both antennae until he reaches the female. To control such pests as the gypsy moth in the northeastern United States, traps are baited with commercially synthesized pheromone to lure males. We mentioned earlier how social insects in the family Hymenoptera make extensive use of pheromones for class and kin recognition (Wilson 1971).

Mammals make extensive use of pheromones; many of the recent studies have been done on rodents. Two general classes of substances that differ in effect have been identified: **priming pheromones** produce a generalized response, such as the triggering of estrogen and progesterone production that leads to estrus; and **signaling or releasing pheromones** produce an immediate motor response, such as the initiation of a mounting sequence. Bronson (1971) suggested that pheromones in mice can be classified by function:

1. Signaling pheromones
   a. Fear substance
   b. Male sex attractant
   c. Female sex attractant
   d. Aggression inducer
   e. Aggression inhibitor

2. Priming pheromones
   a. Estrus inducer
   b. Estrus inhibitor
   c. Adrenocortical activator

In addition, substances excreted in male mouse urine speed up maturation in young females (Lombardi and Vandenbergh 1977). Other urinary products inhibit aggression, increase aggression, stimulate the adrenal cortex, block implantation of embryos, and so on. At this point it is not clear how many different chemicals are involved; different functions may be served by the same pheromone (Drickamer 1989). The sources of these products include the sexual accessory glands and even the plantar tubercles on the mouse's feet.

Many of the substances produced by mammals function as a means of staking out territories or home ranges, much as does bird song. The advantage of pheromones, as mentioned previously, is that the odor may last for many days and nights; this advantage is significant because many animals are nocturnal and visual signals are of little use. Since these substances are often associated with the urinary and digestive systems, eliminative behavior is often highly specialized. Hyena clans mark the boundaries of their territories by establishing latrine areas (figure 11–11). Clan members defecate simultaneously in an area, and then paw the ground. The feces turn white and become quite conspicuous. In the same area or in another area, hyenas engage in *pasting*. Both sexes have two anal glands opening into the rectum just inside the anal opening. When pasting, the hyena straddles long stalks of grass; as the stems pass underneath, the animal everts its rectum and deposits a strong smelling whitish substance on the grass stems (Kruuk 1972).

SOUND

Information about immediate conditions can be transmitted faster by sound than by chemicals. Sound can be produced by a single organ, it can travel around objects and through dense vegetation, and it can be used in the dark. Information can be conveyed by both frequency and amplitude modulation. The best frequency for an animal to use seems to depend on the environment. Brown and Waser (1984) demonstrated that there is a **sound window** for blue monkeys (*Cercopithecus mitis*) living in the forests of Kenya and Uganda (figure

FIGURE 11-11   Activities of the Mungi hyena clan along its boundary

On a particular evening six members of the clan left the starting place or "club" where they had been sleeping and moved toward the boundary with the Scratching Rocks clan, sniffing along the way. Once at the boundary they sought out tall grass stems for pasting and pawing. Two members of the clan also defecated at a site that many others had used previously. After more pasting and pawing, the clan bedded down, having covered about two kilometers in one hour's time. Members of one clan rarely trespass into the territory of another clan.

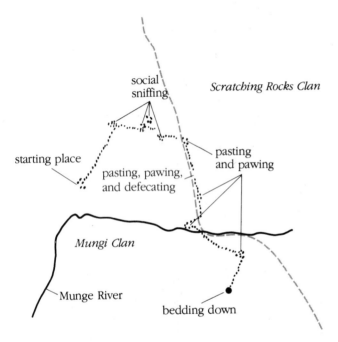

FIGURE 11-12   Sound window in the forest
(a) Background noise levels measured in two East African forests at 9:00 A.M. (b) Excess attenuation as a function of signal frequency. Sound travels farthest in the 100–300 Hz range.

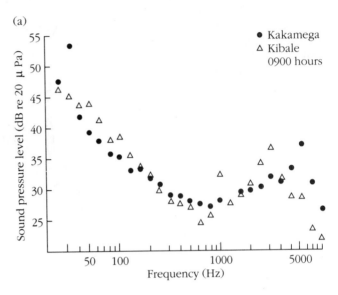

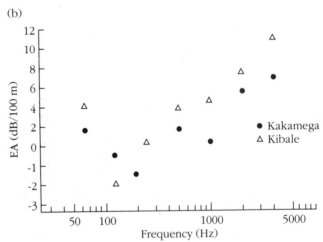

11–12). At around 200 Hz, sounds are attenuated very little and are relatively unaffected by background noises. This frequency corresponds to that of the *whoop-gobble* call produced by the adult male, who has a special vocal sac. Blue monkeys also seem to be unusually sensitive to sounds in this range compared to their more terrestrial cousins, the rhesus monkeys. Howler monkeys (*Alouatta* spp.) in the Neotropical rain forest also signal to other groups with low-frequency calls. Animals with smaller home ranges, such as squirrel monkeys (*Saimiri sciureus*), use higher-frequency sounds, which dissipate rapidly. Such calls serve to maintain contact among group members.

Ultrahigh-frequency sounds are used by a variety of animals, particularly by mammals. The distress calls of young rodents and some of the vocalizations of dogs and wolves are well above the range of human hearing, as are the echolocation sounds of bats. Although bat sounds are used mainly to locate food objects, communication also occurs between predator and prey. Noctuid moths, for example, do not produce sounds themselves, but they possess tympanic membranes on each side of the body that receive sonar pulses from bats (Roeder and Treat 1961). Depending on the location and intensity of sound stimulation, the moth may fly away in the opposite direction, dive, or desynchronize its wingbeat to produce erratic flight.

FIGURE 11-13   Song of humpback whale
The song of a humpback whale can be broken up into units, phrases, themes, songs, and song sessions. Each whale sings its own variation of the song, which may last up to a half hour.

Source: Data from R. S. Payne and S. McVay, "Songs of Humpback Whales," *Science,* 173:585–597, 13 August 1971. Copyright © 1971 by the American Association for the Advancement of Science.

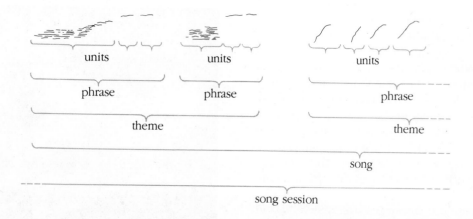

Underwater sound has properties somewhat different from sound in air. Fish and invertebrates produce a variety of sounds, some of which have only recently been investigated. Marine mammals produce clicks, squeals, and longer, more complex sounds that incorporate many frequencies; the short-duration sounds are thought to function in echolocation. Baleen whales (family Mysticeti) produce lower and longer sounds than do the toothed whales, such as dolphins. Payne and McVay (1971) analyzed the sounds of the humpback whale (*Megaptera novaneangliae*), which are varied, occur in sequences of seven to thirty minutes' duration, and are then repeated. The songs have a great deal of individuality, and each whale adheres to its own song for many months before developing a new one (figure 11–13). Researchers have not yet ascribed a clear function to these sounds, but because of the low attenuation of low-frequency sounds in water, they may serve to maintain group cohesion across thousands of miles.

LOW FREQUENCY SOUND AND SEISMIC VIBRATIONS

Substrate-borne vibrations, referred to as *seismic* vibrations, may be used for communication by the white-lipped frogs (*Leptodactylus albilabris*) studied in the rain forests of Puerto Rico (Lewis and Narins 1985). Males

calling from solid substrates make thumps; likewise, females were found to be extremely sensitive to these vibrations.

Other species produce very low frequency sounds that might also have seismic properties. Desert rodents such as kangaroo rats have greatly inflated middle ears, which makes them sensitive to low frequency sounds and vibrations. Bannertail kangaroo rats (*Dipodomys spectabilis*) defend their territories by footdrumming (Randall 1984), producing sounds in the 200–2000 Hz range. Both Asian elephants (*Elephas maximus*) and African elephants (*Loxodonta africana*) use very low frequency rumbles in the range of 14–35 Hz to communicate over distances of several km (Payne et al. 1986; Poole et al. 1988). Such infrasonic sounds have very high pressure levels (over 100 decibels) and suffer little environmental attenuation. The main uses of these calls seem to be long distance coordination of group movements and location of mates.

SURFACE WAVE

Information can be conveyed by patterns of surface vibrations. Males of one species of water strider (*Gerris remigis*) send out ripples of a certain frequency, and receptive females respond by moving toward the source. When a female gets within a certain distance, the male switches to courtship waves (Wilcox 1972, 1979). Wilcox also observed that males generate high-frequency (HF) waves when they are close to another water strider. If return HF waves are not picked up from the second strider, the first attempts copulation. In order to demonstrate the function of HF signals, Wilcox used an ingenious playback method to program females to send out HF signals. He glued a magnet to the female's foreleg and allowed her to move freely inside an electrical coil. When Wilcox played an electrical copy of the male HF signal through the coil, the magnet moved the female's leg and she involuntarily sent out the HF signal. When Wilcox placed a rubber mask over the male's eyes to eliminate possible visual cues, he found that when the male approached a female, he always attempted to copulate when she did not send an HF signal and never attempted to copulate when she did send an HF signal. We might apply this type of playback experiment to studies of other types of substrate-transmitted signals, such as in web communication among spiders.

Males of some orb-weaving spider species use vibrations to court females (Barth 1982). Once at the periphery of a female's web, the male attaches a special mating thread to her web. He then vibrates the web to bring her out for mating. The male is at risk since the resident female is capable of attacking and eating him.

TOUCH

Short-range communication in the form of physical contact is used by many invertebrates whose receptor-covered antennae are the first part of the body to make contact with other objects and organisms. Antennae are used by subsocial insects such as cockroaches, and by social insects such as bees. The honeybee often performs the waggle dance in a dark hive; therefore, much of the information about the type and location of food comes from tactile communication as the workers' antennae contact the dancer, picking up taste cues in the process.

Perhaps the most widespread use of tactile stimuli occurs during copulation. In many rodents, stimulation of the back end of an estrous female produces concave arching of the back and immobility (*lordosis*). In some mammals, vaginal stimulation induces ovulation.

In most primates, grooming is an important social activity (figure 11–14) and seems to function not only in the removal of ectoparasites but also as a "social cement" in the reaffirmation of social bonds. Most grooming takes place between close relatives, but

FIGURE 11-14   Female rhesus monkey grooming offspring
In addition to removing ectoparasites and other foreign matter from the skin and hair, grooming acts as "social cement," solidifying social bonds. Most grooming occurs between close relatives.
Source: Photo by Douglas B. Meikle.

grooming occurs also between nonrelatives in long-term relationships (Sade 1965). In the large, multimale groups characteristic of macaques (*Macaca* spp.) and baboons (*Papio* spp.), grooming between the sexes is mostly confined to the mating season. South American titi monkeys (*Callicebus*), which live in groups monogamously, entwine their tails when resting. These signals may not be complex, but they are no less important than other signals.

ELECTRIC FIELD

Some sharks and electric fish have electroreceptors that they use passively and actively in detecting objects and in communicating socially. Sharks (*Scyliorhinus caniculus*) detect the electric field produced by flatfish prey that are buried in the sand (Kalmijn 1971). In addition to electro-locating objects, electric fish of the African family Mormyridae communicate information about species identity (Hopkins and Bass 1981), individual identity, and sex by modulating the shape of the electric organ discharge (for a diagram of electric discharge see figure 6–2). Moller (1976) demonstrated that members of this family also use electric organ discharges to maintain group coordination in schools. By altering

FIGURE 11-15 Responses of female *Photuris versicolor* to flashes of males

Unmated virgin fireflies of the same species (top row) usually answer only the triple flash of males of their own species. Mated females become *femmes fatales,* answering flashes of different species (middle row). Females mated to sterile males do not become *femmes fatales* (bottom row) and respond significantly less.

Source: Data from J. E. Lloyd, "Aggressive Mimicry in *Photuris:* Firefly Femmes Fatales," *Science* 149:653–654. Copyright © 1965 by the American Association for the Advancement of Science.

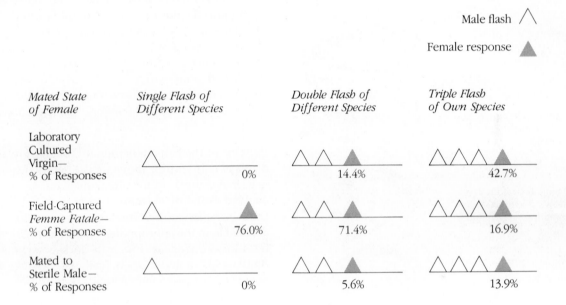

either wavelength or pulse duration, they can communicate threat, warning, submission, and so on (Bullock 1973). The advantages of this sensory mode are that it is useful in dark, murky waters; it can travel around and even through certain objects; and it provides precise information on location.

VISION

The need for a direct line of sight and ambient light limit the use of visual displays, but within social groups, visual displays enable the receiver to locate the signaler precisely in space and time. With a few exceptions, monkeys and apes are social, diurnal primates that rely extensively on visual displays. Primates ourselves, we human observers have studied visual systems more than other systems.

If you live east of the Rocky Mountains, you have probably seen fields and lawns sparkle with flashes of fireflies. These flashes are emitted by beetles of the

family Lampyridae that have specialized photogenic tissue in the abdomen. Such behavior is related in some way to mate attraction; each species has its own flash code. The males' flashes vary in intensity, duration, and interval in a species-specific way, as do the females' responses (Carlson and Copeland 1978). Within a species the flash interval varies, depending on whether the male is searching for a female or courting one he has found. In one particular species (*Photuris versicolor*), the female, once she has mated, may mimic the flash response of females of closely related species, lure the males to her, and then devour them. Such females were aptly termed *femmes fatales* by Lloyd (1965) (figure 11–15). To make matters even more complicated, males of some species of the genus *Photuris* mimic males of other species in order to lure hunting femmes fatales of their own species into a second mating (Lloyd 1980). In other words, a mated female of species *X* who is mimicking the female of species *Y* in order to lure and eat a male of species *Y* is herself lured into another mating by a male

FIGURE 11-16   Graded visual signals in mouthbrooder cichlid fish

An increasing expression of yellow band used in courtship shows in parts (c) through (e).

Source: Data from W. Wickler, "Zur Sociologie des Brabantbuntbarsches, *Tropheus moorei* (Pisces, Cichlidae)," *Zeitschrift für Tierpsychologie*, 26:967–987, 1969.

a. Frightened or submissive fish

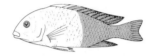

b. Fish in neutral state

c. Beginning appearance of yellow band used in courtship

d. Increasing appearance of yellow band used in courtship

e. Maximum appearance of yellow band used in courtship

of species *X* who is mimicking a male of species *Y*! Copeland (1983) has argued that more data are needed before male mimicry can be assumed.

Fish and some invertebrates are able to change color within seconds by expanding and contracting chromatophores beneath the skin (figure 11–16). The most spectacular in this regard is the octopus; waves of color advance and recede according to the animal's mood. Because of the limitations of visual displays, these displays are usually coupled with other modes of communication, such as audition. For instance, in the song spread, the red-winged blackbird spreads its tail, lowers its wings, and raises its epaulets at the same time as it renders the song.

EVOLUTION OF DISPLAYS

One of the questions asked early on by ethologists was how communicative displays evolved from noncommunicative behaviors. The evolutionary process is referred to as **ritualization.** A behavior pattern may undergo the following changes during ritualization (Eibl-Eibesfeldt 1975):

1. Change in function
2. Change in motivation
3. Exaggeration of movements in frequency and amplitude, but concurrent simplification
4. "Freezing" of movements into postures
5. Stereotyping of the behavior, while keeping frequency and amplitude relatively constant even if motivation varies
6. Development of conspicuous body structures, such as ornamental feathers, enlarged claws, manes, sailfins

One of the behaviors thought to have given rise to displays is **intention movement** (low-intensity, incipient movement). More often, displays seem to have evolved from **displacement activities,** which sometimes occur in conflict situations when an animal is undecided as to the appropriate response to a stimulus. For instance, during courtship, male ducks sometimes preen their wings or touch their feathers. Possibly this behavior is a displacement activity resulting from a conflict between sexual and agonistic behaviors. In mandarin ducks this behavior has become a display involved in courtship. Several conspicuous feathers, or sails, have evolved and are exposed during this sham preening (figure 11–17). Other suggested origins of displays are food exchange, comfort movements, and thermoregulatory patterns (displays that involve feather erection, the original function of which was the regulation of body temperature in birds) (Morris 1956).

Comparisons of closely related species can provide further information about the evolution of displays (see also chapter 19 for a discussion of the comparative method). A common feature of the courtship of pheasants and their relatives, including the domestic chicken, is food-enticing. The male chicken (*Gallus gallus*) scratches several times with its feet and pecks at the ground while calling; if no food is present, he picks up stones as if they were food objects. The hen usually comes running, and the male can then attempt to copulate with her. The ring-necked pheasant (*Phasianus colchicus*) performs the same display. The male impeyan pheasant (*Lophorus impejanus*) bows low with a slightly

FIGURE 11-17   Male mandarin duck with modified primary feather or sail

The mandarin duck provides an example of the evolution of conspicuous morphological traits in the ritualization of courtship display. The male points to the sail with its beak during courtship.

Source: Photo by Michael Hopiak, courtesy of Cornell University Laboratory of Ornithology.

spread tail, and pecks the ground. When the hen approaches and searches for the food, he spreads his wings and tail feathers. The peacock pheasant (*Polyplectron bicalcaratum*) scratches the ground and then bows with wings and tail spread. If he is given food, he will offer it to the female. Finally, the peacock (*Pavo*) male spreads his tail, shakes it, and moves back several steps, then points downward with his beak. The fanned tail arched over his head seems to focus the attention of the hen on the ground in front of him. Young male peacocks food-entice in the "original" form, with scratching and pecking, and develop the ritualized form as they mature (figure 11–18).

This series of behavior patterns shows that the more ancestral, or primitive, form of courtship involves the actual searching for food by the male and the offering of it to the female in order to attract her to him for possible copulation. The most advanced form, which is the furthest evolutionarily from the ancestral pattern, is shown by the peacock (figure 11–18), whose search for food has become a highly ritualized display for mate attraction. The movements are highly exaggerated, and special plumage has evolved. See also chapter 19 for other examples of the evolution of ritualized displays.

FIGURE 11-18   Food calling and courtship in phasianid birds

The rooster shows food calling, or "tidbitting," the original or primitive behavior in (a). In parts (b) through (e), the food is no longer there, but the female is attracted to a spot on the ground by displays of the male. Such comparative series are taken as evidence for the evolution of displays.

Source: Data from R. Schenkel, "Zur Deutung der Balzleistungen einiger Phasianiden und Tetraoniden," *Ornithologische Beobachter,* 53:182–201, 1956.

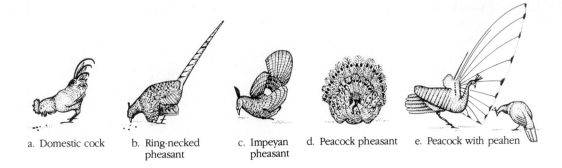

a. Domestic cock   b. Ring-necked pheasant   c. Impeyan pheasant   d. Peacock pheasant   e. Peacock with peahen

HONESTY VERSUS DECEIT IN COMMUNICATION

We now return briefly to the problem with which we began this chapter: Which organism—sender or receiver—benefits from the behavior? Recall that some ethological definitions imply that both sender and receiver must benefit, and that there is evolution toward maximization of information transfer. In this case, displays are assumed to be honest signals.

The sociobiological view of communication is more cynical, assuming that natural selection acts primarily at the level of the individual, in this case the sender. Signals become ritualized so that the sender can control the behavior of the receiver with greater efficiency (Dawkins and Krebs 1978; Krebs and Dawkins 1984). In some cases, information is transmitted to the benefit of both sender and receiver. The honeybee dancer could be thought of as manipulating her sisters. The forager reenacts the flight from the hive to the food through the dance; the recruits amplify the dance into a real flight, and they return with much more food than the forager could have brought back alone. But often the receiver clearly does not benefit, as is the case with the male firefly that is lured to his death by the femme fatale.

This account suggests that deceit should be widespread both between and within different species. Interspecific deceit is rather common, particularly in predator-prey relationships (see chapter 17), but intraspecific deceit seems to be much less common, possibly because the sender and receiver belong to a common gene pool (Dawkins and Krebs 1978). If a mutant "liar" with gene A appears in a population of nonliars (gene a), it may increase its fitness by deceiving conspecifics. As gene A spreads, the likelihood of an attempt to deceive another bearer of gene A increases. If bearers of gene A can recognize and lie only to bearers of gene a, gene A will become fixed in the population. The lying habit would eventually cease, however, because there would be no one left to deceive. Lying is an advantage only when most others tell the truth. Selection also would favor the ability to detect deceit by others, and thus would keep lying relatively rare.

COMPLEX COMMUNICATION

We end this chapter by considering complex communication examples from insects, birds, and mammals that cast doubt on the notion that language separates humans from the rest of the animal world.

FOOD LOCATION IN HONEYBEES

Hours or days may go by before a foraging honeybee (*Apis mellifera*) discovers a sugar or honey solution placed outdoors, but then new bees arrive within minutes. Aristotle thought that the other bees simply followed the forager to the food. However, von Frisch (1967) demonstrated that recruitment takes place even when the forager is not allowed to return from the hive to the food source. He hypothesized that bees obtain information on food location through odor communicated in the hive by the forager's dance. Hive members maintain antennal contact with the dancer's body and taste samples of regurgitated food. But von Frisch's observations later suggested to him that the bees were getting more specific information on location of food sources, and he went on to develop his famous dance-language hypothesis.

By placing food sources at varying distances and angles from enclosed observation hives, von Frisch found that foragers perform a dance on the comb. For food at short distances (from 20 to 200 meters, depending on the strain of bees used), returning foragers perform the round dance, a series of circles with reversals in direction every second or so. For food at greater distances, they perform the waggle dance, a figure eight with a straight run in the middle of the figure (figure 11–19). The forager waggles its body and emits sound bursts during the straight run. The direction of the food relative to the sun is the same as the direction of the straight run relative to gravity. The duration of the straight run increases with distance at the rate of about one complete waggle per 25 meters. The area that the dance occupies on the comb, the duration of each complete figure eight cycle, and the duration of sound bursts are all also correlated with distance to the food source. Von Frisch argued that this symbolic language is used to communicate information about the location of food.

Although these behaviors are highly correlated with food location, how do we prove that the bees actually use this information? Some have argued that since correlations exist, the dance must have a purpose, and that purpose must be communication. Such teleological reasoning has been unacceptable to others, however. First, many species of insects have the ability to transpose an angle flown or walked with respect to the sun into an angle with respect to gravity (Gould 1976). Correlations between waggling or buzzing and distance to food sources exist in species of insects besides honeybees, yet there is no indication that these correlations are part of a language. Second, von Frisch's studies had not eliminated the possibility that odor was solely responsible

FIGURE 11-19 Waggle dance of the honeybee

A foraging honeybee returns after discovering a food source. If the bee dances outside the hive, it waggles or vibrates its body as it passes through the straight run, which points directly toward the source. If it dances on an enclosed vertical comb, it orients itself by gravity and substitutes a point directly overhead for the sun. The angle θ between the sun and the food source is the same as that between a point directly overhead and the food source.

Source: Data from K. von Frisch, *The Dance Language and Orientation of Bees.* Copyright © 1967.

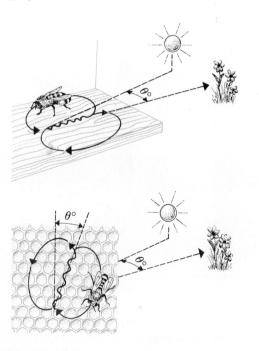

for the target accuracy of the recruits, since odors of specific locales could be transmitted during the dance. Wenner (1967) and others set out to test the olfactory hypothesis further. After many experiments, they concluded that the following foraging rule could account for what was observed: After the dance, recruits leave the hive, and they drop downwind and pick up the odors to which they have been recruited. There were a number of differences between von Frisch's and Wenner's experiments that could have accounted for the discrepancies. In general, Wenner used concentrated sucrose foods with strong odors at short distances from the hive. Von Frisch used low concentrations with weak odors at much greater distances. Dancing would be of less utility in the former situation than in the latter.

One way to prove that the bees actually use the dance to find food is to design an experiment similar in principle to Wilcox's study of the role of high-frequency vibrations in water striders, which we discussed earlier. If the forager could be tricked into a dance giving the wrong information about food location, but gave correct information about odor, we would be able to determine the method of communication by seeing where the recruits ended up. Gould (1976) placed a light in the hive, thereby causing the foragers to orient their dance to it rather than to gravity. When the ocelli (the three simple eyes between the compound eyes) were painted over, the bees foraged and danced normally but were less sensitive to light. When they returned to the hive after discovering a food source, their waggle dance was oriented to gravity instead of the light present in the hive. But the untreated recruits oriented to the light. If recruits used the dance information, they would be misinformed and would go to a place that corresponded to the angle of the dance and the light. If the food source was at an angle of 20 degrees to the right of the sun, the partially blind forager, using gravity, would dance at an angle of 20 degrees to the right of the vertical. But if a light was placed in the hive at an angle of 40 degrees to the right of the vertical, the untreated recruits should end up at a place 60 degrees to the right of the sun—off by 40 degrees. In fact, Gould's bees did just that, demonstrating that they used direction information from the dance (figure 11–20). In other experiments (reviewed in Gould 1976) researchers fed foragers a poison, which caused them to waggle slower than normal. Recruits wound up short of the food supply by the predicted amount, demonstrating that bees also use distance information in the waggle dance.

More recently, European scientists have constructed a "robot" bee that can be programmed to dance, give out food samples, and make sounds like real foragers. The robot can successfully direct recruits (reviewed by Moffett 1990).

One result of these studies has been the demonstration of redundancy in communication systems. Both odors of specific locales and symbolic dance are used in many cases. The communication is multichanneled as well: touch, smell, vision, and sound all play important roles.

LANGUAGE ACQUISITION

Communication using true language has traditionally provided a clear-cut separation of humans from other animals. By *true language* we mean both the use of symbols for abstract ideas and the understanding of syntax, so that symbols convey different messages depending on their relative positions.

FIGURE 11-20   Effect of artificial light in the beehive on recruitment direction

The presence of a light in a beehive causes the dancing foragers to use the light rather than the vertical axis as the reference point. When foragers are partially blinded so that they can see the sun but not the light in the hive, they use the vertical referent, opposite gravity, whereas recruits use the light. Bars denote the number of recruits to six stations within 30 minutes, and arrows denote the angle indicated by the dance in the three different experiments (a, b, and c).

In all cases the food was located at 0 degrees. As the angle of the light was shifted, the dances indicated a new direction, and the distribution of recruits shifted accordingly. These misdirection experiments prove that bees use the direction information in the waggle dance.

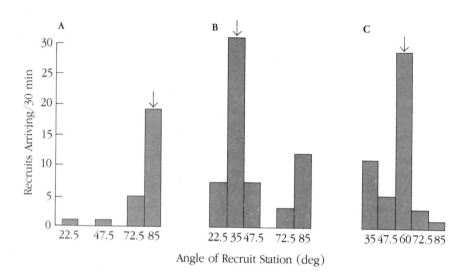

CHIMPANZEES.   Numerous investigators have explored the potential of chimpanzees (*Pan troglodytes*) to learn language. An early subject was the Hayes's (1951) home-reared chimp that was taught to use human language. Its vocabulary consisted of only a few simple words, such as "cup" and "mama." Chimps seem to lack the motor ability to pronounce human sounds, but different approaches have demonstrated that they do not lack the ability to deal with the other aspects of learning complex language. The Gardners (1971; Gardner et al. 1989) used the American Sign Language for the Deaf, and their chimp, Washoe, learned over a hundred words. Another chimp, Sarah, learned to use plastic pieces as word symbols and could communicate with them (Premak 1971).

Using computer-age technology, Rumbaugh and Gill (1976) taught their chimp Lana to press buttons with symbols embossed on them in order to gain access to food and drink from vending machines, or to get human companionship. She had a vocabulary of several hundred words, used verbs and pronouns, and could assemble words in novel ways. Table 11–4 presents a conversation between Lana and an experimenter, Tim Gill.

Lana's conversations were pragmatic; once she obtained her immediate goal of food, drink, or companionship, the conversation ended. Her curiosity was related to her immediate needs, and she showed no interest in extending her knowledge of the world or how things in it work. This pragmatism contrasts with language development in humans, who use language at an early age to gain information about all aspects of the environment.

A controversy resulted when Terrace and his collaborators (1979) asserted that chimps could not really create sentences. They worked with their own chimp, Neam Chimpsky (Nim for short, named after the famous linguist Noam Chomsky), and also reanalyzed the videotapes and films made by other investigators. Nim mastered a respectable vocabulary of sign language words and, like other chimps, used them to convey information to another individual. Nim did use two-sign combinations that were syntactically consistent; for example, he signed the correct "eat banana" more often than "banana eat." But he was apparently imitating his teachers' previous utterances or responding to other cues, rather than creating sentences on his own. The Gardners in response said that Nim was trained in an

TABLE 11-4 Conversation between Tim Gill and the chimpanzee Lana (June 11, 1975)

Conditions: Formal Test. Tim had entered the anteroom with a bowl of monkey chow. Lana had asked that it be loaded into the machine; however, the conditions of the test called for Tim *not* to comply, to load cabbage for vending instead, and to declare that chow (which she had requested) was in the machine. Although Lana might have asked the machine to vend "chow," she did not—[which was] appropriate to the fact that cabbage, and not monkey chow, was in the vendor. She said:

Lana:	Please machine give piece of cabbage.	16:53
	? You *(Tim)* put chow in machine. *(5 times)*	16:54 & 16:55
Tim:	*(lying)* Chow in machine. *(in response to each of the 5 requests)*	
Lana:	? Chow in machine.	16:57
Tim:	*(still lying)* Yes.	16:57
Lana:	No chow in machine. *(which was true)*	16:57
Tim:	?What in machine. *(repeat once)*	16:57 & 16:58
Lana:	Cabbage in machine. *(which was true)*	16:59
Tim:	Yes cabbage in machine.	16:59
Lana:	? You move cabbage out-of machine.	17:00
Tim:	Yes. *(whereupon he removed the cabbage and put in the monkey chow)*	17:01
Lana:	Please machine give piece of chow. *(repeatedly until all was obtained)*	17:01

Conclusion: Lana discerned what had, in fact, been loaded in the machine, did not concur with Tim's assertion that it was "chow," asked that he remove it, and then asked for "chow" when it was loaded for vending.

Source: From D. M. Rumbaugh and T. V. Gill, "The Mastery of Language-Type Skills by the Chimpanzee (Pan)" in *Annals of the New York Academy of Sciences* 280, 562–578. Copyright © 1976. Reprinted by permission.

environment unlikely to produce spontaneous behavior, and that the film segments of their chimp, Washoe, that Terrace analyzed were too short to demonstrate the complexity of communication (Marx 1980).

More recently the Rumbaughs questioned whether Lana and her successors, Austin and Sherman (figure 11–21), were using symbolization in the same way that humans do (Savage-Rumbaugh et al. 1980). Symbolization means the use of arbitrary symbols to refer to objects and events that are removed in time and space. Although chimps may learn to string words together in social interaction routines to attain goals, this accomplishment is not proof that they can do more than associate a word with an object—that is, their language learning does not demonstrate a referential relationship. However, in another paradigm, Sherman and Austin learned how to request tools from one another in order to obtain food that they then shared (Savage-Rumbaugh 1986). Thus, if Sherman was shown a container of food that required a wrench to open, he would punch the appropriate symbol on his keyboard. Austin

then knew a wrench was needed, rather than some other tool, and would hand it to Sherman who could then obtain the food. This is a very clear demonstration that their symbols truly represented the items for which they stood. Hence, it is concluded that they use their symbols symbolically (Savage and Rumbaugh 1986).

PARROTS. We all know that parrots can talk, but we usually assume that they are simply mimicking what they hear, as Terrace argued was the case for sign language learning in chimps. An African grey parrot (*Psittacus erithacus*) named Alex, bought in a pet store in 1977, seems to be demonstrating otherwise. So far he has learned, via spoken English, to identify more than eighty different objects; he can quantify collections of up to six objects; he can identify shapes and colors; and he understands concepts such as "same" and "different" (Pepperberg 1987a,b). Although Pepperberg does not claim that Alex uses true language with its complexities of syntax and grammar, she does believe that he is using words to represent abstract concepts.

FIGURE 11-21   Sherman and Austin at the computer keyboard
Each key of the keyboard has a different symbol that represents a word. The positions of the keys are scrambled frequently to avoid the use of position cues. In order to obtain goals such as food, water, and companionship, chimps must press the keys in the correct order.
Source: Photo by Elizabeth Rubert, courtesy of Yerkes Regional Primate Research Center.

SUMMARY

Communication may be defined as an action on the part of one organism that alters the probability pattern of behavior in another organism in a fashion *adaptive* to either the sender or both the sender and the receiver. Behavior patterns that are specially adapted to serve as social signals are termed *displays*.

Social signals, which vary in fade-out time, effective distance, and duration, convey information by being *discrete* or *graded*. They may be combined to form *composite* signals, and the order in which they appear may affect the information transmitted (*syntax*). *Metacommunication*, communication that alters the meaning of the message that is to follow, is seen among nonhuman animals mainly in conjunction with play behavior.

Communication can be measured by comparing the frequencies of occurrence of each behavioral response against the expected frequencies if the responses were occurring randomly. If the response frequencies are significantly altered, information transfer is taking place. The use of information theory allows us to quantify the amount of communication in the case of discrete signals.

Communication functions in group spacing and coordination; individual, species, and class recognition; reproduction; agonism and social status; alarm; hunting

for food; giving and soliciting care; soliciting play; and synchronization of hatching. Channels of communication include odor (mainly via pheromones), sound, touch, surface vibration, electric field, and vision.

Displays are thought to have evolved from such noncommunicative behaviors as *intention movements* and *displacement acts*. As behaviors change in function and motivation, natural selection produces exaggerated movements and postures, a process called *ritualization*. Conspicuous body structures, such as ornamental plumes on birds or claws on crabs, may have evolved to reduce ambiguity of the message and hence uncertainty of the response. By comparing differences in displays among closely related species, ethologists infer the evolutionary pathways of this process of ritualization.

Some definitions of communication imply that both sender and receiver must benefit and that there is evolution toward maximization of information transfer. An alternate, sociobiological view argues that communication is a means by which the sender manipulates the receiver, who may benefit or may be harmed: the purpose of a display is to persuade, not to inform.

By way of a dance on the comb, honeybees communicate information about the distance and direction of food to fellow workers. Odor cues also seem to be important, but several experiments have demonstrated that the dance information is of primary importance in locating the food. An example of complex visual communication is sign-language learning in the chimpanzee. Although they lack the motor ability to produce the sounds of human language, chimps can acquire vocabularies of several hundred words by means of hand signs or substitute symbols. Chimps can use nouns, pronouns, and verbs to converse about their immediate needs; they can be taught to communicate with each other, but their ability to create sentences and use true symbolism is still being debated. Studies of an African grey parrot suggest the ability to use simple words to identify objects and represent abstract concepts.

Discussion Questions

1. Although we have discussed how communication operates in several discrete sensory channels, most displays involve more than one channel. Discuss examples in which two or more channels are used. Why have such multimedia displays evolved? What methods would you use to demonstrate the function of such displays?

2. Contrast the following two ideas about the evolution of communication systems in animals: (a) Displays evolve so as to maximize information transfer from sender to receiver; (b) Displays evolve so as to maximize manipulation of the receiver by the sender.

3. The table to the right shows the mean number of scent markings made during a ten-minute period by Maxwell's duikers—small, forest-dwelling antelopes—that were in three groups (I, II, and III). What do these data tell us about the function of marking and the social organization of this species? What can you say about the Type A and Type B females?

Group membership	Marking activity when with own group	Marking activity after presence of additional	
		Male	Female
	Males		
I	6.6	15.2	6.1
II	5.8	10.7	6.2
III	4.4	8.6	4.1
	Type A Females		
I	3.5	3.7	18.6
II	3.4	3.1	12.2
III	1.5	0	1.7
	Type B Females		
I	0.06	0	0.09
II	0	0.1	0
III	0.04	0	0.03

Source: From K. Ralls, "Mammalian Scent Marking," in *Science* 171 (1971) 443–449. Copyright © 1971 by the American Association for the Advancement of Science. Reprinted by permission.

Suggested Readings

Halliday, T. R., and P. J. B. Slater, eds. 1983. *Communication.* Animal Behavior, vol. 2. New York: Freeman.
A readable introductory book, emphasizing the sociobiological approach. Chapters on sensory mechanisms, environment, and evolution.

Sebeok, T. A., ed. 1984. *How Animals Communicate.* 2d ed. Bloomington: Indiana University Press.
A massive tome with chapters by specialists on different taxonomic groups.

Smith, W. J. 1984. *Behavior of Communicating.* 2d ed. Cambridge: Harvard University Press.
An in-depth text that covers the entire subject, emphasizing the ethological approach.

Wilson, E. O. 1975. *Sociobiology: The New Synthesis.* Cambridge: Harvard University Press.
Chapters 8, 9, and 10 provide a concise overview of communication, with an explanation of information theory.

12

AGGRESSION

Konrad Lorenz (1966) stated in the introduction to *On Aggression* that his book was about "the fighting instinct in beast and man which is directed against members of the same species." His use of the word instinct points up one facet of the old nature-nurture controversy that has direct bearing on the human behavior. Is aggression a universal property of social animals, including humans, and is it an inborn trait whose expression is inevitable? How can we explain the widespread tendency of animals in a fight to use restraint and not to fight to the death, even when the most aggressive individuals control resources such as food or mates?

In this chapter we define aggression, agonism, and competition; consider forms of aggression; and then consider two ways in which aggression is expressed: the social use of space and the dominance hierarchy. We look briefly at the internal causes of aggressive behavior—genetic, neural, and hormonal—and at external causes, i.e., the role of environment. Next we discuss control of aggression, explore the use of game theory models of conflict, and conclude with implications of the study of animal aggression for understanding human behavior.

AGGRESSION, AGONISM, COMPETITION

DEFINITIONS

Aggression is a complex phenomenon with many functions and many causes, and it may include predatory behavior, in which the animal being attacked is eaten in the process. It has been defined by psychologists as behavior that appears to be intended to inflict noxious stimulation or destruction on another organism (Moyer 1976). The notion of intent is necessary to exclude such destructive behaviors as a person's accidentally stepping on an ant. Use of the word aggression emphasizes offensive behavior. Behavioral ecologists take a more functional approach and consider aggression as a form of resource competition, in which an animal actively excludes rivals from some resource such as food, shelter, or mates (Archer 1988).

A term with a more precise definition is **agonistic** behavior (not to be confused with agnostic behavior!), which is a system of behavior patterns that have the common function of adjustment to situations of conflict among conspecifics. The term includes all aspects of conflict, such as threats, submissions, chases, and physical combat, but it specifically excludes predatory aggression, since as Scott (1972) argued, ingestive behavior is part of a separate behavioral system.

We can list the forms of aggressive behavior as follows (Moyer 1976; Wilson 1975):

- *Territorial*—exclusion of others from some physical space

- *Dominance*—control of the behavior of a conspecific as a result of a previous encounter

- *Sexual*—use of threats and physical punishment, usually by males, to obtain and retain mates

- *Parental*—attacks on intruders when young are present

- *Parent-offspring*—disciplinary action by parent against offspring (mostly in mammals, usually associated with weaning)

- *Predatory*—act of predation, possibly including cannibalism

- *Antipredatory*—defensive attack by prey on predator, such as mobbing

Most of these behaviors involve conflict among conspecifics, and so would also be included under agonistic behavior. These forms of aggression serve very different functions within and between species, and independent regulatory centers in the brain may even have evolved for them. Therefore, the term *aggression* is not a unitary concept. Most of our discussion here concerns agonistic behavior among conspecifics; we will consider interactions between different species in chapter 17.

COMPETITION FOR RESOURCES

Most agonistic behavior involves competition for some limited resource (food; water; access to a member of the opposite sex; or space for nesting, wintering, or safety from predators). Such behavior is distributed throughout the animal kingdom. Even organisms as simple as sea anemones (Phylum Coelenterata) behave aggressively as they use stinging cells (*nematocysts*) against conspecifics; the loser of these encounters, usually the smaller of the two, closes up and can be dislodged from the substrate (Brace et al. 1979).

EXPLOITATION AND THE IDEAL FREE DISTRIBUTION. Competition can be divided into two forms: **exploitation,** in which organisms passively use up resources, and **interference,** in which organisms interact so as to reduce one another's access to, or use of, resources. Suppose that there is a group of twelve ducks on a pond in a city park. Two people, one at each end of the pond, are feeding bread to the ducks. One is feeding the ducks twice as fast as the other. How should the ducks distribute themselves so that each gets the most food? Common sense tells us that if the birds don't guard the access to the bread, and if they have all the information they need about the feeding rates, an average of four should be at the end with the low feeding rate and eight at the end with the high feeding rate. This theoretical pattern is called **ideal,** because it assumes that animals have accurate and complete information about the distribution of resources, and **free,** because individuals are passive toward each other and free to go wherever they can exploit the most resources (Fretwell 1972; Fretwell and Lucas 1970). When just such an experiment was conducted using thirty-three mallard ducks on a lake in England (Harper 1982), the ducks distributed themselves as predicted (figure 12–1).

INTERFERENCE COMPETITION AND RESOURCE DEFENSE. More often it is the case that the individuals are not passive. Some individuals may establish territories and defend the resources, or some may be dominant and control the access of others to the resources. In the duck experiment above, it turned out that even though the results fit an ideal free distribution, some ducks got more than their share of food and became **despots** (Harper 1982). In species of birds, the best habitat quickly gets taken over; younger or less aggressive individuals are excluded and forced to breed in less suitable habitats. For instance, prime breeding habitat for great tits (*Parus major*) is oak woodland. Excluded individuals attempt to breed in hedgerows, where they have lower reproductive success than those in prime habitat. Should a breeder in oak woodland die, it is quickly replaced by a hedgerow resident (Krebs 1971).

SOCIAL USE OF SPACE

The defense of fixed space against members of the same species accounts for much of the agonistic behavior in a host of animals, ranging from limpets to long-billed marsh wrens. This is not to say that all animals defend territories, as might be assumed from reading such books as Robert Ardrey's *The Territorial Imperative* (Ardrey 1966); in fact, the majority of species probably do not. Animals use space in a number of different ways.

FIGURE 12-1   Ideal free ducks
(a) Mean number of ducks at site A plotted against time since start of trial, when patch profitability ratio was unity. The horizontal line is the ideal free prediction: half the 33 ducks at site A. (b) Mean number of ducks at least profitable site plotted against time since start of trial, when patch profitability ratio was 2:1. The horizontal line is the ideal free prediction: one third of the 33 ducks at the least productive site.

FIGURE 12-2   Capture locations of white-footed mice
Figures denote number of captures at live-trap stations spaced 7 meters apart, and dots represent traps where the mouse was not caught. The center of activity indicates the average of the coordinates. Most captures are close to the center of activity; but some mice, such as mouse #5208, seem to use secondary nest sites. Males usually have larger and less exclusive home ranges than females. Ellipses include 95 percent of expected captures.

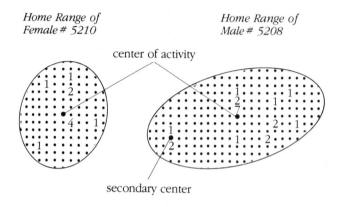

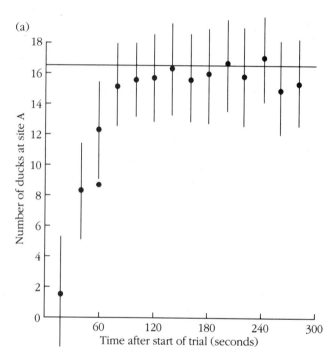

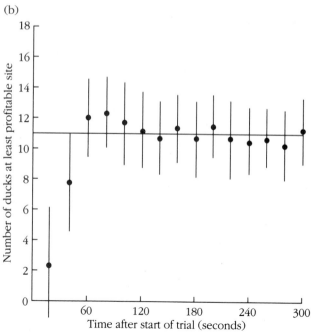

DEFINITIONS OF THE USE OF SPACE

The area used habitually by an animal or group, in which the animal spends most of its time, is its **home range** (figure 12–2). Most organisms spend their lives in a relatively restricted part of the available habitat and learn the locations of food, water, and shelter in this area. We may have difficulty determining the actual boundaries of the home range, since an animal or group may occasionally wander some distance away to a place it will never revisit. To permit comparisons with other studies, we must specify what criteria are being used. For example, the white-footed mouse (*Peromyscus leucopus*) spends most of its time at the nest near the center of its home range. The likelihood of finding it away from that spot decreases according to a normal distribution. Thus we can arbitrarily propose that its home range be defined as that area in which the mouse can be found (captured, radio-tracked, and so on) 95 percent (or some other percentage) of the time. We can compute the home range in this case by calculating the area within which we expect 95 percent of the captures, assuming a normal distribution about the center of activity (95 percent confidence ellipse) (figure 12–3).

The area of heaviest use within the home range is the **core area.** This location may contain a nest, sleeping trees, water source, or a feeding tree. As with home range, the designation of a core area is somewhat arbitrary, but useful in understanding the behavior and

FIGURE 12-3   White-footed mouse in nest box
Nest boxes (left) can supplement live trapping to provide
data on behavior and population dynamics. Note
individually numbered metal tag on mouse's left ear (right).
Source: Photos by Stephen H. Vessey.

**FIGURE 12-4   Home ranges and core areas of nine
groups of baboons in Nairobi Park, Kenya**
Although home ranges overlap extensively among groups
of baboons (*Papio anubis*), core areas overlap little. Shading
indicates core areas.

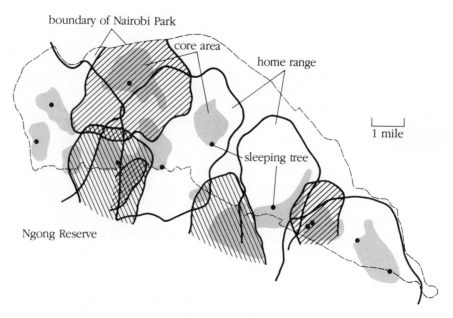

ecology of different species, or the same species in different habitats or at different population densities.
Figure 12–4 illustrates home ranges and core areas of
baboons in Africa.

The minimum distance that an animal normally
keeps between itself and other members of the same
species is its **individual distance.** In birds, it is frequently the distance that one individual can reach to

FIGURE 12-5   Individual distance in young green herons
Each bird maintains a uniform distance from its nearest neighbor.
Source: Photo by Roy Lowe, courtesy of U.S. Fish and Wildlife Service.

peck another; we see this spacing in starlings lined up on a telephone wire, in sea gull nests in a colony, or in green heron (*Butorides virescens*) fledglings on a tree branch (figure 12–5). Individual distance, or personal space, exists for humans as well. For example, when talking to another person, most of us become uncomfortable and back off if the other person comes too close. The amount of personal space required varies from culture to culture and from situation to situation (Hall 1966).

TERRITORY

An area occupied more or less exclusively by an animal or group and defended by overt aggression or advertisement is a **territory.** Although he did not discover the concept, Howard (1920) presented the first modern study of the phenomenon. Working mainly with aquatic birds, he described the defense of an area by a mated pair, pointed out the birds' need to defend a resource in order to breed, and noted the role of territory in population regulation. We treat this last point in chapter 18 on population biology.

To demonstrate territory, we must show that an individual, mated pair, or group has exclusive use of some space, and we must also observe their defense of that area. We can easily make these observations of conspicuous, diurnal species such as sea gulls (figure 12–6), but how do we directly observe defense by a small, nocturnal, secretive, cryptically colored creature such as a white-footed mouse? Not surprisingly, even though millions of these mice have been snap-trapped, live-trapped, and radio-tagged, we still know very little about their social organization.

The size of the territory or home range depends on the size of the animal as well as on the particular resource the animal is defending. The type of territory or home range most common in small mammals and insect-eating birds is relatively large and includes the food supply, courtship, and nesting areas. Its size is an approximate function of the animal's weight and metabolic rate (McNab 1963). For mammals, $Area = 6.76 \, W^{0.63}$, where *Area* equals the expected home range or territory in acres, and *W* equals the body weight in kg. Thus a 20-gram mouse should have an area of 0.57 acres; this figure is not far off the mark. The productivity of the habitat is important, and white-footed mice range farther in less productive habitats. If the area is defended

FIGURE 12-6 California gull defending territory
The bird in the left foreground attacks an intruder. Both parents defend the nest site in this monogamous species.
Source: Photo by Bruce Pugesek.

against conspecifics, we would expect a smaller value. This relationship between home range or territory size and body weight suggests that ultimately what most vertebrates defend is the food source.

Territorial animals spend much time patrolling the boundaries of their space, singing, visiting scent posts, and making other displays. Such behavior would seem to take more time and energy than would simple exploitative competition. However, these displays often have evolved so as to require relatively little energy, and once the territory has been established, the neighbors have been conditioned and need only occasional reminders to keep out. The cost of defense could be less than the benefit of having exclusive use of a resource.

When should an animal establish a territory? The key seems to be **economic defendability** (Brown 1964), such that the costs (energy expenditure, risk of injury, etc.) are outweighed by the benefits (access to the resource). Important are such things as the distribution of the limited resource in space, and whether or not the availability of the resource fluctuates seasonally. A limited resource (food) that is uniformly distributed in time and space is most efficiently utilized if members of the population spread themselves out through the habitat, possibly defending areas. A resource that is spaced in clumps and that is unpredictable might favor colonial living or possibly nomadism.

The relationship between the social use of space and the environment was demonstrated by Orians' (1961) work with red-winged and tricolored blackbirds, which

are sympatric in northern California. The tricolor, responding to a food source that is concentrated and unpredictable, is colonial, nomadic, and monogamous, and the males defend small territories; the redwing is polygynous, and the males defend large areas and space out regularly, responding to a food source that is less concentrated and more predictable. (This study is described more fully in chapter 14.)

TERRITORIAL MODEL. Carpenter and MacMillen (1976) studied a nectar-feeding bird, the Hawaiian honeycreeper *Vertiaria coccinea*, and constructed a model to predict territorial behavior. They argued that in order for territorial behavior to occur, E, the basic cost of living, plus T, the added cost of defending a territory, must be less than the yield to the individual (fraction a of productivity P) if not territorial, plus the extra yield gained by the reduced competition (fraction b of productivity P) if territorial. In other words:

$$E + T < aP + bP.$$

Further development of the model leads to the prediction that at very high levels of food productivity, the birds can get enough food without excluding other birds, so territoriality disappears. Below a certain level of food availability, territorial behavior also should disappear since the resource is no longer worth defending. The birds then should switch to other foods or leave the area. Carpenter and MacMillen were able to estimate the parameters needed to test their model, and the honeycreepers behaved as predicted.

Not all territories include all the resources that the animal needs. Male bullfrogs (*Rana catesbeiana*) defend the sites that are most suitable for development of the larvae (Howard 1978) (figure 12–7). Juvenile lizards defend space that provides a refuge from predators (Stamps 1983). Colonial species may exhibit some territoriality; for example, the tricolor blackbird defends only a small area around the nest. Some species have overlapping home ranges but defend a core area.

LEK. A mating system involving a peculiar type of territory is the **lek** (see also chapter 13). In this case, the only resource that the organism defends is the space where mating takes place. Feeding and nesting occur away from the site. Lek, or arena, systems are characterized by promiscuous, communal mating; the males are likely to have evolved elaborate ornamental plumages, as exemplified by the sage grouse (*Centrocerus urophasianus*) studied by Wiley (1973) and Gibson and

FIGURE 12-7   Male bullfrogs fighting

Males establish territories through contests, with larger males usually winning and controlling the better sites. Females preferentially mate with these males and lay eggs in their territories. Thus the outcomes of male aggression are access to more females and higher reproductive success.

Source: Photo by Richard D. Howard.

Bradbury (1985) (figure 12–8). The same area is typically used year after year. Males arrive early in the breeding season, and with highly ritualized agonistic behavior they stake out their plots. Certain territories and/or displaying males appear to be more attractive than neighboring ones, in the sense that some males do much more breeding than others. Females move through the areas while the males display, mate with one or more males, then leave. While on the lek, the males do little or no feeding; they spend all their time and energy patrolling the boundaries, displaying to other males, and attempting to attract females into their area. Although lek mating systems are rare among animals, they do occur in a wide range of species, from Hawaiian fruit flies (family Drosophilidae) to African hammerheaded bats (*Hypsignathus monstrosus*).

DOMINANCE

When two adult male laboratory mice of an aggressive inbred strain, such as CF1 or BALB/C, are socially isolated for a few weeks and then put together, they begin to fight. One or both may assume a hunched posture and advance with mincing steps; they may also vibrate their tails to produce a rattling sound, and one may roughly groom the other on the lower back. They may then start directing bites at each other's face, shoulders,

FIGURE 12-8   Sage grouse lek in Montana

Males occupy small territories in one of the mating centers of the lek as females aggregate there. Males with more peripheral territories of the lek, away from a mating center, seldom have the opportunity to mate.

Source: Photo by R. Haven Wiley.

and lower back, wrestling vigorously (see figure 3–6). Before long one begins to get the best of things, and the other may try to escape. Once cornered, the newly submissive mouse will rear up, squeak, and box with its forepaws. If the two are separated for a day and then reintroduced, the previous day's victor will establish his superiority more quickly, and the submissive mouse will retire sooner. By including other mice, we can demonstrate that mice individually recognize each other; they remember which mice beat them up previously and which ones they were able to dominate. (Some of this information is communicated via urinary odors.) The establishment of dominance hierarchies requires individual recognition and learning based on previous encounters among the individuals.

DOMINANCE HIERARCHIES

We say that an animal is dominant if it controls the behavior of another (Scott 1966). In another sense, a prediction is being made about the outcome of future competitive interactions (Rowell 1974). If four or five mice that are strangers to each other are put together, several outcomes are possible. Most likely a despot will take over, and all the subordinates will be more or less equal:

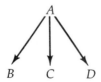

Another possibility is a linear hierarchy, where A dominates B, B dominates C, and so on:

$$A \rightarrow B \rightarrow C \rightarrow D.$$

Sometimes triangular relationships form, where:

In other species coalitions may affect dominance, such that:

but

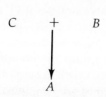

The idea of dominance was first promulgated by Schjelderup-Ebbe (1922), who worked out the peck order in domestic chickens. Dominance hierarchies occur most often in arthropods and vertebrates living in permanent or semipermanent social groups, but they have been reported even in nonsocial sea anemones (Brace et al. 1979). There is much variation in the intensity of dominance and in the frequency of reversals. Peck-right hierarchies are formed when all the aggression goes from dominant to subordinate (table 12–1). In peck-dominance hierarchies only a majority of agonistic acts go from dominant to subordinate. Among closely related species there may be a clear-cut hierarchy in one species and not in the other. For example, African green monkeys (*Cercopithecus sabaeus*) commonly kept in zoos show little or no dominance hierarchy, even when access to some highly prized food is limited; the vervet (*Cercopithecus aethiops*), a close relative, has a pronounced linear hierarchy. Species that are territorial in the wild may change over to a dominance hierarchy when crowded in captivity. Sometimes the dominance rank order is a function of the resource for which the animals are competing; for example, one individual may have first access to water, another to a favored breeding site.

Rowell (1974) and others have argued that the whole concept of dominance should be reassessed because it is largely a product of stress induced by cap-

TABLE 12-1 Method for ranking individuals based on outcomes of dyadic (two-way) interactions
In a hypothetical group of five individuals identified as *A*, *B*, *C*, *D*, *E*, (a) Matrix of wins and losses in which individuals are arbitrarily arranged alphabetically. For instance, *B* was the winner and *A* was the loser in eight contests; *A* never defeated *B*. (b) The order has been arranged to maximize the values in the upper right half of the matrix. Thus the correct order is *C*, *D*, *B*, *E*, *A*. Some reversals did occur, as seen by numbers in the lower left half of the matrix. Thus *E* is dominant to *A* because *E* defeated *A* four times, and *A* defeated *E* only once. This type of rearrangement works only if the hierarchy is linear.

a.

		Loser				
		A	*B*	*C*	*D*	*E*
	A		0	0	0	1
	B	8		0	1	4
Winner	*C*	7	7		6	9
	D	12	8	0		5
	E	4	0	0	1	

b.

		Loser				
		C	*D*	*B*	*E*	*A*
	C		6	7	9	7
	D	0		8	5	12
Winner	*B*	0	1		4	8
	E	0	1	0		4
	A	0	0	0	1	

tivity. Nevertheless, the phenomenon of social control is widespread, particularly among primates, including humans. We saw in chapter 4 that differences in competitive ability provide the means by which natural selection acts. Dominance hierarchies seem to be a common result when socially living organisms engage in competition.

DOMINANCE HIERARCHY IN THE RHESUS MONKEY

To understand the nuances of social control, let us examine in some detail the dominance relationships in a relatively aggressive primate with a well-developed linear hierarchy. By the time infant rhesus monkeys

FIGURE 12-9   Threat and submission in rhesus monkeys
The open-mouthed threat (top) by an adult female rhesus monkey is directed at a monkey out of view to the right. The submissive grimace (bottom) by the young male on the left is in response to the alert posture of the dominant male on the right.
Source: Photos (top) by Douglas B. Meikle and (bottom) by Stephen H. Vessey.

(*Macaca mulatta*) are three to four months old, they have developed most of the threat and submission postures and vocalizations of the adult (figure 12–9). Although fights are not common among infants, by the end of the first year the infants in a group have established a linear peck order (Vessey and Meikle 1984).

DOMINANCE IN FEMALES. Position in the hierarchy is determined by the mother's rank (Marsden 1968; Sade 1967). Adult females are linearly ranked, with sons and daughters generally ranked just below their mothers. The infants get support from mothers and siblings; offspring of lower-ranking mothers soon learn to submit. Sometimes a one-year-old infant chases a two- or three-year-old juvenile, but only if the latter comes from a lower-ranking family. Hierarchies among families are very stable; those observed at La Parguera rhesus colony in Puerto Rico remained stable for more than 20 years. Because female rhesus monkeys never change groups, extensive genealogies develop over the years. As the heads of genealogies die off, the highest-ranking daughters take over. At about puberty (three years), each female becomes dominant over all her older sisters for reasons that are not clear; perhaps the greater support received from her mother gives the young female an advantage in agonistic encounters. If the mother should die, the new dominant female could be as young as three years old.

Dominance in these families manifests itself subtly but constantly at a food source, water source, or resting spot, with dominants supplanting subordinates. Fights break out sporadically in the group for no apparent reason, and family members are quick to support each other. In some groups, the highest-ranking female and some of her daughters collaborate with high-ranking males to break up fights among other families and to oppose threats from outside the group.

DOMINANCE IN MALES. Dominance among males is more complex because males leave the natal group shortly after puberty (three and a half years). While in the natal group, they rank below their mothers. In a new group, of course, family membership is meaningless, and the males must form new alliances. As males move into new groups, dominance relationships are quickly formed. Surprisingly, size and previous fighting experiences do not seem to influence rank directly; as long as they stay in the new group, their rank is positively correlated with seniority: those in the group the longest rank the highest (Drickamer and Vessey 1973). Establishing ties with the central high-ranking females of the group may be of importance in this respect. Males generally move into a group that contains an older brother (half-sib), and brothers may collaborate to stay in the group and attain high rank, thereby increasing their inclusive fitness (Meikle and Vessey 1981).

TABLE 12-2 **Change in spatial relations and agonistic behavior of group's second-ranking male after removal of alpha male**

Days after removal of alpha male	Hours observed	Percentage of time seen at			Fights broken up /hr	Attacks on females/hr
		edge	inside	midst		
−30 to 0	4.2	90	10	0	0.5	0.2
0 to 8	2.2	12	38	50	0.3	2.7
8 to 34	6.0	13	9	78	1.3	1.0
34 to 90	5.8	5	10	85	0.9	0.3

Reprinted by permission from Stephen H. Vessey, "Free-Ranging Rhesus Monkeys: Behavioural Effects of Removal, Separation, and Reintroduction of Group Members," in *Behaviour*, 40(1971) pp. 216–227, E. J. Brill, Leiden.

Although there is no proof that attaining high rank affects mating success in rhesus monkeys, studies suggest that such a relationship exists (Smith 1981), and therefore, there should be some selective advantage to high rank. Furthermore, in times of food shortage when access to food is restricted, high-ranking individuals have a distinct advantage. All group members, even the lowest ranking, probably benefit from a dominance hierarchy because there is less energy spent in scrambling for resources. Also, as we shall see later, the advantages of group living, even as a subordinate, may more than compensate for the stress endured.

Alpha males. Early monkey watchers were much impressed with dominant (*alpha*) males and wrote about their confident walk, upright tail carriage, and sexual exploits; about how they led the group from place to place; and about how they determined the rank of their group in the intergroup dominance hierarchy. The usual method of collecting field data was ad libitum note-taking: the observer simply wrote down behaviors that caught his or her attention. This technique led to biased conclusions because the most conspicuous individuals were the large high-ranking males, and the most conspicuous behavior was aggression. More recent field studies use bias-free techniques such as focal animal observations, in which the observer singles out and observes certain individuals exclusively for a fixed time interval (Altmann 1974).

Much of the mystique of the dominant male evaporated when observers noted that these males are usually the last to get involved in a skirmish, situate themselves safely in the center of the group, and let the lower-ranking peripheral males do most of the fighting. The alpha male has little to do with the group's rank, he does not actually lead the group, and he may not be the most sexually active. The ascription to alpha males of the control role—breaking up fights by charging one or both of the combatants, and protecting the group against serious extragroup threats—does seem to have substance (Bernstein 1966) (table 12–2).

ADVANTAGES OF DOMINANCE

The advantages of being dominant would require little comment if firm data were available for many species. Studies of group-living birds and mammals indicate that the dominant animals are well fed and healthy. Subordinates may be malnourished or diseased and thus suffer higher mortality. Christian and Davis (1964) have reviewed data for mammals showing that low-ranking individuals (frequent losers in fights) have higher levels of adrenal cortical hormones than do dominants. These hormones elevate blood sugar and prepare the animal for "fight or flight." The cost is a reduction in antigen-antibody and inflammatory responses—the body's defense mechanisms— and a reduction in levels of reproductive hormones.

Higher rank is often directly correlated with reproductive success, as in elephant seals (Le Boeuf 1974; Le Boeuf and Reiter 1988) (see table 14–2). But in some cases the highest-ranking male may be so busy displaying that lower-ranking males copulate with the females. In many species, the male dominance hierarchy is age-graded, with younger, lower-ranking males working their way up the hierarchy as older males die off or leave the group. Thus, low rank does not necessarily mean low lifetime reproductive success. Needed studies on lifetime reproductive success are only now being published (Clutton-Brock 1988).

Many popular treatments have emphasized the importance of dominance and aggression in "keeping the species fit," since the survivors of fierce battles are the most vigorous and therefore "improve" the species by passing those traits on. However, the simplest and most likely explanation of aggression and dominance is that individuals benefit. As with other traits, there is an optimum level of aggression that depends on the individual's particular social and physical environment. Individuals that are too aggressive are selected against, as are those that are too passive.

EXTREME FORMS OF AGGRESSION: CANNIBALISM

The intensity of aggression varies widely—from actual physical contact and killing to a subtle threat, such as a direct stare or eyebrow raising to expose the eyelids. So

many cases of violent aggression and killing have been reported that some have questioned whether killing is always accidental, maladaptive, or otherwise anomalous. For example, the Indian langur monkey (*Presbytis entellus*) lives in a variety of habitats. In some areas, social groups of langurs contain only one adult male, five to ten females, and their young. Several observers have reported instances in which a new male came into the group, chased out the old male, and killed some or all of the infants (figure 12–10). One interpretation of such behavior is that the group is socially disorganized by the change in males, and the usual restraints on overt aggression are absent. Once the new male has established himself, he ceases his attack on the young. From the standpoint of the species or group, such behavior is maladaptive and should be rare, which in fact it is. A second interpretation of these events is that infanticide is adaptive for the new male, since he removes the offspring of the presumably unrelated male and causes the females to come into estrus sooner to bear his own offspring (Hrdy 1977a,b). Hrdy's view is consistent with the notion that social behavior results from the action of natural selection on the individual.

FIGURE 12-10   Two female langur monkeys attacking male
The male langur monkey has stolen the infant of a third female (not visible). Males frequently wound or kill infants sired by other males.
Source: Photo by Sarah Blaffer-Hrdy from Anthro-Photo.

We can use the same arguments to explain acts of cannibalism performed by a variety of animals. Observers note that particularly under crowded conditions, rodents often eat their young, usually those already dead. We could interpret such behavior as social pathology, which results from crowded conditions and leads to maladaptive behavior. We should note, however, that under crowded conditions or when food is scarce, the young would probably not survive anyway, so by consuming their young, the parents save their reproductive investment for more propitious times. Rodents and lagomorphs frequently resorb embryos into their own bodies during the development of the embryos in the uterus, with the same effect as cannibalism. In kangaroos most of the development of the young takes place in the pouch, and females can terminate "pregnancy" by simply throwing the young out of the pouch.

Such killing of young is not limited to adults. Among some species of birds, older and larger young sometimes kill their younger and smaller siblings (Mock 1984). Such behavior might be expected only when food is so scarce that the survival of the entire brood is threatened. Cannibalism, whether infanticide or siblicide, seems to fit our definition of aggression; but such acts are relatively infrequent compared to the ritualized combat associated with most competition. (For papers on this topic, see the book edited by Hausfater and Hrdy 1984).

INTERNAL FACTORS IN AGGRESSION

LIMBIC SYSTEM

Most of what we understand about central nervous system control of agonistic behavior comes from studies on mammals, particularly the cat (Flynn 1967). The brain structures involved are part of the limbic system (see figure 6–13). The hypothalamus is involved in defense and escape behavior in animals as diverse as pigeons, cats, monkeys, and opossums. Using lesions, electrical stimulation, and single neuron recordings from specific areas of the brain, researchers have found that different brain sites are responsible for different types of aggression. For example, in cats, electrical stimulation of the ventromedial nucleus of the hypothalamus produces growling, hissing, and attacking with claws (defensive attack). Stimulation of the lateral hypothalamic area produces a biting attack with no defensive elements. Thus, agonistic behavior's involving a single neural

system may be an oversimplification. Other areas of the brain that are involved in aggression are the amygdala of the forebrain, and the central gray in the midbrain. These areas are connected by nerve pathways, and they interact. For example, electrical stimulation of certain areas in the thalamus causes cats to attack rats (Bandler and Flynn 1974). Using special staining techniques, axons in the thalamus were traced to areas in the central gray. Stimulation of those latter sites elicited similar attacks.

We can study the effects of electrical stimulation of specific brain areas in seminatural social groups by the use of radio transmitters (Delgado 1967; Herndon et al. 1979). Monkeys with electrodes implanted in certain parts of the thalamus, hypothalamus, or central gray became aggressive when the electrodes were activated (figure 12–11). In some cases, when stimulation was applied to the hypothalamus, lower-ranking monkeys became dominant as a result (Robinson et al. 1969).

FIGURE 12–11 Radio-stimulated attack in rhesus monkey
Social aggression in rhesus monkeys can be elicited by telestimulation of hypothalamic structures.

Source: J. G. Herndon, A. A. Perachio, and M. McCoy, 1979, Orthogonal relationship between electrically elicited social aggression and self-stimulation from the same brain sites, *Brain Research* 171:374–80. Reproduced courtesy of Yerkes Regional Primate Research Center.

HORMONES

Interacting with the limbic system are neurosecretions and hormones. Epinephrine (adrenaline) and norepinephrine (noradrenaline) are related to physiological arousal. These and other substances, such as dopamine and serotonin, act as neurotransmitters and may affect aggressiveness.

The effect of castration on sexual and aggressive behavior in males has been known since the time of Aristotle. Early work on chickens demonstrated that testosterone made them more aggressive and increased their dominance rank. Some of the hormone changes are rapid; male rhesus monkeys that have been defeated in fights show declines in circulating testosterone within hours (Rose et al. 1975). In many species of mammals, exposure to sex steroids early in ontogeny is a primary factor in the appearance of aggressive behavior in adulthood; thus they have organizational affects as well as activational effects (chapter 7). Female rhesus monkeys that received testosterone before birth and that were tested as juveniles threatened more and played more roughly than did untreated females (Phoenix 1974). In species of mammals where multiple embryos are present in the uterus, hormones produced by one sex may affect the other (vom Saal 1983). When tested as adults, males that were between males in utero were found to be more aggressive than males that were between females. Similarly, females that were between males as embryos fought more than females that were between other females. (See also chapter 7.)

The notion that testosterone is *the* substance causing aggression is much too simplistic, however. Luteinizing hormone (LH), rather than testosterone, increases aggression in some species of birds (Davis 1963). Castrated zebra finches (*Poephila guttata*) receiving estrogens become just as aggressive as those receiving testosterone. Once injected, hormones undergo chemical changes as they are broken down by the body; it appears that the metabolites of these hormones are the substances actually causing the increase in aggression (Harding 1983). In fish, there is evidence that thyroid hormone is involved in control of aggression (Villars 1983), and the adrenal glands have been implicated in dominance-related color changes in reptiles (Greenberg 1983). Invertebrate hormones are much different from those of vertebrates, and much less is known about how they work (Breed and Bell 1983). Roseler et al. (1986) have implicated both juvenile hormone and ecdysteroids produced by the ovaries of female paper wasps (*Polistes gallicus*) in the establishment of dominance hierarchies as they initiate nests in the spring.

In humans, Reinisch (1981) has shown that pregnant women who were treated with synthetic progesterone produced offspring who were more aggressive when tested as adolescents than similar adolescents not treated prenatally. See chapters 6 and 7 for further discussion of neural and hormonal mechanisms.

GENETICS

The synthesis of nerve structures, neurosecretions, and other compounds is under genetic control, and agonistic behavior is shaped by natural selection, as is any other behavior. Artificial selection can lead to significant changes in levels of aggression within just a few generations. For example, Ebert and Hyde (1976) tested wild female house mice for aggressiveness, and by selecting high- and low-scoring mice, they produced two lines: one with highly aggressive females and one with passive females. The unselected control lines were, as expected, intermediate. Domestic laboratory strains of mice differ widely in aggressiveness (Southwick and Clark 1968), as do dog breeds. Siamese fighting fish (*Betta splendens*), fighting cocks, and even crickets have been artificially selected over the years for performance in contests with large sums of money riding on the outcome.

A genetic link to aggression in humans has been postulated to explain the fact that males with two Y chromosomes (XYY males) are found more frequently than expected in maximum security mental institutions, given their frequency in the population at large. However, it turns out that XYY males also tend to be of below-average intelligence, and thus may merely be more likely to get caught. Also there is no indication that they are more violent than are normal males, nor do they have elevated levels of testosterone (Mazur 1983).

EXTERNAL FACTORS IN AGGRESSION

LEARNING AND EXPERIENCE

Although genes control production of the hormones, neurosecretions, and neurons that produce aggressive behavior, researchers generally believe that some factor outside the animal triggers the response. Previous experience can produce semipermanent changes in the expression of agonistic behavior, and animals can be conditioned to win or lose. If we give a naive male laboratory mouse that has been socially isolated for a few weeks a few easy victories over less aggressive mice, he will turn into a fighter mouse and will quickly attack any other male in sight. Likewise, we can "create" a loser by repeatedly exposing a naive male laboratory mouse to highly aggressive mice (Scott and Fredericson 1951).

Interest in primates has centered on the importance of early experience on later levels of aggression. For instance, we know that monkey societies differ widely in the frequency and intensity of aggression. Bonnet macaques (*Macaca radiata*) have a rather loose dominance hierarchy, with little fighting and much friendly interaction. Pigtail macaques (*Macaca nemestrina*) have a more rigid hierarchy, with more fighting and a low tolerance of strangers. Treatment of infants also differs between the two species. Pigtail mothers are very restrictive and do not let others handle or carry the infant as much as bonnet mothers do. The latter tolerate "aunting" by young females. Bonnet infants are cared for by others if the mother is removed; in contrast, pigtail infants become very depressed and huddle in the corner (Rosenblum and Kauffman 1968). Whether this difference in early experience is responsible for the different social structures that characterize the adults of the two species remains to be demonstrated.

Scott (1975, 1976) further emphasized the role of experience and learning in the development of aggression. Although he acknowledged the genetic and physiological bases of aggression, Scott in no way believed that the expression of hostility is inevitable. He emphasized the importance of early experience and learning in the development of aggression and suggested that one of the most important controls of agonistic behavior is passive inhibition. Animals form the habit of not fighting, particularly during critical periods of development. Some evidence in support of this view comes from studies of rats and mice (Scott 1966), and monkeys (Rosenblum and Kauffman 1968). Although it is clear that animals can be trained in the laboratory to be passive and nonaggressive, researchers have not clarified the role of early experience in the restraint of aggression in natural populations.

PAIN AND FRUSTRATION

More direct causes of aggression are pain and frustration. Researchers have shown that noxious stimuli, such as loud noises, foot shocks, tail pinches, and intense heat, cause different lab species to attack a wide variety of objects, from conspecific cage mates to tennis balls (Ulrich et al. 1965). Although referred to as attack behavior, the response often seems to be defensive in

nature; for instance, shocked rats assume the upright "boxing" posture that rats losing a fight normally exhibit. These defensive attacks occur in a laboratory situation in which the victim is prevented from escaping (i.e., the victim is cornered). The full range of attack behaviors seen in the wild is not elicited by painful stimuli.

Researchers have explored the frustration-aggression connection in humans. In some early experiments, researchers allowed children to view a room full of toys, but denied them access to it; when they finally let them into the room, the children frequently smashed toys and fought more than did a control group that was allowed immediate access to the room (Berkowitz 1969; Johnson 1972). Frustration is now considered to be only one of several causes of aggression. Restricted access to resources triggers aggression in most animals. Rhesus monkeys on a reduced diet fought less and became lethargic, but when limited food was placed in a small number of feeders, fighting became intense (Southwick 1967).

SOCIAL FACTORS

Another strong stimulus for aggression is the presence of strangers. In most species, the introduction of a strange animal into an established social group produces the violent reaction of xenophobia, usually in members of the same sex as the stranger (Southwick et al. 1974). The amount of fighting tends to decrease with time, but acceptance of strangers is usually a slow process. A related phenomenon is isolation-induced aggression, in which socially isolated animals become hyperactive and prone to attack other animals (Scott 1966).

Crowding per se tends to increase interaction rates but does not always increase aggression. Overall group size, presence of strangers, or restricted access to resources has a much more powerful effect on aggression than does reducing the space available to an already established social unit.

The relationship between sex hormones, particularly testosterone, and aggression is striking in seasonally breeding species. As the gonads increase in size in response to environmental changes in photoperiod, rainfall, vegetation, and so forth, fighting and wounding increase as well. Most of this increase is related to competition for breeding territories, social rank, or access to females. In some monogamous species (e.g., sea gulls), fighting occurs during the early phases of pair formation before a sexual bond develops. In others, courtship and copulation seem to have aggressive components, as evidenced by the appalling racket mating cats make, which makes the listener wonder whether it is a fight to the death or courtship. Other carnivores also have rather violent mating behavior. When polecats mate, the male grabs the female by the scruff of the neck and drags her back to his nest. Some of this aggression at the time of mating probably stems from the fact that males tend to court females rather indiscriminately, often making advances when the females are not sexually receptive. Some mammals, mainly carnivores and lagomorphs, are induced ovulators, and vigorous courtship plus copulation may be needed to trigger the release of an egg into the oviduct as copulation occurs.

Some evidence has shown that in groups of fish, the initiation of feeding increases intragroup agonistic behavior. Albrecht (1966) proposed that predatory (feeding) and agonistic behaviors, although functionally distinct, are motivationally linked. However, Poulsen and Chiszar (1975) found that receipt of aggression inhibited feeding in submissive bluegill sunfish (*Lepomis macrochirus*). Feeding was unaffected by levels of aggression in dominant bluegills. They concluded that feeding and aggression are independently motivated.

EVOLUTION OF RESTRAINT AND APPEASEMENT

Given the obvious benefits gained by the winners of competitive interactions, one might expect that natural selection would lead to the evolution of highly aggressive individuals with sophisticated armamentaria. However, most individuals show considerable restraint in their use of force. Why is it that animals tend to use restraint?

Displays, discussed previously in chapter 11, are behaviors that have evolved to communicate information. Many displays communicate agonistic behavior, such as a threat or a submission. For instance, in chapter 11, figure 11–8 shows a threatening dog and a submissive dog; the postures are in many ways opposites of one another (Darwin's theory of antithesis). Displays that communicate an animal's submissive intentions may inhibit attack by another. Lorenz (1966), Scott (1966), and others argued that such behaviors are necessary to avoid large-scale killing and wounding, which would be detrimental to the social group or species. Species with effective weapons like large canine teeth or horns are most likely to have evolved such displays, since the potential for inflicting severe damage to opponents is so great.

The species preservation argument implies that natural selection operates at the group or species level. How else could individuals be selected that yield to dominants without a struggle or that fail to go all out for the kill when given the opportunity? In contrast, Tinbergen (1951) pointed out that ritualized contests (displays) would reduce the risk of injury to the aggressor, and that threats take less energy than physical combat and are of advantage to individuals as well as to groups. More recently, behavioral ecologists have applied game theory, a branch of applied mathematics, to understand the evolution of fighting strategies in animals better.

GAME THEORY MODELS

Maynard Smith and Price (1973) independently developed a model of how such displays could be selected for at the level of the individual. One version is the Hawks-Doves game (Parker 1984). In a hypothetical population, all individuals (doves) use only ritualized conventional signals; they display, but always retreat at the first sight of serious conflict and thereby avoid injury. If a mutant fighter (hawk) appears and fights vigorously enough to win each contest, the hawk will be highly successful. If hawk behavior is heritable, hawks will become more common and doves less so in subsequent generations. But as the hawks increase in number, the likelihood that contests will involve two hawks rather than a hawk and a dove also increases. The average benefit of winning (controlling the resource) may be outweighed by the cost of losing (being injured or killed); thus the doves will then do better. The result, presented as a payoff matrix (table 12–3), is that either group does well when it is rare, but is outdone by the other when it is common.

In this situation, the best strategy is a function of what all the others in the population are doing. The relative benefits of winning and the relative costs of losing bring about a stable mixture of hawks and doves, which we refer to as an **evolutionarily stable strategy (ESS)**. An ESS is a strategy, which, if adopted by all members of the population, cannot be successfully invaded by any rare alternative strategy (Maynard Smith 1974). In any population, we can expect either a mixture of individuals that are always highly aggressive and individuals that always display without combat, or individuals that vary their behavior and follow a mixed strategy of being either aggressive or noncombative a certain proportion of the time. The proportion of hawks and doves depends on the benefits and costs of winning and losing.

TABLE 12-3 Payoff matrix for hawks-doves game
Hypothetical costs and benefits have been assigned to show how the game works. With this situation, what would happen if it starts out with all doves? The average payoff is $+15$. But then a mutant hawk could invade and spread. What if we start with all hawks? Here the payoff is -25, so a mutant dove would spread. Thus, neither pure hawk nor pure dove is an ESS. Further analysis shows that there is a stable mixture of 7 hawks and 5 doves, and an ESS results, and the group is resistant to invasion by hawks or doves.

| | *Payoff when opponent is:* | |
	Hawk	Dove
Hawk	$(B + C_{inj})/2$ $= -25$	B $= +50$
Dove	0 $= 0$	$((B + C_{dis}) + C_{dis})/2$ $= +15$

hawk: tries to damage opponent, retreats only if injured
dove: tries to settle amicably by display, retreats if opponent escalates
B = benefit of winning $= +50$
C_{inj} = cost of injury $= -100$
C_{dis} = cost of displaying $= -10$
Source: G. A. Parker. "Evolutionarily Stable Strategies," *Behavioural Ecology: An Evolutionary Approach*, 2d edition. Copyright © 1984 Blackwell Scientific Publications, Ltd, Oxford, England.

Thus, if the cost of losing is low relative to the benefit, a pure hawk strategy is the ESS, whereas if the cost is very high, a pure dove strategy is the ESS. One conclusion we may draw from this model is that noncombative individuals that control their aggression by displaying and retreating rather than escalating conflicts are following an adaptive strategy; they are not necessarily the losers or less fit victims of the aggressive dominants.

More realistic (and complex) models have been developed that include situations where the contestants differ in some quality. Three kinds of asymmetry have been considered: (1) cases where one individual already owns the resource or occupies the area, which leads to home field advantage; (2) cases where contestants differ in resource holding power (RHP), as when one is larger or stronger than the other; (3) cases where the value of the resource differs between the contestants.

Maynard Smith (1976) added a conditional strategy to the Hawks-Doves model called "bourgeois" to deal with case (1). These individuals play a hawk strategy if they are on home ground (or already own the resource in dispute) but play as a dove if it is someone else's territory. Assuming that there are no other asymmetries, bourgeois turns out to be a pure ESS. Male speckled butterflies (*Pararge aegeria*) occupy mating territories on the forest floor; the resident male always wins when intruders enter, and the outcomes reverse if ownership changes (Davies 1978). As predicted by the model, both contestants behave like hawks if they can be tricked into ownership of the same territory.

Asymmetries in RHP and resource value have been manipulated in laboratory studies of spiders. Working with female funnel-web spiders (*Agelenopsis aperta*) Riechert (1984) found that residency was usually unimportant in determining the outcome and that assessment of relative weights occurred early in the contest. Subsequent behavior depended on the results of this assessment. If a spider had a large weight advantage, it immediately escalated the fight, whereas smaller spiders tended to retreat. If they were the same size, then the resident won. Austad (1983) studied contests between male bowl and doily spiders (*Frontinella pyramitella*) over access to a female. He tested his data against a model referred to as the "war of attrition" (Maynard Smith 1974). In this game, two individuals display or fight continuously until one wins; costs are assumed to increase linearly with persistence time. Austad varied resource value by introducing an intruder male at various times while the resident male mated with the female. As more and more of her eggs were fertilized over time, her value to the resident male declined, and he was less inclined to fight. Males varied in RHP in terms of size: larger males usually won. As predicted by the model, fights were longest when males were of similar weight, and also when the female was of greatest value to the resident.

RELEVANCE FOR HUMANS

Whatever its evolutionary origins, the ritualization of agonistic displays into contests instead of struggles has occurred in practically all species with aggressive behavior. As we have seen, killing and wounding do occur;

however, given the weapons available and the competition for resources that probably occurs at some point in the lives of all organisms, serious injury is much rarer than is potentially possible.

Control of aggression in nonhuman animals may have special relevance for us. Lorenz (1966) argued that humans are aggressive for the same reasons as are other animals. However, humans have only recently acquired the means of inflicting serious injury through the use of weapons; we have not evolved the ritualized displays typical of species with such natural weapons as canine teeth or horns. Thus aggression in humans escalates with violent consequences.

Humans also become violent, according to Lorenz, because we have few harmless outlets for aggression—a particular problem in modern society, where so little of our time and energy is expended in subsistence activities. Lorenz, in his hydraulic model of behavior, argued that action-specific energy (in this case for aggression) builds up until it is released in some way (see chapter 10 and figure 10–13). Although other research discredits the idea that specific "energies" are stored in the brain, the discovery that different chemicals are involved in transmitting information to specific parts of the brain could give credence to his theory. If aggression is inevitable, Lorenz argued, then we must find harmless ways to vent it. This cathartic approach suggested to him the importance of ritualized tournaments in which participants and spectators alike can work out their hostilities; for example, sports events. But what about the riots triggered by the sports events themselves?

The idea that we are more aggressive if we are deprived of aggression is not supported by experiments on laboratory animals. For instance, rats trained to kill mice showed no tendency to increase their rate of killing after being deprived of mice for a few days (Van Hemel and Myer 1970). A number of studies on humans have tried to evaluate the effect of the observation of aggression in movies and television on subsequent aggressive behavior. Possible changes, if any, after viewing violence would be either a decrease in aggressiveness through catharsis or an increase in aggressiveness through some type of learning, through a generalized arousal, or through a reduction in inhibitions. The results of television and movie viewing are not clear-cut: some studies have shown increases and some have shown decreases (Feshbach and Singer 1971); but a 1982 report by the National Institutes of Mental Health concluded that violence on television does lead to aggressive behavior in children and teenagers who watch the programs. Not surprisingly, television networks challenged the findings (Walsh 1983). Controlled laboratory experiments indicate that various visual stimuli (for example, comic book violence) do increase the subjects' aggressive behavior (Berkowitz 1969).

Conditioning to situational stimuli may be important in humans, as Berkowitz and LePage (1967) showed. The researchers instructed a subject to play the role of an experimenter trying to teach others a task. When the "subjects" made a mistake, the "experimenter" (i.e., the real subject) was supposed to administer a punishment in the form of an electric shock. The researchers could then measure the aggressiveness of the real subject by the number of shocks he or she delivered. A second variable was the presence or absence of a gun on a table near the "experimenter." The real subject administered more shocks when the weapon was present than when it was absent. Presumably the gun facilitated aggression because of its association with violence.

SOCIAL CONTROL AND SOCIAL DISORGANIZATION

Observations of cichlid fish (*Cichlasoma biosallatum*) (aptly named the Jack Dempsey fish, after the heavyweight boxer) show that when strange fish of the same species are continually introduced into the group, fighting remains at a high level; if group membership is allowed to stabilize, fighting declines to a relatively low level (Scott 1975). Thus, one cause of aggression seems to be social disorganization, as seen in newly formed groups or groups that have been disturbed. When a high-ranking male or female rhesus monkey dies or leaves the group, aggression within the group may increase. New males moving into the central hierarchy attack females and are sometimes mobbed by them in return. The absence of a high-ranking control animal contributes to the increase of aggression. As new relationships are established within the group, and as control animals emerge and begin to break up fights, group aggression decreases (Vessey 1971). The following tactics can be used by any social animal to avoid or deescalate aggression (modified from Marler 1976):

- Keep away
- Evoke behavior that requires proximity, is physically incompatible with aggression, and reduces arousal (e.g., grooming)
- Avoid provoking extreme arousal of frustration
- Use submissive displays (antithesis of aggressive displays)

- Behave predictably

- Divert attack elsewhere

Maintenance of a territory or a position in a dominance hierarchy requires constant reinforcement by songs, threats, or other displays; individual recognition and memory of the results of previous encounters are always involved. Although male macaques and baboons use their enormous canines in fights against other males, once they have formed a stable hierarchy, they can maintain rank without them, as happened in a captive colony of Japanese monkeys in Oregon after experimenters removed the males' canines (Alexander and Hughes 1971).

Aggression in human societies can be traced, in part, to social disorganization and xenophobia. Sociologists use the term community to refer to a social system with stable membership whose members interact extensively among themselves. When group membership is unstable and when relationships and roles are uncertain, alienation and aggressive interactions are frequent (Scott 1976). The high rate of violent crime in inner cities may be spurred in part by unstable family structures.

SUMMARY

Agonistic behavior, defined as social fighting among conspecifics, includes all aspects of conflict, such as threat, submission, chasing, and physical combat, but excludes predation. *Aggression* emphasizes overt acts intended to inflict damage on another, and may include predation, defensive attacks on predators by prey, and attacks on inanimate objects. Many species of invertebrates and nearly all vertebrates engage in some type of agonistic behavior, which may include the formation of a dominance hierarchy, territoriality, combat for mates, or parent-offspring conflict.

Agonistic behavior involves competition, one form of which is *exploitation,* as individuals passively use up limited resources and follow an *ideal free distribution.* A second form is *interference,* where individuals actively defend resources.

Much conflict behavior involves the social use of space. *Home range* is the area used habitually by an individual or group; *core area* is the zone of heaviest use within the home range. Many social animals maintain a personal space around themselves, also referred to as *individual distance. Territory* is the space that is used exclusively by an individual or group and that is defended. Individual territories occur most often when a needed resource is predictable and evenly distributed in space. Some territories include food and water supply, nest site, and mates; other territories contain only one resource that is defended—for example, a place where only mating takes place, as in the *lek.*

Species that live in more or less permanent groups usually develop *dominance hierarchies,* in which individuals control the behavior of conspecifics on the basis of the results of previous encounters. Dominance hierarchies are not always determined by fighting ability; age, seniority, maternal lineage, and formation of alliances with friends have been shown to be important in some mammals.

Agonistic behavior is usually highly ritualized and communicative in nature; killing or wounding is infrequent. Recent observations interpret overt violence as an expression of genetically selfish behavior on the part of an individual rather than as a maladaptive response to abnormal conditions.

Researchers have studied the causes of aggression from both an internal and external perspective. Internal factors include the *subcortical limbic system,* in particular the thalamus, hypothalamus, amygdala, and central gray. Various *neurotransmitters* and *hormones* are linked with the exhibition of different forms of aggression. For example, testosterone is most often associated with aggression, but other hormones are important in producing aggression in some species. Artificial selection in the laboratory for high and low aggression demonstrates a genetic component of aggression upon which natural selection can act. In the laboratory animals can be conditioned to be dominant or subordinate, which leads us to conclude that previous experience and learning affect later aggressive behavior. External factors that influence aggression include early learning and experience, pain and frustration due to restricted access to resources, xenophobia, crowding, and environmental factors that trigger the onset of reproduction.

Some animal behaviorists have held that the widespread use of restraint and display in social species has evolved for the preservation of the group or the species. However, models based on game theory suggest that natural selection acting at the level of the individual can produce an *evolutionarily stable strategy (ESS)*, in which a certain proportion of individuals in a population are aggressive and a certain proportion are passive. The behavioral strategy used depends on the relative benefits and costs of winning and losing, and on what the other members of the population are doing.

Control of aggression is of prime interest because it relates to noninstitutionalized violence in humans. Some researchers have emphasized the inevitability of aggression in humans and have sought ways to divert it into harmless channels. Others have suggested that the modification of the social environment through early experience and avoidance of social disorganization can help to control aggression. Encouraging noncompetitive habits during early experience and creating stable communities may help reduce violent agonistic behavior.

Discussion Questions

1. Distinguish between agonism and aggression.

2. Discuss ways aggression is restrained in animals. Evaluate the arguments put forth to explain the evolution of such restraint.

3. What, if anything, can we learn about aggression in humans by studying aggression in nonhumans?

4. The data below present ranks and maternal lineages among rhesus monkeys at La Parguera, Puerto Rico in June 1969. Part of A Group, which was formed in 1962 when unrelated adult females were placed together in a large cage, is shown. Females

34, 227, 184, and 236 established a linear dominance hierarchy in the order listed; they were then released. The offspring of those females are shown connected to the females with a line; their ages are given in parentheses. Most male offspring left the group at three or four years of age and are, therefore, not shown. The bottom row of numbers and letters shows the monkeys ranked on the basis of agonistic encounters, with those having won the greatest number of encounters on the left, and the least, on the right. Males are in squares. What kinds of rules seem to apply in determining dominance orders in rhesus monkeys? Can you explain how or why such rules come about?

Ranks and maternal lineages among rhesus monkeys

34(13+)	227(13+)	184(13+)	236(13+)
313(6) A9(4) G4(3) 13(2)	268(6) A2(4) E9(3) 12(2)	B2(4) D0(3) F9(2)	269(6) 300(5) D7(4) 19(2)
313-68(1)			K4(1)

34, G4, A9, 313, 313–68, I3, 227, E9, A2, 268, I2, 184, B2, D0, F9, 236, D7, 300, K4, 269, 19

Suggested Readings

Archer, J. 1988. *The Behavioural Biology of Aggression.* New York: Cambridge Univ. Press.
An introductory treatment that considers both how and why aggression occurs throughout the animal kingdom. A good synthesis of ethology, psychology, and sociobiology.

Johnson, R. N. 1972. *Aggression in Man and Animals.* Philadelphia: Saunders.
Basic, introductory text with emphasis on humans.

Moyer, K. E. 1976. *The Psychobiology of Aggression.* New York: Harper & Row.
The psychologist's perspective; deals with neural and hormonal factors.

Southwick, C. H. 1970. *Animal Aggression: Selected Readings.* New York: Van Nostrand Reinhold.
Although getting a bit dated, provides a balanced collection of review papers on agonistic behavior of nonhumans.

Svare, B. B., ed. 1983. *Hormones and Aggressive Behavior.* New York: Plenum Press.
Series of review articles; chapters on invertebrates and each of the vertebrate classes.

13

SEXUAL REPRODUCTION

*A*fter reaching maturity at sea, the Pacific salmon (*Oncorhynchus* spp.) works its way up rapids to reach its birthplace, expends its remaining energy by shedding masses of eggs or sperm, and then dies. We can ask a number of "why" questions about the evolutionary and long-term ecological determinants of the salmon's reproductive behavior. For example, why does the salmon return to its birthplace to reproduce, and not to some other stream? Why does it put all its effort into a single explosive reproductive episode rather than produce smaller numbers of young at intervals? For that matter, why does it reproduce sexually at all? In the first part of this chapter, we look at the ultimate evolutionary forces that determine such things as the existence of sex in the first place, sex ratios, and sexual selection.

We can also discuss reproduction from the point of view of "how" questions. How does the salmon locate its ancestral stream at a certain time of year and arrive at its birthplace? In the second part of this chapter, we briefly examine the proximate, or short-term, mechanisms of sex, such as neural and endocrine control of behavior, and how they are linked to environmental fluctuations. Mating systems and strategies of parental investment are covered in chapter 14.

COSTS AND BENEFITS OF SEX

Many unicellular organisms, plants, and invertebrates reproduce asexually during one part of their life cycle and reproduce sexually at other times. During the sexual phase, the organism will often produce seeds or eggs that are resistant to environmental stress and that are capable of being dispersed. Some sort of genetic recombination occurs in virtually all organisms. Bacteria and blue-green algae, which are among the prokaryotes, recombine genes in a process called transformation. Single-celled algae, such as diatoms, have eggs and sperm that resemble those of vertebrates; however, their courtship behavior is probably not particularly complex. Some organisms—for example, earthworms, snails, and certain species of fish—are hermaphrodites: they possess functional sexual organs of both males and females. Usually hermaphrodites do not self-fertilize, possibly because the new genetic combinations created would not be novel enough to outweigh the costs of breaking up previously successful combinations. Some species of fish and some invertebrates are sequentially hermaphroditic; they change from female to male (**protogyny),** or from male to female (**protandry)** usually once in their lives (figure 13–1).

FIGURE 13-1 Sequentially hermaphroditic coral reef fish

One male in the species *Anthias squamipinnis* defends a territory containing several females. The dominant female in the group (top) changes into a male (bottom) upon the death or disappearance of the male. Protogyny is common in reef fish.

Source: Photo by Douglas Shapiro.

COSTS OF SEX

Why is sexual reproduction so widespread? By reproducing asexually, an organism maximizes its genetic contribution to offspring by creating carbon copies of itself. By reproducing sexually, genotypes are broken up at meiosis (chapter 4), and the genes combine with those of another individual during fertilization. The result is a 50 percent reduction in transmission of genes to the next generation, which is referred to as the **cost of meiosis** (Williams 1975). The energetic cost of raising young is about the same whether the offspring is produced sexually or asexually, so a female would have to produce two sexual offspring for each asexual offspring in order to pass the same number of genes to the next generation. If a female can get a male to share in raising the offspring, the cost of meiosis is reduced. If they share equally, a female can rear two sexual offspring for the same cost as a single asexual offspring, and the cost of meiosis might be reduced to zero (Wittenberger 1981). A further cost of meiosis, however, is that the unique combination of traits in the parent is broken up in sexually produced offspring. This aspect of the cost of meiosis can be reduced if individuals mate with relatives, since they are likely to share some of the same combinations of genes (Shields 1982).

A second cost of sex is sometimes referred to as the **cost of producing males.** A single male produces enough sperm to fertilize the eggs of many females. However, most sperm are wasted and most males never even fertilize one egg (Clutton-Brock 1988); they waste resources that could be used for reproduction in asexual form. A third cost of sexual reproduction is the **cost of courtship and mating,** since it takes energy to secure a partner, and there is often an increased risk of being injured or killed by a competitor or a predator (Daly 1978).

BENEFITS OF SEX

Given these costs, one might wonder how sexual reproduction got started in the first place and how it is maintained in a population. We might expect parthenogenetic (asexual) individuals to "infect" sexual populations and replace the sexually reproducing organisms. What are the possible benefits of sex and how can it be maintained in a population? Most often mentioned is that of **faster evolution.** Fisher (1958) first pointed out that sexually reproducing populations can evolve faster in a changing environment. He stated that the rate of evolutionary change in a population is a function of the degree of genetic variation available. Through sexual recombination, mutations occurring at different loci in different individuals can appear as novel combinations in the offspring. A second, related advantage was noted by Muller (1964) and is referred to as **Muller's ratchet.** Suppose that a deleterious mutation occurs in an individual member of an asexual population. The gene will be passed on to all its offspring. The only way an individual that is free of this mutation can arise is by back (or reverse) mutation, an unlikely event.

Thus the ratchet turns one notch each time a deleterious mutation occurs, and mutations accumulate in the population. However, if sexual reproduction occurs, recombination between two (diploid) individuals with different mutations could produce offspring with neither trait. In this way, harmful mutations can be edited out of a population.

The preceding arguments have emphasized the advantages of sex for the population, or group: faster evolution and the elimination of mutations from the gene pool. Williams (1975) has championed benefits that might accrue to individuals, rather than populations, since natural selection is thought to act primarily at the level of the individual (chapter 4). He noted that in species with both sexual and asexual reproduction, the sexual phase usually occurs before unpredictable changes in environmental conditions begin. Parents produce genetically variable offspring "in the hopes" that a few will be adapted to one of the new environments. Williams uses a **raffle analogy**: If there is only one survivor (winner) in each different habitat, reproducing asexually would be like having tickets that all have the same number printed on them. The chances of picking a winner are much less than if all the tickets have different numbers—the case with sexually produced offspring. From this we might predict that species living in highly variable and unpredictable environments would be more likely to have sexual reproduction; those in stable and predictable environments would be more likely to have asexual reproduction. The emphasis here is variation in the *physical* environment.

Others have argued that *biotic* fluctuations make it even more necessary to produce genetically variable progeny (Bell 1982; Van Valen 1973). For instance, as predators are selected to become more efficient at catching their prey, the prey must continually evolve new ways to avoid being eaten. Such a coevolutionary "race" takes place between competing species. Competition also occurs between pathogens and hosts. Because of their short life spans, bacteria, viruses, and parasites rapidly evolve into new strains. The host produces variable offspring "in the hopes" that some will be resistant. This explanation for the maintenance of sex has been called the **Red Queen** hypothesis because of Alice's encounter with the Red Queen in *Through the Looking Glass*. After a wild run they ended up where they started, and the queen told Alice: "Now *here*, you see, it takes all the running *you* can do, to keep in the same place."

One prediction of the raffle hypothesis is that sexual species should predominate in environments that are highly, unpredictably variable, while asexual species should predominate in habitats that are more stable. In fact, just the opposite is usually the case, and the Red Queen hypothesis seems the more likely explanation. In stable environments, biotic interactions are likely to be more intense, and animals have more predators, parasites, and pathogens to escape from. However, in species that can alternate their mode of reproduction during the year, there is a tendency for reproduction to be asexual in spring and summer, and sexual in the fall before harsh winter conditions set in. Often the zygote is highly resistant to extreme temperature and moisture conditions. These cases seem to support the raffle hypothesis.

Further support for the Red Queen hypothesis comes from a species of snail (*Potamopyrgus antipodarum*) from New Zealand. Those living in lakes, where rates of parasitism are high, are more likely to reproduce sexually than are those from streams, where parasitism is lower (Lively 1987). Also, the percentage of males in different populations (an indicator of the extent of sexual reproduction) increases with the degree of parasitism. The advantage of sexual reproduction here may be to "keep up" defensively with the rapidly evolving parasites.

ANISOGAMY

Whatever the reasons for the evolution and maintenance of sex, we can go on to the question of how sex works. In most animal species there are only two sexes that are anatomically different and that are produced in about equal numbers. The possible genetic combinations for two sexes are astronomical—three or more sexes would add little except confusion in mate selection. In most microorganisms, fungi, and algae, the sexes are anatomically similar, and the gametes are all of the same size (**isogamy**). However, division of labor is the rule in most plants and animals: females produce few large, sessile, energetically expensive eggs, and males produce many small, motile, energetically cheap sperm. This difference in gamete size, called **anisogamy,** may have resulted from disruptive selection (Parker et al. 1972) and has set the stage for many differences in the reproductive behavior of males and females (Trivers 1972). Since females by definition produce a relatively small number of the expensive gametes, they are probably a limited resource for which males compete. A male's reproductive success is usually a function of how many different females he can inseminate, while a female's is a function of how many eggs she can produce.

One of the results of this difference between the sexes is that the reproductive success of males is likely to vary more than that of females. In laboratory populations of fruit flies (*Drosophila melanogaster*), Bateman (1948) demonstrated that nearly all females mated; however, many males failed to mate, and some males mated several times. In other words, the variance in copulatory success was higher for males than for females (the **Bateman effect**). The flies had chromosomal markers so that parents of particular offspring could be identified. Males that copulated most also sired the most offspring. On the other hand, females needed only one mate to produce the maximum number of offspring.

When reproductive success is determined over the entire lifetime, the Bateman effect becomes evident, especially in species such as lions, where male-male competition for access to pride females is intense (figure 13-2). However, in species such as scrub jays, where the sexes form stable pair bonds, no such effect is present.

SEX DETERMINATION AND THE SEX RATIO

We are accustomed to thinking that sex is determined by sex chromosomes, as it is in birds and mammals. However, things are a bit more complicated in other organisms. Some animals have physiological and behavioral control over sex determination. In social insects such as bees and ants (Hymenoptera), males are derived from unfertilized (haploid) eggs and females from fertilized (diploid) eggs. Such a mode of sex determination is called **haplodiploidy.** Early in the season, the queen, after being fertilized, does not release stored sperm when she lays her eggs—and male offspring result. Later in the season she releases sperm when she lays her eggs—all-female broods that become workers result. In many species of turtles, sex is determined by the incubation temperature: those incubated at a low temperature become males, while those incubated at a higher temperature become females (Bull 1980).

Among organisms that undergo a sex change are protogynous fishes such as the sea bass (*Anthias squamipinnis*) studied by Shapiro (1979) on Aldabra Island off the east coast of Africa. The mechanism that controls the fish's change from female to male is incompletely understood; it is triggered, however, by the death or departure of the (usually) single resident territorial male. Shapiro observed that upon the permanent absence of the male, the largest female undergoes a marked change in behavior, engaging in more aggressive displays, as she assumes the male color pattern and reproductive role

FIGURE 13-2 Variance in lifetime reproductive success (LRS)

(a) LRS for African lions. Most females have from 1 to 5 surviving offspring, whereas male reproductive success varies widely. (b) LRS for Florida scrub jays. Note that the variances are similar for both sexes.

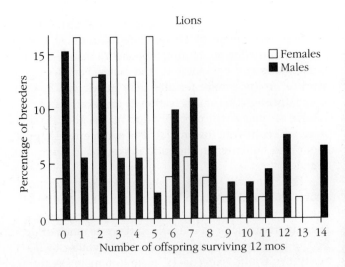

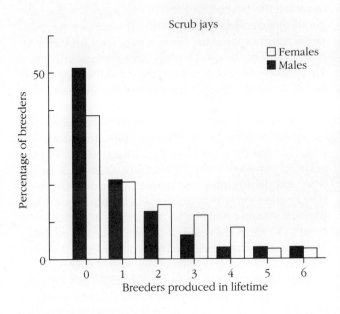

(see figure 13-1). In another species, the saddleback wrasse (*Thalassoma duperrey*), sex change is a function of the relative size of neighbors. When few larger fish (usually males) are present, females change into males (Ross et al. 1983).

The sex ratios of populations of most species tend to be about 1:1 at birth or hatching, but may deviate significantly from equality among adults. The **operational sex ratio** considers only the reproductively active members of the population. Deviations in this ratio can have a large impact on the mating system, since members of the abundant sex will compete for access to the scarcer sex.

Why does the sex ratio tend to be 1:1 at birth? This problem was solved by Fisher (1958) as follows: Consider a population containing more males than females. Some males will be unable to find mates. Females that tend to produce more females than males will be favored through natural selection because all of their female offspring will probably find mates, and the sex ratio will converge toward 1:1. The symmetrical argument also holds when the ratio is skewed in favor of females.

The argument has been developed further by Hamilton (1967) and others, who point out that equality of the **investment** by parents in offspring of each sex is the important factor in the ratio. If twice as much effort by the parents is required to raise male young to maturity, we would expect a 1:2 male to female sex ratio in young that reach maturity. In species of wasps of the genus *Symmorphus*, the sex ratios of offspring were inversely proportional to the amount of food provided by the mother to the developing larva (Trivers and Hare 1976). In those species where more food was provided to female larvae, the ratio was biased toward males, and vice versa.

Deviations from 1:1 also occur in populations that are spatially divided into smaller units. The parasitoid wasp *Nasonia vitripennis* lays eggs directly in the larvae of their host—a blowfly. Developing wasps feed on the host, and the flightless males mate on or near the host. If only one female lays eggs in a host, **local mate competition** occurs as male sibs try to mate with the same sisters (Werren 1983). From an evolutionary perspective, it makes little sense for close relatives to compete with each other because this lowers inclusive fitness (see chapters 4 and 20); in addition, one male can fertilize many females. In such situations, the primary sex ratio is strongly female biased. What if a female lays eggs in a host that has already been parasitized? In these cases, the female somehow can assess the proportion of the total eggs in the host that are hers and adjusts the sex of her eggs accordingly: the fewer that are hers, the more male-biased is the sex ratio (Werren 1983).

The Bateman effect sets the stage for another possible deviation from 50:50, according to a model based on **maternal condition** proposed by Trivers and Willard (1973). The model assumes that parental investment in male and female offspring is the same, and that mothers in the best physical condition produce healthier offspring that are better able to compete for mates or other resources. The reproductive success of males, which is highly variable, should be high if their mothers are in good condition, but low, perhaps zero, if their mothers are in poor shape; female offspring are likely to breed anyway, regardless of their mother's condition. Therefore, Trivers and Willard predicted that a female would produce male offspring if she were in good condition and female offspring if she were in poor condition; they cited data from mammals such as mink, deer, seals, sheep, and pigs in support of their model.

Meikle et al. (1984) found that high-ranking female rhesus monkeys (*Macaca mulatta*) produced significantly more male offspring than did low-ranking females. In fact, these sons did have higher reproductive success than did sons of low-ranking females, mainly because they were more likely to survive to breeding age. Daughters of low-ranking mothers reproduced almost as well as daughters of high-ranking mothers did. Producing daughters is a safe investment; producing sons is a gamble with the possibility of a big payoff in reproductive success. Similarly, in high-ranking social groups of monkeys, the sex ratio was skewed toward males, with the opposite true for low-ranking groups (figure 13–3).

Similar results were reported for red deer (*Cervus elaphus*) (Clutton-Brock et al. 1984). Dominant females have access to the best feeding sites and are able to invest more in their offspring via lactation. Male offspring in this species grow faster than females do, and would seem to benefit more from this greater investment. As adults, males compete intensively to control harems. A female who produces a successful son can achieve more than twice the reproductive success of a female producing a daughter.

However, there are other cases where different results are obtained. In studies of a South African prosimian primate, the bush baby (*Galago crassicaudatus*), Clark (1978) noted that the sex ratio was male biased. Female offspring tended to remain near the mother's home range while males dispersed. Daughters therefore would compete with their mothers and sisters for food. Clark argues that by producing fewer daughters, this **local resource competition** would be reduced. A troop of baboons (*Papio* spp.) living in Amboseli National Park,

FIGURE 13-3   Sex ratio in rhesus monkeys
Proportion of males born in social groups on La Cueva from 1964–1978. Proportions are shown for each year (1976–1978) for groups C, A, and I after they changed ranks in 1975. Note the relationship between proportion of males born and the group's social rank.

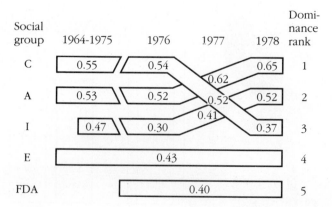

highest among mothers in poor condition. In this way the maternal condition hypothesis could be explained, as females in poor condition would lose their male embryos and possibly recycle until a female is conceived.

SEXUAL SELECTION

The success of an individual is measured not only by the number of offspring it leaves, but also by the quality or probable reproductive success of those offspring. Thus, it becomes important who its mate will be. Darwin (1871) introduced the concept of **sexual selection,** a special process that produces anatomical and behavioral traits that affect an individual's ability to acquire mates. Sexual selection can be divided into two types: **intersexual selection,** in which members of one sex choose certain mates of the other sex and **intrasexual selection,** in which individuals of one sex compete among themselves for access to the other sex. A result of either type is the evolution of sex differences—that is, the sexes become **dimorphic.**

INTERSEXUAL SELECTION

EVOLUTION OF SEXUALLY SELECTED TRAITS. In intersexual selection, individuals of one sex (usually the males) "advertise" that they are worthy of an investment; then members of the other sex (usually the females) choose among them. Most naturalists after Darwin discounted the importance of mate choice in evolution, but it has recently become a popular topic of study. Fisher (1958) used birds as an example. Suppose a plumage characteristic in males gives a reproductive advantage to the bearer. If females prefer those males as mates, those females will have higher reproductive success; most likely their sons will have the trait and their daughters will prefer the trait when choosing a mate of their own. Further development of the trait will proceed in males, as will the preference for that trait in the females, resulting in a **runaway** process. A possible result of this process is seen in the peacock (figure 13–4). Sexually selected traits may become so exaggerated that survival of the males may be reduced. Counterselection in favor of less ornamented males will occur—due, for example, to greater predation on the more ornamented males—and sexual selection will be brought to a steady state.

Kenya, has been studied for many years by the Altmanns (1988). Dominant females produce more daughters than sons, while subordinate females produce more sons than daughters, exactly the opposite found for rhesus monkeys and red deer! The Altmanns point out that daughters, who remain in their natal group for life, share the social rank with their mothers and could benefit from their mother's high rank. Presumably, high-ranking families would get larger and even more powerful as more and more daughters are born. On the other hand, as with Clark's bush babies, sons usually leave the natal troop and are thus not benefited by their mother's rank. In this case, the best strategy for dominant females is to produce sons; for low-ranking females, the best strategy is to produce daughters.

When Williams (1979) reviewed the data for birds and mammals, he concluded that there was no good evidence for facultative adjustment of the sex ratio. The more recent studies cited above seem to demonstrate otherwise. However, more data are needed before these or other models can be accepted or rejected. Nor is it clear how a sex ratio adjustment could occur. In many species of mammals, the sex ratio at conception is male biased. Intrauterine mortality rates are higher for male than for female embryos, and these rates are probably

FIGURE 13-4   Peacock in display

In this extreme example of sexual selection the courting male spreads his tail feathers and shakes them in front of the female. Such plumage would seem to confer little advantage to male survival, but it may indicate to the female the male's genetic superiority.

Source: Photo by Terence A. Gili from Animals Animals.

FIGURE 13-5   Sexual selection for tail length in long-tailed widow birds

(a) There was no difference among the four groups before the tails were altered. (b) After the tails were cut and lengthened, the mating success went down and up respectively. The two kinds of control birds were (I) unmanipulated, and (II) cut and glued back without altering length. Mating success is measured as the number of active nests in each male's territory (Andersson 1982).

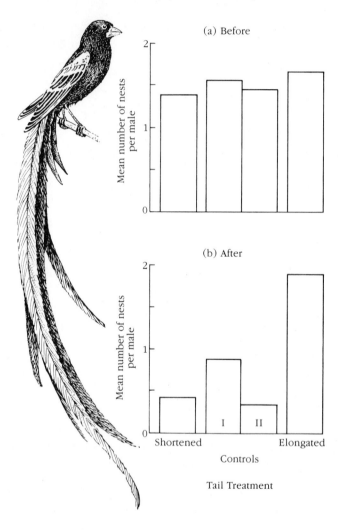

Female preference for male ornamentation has been demonstrated in a small African species of widow bird (*Euplectes progne*). Females prefer males with long tails. When tails are artificially shortened or lengthened, females still prefer those with the longest tails, showing that they are cued by tail length itself, and not some other correlated trait (Andersson 1982) (figure 13-5).

Laboratory studies of mate choice reveal that subtle differences can indeed influence mate choice. Burley et al. (1982) were able to influence choice of mates in a colony of zebra finches simply by using different colored leg bands. Males with red bands and females with black bands were preferred as mates and had higher reproductive success than did males with green and females with blue.

An alternative to Fisher's runaway selection hypothesis is the **handicap** hypothesis proposed by Zahavi (1975). Agreeing that sexual selection can produce traits that are detrimental to survival, he adds that they are both costly to produce and linked to superior qualities in the males. Thus a bird that can survive in spite of

conspicuous and costly-to-produce tail feathers, or a stag that can afford to spend energy on a huge bony growth of antlers, must be genetically superior and a worthy mate. Important to this hypothesis is the notion of **"truth in advertising"**; the male's handicap must be linked to overall genetic fitness. Only in this way will females benefit by picking a male with this handicap. One problem with this model is that when a male with a handicap is picked by a female, not only are his favorable traits passed on to his offspring, but also the genes for the handicap, which should be selected against.

Although Darwin, Fisher, and Zahavi thought of sexual selection as a process distinct from natural selection, some biologists argue that the two are inseparable. Kodric-Brown and Brown (1984) claim that most sexually selected traits are aids to survival, rather than handicaps. Thus a stag with a big set of antlers may be dominant over other males and may have better access to a food supply. They further argue that the sexually selected trait must be a reliable index of the male's condition. Thus, the trait must not be genetically fixed, but be influenced by environmental conditions.

Evidence for this type of role for a sexually selected trait comes from a hypothesis put forth by Hamilton and Zuk (Hamilton and Zuk 1984; Zuk 1984). They argued that sexually selected traits have evolved to indicate an animal's state of health; specifically whether or not it is free from disease or parasitism. They noted that bird species whose males have the showiest plumage are those species whose members are most prone to infestation with blood parasites. If brightness is linked to overall condition, the brighter males would be relatively disease free and thus preferred as mates.

There is other evidence that some aspect of courtship enables females to choose males with "good genes." Male butterflies of the genus *Colias* vary in enzymes involved in glycolysis, the process that provides energy to the flight muscles. Certain genotypes are superior in terms of flight capacity and longevity. In tests of mate choice, the superior male genotypes are favored, especially by older, more discriminating females who had already mated once (Watt et al. 1986).

BOWERBIRDS. As an example of intersexual selection in action, consider the bowerbirds of southeast Asia. Bowerbirds are unique in that they build a structure whose only function seems to be to attract mates. The male satin bowerbird (*Ptilonorhynchus violaceus*) of Australia and New Guinea stations itself alone in the forest, clears a space, weaves the bower from twigs, and decorates it with brightly colored objects (figure 13–6). Gilliard (1963, 1969) reasoned that the bower functions in place of showy plumage, and noted that those species with the fanciest bowers had the drabbest plumage. He argued that such behavior arose to synchronize mating between males and females.

FIGURE 13-6   Satin bowerbird at bower
The male satin bowerbird (*Ptilonorhynchus violaceus*) constructs the bower as part of his courtship ritual to attract a female. Males that build structures with more sticks and brightly colored objects are more successful in attracting mates.
Source: Photo by Richard A. Forster, courtesy of the Massachusetts Audubon Society.

Borgia (1985) assessed the relative contribution of the male satin bowerbird's decorations to his mating success. Human observers and cameras controlled by infrared detection devices monitored activities at the bower. Removing decorations reduced mating success. Aspects of the bower that were positively related to mating success were the number of blue feathers, snail shells, and yellow leaves. The general construction of the bower and the density of sticks in the walls were also important. Borgia suggested that bowers function as "markers," conveying information to females about the male's genetic quality. Dominant males have high-quality bowers that attract the most females.

INTRASEXUAL SELECTION

COMPETITION BEFORE MATING. Intrasexual selection involves competition within one sex (usually males), with the winner gaining access to the opposite sex. Competition may take place before mating, as it does

FIGURE 13-7   Use of antlers in combat between male Pere David deer
Winners of contests become dominant and control access to resources such as mates.
Source: New York Zoological Society Photo.

with ungulates such as deer (family Cervidae) and antelope (family Bovidae). Typically, males live most of the year in all-male herds; as the breeding season approaches, males engage in highly ritualized battles with their antlers or horns. The winners of these battles gain dominance and do most of the mating (figure 13-7). Antlers are better developed in those cervid species where males compete strongly for large groups of females (Clutton-Brock et al. 1982).

It is often difficult to determine which type of sexual selection is operating to produce an observed effect, since members of both sexes may be present during courtship. The red belly of the male stickleback both attracts females and intimidates other males. Similarly, the chirp of male field crickets (family Gryllidae) is used to exclude other males from an area and to attract females (Alexander 1961). As a final example of this problem, you may have noted that we used the antlers of deer to demonstrate the effects of both female choice and male-male competition. Females may incite competition among males and thus maintain some control over the choice of mate. For example, female elephant seals (*Mirounga angustirostris*) vocalize loudly whenever a male attempts to copulate. This behavior attracts other males and tests the dominance of the male attempting to mate (Cox and Boeuf 1977). In response to the female's sounds, the dominant harem master will drive off low-ranking, potentially inferior mating partners.

FIGURE 13-8   Male dung flies fighting over female
The attacking male (on the left) is attempting to push the paired male (on the right) away from the female (only her wings are visible). Sometimes the attacking male succeeds in taking over the female and mating with her; she then continues to oviposit eggs fertilized by the second attacking male.
Source: Photo by G. A. Parker.

Males may fight directly at the time of copulation, as exemplified by the male of the yellow dung fly (*Scatophaga stercoraria*) (Parker 1970a). A male will sometimes attack a mated pair, displace the male and mate with the female (figure 13-8).

COMPETITION AFTER MATING.   Nor does competition among males to sire offspring cease with the act of copulation. Females of many species store sperm and may remate before sperm from the previous mating are used up, creating the possibility of **sperm competition** (Parker 1970b). Sperm competition does not involve individual sperm actually fighting it out to gain access to eggs, rather it involves a selection pressure that has led to two opposing types of adaptation in males: those that reduce the chances that a second male's sperm will be used (*first male advantage*), versus those that reduce the chances that the previous male's sperm will be used (*second male advantage*) (Gromko et al. 1984).

First male adaptations include mate-guarding behavior and the deposition of copulatory plugs, both of which reduce the chance of sperm displacement by a second male. In fruit flies (*Drosophila melanogaster*), first males may also transfer to females anti-aphrodisiac substances that inhibit courtship by other males (Jallon et al. 1981). Female spiders store sperm for long periods and often mate with several males. In laboratory studies of the bowl and doily spider (*Frontinella pyramitela*), Austad (1982) found that the first male's sperm had priority in fertilizing eggs over sperm of subsequent males, as is the usual case in spiders.

Among insects, however, the advantage usually accrues to the sperm of the second (or subsequent) male as in the yellow dung flies studied by Parker (1970b). The last male to mate sired 80 percent of the offspring subsequently produced. The ultimate second male adaptation may be the penis of the damselfly (*Calopteryx maculata*) which has a dual function: it removes sperm deposited in the female by a previous male via a special "sperm scoop" and then replaces it with its own (Waage 1979).

Sperm competition has been suggested in the dunnock (*Prunella modularis*), a small, European sparrow-like bird. When a male's mate is likely to have mated with another male, the first male repeatedly pecks at the cloaca of his mate until she everts it, sometimes ejecting a sperm bundle. He then reinseminates her (Davies 1983) (figure 13-9).

Moths and butterflies (Lepidoptera) produce two kinds of sperm. One kind is the usual type (*eupyrene*) that fertilizes the eggs. The second type (*apyrene*) contains no nuclear material but may comprise more than half the sperm complement. Why should males waste energy on these dud sperm? Silberglied et al. (1984) suggested that apyrene sperm are the result of sperm competition, possibly displacing eupyrene sperm from first males or delaying remating by the female.

A somewhat similar suggestion has been made to explain the relatively large number of deformed sperm in mammals, up to 40 percent in humans, for example (Baker and Bellis 1988). Baker and Bellis argue that these deformed sperm play a "kamikaze" role, staying behind to form a plug to inhibit passage of sperm from a second male. Meanwhile, a small number of "egg-getter" sperm proceed to the oviduct to attempt fertilization. The kamikaze sperm hypothesis has been criticized on several grounds: selection should favor use of seminal fluids rather than sperm to form the plug, and there should be more deformed sperm in ejaculates from species where the female is likely to mate with several males. However, no such relationship seems to exist (Harcourt 1989). Clearly more data are needed to adequately test this intriguing hypothesis.

In primates, the type of mating system is related to the amount and quality of sperm that the male produces. In gorillas and orangutans, male-male competition takes place, but the winners have relatively free

FIGURE 13-9 Cloaca-pecking in dunnocks.
(a) The male first stands behind the female and repeatedly pecks her cloaca, sometimes causing her to eject sperm from a previous copulation. Then he copulates (b). Copulation itself is very brief; the male appears to jump over the female, and cloacal contact lasts for a fraction of a second.

access to females. In chimpanzees, however, several males may attempt to mate with an estrous female. Moller (1988) argues that in this case, competition takes place in the female's fallopian tubes, and the male with the most and best sperm will gain the fertilization. In fact chimps do have large testes compared to the other apes and produce a high quality ejaculate in terms of sperm number and motility.

Following conception, male-male competition may take a different form. In mice the **Bruce effect** operates early in pregnancy: a strange male (or his odor) causes the female to abort and become receptive (Bruce 1966). Among langur monkeys (*Presbytis entellus*) strange males may take over a group, driving out the resident male (see figure 12–10). The new male may then kill the young sired by the previous male (Hrdy 1977). Females who have lost their young soon become sexually receptive, and the new resident male can inseminate them. Infanticide by adult males thus may be viewed as a "remator male" adaptation. Similar findings have been made for lions (Packer 1986).

PROXIMATE CAUSES OF REPRODUCTIVE BEHAVIOR

Before we proceed further with the discussion of mating systems and types of care that parents give their offspring (chapter 14), we will briefly explore the more immediate causes of sexual behavior and the role of the environment.

CLIMATIC-HORMONAL INTERACTION

Most terrestrial species live in seasonally fluctuating environments. The prime variables are temperature and rainfall; temperature is more variable in temperate and polar regions, and rainfall is more variable in tropical zones. Food supply is linked to these variables and, in some cases, may be the factor that actually triggers reproduction. Because of the delay between fertilization and birth (particularly long in mammals), environmental triggers must signal later conditions that will be optimal for birth.

Many species breed on an annual basis; their gonads regress, and sexual behavior is absent during most of the year. In temperate and polar regions, the change in photoperiod usually triggers the onset of reproductive behavior. Photoperiod provides a more reliable cue than temperature. Species with a relatively short span between copulation and birth (e.g., most birds) respond to increasing daylength; species with long gestation periods (e.g., ungulate mammals) respond to decreasing daylength.

In his study on the white-crowned sparrow (*Zonotrichia leucophrys*), Farner (1964) worked out the details of the sparrow's response to changes in daylength. He manipulated daylength artificially to simulate spring conditions and found that it induced recrudescence of gonads and reproductive behavior out of season. He also found that the photoperiod response is regulated internally to some extent, since a refractory period that follows breeding must be completed before changes in daylength can induce another reproductive cycle.

Orians (1960) found that fine adjustment of the timing of reproduction can be made by weather. Once enlargement of gonads has been triggered by photoperiod, unusually warm weather may speed up the reproductive cycle; in some cases it can even cause breeding to occur in the fall. In the northeastern United States, we can hear song sparrows singing on warm days in December; as the days get longer in February, bird singing increases regardless of weather (Nice 1941).

Some animals combine an annual cycle with lunar rhythmicity and breed at a particular phase of the moon, as exemplified by the palolo worm (*Leodice viridis*) of the South Pacific. Over a three-day period, six to eight days after the full moon that occurs near the end of October or the beginning of November, the posterior halves of mature worms break off and swarm at the surface of coral reefs, where they release eggs and sperm (Smetzer 1969). In another example, peaks in conception rates occur among Malayan forest rats just before a full moon (Harrison 1952).

Much of the world is arid or semiarid. Many annual plants in these arid or semiarid environments respond rapidly to unpredictable rains by flowering and setting seeds that will germinate after the next rain. These seeds are a prime source of food, and since their availability is unpredictable and brief, animals must respond quickly to reproduce while an energy supply is plentiful. Some Australian birds remain in breeding condition with enlarged gonads for many months (Marshall 1960). The stimulus of green vegetation, or possibly the rain itself, triggers courtship and a rapid sequence of nest building, egg incubating, and feeding of young.

The hill kangaroo (*Macropus robustus*) is one of many Australian marsupials that practices a form of delayed implantation: courtship and copulation take place, but the fertilized egg remains unimplanted in the uterus (Ealey 1963). In response to rain or an abundance of food, implantation takes place, and within days the tiny young crawls into the pouch and attaches itself to a teat. Kangaroos, therefore, waste no time in getting hormonally primed, in finding a mate, and in copulating when food is abundant. During the weeks of development in the pouch, the mother can respond to unpredictable events, such as a resumption of dry conditions with little food, by simply throwing the young (*joey*) out of the pouch (Low 1978). Although such behavior might seem cruel and wasteful to us, it is not in the mother's interest to invest energy in young that probably wouldn't survive anyway. Natural selection has favored females who invest in young that have a better chance of survival.

FIGURE 13-10 Experiments on pairs of ring doves with environmental and hormonal pretreatments

In all of these experiments, a nest with eggs was introduced on Day 0. (a) If a pair of ring doves is put in a cage with a nest and eggs, they begin incubating 5 to 7 days later. (b) When the pair is put into the cage with a partition between male and female at Day −7, there is no change and it still takes 5 to 7 days for incubation to begin after introduction of a nest and eggs and removal of the partition. (c) If the pair is introduced with nesting material at Day −7, incubation begins immediately upon introduction of a nest with eggs. (d) The absence of a nest in the pretreatment causes about a one-day delay in the onset of incubation. (e) Pretreatment with progesterone causes immediate incubation when pair, nest, and eggs are introduced on Day 0.

Source: D. S. Lehrman, "Induction of Broodiness by Participation in Courtship and Nest-Building in the Ring Dove," 51:32–36, and "Effect of Female Hormones on Incubation Behavior in the Ring Dove," 51:142–145 in *Journal of Comparative & Physiological Psychology,* 1958. Copyright © 1958 by the American Psychological Association.

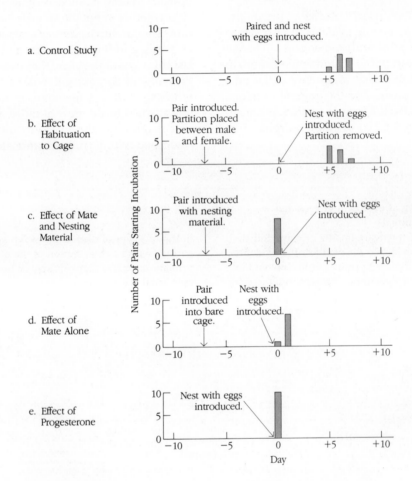

SOCIAL-HORMONAL INTERACTION

Once animals have been brought into a general state of sexual readiness through climatic-hormonal interactions, shorter-term events coordinate the behavior of both sexes through social-hormonal interactions. One of the most detailed studies of the nature of these interactions is the work of Lehrman (1958a,b) on ring doves (*Streptopelia risoria*), which was also discussed in chapter 7. In one series of experiments, Lehrman placed sexually experienced birds in a cage, presented them with a nest and eggs (figure 13–10), and then looked for incubation behavior. Birds that he tested singly failed to incubate at all. The bird's habituation to the cage without contact with a mate did not increase its readiness to incubate; the bird's association with a mate for seven days prior to nest and egg presentation led it to incubate promptly. Lehrman concluded that association with a mate—involving the sounds and postures of courtship, and to a lesser extent, the presence of nesting material—brings about the physiological changes that lead to incubation. When he gave birds daily injections of progesterone beginning seven days before he presented them with a mate, nest, and eggs, they incubated immediately. This behavior suggests that pairing and courtship cause production of progesterone, although other hormones, such as estrogen and prolactin, are also involved.

OVULATION, COPULATION, FERTILIZATION.   In fish, amphibians, and birds, the female ovulates in response to both social and nonsocial stimuli and engages in sexual activity for a limited part of the reproductive season. In mammals, the period of sexual receptivity at the time of ovulation is called **estrus,** and the interval from one estrus to the next is called the **estrous cycle.** Some mammals, like the fox (*Vulpes fulva*), are **monoestrous**—that

is, the female is receptive for a few days once each year—others cycle regularly throughout much of the year. Variation also occurs in the stimuli necessary to cause ovulation. **Induced ovulators,** such as rabbits, cats, and mustelids such as mink (*Mustela* spp.), require copulation for the release of the egg(s) from the ovary. Induced ovulation is characteristic of felids, with an extended estrous period of six to seven days and relatively brief courtship. With the exception of the lion (*Panthera leo*), felids are solitary with no lasting bond between the sexes (Kleiman and Eisenberg 1973). **Spontaneous ovulators,** such as canids and primates, release eggs whether or not the animals have copulated. Among canids, courtship begins several weeks before the onset of estrus, and bonds between male and female tend to last throughout the year (Kleiman and Eisenberg 1973). Primates undergo regular periods of **menses** during which the lining of the uterus is shed. The interval from one menses to the next is called the **menstrual cycle.** (For an outline of the hormonal and histological changes associated with estrus in mammals, see figure 13–11.)

FIGURE 13-11   Timing of events in estrous and menstrual cycles of mammals

In the closeup of part of the ovary (a), the growing follicle contains an egg that is shed into the oviduct after ovulation. The follicle then transforms itself into a corpus luteum that produces the hormone progesterone. Estrogen, produced by the ovary, and progesterone cause the thickening of the uterine lining. Part (b) shows a cross section of the uterine lining of a typical mammal through an estrous cycle and part (c) shows a cross section of the uterine lining of a primate, such as a monkey, ape, or human, through a menstrual cycle.

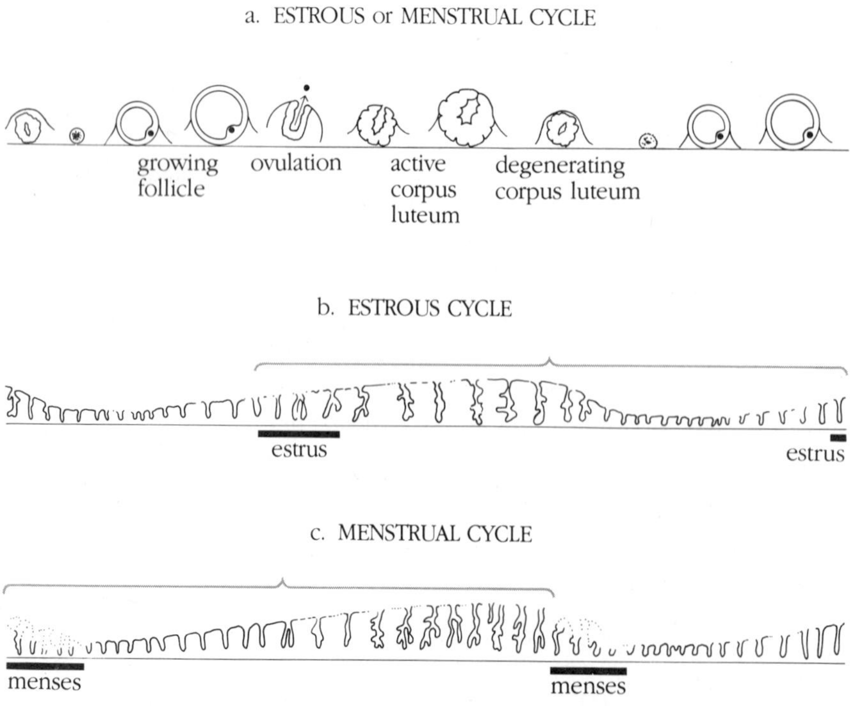

a. ESTROUS or MENSTRUAL CYCLE

growing   ovulation   active   degenerating
follicle   corpus   corpus luteum
                              luteum

b. ESTROUS CYCLE

estrus   estrus

c. MENSTRUAL CYCLE

menses   menses

Dewsbury (1975) studied the mechanics of copulation in mammals, particularly rodents, in detail. Patterns within and between the individuals of a species are highly stereotyped and vary in the following four respects: (1) the presence or absence of a mechanical tie, or lock, between the penis and vagina, in which the penis becomes engorged inside the vagina and the pair is temporarily unable to separate; (2) single or multiple thrusts during each insertion; (3) single or multiple intromissions preceding ejaculation; (4) single or multiple ejaculations in a single episode (figure 13-12 and table 13-1). The patterns are not restricted to large taxonomic units: carnivores such as wolves and dogs lock during copulation, but so do some rodents. We have only begun to explore the ecological and evolutionary significance of these differences. Rodents that lock, such as wood rats and grasshopper mice, are vulnerable to predation while locked together. However, such species have elaborate nests or burrows, where predation

is probably minimal. The lock may function to stimulate the female for pregnancy or it may be a product of sexual selection that reduces the possibility of another male's copulating with the female.

The stimuli that induce actual copulation vary tremendously, of course, but in many species there is a linear sequence of events by each partner in turn (see figure 13-13). If the sequence is broken at a certain point—for example, by an inappropriate response or an interruption—the female ceases to be receptive, and the pair either breaks up or starts over. A classic example of such a chain is that of the three-spined stickleback (Tinbergen 1951) (see chapters 2 and 9). Such complex reproductive patterns suggest a number of possible functions: (1) to coordinate physiological and behavioral events, such as nest building and egg formation in birds; (2) to provide an opportunity for assessment of certain features of the mate that might indicate potential reproductive success; or (3) more generally, to

---

FIGURE 13-12 Classification scheme for male mammalian copulatory patterns
Species vary in the presence or absence of the following processes: copulatory lock, thrusting, multiple intromission, and multiple ejaculation. Varied combinations of these processes yield sixteen possible patterns of copulation.

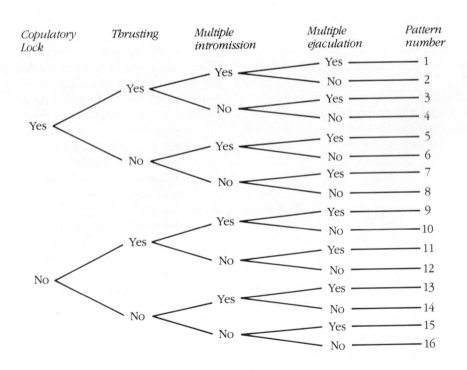

TABLE 13-1   **Examples of different copulatory patterns in mammals**

Common name	Scientific name	Lock	Thrust	Multiple intromissions	Multiple ejaculations	Pattern
dog	*Canis familiaris*	yes	yes	no	yes	3
wolf	*Canis lupus*	yes	yes	no	yes	3
golden mouse	*Ochrotomys nuttalli*	yes	no	no	yes	7
house mouse	*Mus musculus*	no	yes	yes	yes	9
montane vole	*Microtus montanus*	no	yes	yes	yes	9
rhesus macaque	*Macaca mulatta*	no	yes	no	yes	11
bonnet macaque	*Macaca radiata*	no	yes	no	yes	11
meadow vole	*Microtus pennsylvanicus*	no	yes	no	yes	11
Norway rat	*Rattus norvegicus*	no	no	yes	yes	13
Mongolian gerbil	*Meriones unguiculatus*	no	no	yes	yes	13
bison	*Bison bison*	no	no	no	yes	15
black-tailed deer	*Odocoileus hemionus*	no	no	no	no	16

From D. A. Dewsbury, "Patterns of Copulatory Behavior in Male Mammals," in *Quarterly Review of Biology*, 47:1–33. Copyright © 1972 Stony Brook Foundation, Inc., Stony Brook, NY. Used with permission of the *Quarterly Review of Biology*.

assure mating between conspecifics, thus avoiding the production of infertile hybrids. Courtship patterns that prevent mating between closely related species may thus lead to the formation and maintenance of species.

Methods of fertilizing eggs vary among species. Frogs and many fish release sperm over the eggs as the eggs are laid; the females of some salamander species pick up a spermatophore that is deposited on the substrate by the male; all birds and mammals fertilize internally. In addition to ensuring that the eggs become fertilized, internal fertilization gives the female control over which male becomes the parent. Males, however, can be less certain of paternity with internal fertilization than with external fertilization. As we shall see in the next chapter, this difference may explain why parental care by males is more common in fish than in mammals.

To conclude this discussion of proximate causation, we consider a few examples that illustrate different approaches to the study of sexual behavior, and end with a description of a well-studied primate, the rhesus monkey. Other examples, including one of a reptile, can be found in chapter 7.

TURKEY COURTSHIP. Hale et al. (1969) described the complex courtship chain in the domestic turkey (*Meleagris gallopavo*) that is similar in principle to the one that Tinbergen (1951) made famous for the three-spined stickleback (chapter 9). In response to the male turkey's display of strutting with his tail fanned, a receptive female crouches; as the male approaches and mounts her, she raises her head and he treads on her back, orienting himself to her head; she then raises her tail, and in response to pressure on her tail by the male, she everts

her oviduct. Up to this point, the sequence can be stopped and restarted at any point (figure 13-13). Once the hen everts her oviduct, the sequence must return to the beginning if it is interrupted, whether or not the female has come in contact with the male's everted cloaca and whether or not the male has inseminated her. (Most birds lack a penis; they transfer sperm via cloacal contact.) The female jumps up, ruffling her feathers, and may run away. The position of the hen's head during mounting and copulation is crucial to the male's orientation; if we present the male with a model whose head is at the side of its body, the male mounts sideways!

CAT SEXUAL BEHAVIOR. The interplay of neural and hormonal factors have been studied in the sexual behavior of domestic cats (*Felis domesticus*). The peaks of estrus, which can be altered by modification of the photoperiod, are from mid-January through March and from May through June in northern latitudes. Their notorious, rather violent courtship behavior that is sometimes difficult to distinguish from fighting may promote follicle maturation in the female, and the small spines on the cat's penis provide stimulation to induce ovulation (Fox 1975).

By the fourth month of age, the cat secretes androgens that stimulate development of the penile spines, but the male cat first exhibits mating behavior at eight to nine months of age. If we castrate the cat prior to puberty at four months, it never exhibits mating behavior. If we inject the cat with testosterone, typical mating behavior appears. If we castrate the cat after puberty, the cat exhibits a variable decline in sexual behavior, depending on the amount of sexual experience

FIGURE 13-13 **Sequential mating behavior in domestic turkeys**

Arrows denote the path that the sequence follows. For instance sexual display by the male is followed in the female by either avoidance (in an unreceptive female) or sexual crouch. Sexual crouch is followed by approach of the male, and so on. Once the female everts her oviduct, receptivity is terminated whether or not insemination has taken place.

Source: Data from E. B. Hale, W. M. Schleidt, and M. W. Schein, "The Behavior of Turkeys," in *The Behavior in Domestic Animals,* 2d ed. Copyright © 1969 Williams & Wilkins, Baltimore, MD.

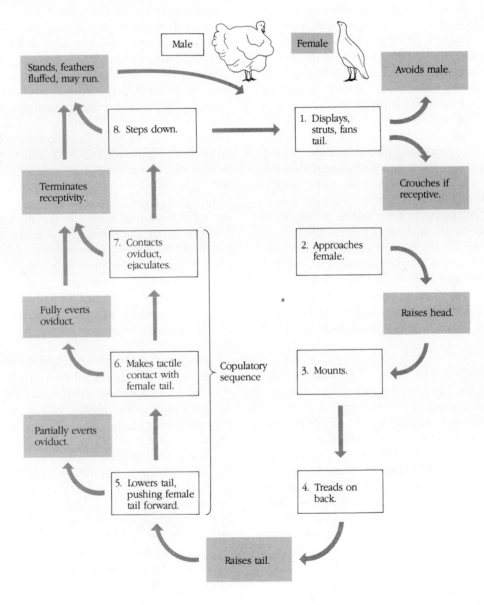

the male had before castration; in males with extensive experience, frequency of copulation and latency to copulate decrease only slightly (Rosenblatt and Aronson 1958).

Aronson and Cooper (1966) investigated the role of sensory input in male feline sexual behavior. When they desensitized the glans penis by cutting the nerves, the cats did not decrease their sexual activity, but they were so disoriented they were unable to achieve intromission.

OCTOPUS SEXUAL BEHAVIOR. Although much of our understanding of hormonal and neural correlates comes from studies of vertebrates, increasing effort is being devoted to invertebrates. For example, in the octopus (*Octopus hummelincki*) the endocrine glands that control both reproduction and senescence are the optic glands in the orbital sinus. Normally the female octopus spawns once in her life; she eats little while caring for the eggs and dies shortly after they hatch. When Wodinsky (1977) removed the optic glands after the octopus had laid her eggs, the female stopped brooding the eggs, ate, gained weight, and lived up to five months longer. Other invertebrates and fish exhibit this kind of control, which demonstrates the relationship between short-term control of sex and the control of the entire life cycle. It should not be too surprising to find that a single system is responsible for determining the age at which an organism reproduces, how often it does so, and when it dies; the timing of these and all other life history events is subject to natural selection, just as are morphological traits. Further discussion of the relationship between life-history traits and the environment is in the next chapter.

RHESUS MONKEY REPRODUCTION. The rhesus monkey (*Macaca mulatta*) has been used widely as a biological model for increasing our understanding of human reproduction. Both rhesus monkeys and humans belong to the order Primates and the class Mammalia. The similarities that exist between the two species, such as the menstrual cycle, are no doubt due to common ancestry and are thus homologous. Other similarities, such as the striking resemblance in hands, may be due to convergent evolution, since both are highly adaptable "generalists" that evolved in similar tropical environments. But there are some important differences as well. Rhesus monkey females have a well-defined period of estrus, or sexual receptivity, around the time of ovulation, and they mate during a discrete five-to-six month breeding season. Human females are sexually receptive at all times

of the cycle and at all times of the year; ovulation is not detectable by human males and seldom by the ovulating female herself. Rhesus monkeys are highly sexually dimorphic—the male has large canine teeth and weighs about 50 percent more than the female. As we might expect from this difference, rhesus monkey society is polygynous, with a small number of males monopolizing many females. Humans are less dimorphic and probably less polygynous (see chapter 20).

The annual mating cycle of rhesus monkeys, which is very consistent and which may be linked to photoperiod, begins in the fall in their native habitat, India. The monkeys live in stable social groups of from twenty to more than one hundred males, females, and young. Rhesus monkeys were introduced on islands off the coast of Puerto Rico; one colony, Cayo Santiago, off the east coast, and the other, La Parguera, about a hundred miles away from Cayo Santiago at about the same latitude. La Parguera monkeys began mating in October, those in India began in September, and those at Cayo Santiago began in July. The two Puerto Rican colonies were on the same photoperiod, and temperatures differed little. The main difference, however, was in the timing of the rainy season. At Cayo Santiago rains began in late spring, in India they start in early summer, and at La Parguera they began in late summer (figure 13–14). Following the onset of the rains, the vegetation became lush—mating followed within a few weeks. Experimenters have not been able to determine whether the mere presence of rain and green vegetation is sufficient to trigger mating behavior or if there is some nutritive component, such as hormones produced by plants, that starts the cycle (Vandenbergh and Vessey 1968).

Under constant environmental conditions in the laboratory, rhesus monkeys breed year round; but when they are given access to outside runs, they typically do not mate during the hottest part of the summer. Females normally conceive once each year and give birth about 23 weeks later. Some degree of endogenous control is involved, because females that lose infants from the previous year begin their estrous cycles earlier than do those with infants. Females with ligated oviducts, normal hormonally but unable to become pregnant, stop and start cycling along with the rest of the female population (Vessey and Marsden 1975).

An increase in redness of the skin around the perianal region in both sexes is indicative of the onset of mating. The redness pales in castrated adult males, and the redness is restored when these males are treated with testosterone. If males are injected with estrogen, the sex color more closely resembles that of the female rhesus

Siblicide in great egrets (*Casmerodius albus*)
Two egret chicks exchanging blows as a parent
passively (chapter 12) stands by. Parents rarely
interfere in such acts that usually lead to the death of
younger and smaller chicks.
Source: Photo by Douglas Mock.

PLATE 6

Caregiving by a non-parent in dwarf mongooses (*Helogale parvula*)

All members of a group help to raise young, even though the offspring are usually produced by a single, dominant pair (chapters 14 and 20).

Source: Photo by Scott Creel.

PLATE 7

FIGURE 13-14 **Relationship between rainfall and reproduction in three populations of rhesus monkeys** Cayo Santiago and La Parguera, Puerto Rico, are only 100 miles apart and at the same latitude. The timing of the rainy season differs at Cayo Santiago, La Parguera, and in the rhesus monkey's native habitat in India. Mating begins about two months after the onset of the rainy season at each location. The small mating period that occurred in spring in one localized area in India is atypical.

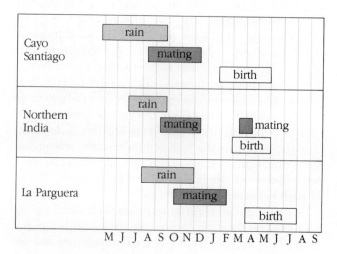

M J J A S O N D J F M A M J J A S

monkey (Vandenbergh 1965). Other hormones, such as progesterone, may be active in the sex skin color of the female rhesus monkey, and Phoenix (1977) reported that both testosterone and estrogen are essential for maximum sexual performance in females. Social factors are involved in the timing of mating; injecting castrated females with estrogen outside the regular breeding season brings intact males into breeding condition at the "wrong" time of year (Vandenbergh 1969). In one experiment, Vandenbergh and Drickamer (1974) seeded a free-living group at La Parguera with two of these injected females four months before mating should have begun. As evidenced by sexual behavior and births, the females in the seeded group bred early that year, compared with females in nonseeded groups.

As the mating season approaches, rhesus monkeys begin consort behavior. A male and a female form a temporary pair bond and engage in following and grooming activities. Males sometimes masturbate early in the breeding season; this behavior suggests that they

come into breeding condition before females. Researchers have done little work to answer the important question of which sex initiates the consort behavior. Frequently the male begins to follow the female, but a female commonly approaches and presents her perianal region to the male as an invitation to mount. Rarely, an estrous female actually mounts the male before presenting.

Presumably, the color of the sex skin is a cue for sexual receptivity; trained observers can even document the color change within the female's cycle, and the male's sex skin reddens noticeably when he is involved in an intense consort. Michael and Bonsall (1977) collected evidence that suggests the importance of odor as a sexual cue in rhesus monkeys. Under the influence of ovarian sex hormones, a bacterial system in the vagina produces an odoriferous substance that induces the male to mount. This substance, called copulin, is a mixture of short-chain fatty acids, such as acetic acid and butanoic acid. If a male has plugged olfactory tracts, he will not mount an estrous female; a synthetic mixture of copulin applied to the perianal region of a nonestrous female causes an intact male to mount. In this case, he usually gets slapped by the female since she does not react systemically to the locally applied copulin.

Goldfoot and his colleagues (1978) reported contradictory results. After testing three male-female pairs for several estrous cycles, they made the males permanently anosmic and found that the males' sexual behavior was not affected. Although we do not know the reasons for the contradictory results, olfactory cues are apparently not absolutely necessary for the sexually active male.

There is some evidence that odors could influence reproductive behavior in humans. Acids similar to those present in monkeys are found in the human vagina, and human subjects could smell the differences among samples from different stages of the menstrual cycle (Michael and Bonsall 1977). In another study Doty (1981) reported that human vaginal odors were scored by both men and women as least unpleasant at about the time the donors were ovulating.

The female rhesus monkey usually presents before the male mounts. Sometimes the male indicates his readiness to mount by placing a hand on her hip. When the male mounts, he places his hands on her lower back and grasps her ankles with his feet; she thus supports all his weight (figure 13-15). Once mounted, the male usually attains intromission rapidly since the presence of a bone (baculum) in the penis reduces the time needed for erection. He thrusts rapidly for about three

FIGURE 13-15   Copulation in rhesus monkeys
Males form temporary bonds with estrous females and
follow, groom, and copulate exclusively with them for two
or three days.
Source: Photo by Douglas B. Meikle.

seconds, then dismounts. The pair may groom for a
minute or two before he mounts again. These mounts
tend to come closer and closer together until ejaculation, which is seen as a "freezing," occurs. Some males
grimace or emit a bark vocalization during ejaculation.
Usually at this point the female looks back, reaches out,
and grabs one of the male's legs; this clutching reaction
is thought to indicate orgasm in the female (Zumpe and
Michael 1968). The existence of orgasm in nonhuman
females is of interest, since in most animals, the female

plays a rather passive role in copulation. Its existence is
difficult to prove, but Goldfoot and his colleagues (1980)
reported that stump-tailed macaque (*Macaca arctoides*)
females undergo uterine contractions and sudden increases in heart rate during the clutching behavior. For
a more detailed discussion, see Chevalier-Skolnikoff
(1974).

After ejaculation, the male dismounts, and the pair
may groom or rest. Mounting may resume in an hour
or so, or the consort may break up. Rapid coagulation
of the semen forms a plug in the vagina, perhaps reducing loss of sperm from the vagina, and the likelihood that another male will copulate with the female.

Consorts persist only two or three days; the pair
sleeps together, but little sexual activity takes place at
night except during a full moon, and a female consorts
with an average of three males during a nine-to-ten-
day estrous period. Since the cycle length is about four
weeks, the female will be receptive again in two to three
weeks. About half the time, the female does not conceive during the first cycle; most females cycle about
three times during the five-month breeding season. For
some reason, many females cycle once or twice after they
have conceived; the hormonal and evolutionary causes
of such behavior are puzzling.

Some females seem to improve their status in the
group while they are consorting with a high-ranking
male. Early experimental work with macaques and baboons emphasized the fluidity of female dominance, but
it is now known that the female hierarchy is much more
stable than that of the male (Sade 1967). Consorting pairs
often do team up against others, but the consort is short
lived and has no long-term effect on the female's status.
We will consider birth and parental care in the rhesus
monkey at the end of the next chapter.

SUMMARY

Asexual reproduction, by which an organism produces
exact copies of itself in the next generation, might seem
the most likely outcome of natural selection since it
maximizes passage of one's genes into the next generation. Most organisms, however, reproduce *sexually* at
some point in their life cycles, incurring the *costs of
meiosis* and of *finding a mate.* The benefit of increased
variability in an ever-changing and unpredictable environment apparently outweighs the costs. Many invertebrates have both sexual and asexual reproduction;

some animals have functional male and female sex
organs; and some fish and invertebrates change sexes
during adult life.

Natural selection tends to maintain a 1:1 ratio of the
two sexes. The gametes of the two sexes differ anatomically, with the female producing few, large, sessile eggs
and the male producing many, small, motile sperm. In
both vertebrates and invertebrates, females may adap-
tively adjust the sex ratio of their offspring in order to
produce the sex that will have the highest reproductive
success.

Sexual selection affects anatomy and behavior at the time of mating. *Intersexual selection* involves choices made between males and females, with the females usually choosing the males. In *intrasexual selection*, competition takes place between members of one sex (usually the male), with the winner gaining access to the opposite sex. Sexual selection may lead to the evolution of elaborate secondary sexual characteristics, particularly in males. The evolution of such traits is thought to have come about either by (1) a process of *runaway selection*, where the trait continues to evolve until countered by natural selection, or (2) a *linkage* of the trait *with superior genes* in the male, so that females selecting males with the trait will have higher reproductive success.

Many species breed on an annual basis, with recrudescence of gonads triggered by changes in photoperiod. Weather also affects the reproductive cycles of some species; for example, in dry environments rainfall may be the cue for the onset of the reproductive cycle in certain species. A few species are affected by lunar cycles. Around the time of fertilization, many sexually reproducing animals engage in behavior that insures that sperm come into contact with eggs (either internally or externally). The details of courtship and the mechanics of copulation are species-specific, and they function to assure mating between conspecifics, to coordinate the physiological and behavioral events leading to production of offspring, and to assess the potential fitness of the mate.

Discussion Questions

1. Males typically compete for access to females and often obtain more than one mate, while females are particular about whom they mate with. In some species, however, the male supplies the female with large "gifts" in the form of captured prey or large, protein-rich sperm packets. What differences might you expect to see in the reproductive behavior of such species when compared with the typical pattern?

2. When we see differences between the sexes, we usually assume that they have come about by sexual selection. Can you think of selection pressures other than mate competition that might produce such differences?

3. The data in the table below relate to ring doves. Female ring doves were placed in a cage. Sounds from the breeding colony were played into some of the cages and not into others. Males were placed on the other side of a glass partition in the cages of the females so that the males and females could view each other but not make contact. Some of these males were castrated, others were intact. The numbers given are the median ranks of ovarian development in the females; the higher the number, the greater the development. All differences between groups were statistically significant except the 40.5 versus 44.5 rankings. Discuss these results in terms of the social-hormonal factors affecting reproduction and relate them to the data in figure 13–10.

Median ranks of ovarian development

	With castrated male	With intact male	Combined groups
With colony sound	40.5	55.5	46
Without colony sound	15.5	44.5	28
Combined groups	24.0	49.5	

From D. Lott, S. D. Scholz, and D. S. Lehrman, "Exteroceptive Stimulation of the Reproductive System of the Female Ring Dove (*Streptopelia risoria*) by the Mate and by the Colony Milieu," in *Animal Behaviour*, 15:433–437. Copyright © 1967. Reproduced with the kind permission of Baillière Tindall, London, England.

Suggested Readings

Daly, M., and M. Wilson. 1983. *Sex, Evolution, and Behavior,* 2d ed. Boston: Willard Grant.
Basic text that covers the whole spectrum of sex, with emphasis on ecological and evolutionary aspects. Several chapters on evolution of human sexuality.

Maynard Smith, J. 1984. The ecology of sex. In *Behavioural Ecology: An Evolutionary Approach,* 2d ed., ed. J. R. Krebs and N. B. Davies. Oxford, England: Blackwell Scientific Publications, Ltd.
Short, readable review of current theories on how sexual reproduction is maintained in populations.

Michod, R. E., and B. R. Levin. 1988. *The Evolution of Sex.* Oxford, England: Blackwell Scientific Publications, Ltd.
A collection of papers by researchers who have made major contributions in this area. Heavy going in spots, but papers by Ghiselin and Williams provide good summaries of this highly controversial subject.

Naftolin, F., and E. Butz, eds. 1981. Sexual dimorphism. *Science* 211:1263–1324.
Series of articles reviewing the bases of sex differences. Deals with proximate mechanisms.

14

MATING SYSTEMS AND PARENTAL CARE

*T*he previous chapter dealt with the problem of the evolution and maintenance of sexual reproduction, with how the sexes come to differ in many species, and with the proximate factors controlling reproduction. Here we extend the discussion to include the diversity of mating systems found in animals and associated patterns of parental care.

MATING SYSTEMS

Anisogamy prevails in nearly all animals, with females investing more into each egg than males invest in each sperm (chapter 13). According to Trivers (1972), this difference sets the stage for male-male competition for access to females, and for attempts by males to mate with more than one female, a condition referred to as **polygyny.** Polygyny results in greater variation in the reproductive success of males than of females: For each male that fertilizes the eggs from a second female, another male is likely to fertilize none, as we saw earlier for lions. We have also seen that sexual selection tends to act more strongly on males than on females. However, not all species are polygynous.

In trying to evaluate the adaptive significance of differences in mating systems, we must look at ecological factors as well as historical ones. For example, group size may be related to predator pressure and food distribution. In the open plains, where large predators are present and food is widely distributed, omnivorous primates and grazing mammals such as ungulates live in large groups, in which mating with several members of

FIGURE 14-1   The influence of the spatial distribution of resources (food, nest sites) or mates on the ability of individuals to monopolize those resources.
Dots are resources and circles are defended areas. Uniform distribution of resources on the left offers little opportunity for monopolization. Monogamy is the likely mating system here.

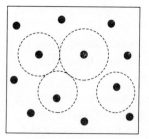 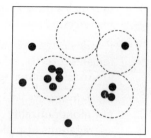

Uniform distribution.
Little polygamy potential

Clumped distribution.
High polygamy potential

the opposite sex is likely for both males and females. In densely forested areas, where communication over long distances is difficult, small family units and monogamy prevail. Figure 14–1 illustrates how the spatial distribution of resources (food, nest sites, or mates) might influence the type of mating system.

Most classifications of mating systems are based on the extent to which males and females associate (bond) during mating. **Monogamy** refers to association between one male and one female at a time. **Polygyny** refers to association between one male and two or more females at a time. **Polyandry** refers to association between one female and two or more males at a time. **Promiscuity** refers to the absence of any prolonged association and to multiple mating by at least one sex. The term **polygamy** is more general, and incorporates all multiple-mating and nonmonogamous-mating systems. One problem with classification systems based on bonding is that a judgment is needed to determine what constitutes an association. Does it include parental care? What about species, such as many nonhuman primates, that live in year-round social groups? A prolonged association exists, but during breeding females mate with several males, and males may mate with many females. Is this polygyny, polyandry, or promiscuity?

Emlen and Oring (1977) developed an ecological classification of mating systems that may eliminate the need to make these judgments because it is based on the ability of one sex to monopolize or accumulate mates, and emphasizes the ecological and behavioral potential for monopolization. Although developed primarily for birds, the classification of mating systems seems generally applicable to most vertebrate and insect taxa.

- **Monogamy.** Neither sex is able to monopolize more than one member of the opposite sex.
- **Polygyny.** Males control access to more than one female.
  - *Resource-defense polygyny.* Males control access to females indirectly by monopolizing critical resources.
  - *Female-defense polygyny.* Males control access to females directly, usually because females are grouped for other reasons.
  - *Male-dominance polygyny.* Mates or resources are not monopolizable; females select mates from aggregations of males, as in leks, based on the quality of the male's display or his territory.
  - *Scramble-Polygyny.* Males actively search for mates without overt competition.
- **Polyandry.** Females control access to more than one male.
  - *Resource-defense polyandry.* Females control access to males indirectly by monopolizing critical resources.
  - *Female-access polyandry.* Females do not defend resources essential to males, but they interact among themselves to limit access to males.

FIGURE 14-2   Monogamous pair of California gulls with young
There is little sexual dimorphism in these male and female gulls (*Larus californicus*) and both sexes care for the young.
Source: Photo by Bruce Pugesek.

MONOGAMY

In monogamous systems, neither sex is able to monopolize more than one member of the opposite sex. When the habitat contains scattered renewable resources or scarce nest sites, monogamy is the most likely strategy. If there is no opportunity to monopolize mates, an individual will benefit from remaining with its initial mate and helping to raise the offspring. The formation of long-term pair bonds also seems advantageous because less time need be spent finding a mate during each reproductive cycle. Long-lived birds such as sea gulls that breed with former mates have higher reproductive success, probably because of less aggression between mates and greater synchronization of sexual behaviors (Coulson 1966). About 90 percent of all bird species are monogamous (figure 14–2). Another factor promoting monogamy could be predation risk. Some species live in small social units and behave secretively in order to reduce the chances of being eaten.

POLYGYNY

In polygynous systems, individual males have access to more than one female. In **resource-defense polygyny,** males defend areas containing the feeding or nesting sites critical for reproduction, and a female's choice of a mate is influenced by the quality of the male and of

FIGURE 14-3 Reproductive success of female as a function of quality of male territory

One curve is for a monogamous pair of birds and another for a female who has to share the territory with another female. As territorial quality improves, so does the female's reproductive success. If territories vary enough from one male to the next, a female may have greater reproductive success by joining an already mated male with a good territory, a process that leads to polygyny. Here, a female mating with a male that is defending a territory of quality (a) would have the same reproductive success (c) as a female joining an already mated male in a superior territory of quality (b). For a test of this and several other models of polygyny in redwings see Lenington (1980).

Source: Data from G. H. Orians, "On the Evolution of Mating Systems in Birds and Mammals," *American Naturalist,* 103:589–603, 1969.

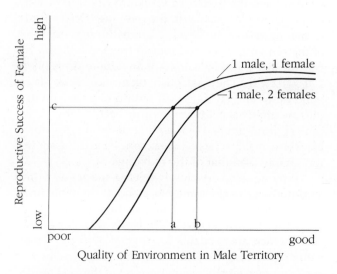

Quality of Environment in Male Territory

his territory. Territories that vary sufficiently in quality may attain the **polygyny threshold,** the point at which a female may do better to join an already mated male possessing a good territory than an unmated male with a poor territory (Orians 1969). Thus some males may get two or more mates while others get none (figure 14–3). Typically males do not provide parental care and usually delay breeding until later in life, when their ability to obtain a good territory has increased.

Many species are probably facultatively (i.e. optionally) polygynous. In habitats where feeding or nesting resources cannot be monopolized by males, monogamy is likely, while habitats with defensible clumped resources would favor polygyny. Thus we may expect to find variation in the mating system of a single species.

Female-defense polygyny may occur when females are gregarious for reasons unrelated to reproduc-

tion. Some males monopolize females and exclude other males from their harems. In many species of seals, the females haul out on land to give birth, and they mate soon after. The females are gregarious because there are a limited number of suitable sites, and the males monopolize the females for breeding. Intense competition among males results in marked sexual dimorphism and a large variance in male reproductive success.

If males are not involved in parental care and have little opportunity to control resources or mates, **male dominance polygyny** may occur. If female movements or their areas of concentration are predictable, the males may concentrate in these areas and pool their advertising and courtship signals. Females then select a mate from the group of males. These areas where males congregate and defend small territories in order to attract and court females are called *leks* (see figure 12–8). Females select a mate, copulate, then leave the area and rear their young on their own. Older, more dominant males may occupy the preferred territories and/or have the most attractive displays, and thus do most of the copulating.

In the absence of territory or dominance in some species, a **scramble polygyny** takes place as the males try to mate. Female wood frogs (*Rana sylvatica*) congregate in small temporary ponds, often during a single night in early spring. Large numbers of males rush about attempting to mate with fecund females (Berven 1981), sometimes dislodging smaller, already-mating males (Howard 1988). Under almost opposite conditions, where females are widely dispersed, the same thing may happen. For instance, male thirteen-lined ground squirrels (*Spermophilus tridecemlineatus*) with little evidence of dominance or territoriality actively search out estrous females (Schwagmeyer 1988).

POLYANDRY

In polyandrous systems, females control access to more than one male. Because female investment in eggs exceeds that of males in sperm, polyandry is rare. In most cases, females provide parental care while males seek new mates. If food availability at the time of breeding is highly variable, or if breeding success is very low due to high predation on the young or the eggs, females may have to produce many offspring. In birds, male incubation is common; a few cases of polyandry in which males do all the incubating and females lay multiple clutches have been documented (figure 14–4).

FIGURE 14-4   Male red-necked phalarope with eggs
In this polyandrous species the male cares for the eggs
and young while the female seeks additional mates.

Breeding sites of the American jacana (*Jacana spinosa*), a large wading bird found in Central and South America, are limited and are divided into small territories by males (Jenni 1974). Female jacanas control superterritories that may encompass the nesting areas of several males. Frequently, several males incubate the clutches of one female, and she provides replacement clutches for them if nests are lost through predation, as often happens. Breeding females are 50 percent larger than the males, who they dominate; and these females provide little parental care. In this reversal of polygyny, the females specialize only in egg production.

Females of the migratory spotted sandpiper (*Actitis macularia*) compete for control of breeding territories, and males provide most of the care of young (Oring and Lank 1982). Females arrive on the breeding grounds before the males and are **philopatric** (they return to the place where they were born). However, the vast majority of birds are either monogamous or polygynous and the reverse is generally true: males arrive first to establish territories and are more likely than females to return to the natal site to breed (Greenwood 1980).

ECOLOGY AND MATING SYSTEMS

A good example of the way in which mating systems are related to resource distribution is provided by Orians' comparative study of blackbird social systems (Orians 1961). The red-winged blackbird (*Agelaius phoeniceus*) is usually polygynous; a male defends a territory containing two or three females. The male arrives three to four weeks before the females (frequently at the same site as the previous year), mates, and then expends a considerable amount of energy defending his territory until the young are fledged. Males rarely help to raise the young; females do all of the nest building and incubating, and nearly all of the feeding of the young, as is typical of polygynous species.

In parts of northern California, a very closely related species, the tricolored blackbird (*Agelaius tricolor*), is **sympatric** (coexists) with the redwing (Orians 1961); its mating system, however, is quite different from that of the redwing. Male and female tricolored blackbirds pair off in a nomadic colony of anywhere from 100 to 200,000 birds. They establish territories, find mates, build nests, and lay eggs—all within one week. Activities are highly synchronous within each colony. As we can see from figure 14–5, both sexes of the tricolored blackbird make a large investment—but in a shorter time frame than that of the redwings.

Why do two very closely related species do things so differently in the same place and at the same time? The answer seems to be related to their energy source. Redwings have a relatively stable diet of seeds and insects, which are available for several months. A male redwing defends the same territory several years in a row, and each territory contains most of the food needed to support the females and the young. In contrast, tricolors go out from the colony on mass feeding flights, possibly to assess concentrated food sources; and they attack rice and other grain fields at the time of seed maturity. Thus the tricolor mating system seems designed to take advantage of an ephemeral, but rich and concentrated, food source. When tricolors locate such an energy supply, they move in, establish a new colony, and reproduce before the source is gone. A similar strategy is used by African weaver finches (*Quelea* spp.), which move all over the continent, attack rice and wheat fields, breed quickly, and then move to another ripening area (Ward 1971). Because of their nomadism, these serious economic pests have been very difficult to control.

FIGURE 14-5 **Time expenditure for pair of red-winged blackbirds and tricolored blackbirds during breeding season**
The reproductive effort is spread out over more than four months in the redwing (a), but it is completed in about half that time by the tricolor (b). Tricolored blackbirds engage in mass feeding flights to concentrated food sources away from the breeding colony.

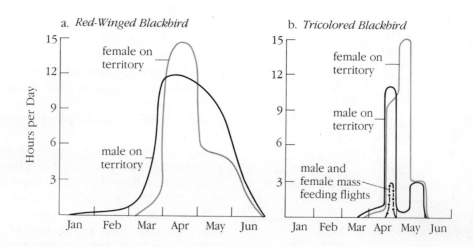

PARENTAL INVESTMENT

Once eggs are laid or young are born, what may the parents do to improve the chances of their offspring to survive? We can define **parental investment** as any behavior toward offspring that increases the chances of the offspring's survival at the cost of the parent's ability to rear other offspring (Trivers 1972, 1974).

WHICH SEX SHOULD INVEST?

Because of anisogamy, an egg requires a greater investment of energy than does a sperm; thus male and female strategies are expected to differ. Since eggs are likely to be limited in number, we expect that males will compete for the opportunity to fertilize them and thus will be subject to sexual selection. A female is likely to mate, but given her already large investment, her ability to invest further may be limited; thus she will be particular about which male fertilizes her precious eggs. Males, on the other hand, will try to inseminate as many females as possible. Mating systems in which the male mates with more than one female should be the most common (figure 14–6) (Trivers 1972).

However, just because only one sperm fertilizes an egg, millions are generally required in each ejaculation to ensure fertilization—even of a single egg (Dewsbury 1982). Also, there is a limit on the number of times most males can ejaculate within a certain period. Remember also the evidence in chapter 13 that sperm competition in males of some species leads to selection for increased sperm production. Thus a male's investment in sperm is not necessarily trivial, and he too can be expected to be somewhat choosy about his mate. In some species of insects, the male contributes nutritive substances in addition to sperm and is the choosier sex (Gwynne 1981). Nevertheless, it is generally assumed that in most species, the female's investment in gametes is greater than the male's.

FIGURE 14-6   Parental investment and reproductive success as a function of number of offspring produced
Because the parental investment (cost) for females usually rises more steeply than for males, the optimum number of offspring (highest reproductive success at lowest cost) for females is less than for males. Males, consequently, may seek more than one mate to attain maximum net reproductive success. Broken lines indicate maximum male and female net reproductive success.

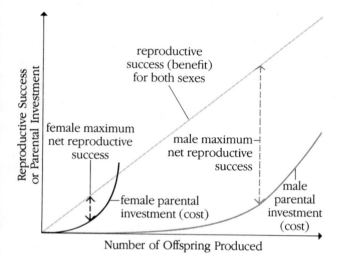

INVESTMENT PATTERNS IN FISH, BIRDS, AND MAMMALS. Other factors besides anisogamy seem to affect the contribution of each sex to parental care. Certain taxonomic groups of animals seem to be predisposed to a particular pattern. Thus, birds tend to be monogamous, and both parents care for young. Mammals are generally polygynous and the male contributes little to raising the offspring. Most species of fish do not care for their young at all; however, among those species that do, it is usually only the male.

What factors could be responsible for these taxonomic differences? Trivers reasoned that confidence of parentage might explain the results (Trivers 1972). In species with internal fertilization and where sperm competition could take place, the male might be inclined to desert and seek additional mates. The female is confident of her genetic relationship to the offspring, so she invests further in it. Another possibility, suggested by Williams, is that parental care evolved in the sex that is most closely associated with the embryos (Williams 1975). Where eggs are fertilized internally, the evolution of embryo retention and live birth might be more likely, followed by further care of the young by the mother. Where eggs are laid and fertilized externally in a male's territory, the male becomes most closely associated with them and additional paternal care is likely. Table 14-1 illustrates that internal fertilization favors female care; and external fertilization, male care, as both these theories predict (Gross and Shine 1981).

TABLE 14-1   Parental care and fertilization mode in teleost fishes
The table shows number of families; a single family may appear in more than one category, but is not listed under 'no parental care' unless care is completely unknown in the family.

Parental care by	*Number of families when fertilization mode*	
	Internal	*External*
Male	2	61
Female	14	24
Neither	5	100

Source: Data from Gross and Shine 1981.

Still other factors may predispose a taxon to a particular pattern. In the case of birds, males can incubate eggs and feed young—in the case of pigeons, they can even provide crop milk. In mammals, however, gestation and milk production are restricted to the female, and there is relatively little the male can do to provide direct care for the young. In mammals whose young are relatively advanced at birth (precocial), the opportunities for male investment are even lower, and males compete for multiple mates more than they do in species whose young are immature at birth (altricial), and whose males and females can more equally invest (Zeveloff and Boyce 1980).

TYPES OF PARENTAL INVESTMENT. In polygynous systems, competition among males can be intense. For example, mating of elephant seals takes place in colonies. Males establish a dominance hierarchy, and only the high-ranking males breed. Le Boeuf (Le Boeuf 1974; Le Boeuf and Reiter 1988) observed that typically less than one-third of the males copulate at all, and the top five males do at least 50 percent of the copulating (figure 14–7 and table 14–2). Beyond the sperm, there is no male investment in offspring, as evidenced by the males'

FIGURE 14-7 Sexual dimorphism in elephant seals
Among a herd of females, two males fight to establish dominance. Males differ strikingly from females, are about three times larger, possess an enlarged snout, or proboscis, and have cornified skin around the neck. In this highly polygynous species males invest nothing in their offspring other than DNA from sperm.
Source: Photo by Burney J. Le Boeuf.

TABLE 14-2   Number and percent of male elephant seals copulating during consecutive breeding seasons

	1968	1969	1970	1971	1972	1973
Number of males present	103	120	125	136	146	180
Number of males copulating	14	17	32	41	51	62
Percent of males copulating	14	14	26	30	35	34
Percent copulations by the five most active males	83	92	69	65	53	48
Number of females present	193	243	311	352	408	470

From B. J. Le Boeuf, "Male-Male Competition and Reproductive Success in Elephant Seals," in *American Zoologist*, 14:163–176, 1974. Copyright © 1974 by the American Society of Zoologists, Thousand Oaks, CA. Reprinted by permission.

FIGURE 14-8   Egg brooding by male water bug (*Abedus herberti*)
Note newly hatched nymph nearby. The female lays the eggs directly on the back of the male.
Source: Photo by John Cancalosi.

possible trampling of their own pups when striving to inseminate females. The males have no way of knowing which young are their own, since pups are born a year after copulation.

In a few species the situation is reversed, and the male does most of the caring for young. Since the males' investment in offspring becomes larger than the females', we would expect females to compete for and try to attract males, rather than vice versa. Indeed, this reversal of the typical roles occurs in fish such as pipefishes and sea horses, in polyandrous birds such as phalaropes, jacanas, tinamous, and sandpipers, and in some insects (figure 14-8).

Another type of investment occurs when the male provides nourishment to the female at the time of copulation, as in prenuptial feeding in birds. In some orthopteran insects, more than 25 percent of the male's weight may be transferred to the female in a spermatophore (figure 14-9). This protein-rich meal increases the number and size of eggs produced by the female (Gwynne 1984). In cases where the male's investment in the spermatophore exceeds that of the female in eggs, males should become the choosy sex and females should compete for access to them. Gwynne (1981) found that male Mormon crickets (*Anabrus simplex*) mated more often with heavy females that contained more eggs. Females, on the other hand, competed aggressively for access to singing males.

In many species, both sexes care extensively for young, and the pair is monogamous; this is usually the case in birds. However, the investment in offspring is not equal at all times (figure 14-10). It may be advantageous for one partner to desert the other and find a new mate if the remaining partner's investment is greater and the offspring are likely to survive anyway. Because of its large investment, the remaining partner has the burden of caring for the young (Trivers 1972).

But should a parent continue to invest just because it has made a previous commitment in terms of time and energy? According to Dawkins and Carlisle (1976), such an individual would be committing the "Concorde fallacy," named in honor of the supersonic transport that was completed even though a profitable return was unlikely, because of a large previous financial investment. It has been argued that organisms should behave so as to increase future reproductive success regardless of prior investments in offspring. However, Trivers' concept of parental investment actually does consider future ability to invest. If one parent has already made a large commitment, that parent may be physically unable to begin another breeding cycle; its reproductive success is thus increased if it remains with its young.

FIGURE 14-9   Mormon crickets with spermatophore
A female Mormon cricket is shown (a) mounting a male, (b) with attached spermatophore from male, and (c) consuming the spermatophore.

Source: Gwynne, D. T., 1981, Sexual differences theory: Mormon crickets show role reversal in mate choice, *Science* Vol. 213, 14 August, pp. 779–80, Fig. 1.

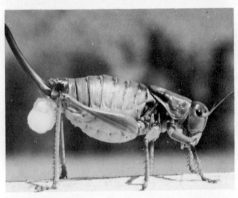

FIGURE 14-10   Parental investment in offspring by male and female
In this hypothetical example we find that the male establishes and defends a territory, so his cost is initially higher; but the female invests heavily in eggs, and her cost soon exceeds that of the male. The male then incubates the eggs, and his cumulative investment exceeds hers until termination of parental care. Differences in investment may affect such behaviors as mate desertion.

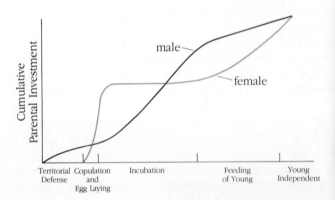

PARENTAL CARE AND ECOLOGICAL FACTORS

The extent of parental care varies tremendously in the animal kingdom. The amount of care given generally increases as the complexity of the organism increases. While many aquatic invertebrates simply shed eggs and sperm into the water, primate care may last for several years, which may amount to 25 percent of the offspring's life span. The kinds of parent-offspring relationships are varied and not simple to describe, however. Birds display a wide range of care techniques—from burying their eggs in rotting vegetation that provides heat for incubation (thus freeing them), to sitting on the eggs without feeding until they are hatched.

The term **reproductive effort** is used to denote both the energy expended and the risk taken for breeding, which reduce reproductive success in the future. Finding mates and caring for young take extra energy, and parents may run a greater risk of predation. Individuals are faced with the decision—conscious or otherwise—of whether to breed now or wait until later. If the choice is to breed now, should some effort be spared for another attempt later?

What environmental factors influence the investment that parents make in their young after birth? Ecologists have attempted to relate environmental conditions to parental care. For example, species adapted to stable environments have a tendency toward larger body size, slower development, longer life span, and having young at intervals **(iteroparity)** rather than all at once **(semelparity)**. Typically, individuals of these species will occupy a home range or territory (see chapter 12). These stable conditions favor production of small numbers of young that receive extensive care and thus have a low mortality rate. Such species are said to be **K selected,** in reference to the fact that populations are usually at or near K, the carrying capacity of the environment. Intraspecific competition is likely to be intense, and the emphasis is on producing high quality offspring, rather than high quantity.

Species that are adapted to fluctuating environments have high reproductive rates, rapid development, small body size—and need little parental care. Their populations tend to be controlled by physical factors, and their mortality rate is high. Such species are said to be **r selected,** where r refers to the reproductive rate of the population. Pacific salmon, who reproduce far upstream from feeding areas, must expend a great deal of energy before they can breed. Once they incur the great cost of migrating upstream, they breed explosively and die; in no other way can the benefit of reproduction become greater than the cost.

Other species, which do not have such a high initial cost before breeding, may defer reproduction or spread it out over time. In environments where survival of offspring is low and unpredictable, parents may "hedge their bets" and put in a small reproductive effort each season. The California gull (*Larus californicus*) uses such tactics; the birds live fifteen years or more but rear only one or two chicks per year (Pugesek 1981, 1983). As the parents age, they increase their effort, laying more eggs, feeding the chicks more food, and defending them more vigorously, possibly because the parents' chances of surviving another year become smaller.

Note that the predictions of "bet-hedging" contradict those of r and K selection. In unstable, unpredictable environments r selection, which favors high reproductive rates, should be important. However, bet-hedging theory suggests low reproductive rates, and spreading reproductive effort across many breeding seasons. More data are needed from natural populations before we can resolve this apparent contradiction.

Prolonged dependency and extensive parental care are also favored when a species—for example, large mammalian carnivores such as the felids and canids—depends on food that is scarce and difficult to obtain. Much effort is spent searching for prey, and in some species, cooperation is needed for the kill. During the prolonged developmental period, the young benefit from a considerable amount of learning through observation of parents and through play.

The Old World monkeys and the great apes have the longest period of dependency. Typical of these species is an infancy of eighteen to forty months and a juvenile phase of six to seven years, making up nearly one-third of the total life span. The reason for this prolonged dependency may be related to their flexible behavior that is shaped largely by learning. The complexity of monkey and ape social systems and the importance of kinship depend on a knowledge of individuals and an extensive behavioral repertoire.

PARENT-OFFSPRING RECOGNITION

One practical problem faced by parents is being able to recognize their own offspring. We predict that mechanisms of recognition will evolve when there is a risk of misdirecting parental care toward nonrelatives. Among gulls and terns that nest on the ground in dense colonies, the precocial chicks begin running around several days after hatching. At about this time, the parents learn to recognize their own chicks (Tinbergen 1960). Thus, ring-billed gull (*Larus delawarensis*) parents will accept chicks from other parents until the chicks are about five days old; after this time the acceptance rate declines sharply (Miller and Emlen 1975). Another species of gull, the kittiwake (*Rissa tridactyla*), nests on tiny ledges on steep cliffs. For obvious reasons, chicks don't leave the nest until they are several weeks old and can fly. Mechanisms of recognizing chicks have not evolved in this species: parents will adopt older chicks, even those of other species (Cullen 1957).

FIGURE 14-11 Calls of young cliff and barn swallows
Note the greater complexity of the cliff swallow calls. Cliff
swallows nest in large colonies, where the chance of a
parent's misdirecting parental care is high.

Cliff Swallow   Barn Swallow

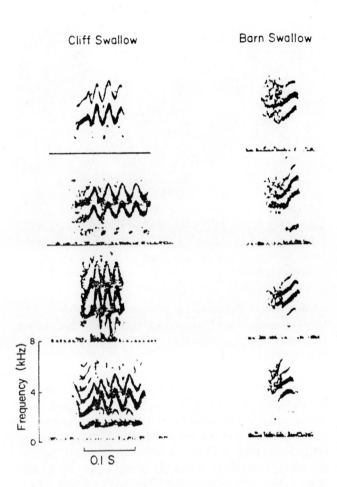

Frequency (kHz)

8

4

0

0.1 S

The cues used to recognize offspring (or other kin)
vary with the taxonomic group (see chapter 11 for more
on kin recognition). Vocalizations are thought to be important in birds, as parents learn to recognize the calls
of their own young. Some species of swallows produce
distinctive calls, and each young has its own "signature" (Beecher 1982; Beecher et al. 1986). Other species
do not have such signatures (figure 14–11). The two
species' social organization differs: rough-winged (*Stel-
gidopteryx serripennis*) and barn (*Hirundo rustica*) swallows nest by themselves or in small groups, and it is

unlikely that a parent would confuse her young with
those of other parents. Bank (*Riparia riparia*) and cliff
(*Hirundo pyrrhonota*) swallows live in large colonies,
where breeding is highly synchronized, and there is a
good chance that parents could misdirect parental behavior, as young are all of similar age and often return
to the wrong burrow entrance. Information analysis
(chapter 11) confirmed that the calls of the colonial species contained more information than those of the solitary species (Beecher 1982).

PARENT-OFFSPRING CONFLICT

Anyone who has raised children or grown up with a
sibling has observed frequent disagreement between
parent and child. It is not unusual to see a mother rhesus
monkey bat her ten-month-old infant or raise her hand
over her head, thereby pulling her nipple from its
mouth. Frequently the infant responds by throwing a
"temper tantrum." We can view much of the conflict as
a disagreement over the amount of time, attention, or
energy the mother should give to the offspring (i.e., the
infant wants more than the parent wants to give).

Although we often interpret such conflict in
humans as maladaptive, related to psychological problems of parent or child or to some negative cultural influence, work with nonhuman primates suggests that
conflict is part of the weaning process and is necessary
for the infant to become an independent functioning
member of the social unit (Hansen 1966). It is not clear
from this interpretation why the infant should protest
so vigorously.

Another hypothesis, based on the coefficient of relationship (chapter 4) and on parental investment, is that
the conflict arises because natural selection operates differently on the two generations (Trivers 1974). From the
mother's standpoint, she should invest a certain amount
of time and energy in her offspring and then wean the
young and invest in new young. When the cost to the
mother in fitness exceeds the benefit, she should reject
the young. From the offspring's standpoint, however,
the offspring will profit from continued care until the
cost to its mother is twice the benefit (since the offspring shares half of its mother's genes). When the cost
is twice the benefit to the mother, then the offspring's
own inclusive fitness will start to decline. At this point,
the offspring should leave the mother and allow her to
increase the offspring's inclusive fitness by having more
offspring (figure 14–12).

FIGURE 14-12   Parent-offspring conflict: Ratio of cost to mother and benefit to offspring as a function of age of offspring

When the offspring is young, the ratio is less than 1, and both the mother and offspring gain in terms of fitness from the relationship. Above 1, the fitness of the mother begins to decline, and so she should try to cease caring for this offspring and invest in new offspring. Between 1 and 2 is a zone of conflict (shaded area) because the fitness of the offspring is still increasing. Above 2, the fitness of the offspring also declines, and the offspring willingly becomes independent.

Source: Data from R. L. Trivers, "Parental Investment and Sexual Selection," in *Sexual Selection and the Descent of Man,* edited by Bernard Campbell. Copyright © 1972 by Aldine de Gruyter, Hawthorne, NY.

FIGURE 14-13   Cattle egret chicks fighting
Due to asynchronous hatching of eggs, the chicks differ in age and size. Here the oldest chick is delivering a blow to a younger sibling. Siblicide is common in this species, and the parents do not intervene.

Alexander (1974) challenged this idea; he argued that selection will work *against* behaviors of the off-spring that allow them to cheat and to receive more than their share of care from the parent. Such offspring will pass those traits on to their own offspring, who will cheat them in turn and lower their fitness as parents. Thus, parents will win in cases of parent-offspring conflict. Alexander coined the term **parental manipulation** to refer to cases where a parent selectively provides care to certain offspring at the expense of others.

Conflict is not always limited to parents and off-spring. Siblings may engage in deadly combat, a process studied by Mock (Mock 1984; Mock and Ploger 1987) in great egrets (*Casmerodius albus*) and cattle egrets (*Bubulcus ibis*). **Siblicide** is when young kill each other outright or force each other out of the nest (figure 14-13). One cause of it seems to be the size of the food objects brought by the parents. Small fish are monopolizable by one chick and promote aggression and

dominance. In contrast to great egrets, great blue herons (*Ardea herodius*) feed their chicks large pieces that cannot be monopolized; siblicide is infrequent in this species (Mock 1984).

A second factor that increases the chances that young will dominate and even kill each other is hatching asynchrony. When incubation is begun after the first egg is laid, chicks hatch at different times. The chick from the first egg laid hatches first and gets a head start. That chick is likely to kill younger siblings and get more than its share of food. Clearly it is not in the chicks' best interests to run the risk of being bludgeoned or thrown out of the nest by their older siblings. Why does this behavior persist? Mock and Ploger (1987) experimented with cattle egrets by moving newly hatched chicks from one nest to another in order to create artificial synchronous and asynchronous broods. They found that chicks in synchronous broods fought more, demanded more food, and survived *less* well than did the asynchronous broods. Thus, although hatching asynchrony seems to work against the well-being of all but the oldest chicks, it is in the parent's interest to create inequalities in chick size. In this form of parental manipulation of offspring, brood size adjusts to the optimum number of chicks for the food available.

PARENTAL CARE IN THE RHESUS MONKEY

At the end of the previous chapter, we described mating behavior in the rhesus monkey as an example of a well-studied primate. Now we continue with an account of early development. After a five-and-a-half-month gestation period, birth occurs at night, when the group is stationary. Although few normal births have been witnessed in natural populations of rhesus monkeys, there is no evidence that other members of the group assist in the delivery. From the time of birth, the infant clings unassisted to the belly of the mother. The mother expels the placenta within an hour and usually consumes it, as do most mammals. Both nutritive and antipredator functions have been ascribed to this behavior.

The infant begins moving away from its mother within a day or two, and by ten weeks of age, it spends half its time away from her. Initially, the mother restrains the infant by pulling its leg as it struggles to escape; later, however, the mother initiates the separations as she begins to wean her infant (Berman 1980; Vessey and Meikle 1984).

The infant's attachment to the mother is very strong; it clings to her ventral side with a nipple in its mouth for the first year. A mother with a newborn youngster is a focus of attention for family members and other females, who gather around the mother, groom her and make "girn" vocalizations. Clearly, their intent seems to be to touch and handle the infant. By far the most persistent are the two- to four-year-old nulliparous females, who may kidnap the infant during the first few weeks (figure 14–14). A kidnapped infant emits a distinctive "lost" call that is recognized by its mother. Females that surgically have been made incapable of conception may refuse to surrender a kidnapped infant, but intact females usually surrender it within hours. After a few weeks, the infant seems to lose some of its appeal and can usually escape under its own power (Vessey and Marsden 1975).

The amount and type of socialization that infants receive varies extensively among primates. Rhesus macaque mothers are relatively selfish with their infants; bonnet macaque (*Macaca radiata*) and langur (*Presbytis entellus*) mothers allow their infants to be passed around the first day. Such differences seem to be correlated with overall levels of aggression and dominance in the group. Simonds (1965) noted that bonnet macaques are less aggressive and have more relaxed dominance interactions

FIGURE 14-14   Infant rhesus monkey handled by immature female "aunt"
Such attempts at care giving may prepare immature females to care for their first infant, but the infants seem to gain little.
Source: Photo by Dennis Ferguson.

than do rhesus macaques. Rosenblum and Kaufman (1967) conducted a detailed laboratory study comparing bonnet with pigtailed macaques (*Macaca nemestrina*). When the experimenters removed the mothers of pigtailed infants from small social groups, the infants showed acute separation distress. Infants of the less-restrictive bonnets tolerated the absence of their mothers with much less trauma. Unfortunately, the relationship between early social experience and later social organization tells us nothing about cause and effect. Does the high level of aggression in rhesus monkey groups result from the restrictive and punitive experience in early life, or does it cause it?

By the end of the first year, the infant rhesus monkey, which ranks just below its mother and siblings in the dominance hierarchy, has established a network of social relationships; it knows its own status and that of juveniles, adult females, and its peers (Sade 1967). Male and female infants behave similarly in the first year, although male infants tend to engage in a bit more rough-and-tumble play than do females, and Mitchell and Brandt (1970) provided some evidence that mothers hit their male infants more often than their female infants, and move away from males more often. One big difference is that males begin mounting other infants as early as eight weeks of age; this pattern is well developed by the end of the first year. Males deprived of social interactions during early life are incapable of properly mounting and inseminating a female later in adulthood (Harlow 1965). For contradictory evidence, see Meier (1965).

By the end of the first year, conflict between mother and infant is substantial. Although the infant can feed itself under optimal conditions by six months of age, it still nurses at the end of one year and still clings ventrally, with a nipple in its mouth, 5 to 10 percent of the time during the day and most of the time at night. The female gives birth once a year and rarely allows the infant on the nipple once the new infant is born. She weans the yearling by withdrawing the nipple, pushing the yearling away, or hitting it. The yearling invariably jerks its head, makes a "geck" sound, and may throw a full-fledged temper tantrum.

BEHAVIOR DEVELOPMENT

After the first year, the behaviors of the sexes become more and more different. Males spend increasing amounts of time with their male peers at the periphery of the group. One- and two-year-old males sometimes sleep together at night (Vessey 1973). But even as three- and four-year-olds, they often come back to groom their mothers and siblings. At about the age of three and a half, their testes descend, and they may visit the periphery of another social group in the breeding season. By seven years of age, they will have moved to a social group different from the one they were born into (Drickamer and Vessey 1973). Females never leave their natal group, but spend increasing amounts of time grooming, handling, and carrying infants until their first young is born when they are about four years of age. As is typical of polygynous species, most males engage in little sexual behavior until well past maturity, although both sexes mature at about the same time (three-and-a-half years).

SUMMARY

Ecological factors, such as food distribution and predator pressure, affect group size and thus the type of mating system. *Monogamy*, in which one male and one female form a pair bond for one or more breeding seasons, occurs when neither sex is able to monopolize more than one member of the opposite sex. In *resource-defense polygyny*, males defend areas containing feeding or nesting sites critical for reproduction and thus gain access to more than one female. If females are gregarious, as in *female-defense polygyny*, males may form harems. In *male dominance polygyny*, males may concentrate in an area and display to attract females for mating. Occasionally females monopolize males, as in *resource-defense polyandry* or *female-access polyandry*, but such cases of polyandry are restricted to relatively few species.

Parental investment is any behavior toward offspring that increases the chances of the offspring's survival at the cost of the parents' ability to rear other offspring. Because the female's initial investment in eggs is greater than the male's in sperm, her ability to invest in future offspring is often less than his. Therefore, the female is typically particular about the male she mates with; the male tries to mate with as many females as possible. However, many other factors besides initial investment in eggs and sperm affect the decision about how much each sex invests in its offspring.

The more complex the organism, the greater the amount of parental care given to the young. As species adapt to stable environments, their body size tends to become larger, their life span increases, and they tend to have young at intervals (*iteroparity*), rather than all at once (*semelparity*). These conditions favor production of small numbers of young, which receive extensive care and thus are subject to a reduced mortality rate. An organism's dependence on food that is scarce and difficult to obtain also favors prolonged dependence of the young and extensive parental care. When survival of offspring is low and unpredictable, parents may bet-hedge, producing few offspring each year.

In species with parental care, mechanisms have evolved to reduce the chances of misdirecting parental investment to unrelated offspring. Such mechanisms are particularly well developed in species with highly mobile offspring and in species that breed in large groups.

Parent-offspring conflict has traditionally been viewed as maladaptive or as a necessary part of the weaning process. More recent ideas point out that natural selection operates differently on the two generations, so it is in the offspring's interest to receive more care than the parent is willing to give. Parents usually win such conflicts and *manipulate offspring* to maximize their own reproductive success.

As shown in some detail in the rhesus monkey, the physical and social environments interact with the neural-hormonal state at all levels to regulate development and parent-offspring relationships.

Discussion Questions

1. Why is polyandry so uncommon? What sorts of ecological and phylogenetic circumstances might favor polyandry in animals?

2. Discuss the possible relevance to humans of Trivers' concept of parent-offspring conflict.

3. The graph to the right shows the relationship between harem size, or the number of females that a male monopolizes, and the ratio of male-to-female body length, a measure of sexual dimorphism, in a variety of species of seals. (a) What is the most likely cause of this relationship? (b) What prediction could you make about the breeding habitat occupied by a species at the lower left of the curve versus one at the upper right?

From R. D. Alexander, et al., "Sexual Dimorphisms and Breeding Systems in Pinnipeds, Ungulates, Primates and Humans," *Evolutionary Biology and Human Social Behavior: An Anthropological Perspective*, edited by N. A. Chagnon and W. Irons. Copyright © 1979 Duxbury Press. Reprinted by permission of the editor.

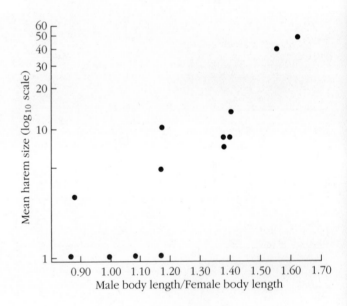

Suggested Readings

Barash, D. P. 1982. *Sociobiology and Behavior.* 2d ed. New York: Elsevier.
An easy-to-read account with good coverage of mating systems and parental care.

Thornhill, R., and J. Alcock. 1983. *The evolution of insect mating systems.* Cambridge: Harvard University Press.
Reviews insect reproductive behavior from an evolutionary perspective, with particular emphasis on sexual selection theory. Demonstrates the advantages of working with insects with their great diversity, short generation times, and high observability.

Trivers, R. L. 1972. Parental investment and sexual selection. In *Sexual Selection and the Descent of Man 1871–1971*, ed. B. Campbell. Chicago: Aldine.
This book chapter sets forth the theory of parental investment upon which much of the recent research in this area rests.

PART FIVE

BEHAVIORAL ECOLOGY: ENVIRONMENTAL PROCESSES

15

MIGRATION, ORIENTATION, AND NAVIGATION

In the previous section on behavioral ecology, we examined how animals interact with one another through communication, aggression , reproduction, and parental behavior. In part 5 we will examine the ways animals interact with their abiotic and biotic environment. Among the problems organisms face in trying to survive and reproduce are (1) finding their way about over both long and short distances (chapter 15), (2) selecting a suitable place to live and reproduce (chapter 16), (3) finding food and avoiding being eaten by others (chapter 17), and (4) responding to changes in population density (chapter 18).

Many species of birds—from tiny hummingbirds and warblers to cranes and hawks—migrate over water and land southward from temperate regions, often to the same location each winter; they then migrate back to the same location and even the same nesting tree each spring. Other animals undertake seasonal movements. For example, the elk in the western United States forage at higher elevations in the summer, but retreat to lower meadows and valleys for the winter. Many animals disperse from natal sites, and most animals move about every day as they feed, nest, and search for mates. How do these animals find their way?

In the first part of the chapter, we look at **migration**—a periodic movement from one location and climate to another location and climate—and its evolutionary history. In the second part of the chapter, we examine the means of orientation and navigation that have evolved; in addition, we will discuss the cues that animals use to find their way not only in long-distance migrations, but also within their home areas. We define **orientation** as the way in which an organism positions itself in relation to external cues. **Navigation** is the process by which an animal uses various cues to determine its position in reference to a goal as it moves about from place to place. **Homing,** which involves the use of navigation, is the ability of an animal to return to its home site or locale after being displaced. **Piloting** can be defined as the animal's use of familiar landmarks to find a goal or direction.

MIGRATION

BIRDS

Most of us are familiar with the fall and spring migrations of various bird species. The honking of geese and their V-shaped formations winging overhead herald the true arrival of fall each year, just as the appearance of certain warbler species signals the onset of spring (figure 15-1). We can ask four questions about migratory flights: (1) Which species migrate, from where to where, and when? (2) What external cues and internal physiological events trigger migration? (3) What is the evolutionary history of migratory behavior? and (4) What types of cues do birds use to guide them in their long-distance flights? We examine the first three questions now and treat the last question in the second half of the chapter.

FIGURE 15-1   Migrating knots
Each fall many thousands of birds of many species desert their summer residences in the north temperate zones to fly southward. Some fly in small groups; others, such as knots (*Calidris canutus*), fly in large flocks numbering in the thousands.
Source: Photo by Allan Cruickshank/Photo Researchers, Inc.

DESCRIPTION OF MIGRATORY BEHAVIOR. Information about the species of birds that migrate, the locations of their summer and winter ranges, the routes they follow in flight, and the speed at which they travel over long distances is gathered through several techniques. A common method is banding, in which researchers fit small, colored or numbered cylinders of metal or plastic on birds' legs. Researchers capture the birds in special nets or in traps (see chapter 3), place bands on the legs, and make careful records of the species, sex, age, and place of capture. When the same bird is later caught, shot, or found dead in another location, researchers hope the band will be noticed and the information reported. In the United States, the U.S. Fish and Wildlife Service is the clearinghouse for securing permission to band birds and for reporting all information pertaining to banding.

From accumulated bird-banding records, we can draw a picture of migratory species and collect vital facts about their travels. For example, from banding data, we have learned that many duck species, including teal (*Anas discors*) and mallards (*Anas platyrhynchos*), spend the summer months nesting in marshes and lakes in the northern United States and Canada. These birds fly to the southern states for the winter. Examples of spectacular migration speeds have come from banding data (see Griffin 1974); for instance, a sandpiper (*Actitis* spp.) flew 3,800 kilometers from Massachusetts to the Panama Canal Zone in nineteen days (an average speed of 200 kilometers per day), and a very small (10 grams) lesser yellowlegs (*Tringa flavipes*) traveled 3,100 miles from Massachusetts to the island of Martinique in the Caribbean in just six days (a rate of about 500 kilometers per day).

Radar is used in the study of flock migration to approximate the number of birds, their speed, their altitude, and their flight path. We can also focus a telescope on the moon; the birds' passage across the lighted background allows us to estimate the numbers of birds, and to identify some of the species, using the silhouettes. In addition, small radio transmitters with antennas that trail to the rear with the tail feathers can be affixed to the backs of birds. The signals can be picked up by receivers at varying distances from the bird, depending upon the power of the telemetry device. These signals can be used to track movement patterns and to ascertain flight speeds.

TRIGGERS. The timing of migration is probably under the control of endogenous biological clock mechanisms that are set by and adjusted to external stimuli. Two processes appear to be involved: a phase of preparation for migration, and the actual triggers that initiate migration. Of all the stimuli in the birds' environment, the most prevalent and consistent is daylength or photoperiod. The annual cycle of increasing daylength in the spring and declining daylength in the fall is consistent from year to year. Thus, photoperiod is the birds' best cue for initiating preparations for a fall migration.

The preparation phase is characterized by two major features: increasing fat deposition and migratory restlessness (*Zugunruhe*). The metabolic system of most migratory birds goes through two cycles each year, one in the fall and another in spring. At these times, the birds add large amounts of fat, the food reserves that they need for energy during flight. For the bird species whose pre- and postmigratory body weights have been determined, the average weight loss during migration is 30 to 40 percent. Species that make long-distance flights over water or that fly nonstop from the summer residence to the winter resting ground store virtually all of the necessary energy for the flight before starting. Among the birds that employ this strategy are some warbler species (family Parulidae) that migrate over the Caribbean islands nonstop to and from Cape May, New Jersey or across the Gulf of Mexico (Lowery 1946; Gauthreaux 1971). Spring migrants that fly north across the Gulf from the Yucatan peninsula or further south in Central America may have severely depleted fat reserves when they arrive on the Gulf Coast of the United States (Moore and Kerlinger 1987). Migrants with heavily depleted fat spend longer periods replenishing their food reserves (up to seven days) than do birds that have made the journey with lower energy cost. Other species, like the white-throated sparrow (*Zonotrichia albicollis*) (figure 15–2), store some fat, but make frequent stops to replenish their energy supply.

FIGURE 15-2 White-throated sparrow
The white-throated sparrow provides an example of a migratory species that only stores part of its needed food reserves before initiating its flight. These sparrows make several stops along the route for additional food.
Source: Photo by G. Ronald Austing.

Control of metabolism is centered partly in the hypothalamus and pituitary gland. Hormones secreted by the pituitary gland affect the metabolism of foods and the accumulation of fat reserves (Meier 1973). After the fall migration, the pituitary apparently enters a refractory phase, but reawakens in the spring in time to trigger both the deposition of fat for migration and the production and secretion of sex hormones. The latter signal is to prepare for breeding after the birds return to their summer home. Earlier, observers suggested that because changing photoperiod stimulated both migration and changes in sexual activity, these two activities were interrelated. Experiments with golden-crowned sparrows (*Zonotrichia atricapilla*) (Morton and Mewaldt 1962) indicated that castrated birds still develop fat deposits; the two processes may involve some of the same hormones and triggers, but they are not directly linked.

A second characteristic of the preparation phase is migratory restlessness, or *Zugunruhe*. As birds approach the time of migration, they show increased activity levels. Automated perches placed in cages with captive birds measure the birds' movements, as shown in figure 15–3 (Farner 1955; Gwinner 1986). Restlessness is related to daylength and may also be affected by other weather conditions, such as storms. Birds exhibit more restlessness under conditions that are favorable for migratory flights.

FIGURE 15-3 Zugunruhe in male white-crowned sparrow

During the molting period, prior to the onset of Zugunruhe, or migratory restlessness (a), the sparrow exhibits a different pattern of activity than it does after the onset of Zugunruhe (b). Lined bars represent the twilight, shaded bars indicate periods of darkness, and the open bars are daylight hours.

Source: D. S. Farner, "The Animal Stimulus for Migration: Experimental and Physiologic Aspects," in *Recent Studies in Avian Biology,* edited by A. Wolson. Copyright 1955 University of Illinois Press.

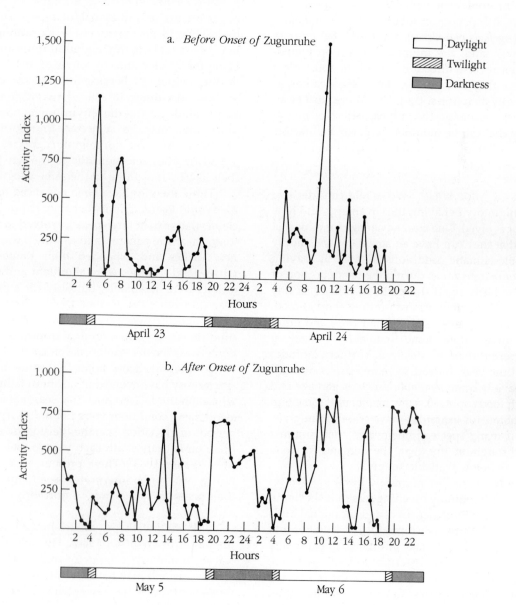

Thus, some type of endogenous rhythm involving neuroendocrine mechanisms appears to regulate the processes of fat deposition and the onset of migratory restlessness (Meier 1973; Gwinner 1986). The initiation of migration is triggered by the accumulation of sufficient fat deposits, which is possibly monitored by neural and hormonal systems; and it is triggered by the appearance of favorable weather conditions. Once initiated, some migrations are all-or-none phenomena, in which there is no turning back; for example, the nonstop flights of golden plovers (*Pluvialis dominica*). For other species, like American robins (*Turdus migratorius*), blackbirds (*Agelaius phoeniceus*), and those that migrate primarily over land, the migration may stop temporarily or even reverse direction toward the south for a short while if they encounter unfavorable weather, as they frequently do in the spring. Thus there must be an internal clock regulating the annual pattern of migratory activity that can be modified by peripheral factors such as the weather or the energetic condition of the bird (Gwinner 1986).

EVOLUTION OF MIGRATION. We should consider two aspects of migratory behavior: the advantages and how the behavior evolved. One clear advantage for birds that migrate is that they can have an adequate food supply and favorable climatic conditions all year. By leaving the north temperate zones during the harsh portion of the year, the birds avoid the perils of cold or stormy weather and the constant risk of running short of food. Some bird species have adapted their diets and other behavior—for example, some huddle for warmth at night—to permit them to remain in northern latitudes throughout the year. Indeed, as more sources of artificial food have become available, (such as garbage and feeders with food supplies), a few members of some bird species—robins, for example—have ceased to migrate. However, if many species remained in their summer locations throughout the year, the demand for food would far outweigh available supply.

Other advantages for migratory birds involve reproduction. Although the breeding season is shorter in the north, daylength is increased, and individual birds are able to concentrate their reproductive activities into fewer days. They can also take advantage of seasonally rich food supplies at the northern latitudes. In addition, the period of mating and rearing of young is a period when adults (of many species) and their offspring (either eggs or pre-fledging young) are most vulnerable to predation. An advantage of northbound migration is the reduction in the amount of time spent in these activities, which means a reduction in the risk of death or unsuccessful reproduction. One way they reduce predation is by effectively swamping the predator; because so many potential prey are present at one time, predators can take only a small portion of the available prey. Also, because the breeding season is short, the probability of reproductive synchrony is enhanced.

Migration has at least two significant consequences for the genetics and evolution of birds. Geographic dispersal is facilitated because of migration. As a result of migratory flights, individuals of a species may establish themselves in new areas, either by settling down in a new location along the migration route, or possibly by being forced off course by winds or inclement weather during migration. Because individuals of migratory species must reside in at least two different environments (and the interim environments during migration), they may be subjected to rigorous selection pressures. However, nonmigratory species are also subject to rigorous selection due to the harsh conditions they face by remaining in the same geographical area.

Three theories have been advanced to account for the evolution of migratory behavior in birds. One theory claims that certain bird species evolved in southern latitudes near the equator, and as time passed, they filled new niches and expanded their ranges northward where food supplies were abundant and climatic conditions were acceptable, but only for a portion of the year (Lincoln 1950; Graber 1968). These species may have found it necessary to return to more favorable conditions nearer their original tropical and subtropical homelands for the winter season each year.

A different theory holds that some migratory bird species may have evolved in northern latitudes at a time when conditions were such that seasonal movements to avoid harsh conditions were not necessary. Subsequent periods of glaciation (e.g. the Pleistocene epoch) forced these birds to fly south each year to seek better conditions (Sauer 1963). Their present annual migrations northward can be interpreted as attempts to return to their ancestral homelands in northern climes to breed each year.

Still another theory relates bird migration to continental drift (Wolfson 1948). The subtropical or tropical evolutionary origins of many bird species in Gondwanaland may have been followed by migratory movements out to the newer land masses as continental drift occurred.

MAMMALS

Among small mammals, the best-known migrants are bats of the order Chiroptera. Much of what we know about bat migrations has been derived from bat tagging, a technique similar to bird banding (see Griffin 1970). Some bats that live in more temperate climates hibernate during the winter months; others, including some species that also hibernate, migrate southward to warmer climates. The distances of these migratory movements vary from a few hundred to over one thousand kilometers. For instance, little brown bats (*Myotis lucifugus*) move from the New England states in a southwesterly direction to areas of Pennsylvania and adjacent states (Davis and Hitchcock 1965); and in western North America, hoary bats (*Lasiurus cinerea*) move from summer ranges in northern coniferous forests southward into central California and Mexico.

Among the larger mammals, some ungulates migrate; perhaps the most spectacular of these migrations is that of the Arctic caribou (*Rangifer tarandus*). Enormous herds of caribou migrate between winter ranges in Canada's more central latitudes, and summer ranges in the northern Arctic (figure 15–4) (Murie 1935; Banfield 1954; Lent 1966). The summer ranges are north of the timberline, where the caribou subsist on grasses and lichens. With the fall migration southward, they return to the shelter and food supplies provided by the forest.

Certain marine mammals also exhibit patterns of seasonal migration; for example, gray whales (*Eschrichtius gibbosus*) (figure 15–5) and harbor seals (*Phoca vitulina*) (Lockyer and Brown, 1981). Gray whales spend summers in the Arctic Ocean and North Pacific Ocean; in the winter they inhabit shallow waters off the coast of Mexico. Beginning in late January, gray whales calve in the warm waters off the coasts of Baja California and Mexico. Interestingly, the adult whales do not appear to feed either during the migration south or while they are in the breeding area. Instead, they utilize the fat deposits they accumulated the previous summer in the food-rich Arctic waters. In March or early April, the whales begin the long trip back north.

FIGURE 15-4   Migration routes of caribou
From H. L. Gunderson, *Mammalogy*. Copyright © 1976 by McGraw-Hill, Inc., New York, NY. Reproduced with permission of McGraw-Hill, Inc.

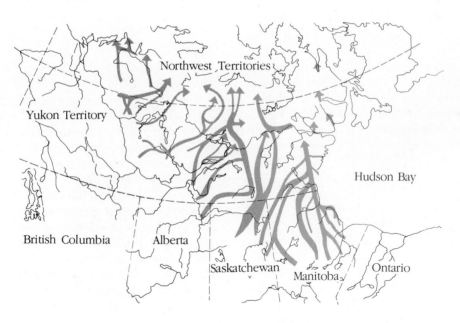

FIGURE 15-5   Gray whale migrating up the California coast

Gray whales spend the winter months in warm lagoons off Baja California and other parts of the Mexican coast. They migrate northward each spring and summer to the Arctic and North Pacific oceans.

Source: Photo by Steven Katona.

FIGURE 15-6   Monarch butterflies (*Danaus plexippus*) at overwintering site

The monarch butterfly is a seasonal migrant, flying south to Mexico or southern California for the winter. The butterflies that return north the following spring are the result of several generations of breeding en route.

Source: Photo by Lincoln P. Brower.

Orr (1970) stated that three categories of factors—alimental, climatic, and gametic—help to explain the evolution of migration in mammals (and quite possibly in many other classes of organisms, too). Alimental factors pertain to diet and available food supply; the need to obtain sufficient food energy underlies the evolution of migratory behavior in species like the caribou. Climatic factors are those that involve the ambient conditions, such as temperature and rainfall. Gametic factors appear to be important in some species; in, for example, the gray whales just described and the Pacific salmon (*Oncorhynchus* spp.). Food and environmental conditions do not explain their migratory behavior. Rather, the migration seems to bring the sexes together for the purpose of mating, and it provides for a place for birth and early development of young that will enhance successful reproduction (see Dingle 1980).

INVERTEBRATES

Monarch butterflies (*Danaus plexippus*) of the eastern United States fly south to sites in Mexico for the winter; and monarchs from the west coast move to locations along the central and southern California coast (figure 15–6). Unlike birds and mammals, monarchs do not restrict their breeding activities to their summer residences in northern locations. Adult butterflies migrate

southward in the fall and overwinter in large aggregations in trees (Zahl 1963; Brower et al. 1977; Calvert and Brower 1981). Those that survive until the following spring breed at or near the overwintering site. Then as the monarchs move northward, they complete up to three additional generations. Thus the residents of some summer locations are the third or fourth generation descendants of the butterflies that migrated south the previous fall.

Locusts (family Acrididae) exhibit a periodic movement type of insect migration. *Hoppers* (young locusts) that develop in low-density populations exhibit a low level of activity, and after metamorphosis into adult locusts, they are normally rather solitary and do not engage in sustained flights. When young hoppers develop in dense populations, they tend to be more active and gregarious; when they become adults they remain more gregarious and engage in longer flights (Johnson 1969). The character of locust populations varies between the sedentary and migratory phases, depending upon the interaction of social density with environmental triggers such as food availability and climate. When food supplies diminish, the physiology of the developing insects changes. This change (into migrators) combined with an increased population produces large swarms of migratory adult forms that may number in the millions. These swarms then set forth on long flights, and on the way, they consume prodigious quantities of vegetation.

OTHERS

Other organisms, representing all groups in the animal kingdom, exhibit migratory behavior (Dingle 1980). Pacific Ocean salmon migrate upstream to the place of their birth to spawn. Certain eels (*Anguilla rostrata*) live much of their lives in fresh waters in North America and Europe and migrate to the Sargasso Sea region of the Atlantic each year for breeding. Many frogs from North America's fast-flowing freshwater streams migrate to lakes or ponds to breed in an environment where egg deposition and fertilization are more practicable. Among reptiles, the desert tortoise migrates between summer and winter habitats.

ORIENTATION AND NAVIGATION

EARLY WORK

One of the early investigators of animal orientation, Jacques Loeb (1918), theorized that asymmetrical stimulation of an animal's sensory organs results in differential contraction of muscles on the opposite side of the animal until the symmetry is restored for both sense organs and muscle actions. Other investigators (e.g., Mast 1938) concluded that not all animal orientation could be fitted to Loeb's sensory organ-muscle scheme, and they found that different organisms possessed different systems of orientation.

Later, Fraenkel and Gunn (1940) summarized the work on animal orientation up to that date and defined general classes of orienting reactions. **Kineses** are random locomotion patterns in which there is no orientation of the organism's body axis to the source of stimulation. **Taxes** are directed reactions involving (in a single stimulus situation) an orientation of the long axis of the body in line with the stimulus source. Movements toward the stimulus are positive taxes; movements away from the source of stimulation are negative

taxes. For instance, planaria (*Planaria* spp.) exhibit a negative *phototaxis*—movement away from a light source. Extensive studies have revealed other, often complex systems of orientation and navigation in different organisms, only a few of which we examine here.

BIRDS

In the last three decades, a variety of studies have been conducted on orientation and navigation of birds. Many of the hypotheses that have been advanced for birds are potentially applicable to other animals, such as insects and mammals (see reviews by Mathews 1968; Griffin 1974; Emlen 1975a; Able 1980). Navigation requires both a compass to provide directional information and a map to provide the animal with information on its position relative to home or some other goal. As illustrated in the following summaries, considerable evidence exists for the compass, but little for the map (Keeton 1969, 1970).

TOPOGRAPHIC FEATURES. Use of familiar topographic features is certainly a factor in navigation, but it is probably of secondary importance compared with other cues. Nonetheless, many diurnal and nocturnal migratory birds are influenced by topographic features and may utilize them. Birds congregate near and fly along coastlines or river valleys; and they may use landmarks for piloting, particularly in home areas at both ends of the migration, where they are more familiar with specific topographic features. However, the use of topography as a guidance system has inherent drawbacks. How do first-time migrants who have never learned the landmarks find their way? What if a storm blows birds off the normal route into areas they have never traversed before? Also, visual landmarks alone cannot provide the proper orientation over long distances. Both landmarks and certain topographic features could be used in combination with some type of compass mechanism to provide sustained orientation during longer movements. In familiar terrain, landmarks may suffice without any compass (Able 1980).

Landmarks may play a critical role in finding particular stopover locations, as in the case in migrating waterfowl (Bellrose 1964, 1971). Evidence from radar studies of bird migration, however, has shown that nocturnal migrants in different parts of the country ignore most topographic features (Gauthreaux 1971; Richardson 1972; Drury and Nisbet 1964). Studies in which homing pigeons (*Columba livia*) have been fitted with frosted lenses over their eyes (Schmidt-Koenig and Schlichte 1972; Schmidt-Koenig and Walcott 1973) reveal that upon their release from a new location, when they can locate the sun, even when the visibility is only three meters, they still can find their way home—or very close to home. When investigators used airplane tracking and ground radar to follow the flight paths of homing pigeons with frosted lenses, they found that the paths flown by the birds were generally in the direction of home. Thus we can conclude that the pigeons' navigation system does not require detailed vision and can lead birds that have been released 13 to 20 kilometers from home to within a short distance of the loft.

SUN. The most prominent regular cue for diurnal migratory birds is the sun. Nocturnal migrants may also utilize the sun by taking a bearing at sunset and using that bearing to fly at night (Able 1982). The early studies of bird navigation devoted considerable attention to the sun (see Keeton 1974). Kramer (1950,1951) first showed that European starlings (*Sturnus vulgaris*), placed in an outdoor cage with the sun visible, exhibited migratory restlessness in the appropriate direction in the spring and fall (figure 15–7 *a,b*) . When Kramer used mirrors to alter the apparent position of the sun, the pattern of the starlings' migratory restlessness shifted direction in a predictable manner (figure 15–7 *c,d*). As we discuss later, other organisms (e.g., insects, turtles) also appear to use some form of sun-compass navigation system.

How do nocturnally migrating birds use the sun for orientation? The mechanism by which birds use the sun near sunset as a bearing has been explored in some detail by Able (1982, 1989; Able and Able 1990a). Using Savannah sparrows (*Passerculus sandwichensis*), Able has demonstrated that the birds can use the polarization patterns of light from the sun as an indicator of its direction and can take their bearings from such cues. Furthermore, this process is learned in young Savannah sparrows (Able and Able 1990a). Groups of hand-raised young birds were provided with controlled experience of the daytime sky; conditions manipulated were sun azimuth, skylight polarization patterns, and magnetic directions. In their first autumn, the birds were tested in Emlen funnels (figure 15–8) during the period between sunset and the first appearance of stars. The birds learned to perform migratory orientation based primarily on polarized light patterns.

After the initial discovery of sun-compass orientation in birds, theories were put forth to explain how birds use the sun cue (see Mathews 1968). Some investigators theorized that birds use the sun only to gain a

FIGURE 15-7   Starling orientation experiment
Each circular diagram illustrates the orientation of diurnal spontaneous migratory activity in a caged European starling. Experimenters caged the bird in an outdoor pavilion with six windows and conducted tests during the migratory season (a) under clear skies, (b) under overcast skies, (c) with the sun's image deflected 90° counterclockwise by mirrors, and (d) with the sun's image deflected 90° clockwise by mirrors. Arrows denote mean direction of activity, and each dot within the circles represents ten seconds of fluttering activity. The dotted lines indicate the direction of light coming from the sky.

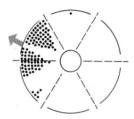

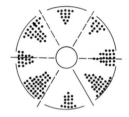

a. Clear skies

b. Overcast skies

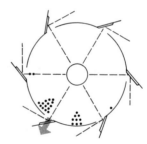

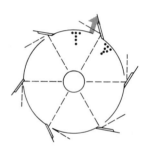

c. Sun's image deflected 90° counterclockwise by mirrors

d. Sun's image deflected 90° clockwise by mirrors

compass bearing to head in a particular direction. Others theorized that they use true navigation—the ability to orient toward a goal regardless of its direction and without the use of familiar landmarks (Griffin 1955).

One way to demonstrate the use of a sun compass is to show whether birds maintain a constant orientation at different times of the day even though the sun's position varies as it traverses the sky. Data from some diurnal migrants provide support for this alternative. Clock-shift experiments have provided a more rigorous test of sun-compass orientation. First, artificial shifting

FIGURE 15-8   Circular cage for orientation experiment
A special cage (called an Emlen funnel) enables researchers to study the pattern or direction of restlessness in migrating birds. A bird stands on the ink pad at the bottom of the funnel-shaped cage where it can view the stars overhead through a wire mesh top. Each jump it makes as if to fly is recorded on paper on the sides of the funnel. (The cage is depicted here in cross-section.)

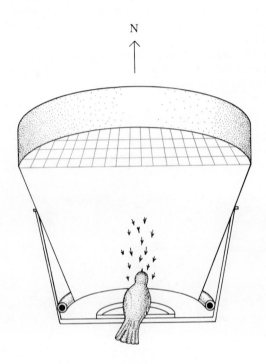

FIGURE 15-9   Clock-shifted pigeon experiment
Experimenters released pigeons that were clock-shifted six hours behind time, at noon at a site 100 miles south of their home loft. The birds' internal clocks indicate that it is 6:00 A.M.; so if they are using sun-arc navigation, sight of a noon sun should lead them to determine their location as thousands of miles east of home, and they should depart westward. They actually depart eastward. This experiment and others support the contention that pigeons use the sun as a directional compass, but not as a "map".

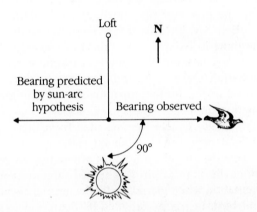

STELLAR CUES.   What about birds that migrate at night? When nocturnally migrating birds (e.g. blackcaps [*Sylvia atricapilla*], lesser whitethroats [*Sylvia curruca*], both found in Europe) are placed in outdoor cages with a view of the clear night sky, they exhibit migratory restlessness in a direction appropriate to the seasonal migration (Kramer 1949, 1951; Sauer 1957). Sauer exposed birds to planetarium skies, which permitted him to manipulate star patterns experimentally. Not only were the results of outdoor studies confirmed, but when he shifted the planetarium's star patterns 180 degrees, the direction of the birds' *Zugunruhe* also shifted. Under cloudy skies or with no stars visible, and with only diffuse, dim illumination in the planetarium, the birds exhibited random orientation. Sauer's results from warblers (*Sylviidae* spp.) have been confirmed in both field and laboratory experiments by Emlen (1967a,b) using indigo buntings (*Passerina cyanea*). Emlen's technique is innovative: The bird stands on an ink pad at the bottom of a funnel-shaped cage (figure 15–8). The funnel sides are covered with paper; the top is covered with a wire mesh screen that permits the bird to see the

of the birds' internal clocks six hours fast (by housing them in an environmental chamber with a controlled light cycle) resulted in predictable 90 degree directional changes in the initial bearing of birds heading for the home site. This finding is consistent with the notion that birds use the sun as a simple compass. As a further test, birds were clock-shifted six hours slow and released at a site 160 kilometers south of the home loft (figure 15–9). If the birds were using true sun navigation, they should have decided that they were 6,400 kilometers east of home and headed directly west. However, the birds headed directly east, as would be predicted if they were using the sun only as a compass, and not as both a map and a compass (Keeton 1974).

The sun, then, is an important directional cue for diurnal migrant birds and for homing pigeons. Evidence in support of the hypothesis that the sun can be used as both compass and map is lacking.

overhead sky. When the bird jumps up against the sides of the funnel in its restlessness, it leaves marks on the paper. Investigators can turn the record of these marks into a vector diagram for analysis.

In additional studies, Emlen (1970, 1975b) rotated the night sky in a planetarium. He conducted an experiment with three groups of young indigo buntings. Individuals in group 1 were raised in a windowless room with only diffuse light. Individuals in group 2 were allowed to see the normal night sky in the planetarium, with a normal rotation of the heavenly bodies around the pole star (the North Star or Polaris) once every other day. Birds in group 3 were raised the same way as those in group 2, except that the heavenly bodies were rotated around Betelgeuse, a bright star in the constellation Orion. The indigo buntings were later measured in the funnel apparatus for their migratory orientation under planetarium skies.

Two major conclusions resulted from this experiment. First, exposure to stellar sky patterns is necessary for normal southward migratory orientation in young buntings. Birds in group 1 exhibited random patterns of orientation when placed under the normal night sky. Second, birds in group 3 oriented 180 degrees away from Betelgeuse—as if headed south, using that star to define the southerly direction. Early experience thus plays a critical role in determining the migratory orientation of buntings, and they may use the sky pattern they learn at this time throughout life. These findings also help account for the evolution of stellar-cue orientation in spite of the change in the earth's magnetic poles that occurs about every thirteen thousand years that causes the positions of star patterns to change in the sky. Birds do not inherit a star map or knowledge of a specific star pattern. Rather, they inherit a predisposition to learn the sky pattern they see when they are very young, and then they use that sky pattern as the basis for orientation during migratory flights.

As was the case in our discussion of the compass/map role of the sun, we are again faced with the question of whether birds are using the stars merely as a compass or whether they are capable of true navigation using stellar configurations (figure 15-10). Use of the stars as a compass differs from use of the sun because the night sky contains many more potential cues. Because the stars shift (and do so at varying speeds, depending on their position), birds must compensate, as has been shown by Emlen (1967b). To use true star navigation, the birds would have to use several star patterns and would need several compensation rates for these different groupings (see figure 16–10). Emlen

FIGURE 15-10   Measurement of migratory restlessness with vectors

Vector diagrams depict the results of studies in which migratory restlessness is measured either by an observer, with the bird in a circular cage, or automatically via the use of the funnel apparatus shown in figure 15.8. Within each circle, the length of a vector (directional line) indicates the amount of migratory restlessness activity oriented in that particular compass direction. Here data for three birds studied by Emlen illustrate that during three two-hour time blocks at night, the orientation of indigo buntings does not shift. This indicates that the birds are capable of time compensation to account for the movement of the pattern of stars in the sky in an apparent circle.

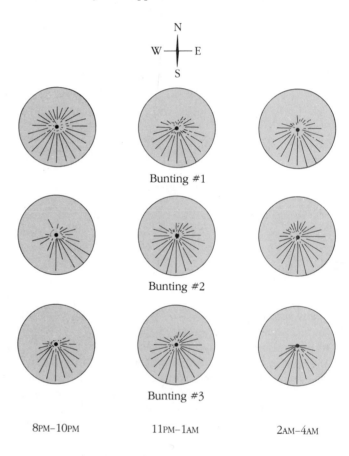

Bunting #1

Bunting #2

Bunting #3

8PM–10PM   11PM–1AM   2AM–4AM

(1975a) and Able (1980, 1982) have suggested that although no data as yet demonstrate star navigation convincingly, data do exist to support the hypothesis that birds use the stars as a directional or compass mechanism.

METEOROLOGICAL CUES. How does the weather influence migration? If we watch birds in the spring or fall, we find that migratory activity in a particular region tends to be concentrated into a few days or nights. Birds are capable of responding to favorable weather conditions (Drury and Keith 1962; Kreithen and Keeton 1974a). Data support the conclusion that birds often fly downwind on their migratory flights (Bellrose 1967; Richardson 1971; Bruderer and Steidinger 1972). Do birds select favorable winds *after* they have oriented themselves using other types of cues? or, as some (e.g., Able 1974) have suggested, Might the birds be using the wind itself as a directional cue?

Further experiments (Able 1982; Able et al. 1982) involved monitoring the natural nocturnal migratory flights of individuals of several *Passerina* species under various weather and wind conditions. When migrants could observe the sun near sunset, or the stars, they flew in the appropriate migratory direction regardless of wind direction. If observations were made under totally overcast skies with no view of sun or stars possible, birds often flew downwind, sometimes in an inappropriate direction. When migrant white-crowned sparrows (*Zonotrichia leucophrys*) were fitted with frosted lenses and released from balloons aloft, they headed downwind, even though that direction was sometimes seasonally inappropriate. These results indicate the visual cues' importance in determining appropriate migratory direction; when deprived of visual cues, some species can use wind direction for orientation.

OLFACTORY CUES. In addition to the foregoing items, we should also note that several investigators have suggested that olfactory cues, primarily odors carried on the wind, may aid orientation, particularly in homing pigeons (Papi et al. 1972, Papi 1982; Walraff 1986). The most direct test of such a hypothesis would involve removing the olfactory capacity and then testing for orientation and homing behavior. The olfactory capacity has been removed by three different techniques: (a) nerve transection; (b) plugging the nostrils with plastic tubes that permit air passage, but preclude contact with the nasal mucosa; and (c) using local anesthesia (Walraff 1986). When the olfactory sense is blocked or when nerves are transected, its influence on orientation can be determined. However, some behavioral effects can result that are not related to the homing ability being tested. Therefore, unequivocal evidence of the role of olfaction is difficult to show. Indeed, studies by Keeton and Brown (1976), Keeton, Kreithen, and Hermayer (1976), Schmidt-Koenig and Phillips (1978)

and Hartwick, Kiepenheuer, and Schmidt-Koenig (1978) all provided data demonstrating little or no effect on homing behavior when the nasal passages were blocked or a local anesthetic was used on the nasal epithelium.

GEOMAGNETIC CUES. Evidence from several sources has accumulated that supports a role for geomagnetic cues in orientation and navigation in some birds. After failing to determine whether birds could even sense the 0.05 gauss geomagnetic field of the earth (e.g., Emlen 1970; Kreithen and Keeton, 1974b), Brookman (1978) conditioned pigeons to magnetic fields in a laboratory experiment. Measurement of *Zugunruhe* in European robins (*Erithacus rubecula*) provided a second line of evidence. Birds were placed in a cage surrounded by Helmholtz coils, which provide an artificial magnetic field. Wiltschko and Wiltschko (1972; and Wiltschko 1972) used this apparatus to demonstrate that the birds were not responding to the polarity of the horizontal component of the magnetic field; they did not shift the direction of their migratory restlessness when horizontal polarity was shifted. However, they did reverse direction when the vertical component was reversed; and they reversed the direction of orientation in a predictable manner. In an artificial magnetic field with a zero vertical component and strong horizontal component, the activity was random.

Other evidence for the role of magnetic fields in orientation comes from the work of Southern (1969, 1972) on ring-billed gulls (*Larus delawarensis*). Gulls in cages exhibited a strong tendency to walk in a southerly direction, except when storms produced temporary aberrations in the earth's magnetic field. Also when gulls with small magnets affixed to their backs were taken some distance from home and then released, they displayed a random pattern of dispersion. Control gulls without the magnets exhibited normal and consistent directional headings to the south. Additional confirmation of the role of geomagnetic cues was reported by Keeton (1971) and Larkin and Keeton (1976), who used bar magnets attached to the wings of homing pigeons.

Investigators have utilized radar and airplane tracking to monitor the behavior of homing pigeons fitted with Helmholtz coils on their heads (Walcott 1972, 1977). Further, Walcott, Gould, and Kirschvink (1979) reported the existence of an organ along the midline of the pigeon's brain that contains magnetite granules. This organ may prove to be a source of sensitivity to geomagnetism in the pigeon.

Recently, Able and Able (1990b) used Savannah sparrows to test early geomagnetic orientation. Birds reared in outdoor cages exposed to the daytime or night sky did orient in the proper direction, whereas control birds reared in the laboratory did not orient properly when tested in an Emlen funnel apparatus. Further, when birds were raised outdoors in the cages described above, and were permitted to observe the daytime sky, the night sky, or both, if the direction of the apparent geomagnetic force was changed to east-southeast, the birds all oriented in a northeast-southwest direction—significantly different from control birds' orientation. Able and Able proposed that the geomagnetic compass of the birds is calibrated early in life and that celestial rotation provides the basis for these adjustments to the compass.

As both Emlen (1975a) and Able (1978) have noted in their reviews of bird orientation and navigation, tremendous progress has been made since the 1960s in new field and laboratory techniques for studying the question and also in overall knowledge of what cues may be important. It is noteworthy that the pace of new findings in bird orientation in the 1980s, after the major developments of the previous decade, has slowed somewhat. It is clear from much of the foregoing discussion that many bird species possess multiple systems for orientation; the use of available cues apparently follows some type of hierarchical scheme. The mechanism employed depends on the bird's preferences for cues and the prevailing weather conditions at its location.

MAMMALS

Having thoroughly explored orientation mechanisms and cues in birds, we might now wonder whether similar cues are used by mammals. Studies on orientation and navigation in mammals have concentrated on small rodents and bats. In many studies on rodents, experimenters displaced the animals and observed their return, measuring the number of successful animals, their speed, and their orientation when they left the displacement site. In some studies, experimenters attached reflector tape or radio transmitters to the animals to plot their courses.

Rodents have an area, called the home range, where they carry out their daily activities. Once we have determined the home range, we can displace the rodent and monitor its homing behavior until it returns to its home range and is recaptured (see review by Joslin

1977a, which includes a number of excellent comments on methods problems encountered in attempting such homing experiments). Data from several species of mice indicate that when displaced, these animals can find their home range again, some from a considerable distance (up to 30 kilometers) and with speeds up to 300 meters/hour. In most studies, the rodents are usually displaced only up to 500 meters away, and may take several days to return to the home range and reenter a trap (see Joslin 1977a). The return rate, or the percentage of mice that successfully return home, varies with the species and the nature of the habitat. Overall figures indicated return rates ranging from 3 percent to over 80 percent. In general, the further from home the test animals are displaced, the lower their percentage of successful return.

Based on present data, the most plausible explanation of homing mechanism in rodents is **piloting**—that is, location by using familiar landmarks or terrain—combined with random search when the displacement is for greater distances (Joslin 1977a,b). An animal learns about its habitat in several ways, including exploration of the home range. Some exploratory forays may take the rodent a distance beyond the home range. Rodents also learn about the habitat during dispersal when the young move away from the natal site to establish home ranges of their own. Home ranges differ greatly in size between species and even within the same species, from several hundred square meters for voles to several square kilometers or more for deermice and squirrels. For example, a rodent whose home range has an approximate diameter of 200 meters may also know something about the terrain and its landmarks for 200 meters or more in all directions around the perimeter of that home range. Thus, when a rodent is displaced for a short distance, it may be in an area it already knows something about. This larger area is sometimes referred to as the *life range*. Alternatively, when it is displaced for a greater distance, the animal may exhibit random, or possibly, systematic searching behavior until it encounters familiar terrain and can again rely on piloting to find the home area.

Little research has been done on how rodents orient within their home ranges. However, Drickamer and Stuart (1984) investigated the movements of deermice (*Peromyscus* spp.) by mapping and analyzing the patterns of tracks left by the deermice on fresh snow (figure 15–11). Drickamer and Stuart used traverses (each segment of the path from tree to tree or from tree to log) to examine the potential use of cues. They found that

FIGURE 15-11   Deermice navigating forest floor
Deermice (left) utilize trees in their environment as cues for piloting their way around the top of the snow cover on the forest floor. Many of their movements are from one tree to another (right).
Source: Photo by Lee C. Drickamer.

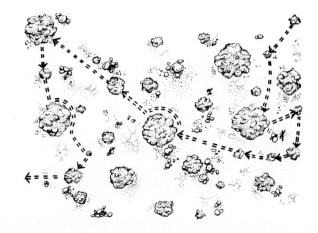

86 percent of all traverses were oriented from one tree to another tree, and covered distances ranging from one to thirty meters. The concentration of tracks suggests that vertical tree trunks are used for orienting. A significant positive relationship ($r = .51$) between the length of the segment traveled by the mouse and the diameter of the tree to which it was headed further suggests that larger trees serve as more conspicuous objects on the horizon for orientation. Laboratory investigations (Joslin 1977b) provide further support for the hypothesis that deermice may use conspicuous vertical cues on the horizon for orientation.

Bats use an echolocation system to orient to objects in the environment, and to capture prey. Bats send out high-frequency sounds that bounce off objects in the environment and return as echoes, which the bat uses to control its flight pattern (Griffin 1958, 1974). Griffin and Galambos (1941) investigated whether bats use the high-frequency sounds to avoid objects when navigating through their environment. They tested bats in a large room in which a barrier of wires was set up. The observers scored "hits" and "misses" as bats flew back and forth past the wires. Data from these experiments (table 15-1) show clearly that any treatments that impaired the bat's ability to send out high-energy sounds or receive the return echoes were associated with a lower percentage of misses. In other words, when the experimenter hindered its capability for using echolocation, the bat was more likely to touch or strike the wires.

To test whether bats use visual cues, experimenters blindfolded bats and then moved them eight kilometers away from the roosting cave (Mueller and Emlen 1957). The number of bats returning and the speed of their return to the cave (where they were caught in a large net set up across the cave's entrance) were the same for blindfolded bats as for bats with unimpaired vision. We should note, however, that in experiments with blindfolds there have been varying techniques (for example, blindfolds or goggles), varying results from different experimenters (Mueller 1966; Barbour et al. 1966), and varying interpretations of the results. At longer distances, vision appears to play a role in the orientation and navigation of bats (Williams and Williams 1967, 1970; Williams, Williams, and Griffin 1966). We know that bats can see with their eyes; studies of optomotor responses and experimental training of bats to respond to visual patterns have revealed visual acuities of several degrees (Suthers 1966; Chase and Suthers 1969). We will need improved experimental techniques and designs to learn more about navigation in bats.

TABLE 15-1   Bat navigation by echolocation

Bats (mostly little brown bats, *Myotis lucifugus*) were tested for their ability to navigate after various impairments of hearing, sound production, and vision. The test room contained 16-gauge wires hung vertically with a 1-foot (0.3 m) spacing interval. When bats flew near the wall, floor, or ceiling, their flights were not counted. The scores in the table represent percentages of flights through the wires during which the bats did not touch the wires. Specific controls used for each experiment are indicated in the "experimental treatment" column. A chance score was calculated as about 35 percent misses.

		Experimentals		Controls	
Number of bats	Treatment	Number of flights	Average percentage misses	Number of flights	Average percentage misses
28	Both eyes covered (controls untreated)	2,016	76	3,201	70
12	Both ears covered (controls untreated)	1,047	35	1,297	66
9	Ears and eyes covered (controls with only the eyes covered)	654	31	832	75
8	Glass tubes in ears and closed (controls with the same tubes in ears but open)	580	36	636	66
12	Both ears covered (controls with one ear covered)	853	29	560	38
6	Eyes and one ear covered (controls with only the eyes covered)	390	41	590	70
7	Mouth covered (controls with eyes covered or intact)	549	35	442	62

From D. A. Griffin and R. Galambos, "The Sensory Basis of Obstacle Avoidance by Flying Bats," in *Journal of Experimental Zoology*, 86:481–506. Copyright © 1941 Alan R. Liss, Inc., New York, NY. Reprinted by permission of Wiley-Liss, a division of John Wiley and Sons, Inc.

AMPHIBIANS AND REPTILES

Can other vertebrates use some of the same cues for orienting that have been found in birds and mammals? Orientation has been studied experimentally in southern cricket frogs (*Acris gryllus*) and northern cricket frogs (*A. crepitans*) (Ferguson, Landreth, and Turnispeed 1965; Ferguson, Landreth, and McKeown, 1967). Captured frogs were placed in a large circular plastic pen in the water; the frogs could see no landmarks on the horizon. When southern cricket frogs were released in the pen under sunlit skies, they swam in the direction that would have taken them to land had they been released off the shore at their home pond. The results were the same whether they could see the sky overhead or not, whether they were released under starlit skies, or whether they were released under moonlit skies.

When researchers placed the frogs in the pen on a moonless night with only stars visible in the sky, the frogs oriented in two opposite directions; some headed toward the home shore and others headed directly away from home. When experimenters released the frogs during the twilight period after sundown but before the stars or moon were visible, or under overcast conditions, the frogs oriented in a random fashion. Thus, it appears that for proper orientation in the cricket frogs, learned shore position, a view of some celestial cue, and a biological clock in phase with the local time are all necessary.

Cricket frogs held in darkness for thirty hours to seven days exhibited a decline in orientation accuracy that culminated in random orientation after seven days. "Dephased" frogs reestablished proper orientation when experimenters exposed them to normal dark/light

FIGURE 15-12   Female green turtle on nesting beach
Sea turtles have been studied extensively by tagging. The turtles return periodically to the same beach to lay their eggs.
Source: Photo by Archie Carr.

cycles or to daily fluctuations of temperature and humidity. These frogs appeared to use a learned shore position, celestial cues, and a timing mechanism to orient themselves properly to the home shore. Similar results have been obtained for northern cricket frogs.

Carr (1967) summarized a number of investigations on orientation in green turtles (figure 15–12). Female turtles (*Chelonia mydas*) deposit their eggs on sandy beaches on certain islands; they lay about a hundred eggs on each trip to the beach, and make three to seven trips spaced twelve days apart. In nonreproductive periods, the turtles live in areas with abundant turtle grass along the coasts of continental land masses. In one study (Carr 1967), turtles were banded at Ascension Island (figure 15–13). Young turtles that hatched on the island floated on currents that took them to the coast of South America. Adults had to swim against the same current to reach Ascension Island to reproduce. How did they find their way? The prevailing current may have certain features that allow the turtles to distinguish it from surrounding waters. This current could provide a general orientation, but not the precise information needed to target a small land mass in the midst of many square miles of ocean. Studies using green turtles have provided tentative data to support a hypothesis of sun-compass orientation (Carr 1965, 1967; Ehrenfeld and Carr 1967).

When the mysteries surrounding the movements of green sea turtles were first being unraveled, Carr and Coleman (1974) hypothesized that the turtles elected to return to the beaches of Ascension for breeding because those locations had once been much closer to South America; over millions of years, continental drift had moved the island further from the continent. Recently, molecular techniques have been used to disprove this notion (Bowen, Meylan, and Avise 1989). A particular type of DNA in animals, mitochondrial DNA (mtDNA) is passed only from mothers to their offspring; this permits investigators to track female lineages. By examining the maps of mtDNA for differences, it is possible to estimate how long ago two populations became separated: The more divergence between their mtDNA make-up, the further back in time they diverged. The mtDNA data do indicate some divergence among the three populations, confirming the notion that green sea turtles do return to their natal site for breeding. However, comparisons of mtDNA from turtles taken at three breeding locations—Ascension Island, Florida, and Venezuela—revealed that the population that uses Ascension Island for egg laying has a degree of difference small enough that any divergence between the groups of turtles took place on the order of 20,000–40,000 years ago, not forty million years ago as postulated by the continental drift hypothesis. Thus, the patterns of migration cannot be explained by continental drift, and other explanations must now be sought and tested.

FISH

We have defined homing behavior as the use of landmarks and navigation to return to a homesite. An animal may become displaced from its home by weather, or it may return to a home location in the course of the life cycle, as do Pacific salmon. Alternatively, researchers may artificially move an organism from its homesite to another location to investigate the cues and means of navigation it employs to return home.

Female Pacific salmon lay eggs in stream beds among the cracks and crevices in the rocks, where male salmon fertilize them by depositing sperm over the eggs. After hatching, the fry develop in the home stream during the summer and then migrate to the ocean,

FIGURE 15-13   Turtle tagging project on Ascension Island

The number of turtles tagged at each nesting beach location is shown by a figure. Solid triangles indicate mainland Brazil locations where researchers recovered turtles. Open triangles on nesting beach sites show turtles that returned after three years. Solid hexagons denote turtles that returned in the fourth year to nest, possibly having made two round trips to Brazil. Arrows indicate prevailing ocean currents.

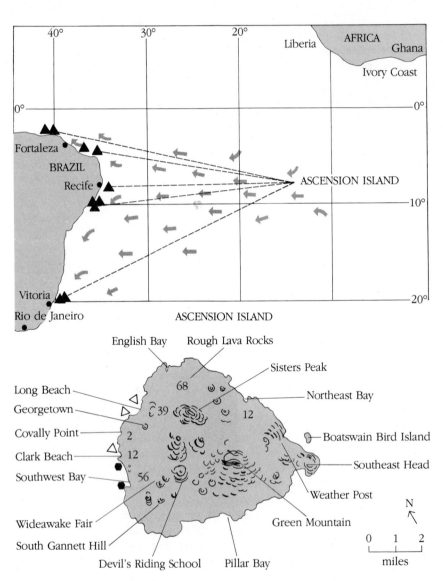

FIGURE 15-14   Pacific salmon migrating upstream
After spending much of their adult life in the ocean, Pacific salmon migrate upstream to breed in the location where they were hatched. The mechanism by which they re-locate their natal stream appears to involve chemical cues in the water.
Source: Photo by Ronald Thompson from Frank W. Lane.

where they mature. Some three to five years later (depending on the species), most of the surviving adults return to spawn and die in the stream where they hatched (figure 15-14).

How do the salmon find the stream where they hatched? The results of numerous experiments conducted over many years (see Hasler 1960) indicate that young salmon fry are somehow imprinted with the odor of the natal stream. When adult salmon migrate upstream, they follow the odor or pattern of odors from the home stream, making correct turns at each junction between streams until they arrive back at the natal stream. The specific nature of the olfactory stimuli involved in this process have not yet been clearly identified.

INVERTEBRATES

Migrating locusts (*Locusta migratoria*) flying in swarms give the appearance of oriented directional movement. Studies of locust swarms have revealed that wind exerts a displacing effect on individuals and thus, over time, on the swarm. Within the swarm there are groups that fly in directions different from that of the swarm as a whole. When these subgroups reach the periphery of the swarm, they turn back inwards toward the main flow of the swarm (Rainey 1959, 1962; Waloff 1958).

The way honey bees (*Apis mellifera*) communicate the location of food sources to one another in the hive and then make foraging flights to the food has been studied by a number of investigators (von Frisch 1967; Lindauer 1961; Gould 1975, 1976; Wenner 1971, 1974). Chapter 11 on communication covered this topic in some detail.

Bees and some other insects, including ants, can receive and interpret information about polarized light (von Frisch 1967). The plane of polarization can provide an axis for orientation for these animals. It is also possible that the perception of polarized light may enable some animals to locate the position of the sun when the sky is partially overcast. The use of polarized light for orientation occurs in a variety of other vertebrates, including fish (Waterman and Hashimoto 1974) and salamanders (Adler and Taylor 1973).

Many of the orientation cues discussed in this chapter with regard to birds and mammals have also been demonstrated in invertebrates. Among these are sun-compass orientation in octopodes, spiders, beetles, crickets, and ants (Waterman 1966; Beugnon 1986); geomagnetic force cues in bees; and landmarks in a variety of species (see Able 1980). In addition, data have been reported that suggest beachhoppers (*Talitrus* spp.) use lunar orientation to guide their nocturnal movements (Papi and Pardi 1953; Scapini 1986).

Spiders constructing webs present an interesting opportunity to examine orientation in a smaller space than some of the examples we have provided thus far (figure 15-15). How is it that the spider is able to produce such a complex and intricate web? What orientation cues are involved? Studies on several orb-weaving spiders that build vertical webs (*Araneus diodematus* and *Leucauge mariana*) have shown that these arachnids use at least three types of cues for orientation: light, gravity, and the threads or lines of the web they are constructing or moving about (Crawford 1984; Eberhard 1988). The spiders use gravity to determine their direction within the vertical plane, light to discriminate between the sides of the web, and a spatial memory of distances and directions on various lines involved in the web's construction.

FIGURE 15-15   Orb-weaving spider and web
Some spiders, like the orb-weaving spider, *Araneus diadematus*, shown here, build vertical webs. These spiders apparently use light, gravity, and the threads of their webs for orientation.
Source: From Peter Witt.

SUMMARY

Many species of animals, including birds, mammals, and insects, migrate seasonally to areas with more favorable conditions in order to avoid the harsh weather conditions and diminished food supplies of the winter months.

The routes and dynamics of bird migration can be studied by direct observation, use of radar, and telescope observation of the birds that fly in front of the moon. The triggers for migration include factors that stimulate fat deposition and migratory restlessness, or *Zugunruhe*, and the actual signals that initiate migratory movements.

The *advantages* of migration for birds include an adequate *food supply*, the concentration of activities related to *reproduction* during the longer summer days in northern latitudes, and the *synchrony between sexes* that may ensure a higher probability of successful reproduction. Migration may affect *geographical dispersal* in birds and may have evolved as a seasonal movement northward and southward from sites of evolutionary origin in or near the tropics. Alternatively, some migratory species may have evolved in northern climates and have adopted seasonal patterns of movement in response to shifts in climate.

Some bats and caribou migrate for *climatic* or *alimental* reasons; gray whales apparently migrate for *gametic* reasons. Monarch butterflies have a unique long-distance migration—the returning migrants in the spring are fourth-generation descendants of those that flew south. Locusts have *sedentary* and *migratory phases* and exhibit periodic movements.

Investigators have studied homing pigeons and other bird species to determine the mechanisms of orientation and navigation. Some bird species are capable of using the sun or stars as a *compass* for orientation. In many instances, birds appear to use topographic landmarks for *piloting*. Investigations have shown that *meteorological cues* can influence migratory flight. A number of reports have provided evidence for the role of *geomagnetism* in bird orientation. Birds appear to have several orientation systems at their disposal; which system is used depends on prevailing conditions.

Homing experiments with small rodents illustrate their use of *piloting* as a means of navigation. Among mammals, extensive studies have been carried out on the *echolocation* system used for navigation in bats, which is supplemented by vision. The sun, landmarks, and geomagnetic forces are also used by some invertebrates for orientation. The life cycle of the Pacific salmon demonstrates homing behavior based on *chemical cues*, and studies of cricket frogs and sea turtles have suggested that the sun or stellar cues are used as a compass for orientation.

Discussion Questions

1. Much of the work on orientation and migration by birds has produced a number of new techniques and methods. Develop an experiment to investigate orientation in a nonavian animal and apply some of the techniques used with birds.

2. In addition to the reasons discussed in the chapter, what are the advantages and disadvantages of migration for organisms of different species?

3. What comparisons or generalizations can you make about the cues used for orientation and navigation by the animals discussed in this chapter?

4. Recent studies support the hypothesis that birds can use geomagnetic cues for orientation. Data from Baker (1980) suggest that the same may be true for humans. The data presented below are taken from a hypothetical study similar to a study conducted by Baker. Individuals were blindfolded, driven 10 kilometers from home, and released one by one (just like homing pigeons). The initial orientation bearing for these individuals at two different sites is recorded here (each dot represents the initial bearing for one individual). Fifteen people were tested at each site. What conclusions can you draw about geomagnetic cues and orientation by humans? What methods problems might you encounter in conducting such an experiment?

5. Recent information based on molecular techniques has helped answer questions about the homing behavior of green sea turtles. What other questions about homing behavior in animals could you test using these molecular techniques?

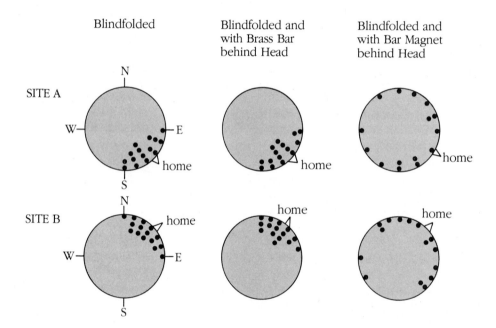

Suggested Readings

Baker, R. R. 1978. *The Evolutionary Ecology of Animal Migration.* New York: Holmes and Meier.
This huge tome covers all aspects of animal migration. Detailed examples and a large selection of references make this a useful resource, though some of the material and presentation remain controversial.

Emlen, S. T. 1975. Migration, orientation and navigation. In *Avian Biology,* vol. 5, eds. D. S. Farner and J. R. King. New York: Academic Press.
Summary with historical development of bird orientation theories. Solid presentation of experimental approaches and techniques by a respected authority in this field.

Gauthreaux, S. A. 1980. *Animal Migration, Orientation and Navigation.* New York: Academic Press.
An edited volume of review articles covering many aspects of these topics. Good mix of mechanistic and evolutionary perspectives.

Gould, J. L. 1980. The case for magnetic sensitivity in birds and bees (such as it is). *Amer. Sci.* 68:256–67.
Very readable. Touches on information relating magnetic cues to the orientation of animals.

Orr, R. T. 1970. *Animals in Migration.* London: Macmillan.
Engaging summary of migratory phenomena in all forms of animal life. Well illustrated and readable.

Schmidt-Koenig, K., and W. T. Keeton, eds. 1978. *Animal Migration, Navigation, and Homing.* New York: Springer-Verlag.
Edited by two major contributors in this research area. Experts treat all theories of orientation and navigation. Best source of references on this topic for the beginning student. Most papers are readable without extensive background.

Schone, H. 1984. *Spatial Orientation.* Princeton, NJ: Princeton Univ. Press.
A thorough summary of all aspects of orientation. Wide taxonomic coverage, good definitions of terms, sensory factors, and a large number of cues to which animals orient, including those we have covered in the chapter, but also such cues as water currents, heat etc.

16

HABITAT SELECTION

The earliest humans no doubt were aware that different kinds of plants and animals live in different habitats. The surface of our planet consists of a mosaic, or patchwork, of habitat types, and the distribution of organisms is neither uniform nor random. What causes these differences?

Plants are generally dependent on natural agents for dispersal, such as water or air currents or other organisms. The result is an essentially random dissemination of plant individuals; few ever reach environments conducive to survival and reproduction. As we saw in the previous chapter, animals have well-developed locomotive abilities at least during some point in the life cycle, and they play more active roles in finding places to live. **Habitat selection** can be defined as the choosing of a place in which to live. This definition does not imply that the choice is necessarily a conscious one, or that individuals make a critical evaluation of the entire constellation of factors confronting them. More often the choice is an automatic reaction to certain key aspects of the environment.

This chapter concerns the **distribution** of species (the presence or absence of a species in a particular habitat) and the **dispersal** of the individuals of a species within the habitat. We first consider how dispersal ability, other organisms, and physical and chemical factors restrict habitat use. Next we examine the choice of breeding sites and why so many animals disperse from their place of birth. We then look at the role of proximate factors in habitat selection—the environmental and social cues that influence the choice of a place to live. Finally, the roles of genetics, early experience, learning, and tradition in the development of habitat

FIGURE 16-1 Typical relationship between distribution in space and density for a species
Areas of optimal habitat support the highest densities and are surrounded by suboptimal habitats with lower densities. This distribution pattern contrasts with that shown in Figure 16–4.

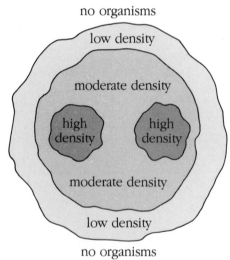

preferences are examined. The question of what determines the **abundance** of a species is considered in chapter 18.

When examining the distribution and abundance of most species, we often find areas of high population density near the geographic center of the species' range; abundance decreases outwardly (figure 16-1). The **fundamental niche** (the multidimensional space that a species occupies under ideal conditions with no

TABLE 16-1   Transplants of heathland ant colonies in England

Ant colonies of *Lasius niger*, which naturally occur in low, wet areas, were moved either to a similar habitat as a control or to higher, drier sites. Each transplanted colony was observed for five years.

Location of transplant	Years survived						Total number of colonies transplanted
	0	1	2	3	4	5+	
Control							
Lasius niger zone 11 m level	1		1			4	6
Dry heath							
13 m level			1		1	3	5
17–20 m level	6	3				1	10
25 m level	1	1	1				3

Source: Data from G. W. Elmes, "An Experimental Study on the Distribution of Heathland Ants," *Journal of Animal Ecology*, 40:495–499, 1971.

competition) is often much larger than the **realized niche** (the space occupied under real world conditions that involve competitors, predators, disease, and so forth).

PRESENCE OR ABSENCE OF SPECIES: FACTORS RESTRICTING HABITAT USE

If a species occupies an area and reproduces there, we know that all its needs are met and that it can compete with other species successfully. A useful way to identify the factors affecting the distribution of a species is to determine why it is absent from a place. We can examine possible reasons one by one, by using **transplant** experiments, by giving individuals a choice of artificial habitats in the laboratory, or by building enclosures in the field that encompass different habitats.

In a transplant experiment, organisms are moved to a new environment, and their survival and reproduction are monitored. Because organisms sometimes survive but fail to reproduce, a long-term project covering several generations is necessary. In a study of the distribution of heathland ants in England, Elmes (1971) dug up colonies and moved them to sites that differed in temperature and in moisture content. He moved eighteen colonies of one species (*Lasius niger*) that normally inhabits low, wet heathland to higher, drier sites, and moved six to other low, wet areas as controls. Elmes monitored their survival over a five-year period and found that the higher the colony was moved, the less likely it was to survive (table 16–1). Those at the 13 m level did as well as the controls at 11 m. One of the main factors that affected survival was competition with other

species that normally inhabit the higher, drier part of the heath. This experiment demonstrates the need for controls in research and the need for long-term monitoring of transplants.

If a transplant is successful, two possible factors may explain why a species is not found naturally in the transplant area: (1) the area is inaccessible because the dispersal ability of the organism is limited, or (2) the organism fails to recognize the area as a suitable habitat (referred to by Lack [1933] as the "psychological factor"). If a transplant fails, the causal factors may be the presence of other species (competitors, parasites, pathogens) or physical and chemical factors (temperature, pH, and so on). When trying to understand why a species is absent from an area, we can thus proceed through a series of steps (Krebs 1985), as shown in figure 16–2.

DISPERSAL ABILITY

Is a species absent from a place because it cannot get there? Many cases of successful introduction by humans have demonstrated that locomotive abilities adequately explain a species' absence. The European starling (*Sturnus vulgaris*) originally was found in most of Europe and Asia. After several unsuccessful introductions of small numbers of starlings into the United States, eighty pairs were released in New York City's Central Park in 1890 by Eugene Scheifflin of the Acclimatization Society. The "goal" of this group was to familiarize Americans with all the birds in Shakespeare's plays (Miller 1975). Within fifty years, starlings had reached the West Coast, and today they are probably the most numerous

FIGURE 16-2 Methodological approach for studying the geographical distribution of a species
A useful way to proceed in the analysis of a species' distribution is to determine why that species is *not* in a particular place. Four basic factors are listed; behavior may be involved in each and the factors may interact.
Source: Data from C. J. Krebs, *Ecology: The Experimental Analysis of Distribution and Abundance.* 3d edition. Copyright © 1985 Harper & Row Publishers, Inc., New York.

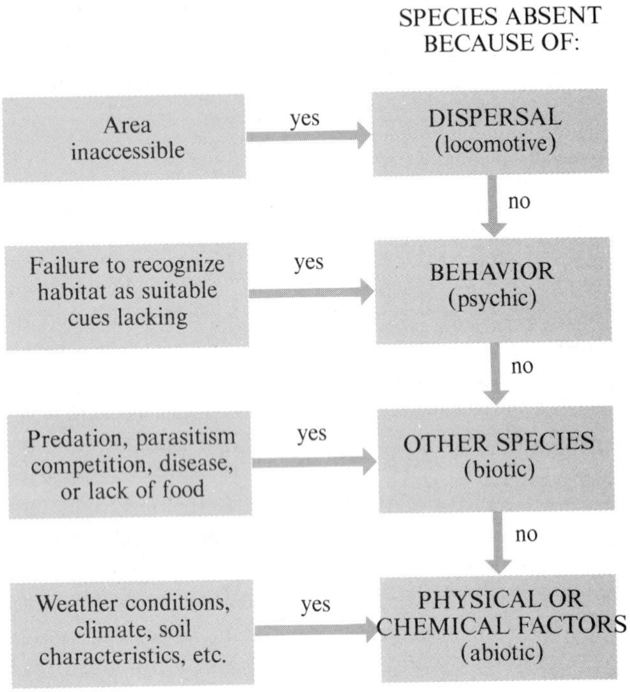

SPECIES ABSENT BECAUSE OF:
Area inaccessible
yes
DISPERSAL (locomotive)
no
Failure to recognize habitat as suitable cues lacking
yes
BEHAVIOR (psychic)
no
Predation, parasitism competition, disease, or lack of food
yes
OTHER SPECIES (biotic)
no
Weather conditions, climate, soil characteristics, etc.
yes
PHYSICAL OR CHEMICAL FACTORS (abiotic)

FIGURE 16-3 Equilibrium numbers of species on islands as a function of immigration and emigration rates
In (a) an island far from a colonizing source or mainland should equilibrate with fewer species, E_f, than an otherwise identical near island, E_n. In (b) a large island equilibrates with more species, E_l, than a small one, E_s, at the same distance from the mainland.

a.
immigration
extinction
near
far
Rates Increasing
E_f
E_n
Species Increasing
b.
immigration
extinction
large
small
small
large
Rates Increasing
E_s
E_l
Species Increasing

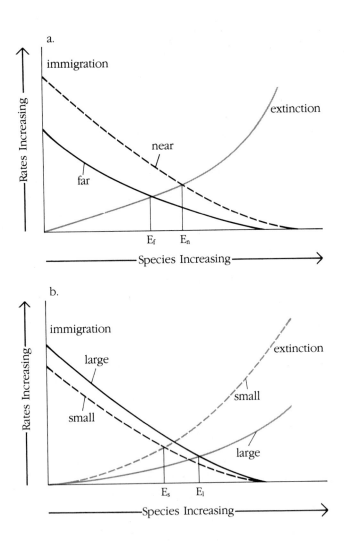

bird species in the country. Starlings nest in tree cavities and are more aggressive than most native species; they will even evict larger species such as flickers (*Colaptes auratus*) from nest holes. The eastern bluebird (*Sialia sialis*) is one native species that now occupies only a fraction of its former range partly because of its unsuccessful competition with the starling. Although they are insectivorous during the summer, starlings are generalists and switch to seeds in the winter; many native insectivorous species must migrate south to find insects. Other aspects of the starlings' habits have led to their success in modern industrialized countries: a tolerance of loud noises and air pollution; a willingness to roost and perch in various places, from trees to bridge supports; and a preference for feeding in grassy areas.

Distributions on islands offer a further test of dispersal powers. MacArthur and Wilson (1967) developed a model that predicts a dynamic equilibrium of the number of species on islands. Although the species change through time, the total number of species remains constant. The immigration rates of new species are affected by island size; smaller islands are smaller targets and therefore have lower rates. Also important is distance from the colonizing pool of species on the mainland; islands farther away have lower rates (figure 16-3). Thus small islands equilibrate at fewer species than large islands, and distant islands equilibrate at fewer species than near islands.

Ecologists sometimes characterize species as being *r*- or *K*- selected; *r* refers to the rate of population increase, and *K* refers to the number in the population at the upper limit, or carrying capacity, of the environment. Species that are *r*-selected have high reproductive rates, rapid development, and great powers of dispersal (MacArthur and Wilson 1967); they live in unstable environments in which recolonization of areas is necessary. Diamond (1974) studied the birds of New Guinea and nearby islands, and found that certain species were always the first to recolonize islands that had lost their fauna due to volcanic explosions or tidal waves; he referred to these *r*-selected species as "supertramps". *K*-selected species have low reproductive rates, slow development, and limited powers of dispersal; characteristically inhabitants of stable environments, these species may fail to colonize new areas separated by relatively small barriers. One island Diamond studied was separated from New Guinea by only 10 m of water, yet it had only half the New Guinea species expected on the basis of the availability of comparable habitats in the two areas.

BEHAVIOR

Sometimes behavior patterns keep species from occupying apparently suitable habitats. Two closely related species of British birds, the tree pipit (*Anthus trivialis*) and the meadow pipit (*A. pratensis*), are both ground nesters and eat similar types of food, but the tree pipit is absent from many treeless areas that the meadow pipit inhabits. The reason has to do with behavior associated with song: the tree pipit ends its aerial song on a perch, such as a tree or pole; the meadow pipit ends its aerial song on the ground. Thus the tree pipit is excluded from areas that it could otherwise occupy because of a specific behavior pattern (Lack 1933).

Female mosquitoes of the genus *Anopheles* (many species of which transmit such diseases as malaria) are very particular about where they lay their eggs. In the southern part of India, *Anopheles culifacies* eggs and larvae are found only in new rice fields, where the plants are less than twelve inches high; however, eggs transplanted to old fields have yielded normal numbers of larvae and adults, and other species of *Anopheles* lay their eggs in these mature fields. Russell and Rao (1942) demonstrated that the mechanical obstruction of the rice plants inhibited *A. culifacies* females from laying eggs. Glass rods or bamboo strips "planted" in the water had the same effect; shade was not a factor. *A. culifacies* females oviposit while they are on the wing, performing a hovering dance two to four inches above the water; possibly the obstructions interfered with this oviposition dance. However, in one experiment researchers partially submerged a box with no lid in the shallow water; although the box did not directly interfere with oviposition, few eggs were laid.

Why should a species not take advantage of a suitable habitat? One possibility is that the habitat is not actually suitable, perhaps because of competition, predation, or other factors the scientist may have failed to detect. Or such habitats may not have been suitable in the past: If organisms responding to certain environmental cues in previous optimal habitats left more offspring, their genetically influenced behaviors would become widespread and persist. New environments, although suitable, may not contain those cues and therefore are not utilized.

INTERACTIONS WITH OTHER ORGANISMS

Do other organisms keep a particular species out? Even if an individual can and "wants" to get to a place, other factors may prevent its becoming established. These factors could be predators, parasites, disease agents, allelopathic agents (poisons or antibiotics), or competitors. Demonstrating conclusively that one species prevents an area from being colonized by another is difficult, but that has been indicated with experimental and observational data. For instance, an elaborate series of experiments by Kitching and Ebling (1967) showed how mussels (*Mytilus edulis*) were kept out of protected bays along the coast of Ireland by three species of crabs and one species of starfish. Where the coast was unprotected, crabs were restricted by wave action, and small mussels could survive. In sheltered waters, mussels survived only in areas such as steep rock faces that the predators could not reach. Kitching and Ebling proposed that four criteria must be met before we can conclude that a predator restricts the habitat of its prey: (1) if they are protected from predators, prey will survive when transplanted to a site where they normally do not occur; (2) if the distributions of prey and predator are negatively correlated; (3) if the predator is observed eating the prey; and (4) if the predator can be shown to destroy prey in transplant experiments.

The more similar two species are, the more likely they are to compete intensely and thereby restrict each other's distribution, but once again positive proof is difficult to obtain. Most often we rely on presumptive evidence from comparisons of the closely related species'

FIGURE 16-4   Influence of other species on distribution of warblers

Abundance is measured as a percentage of all bird individuals observed. As one sample from the side of a mountain in New Guinea shows, *Crateroscelis murina* reaches maximum abundance at 5,400 feet and is abruptly replaced by *C. robusta*. This distribution can be compared with that in Figure 16-1.

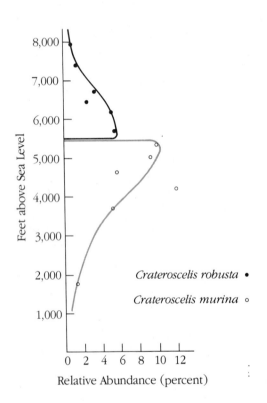

FIGURE 16-5   Altitudinal ranges of species of birds on New Guinea mountains and surrounding islands

In (a) three similar congeneric flower peckers, *Melanocharis nigra*, *M. longicauda*, and *M. versteri*, occupy nonoverlapping areas up to about 11,000 feet on Mt. Michael. On Mt. Karimui, which is smaller and more isolated than Mt. Michael, *M. longicauda* is absent. In (b) two congeneric lorikeets, *Charmosyna placentis* and *C. rubrigularis*, occupy an area on Mt. Talawe, but only *C. rubrigularis* colonized Karkar Island. Altitudinal ranges in all cases are nonoverlapping and are larger in the absence of other species.

Source: From J. M. Diamond, "Ecological Consequences of Island Colonization by Southwest Pacific Birds," in *Proceeding of the National Academy of Sciences USA*, 67:529–536. Copyright © 1970. Reprinted by permission of the author.

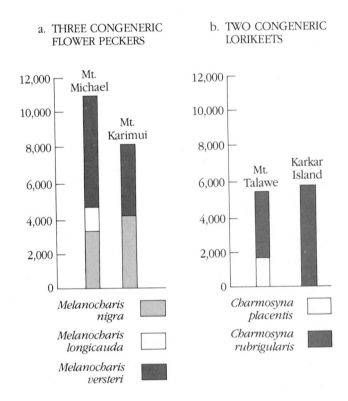

distributions in several habitats. Although a species is usually less numerous at the edge of its distribution than at the center (as shown in figure 16-1), it is sometimes abruptly replaced by a close relative, with both species at maximum density at the interface (figure 16-4). If interspecific competition restricts distribution, we would expect one species to extend its range in the absence of the other. This phenomenon was demonstrated by Diamond (1978) with bird distributions on mountain tops in New Guinea, where a second, closely related species

may be absent due to its inability to disperse (figure 16-5). In these cases, the single species occupied a much larger altitude range than it did on islands where both were present.

Competitive exclusion was actually witnessed by Orians and Collier (1963) when colonial tricolored blackbirds (*Agelaius tricolor*) moved into a marsh already occupied by red-winged blackbirds (*Agelaius phoeniceus*). After the invasion, the redwing territories were restricted to the periphery. Competition can also be studied experimentally by removing one species and noting changes in nearby competitors, or by introducing closely related species and monitoring the success of each, as Vaughan and Hansen (1964) did with two species of pocket gophers (*Thomomys bottae* and *Thomomys talpoides*). Slight differences between species in dispersal powers and environmental tolerances led to one or the other species' winning out.

PHYSICAL AND CHEMICAL FACTORS

If a transplant experiment fails, and no evidence exists that biotic factors have eliminated the species, some combination of physical and chemical factors may be involved. Each organism has a range of tolerances for these factors, and much of its behavior is directed toward staying within these limits. Temperature and moisture are the main factors that limit the distribution of life on earth, but physical factors (such as light, soil structure, and fire) and chemical factors (such as oxygen, soil nutrients, salts, and pH) may be important as well. Most research on the effects of physical and chemical factors on organisms is done by physiological ecologists and will not be discussed further here. The effect of these factors on reproduction was treated in chapters 7 and 13.

DISPERSAL FROM THE PLACE OF BIRTH

The discussion thus far has dealt with the problem of habitat selection at the species distribution level. Animals may also make decisions about whether to remain at (or return to) the natal site or to disperse to other breeding locations. In most species of birds and mammals, members of one sex tend to disperse, while members of the other sex are **philopatric,** breeding near the place where they were born. Among mammals, it is usually the males that disperse, while among most species of birds (perching birds of the order Passeriformes), the opposite is true (Greenwood 1980). The reason for this difference may be that most bird species are monogamous, and the male defends a territory that contains resources vital to him and his mate. It is probably easier for a male to establish such a territory and attract a mate in or near his natal site, where he is familiar with the location of resources and/or predators. On the other

hand, many species of mammals are polygynous. The females form the stable nucleus and the males attempt to maximize their access to them, frequently moving from one group to another (Greenwood 1980).

INBREEDING AVOIDANCE HYPOTHESIS

The ultimate cause of dispersal from the natal site has been argued by many to be inbreeding avoidance. The costs of inbreeding, referred to as **inbreeding depression,** have been documented in many laboratory and zoo populations, but have only recently been studied in natural populations. For example in African lions (*Panthera leo*) males from a small, inbred population showed lower testosterone levels and more abnormal sperm than did males from a large, outbred population (Wildt et al. 1987) (figure 16–6).

If one or the other sex disperses, there will be less chance of matings between related individuals. Among black-tailed prairie dogs (*Cynomys ludovicianus*), young males leave the family group before breeding; females remain. Also, adult males usually leave groups before their daughters mature (Hoogland 1982). Among primates such as vervet monkeys (*Cercopithecus aethiops*) and baboons (*Papio anubis*), males usually leave the natal group at sexual maturation or shortly after. They usually transfer to a neighboring group with age peers or brothers (figure 16–7). Several years later they may again transfer alone to a third group. Cheney and Seyfarth (1983) argue that this pattern of nonrandom movement followed by random movement minimizes the chances of mating with close kin. Packer (1979) reported that a male baboon, after failing to disperse at sexual maturity and after mating with relatives, sired offspring with low survival rates compared to the offspring of outbred males. Thus, in this case there seems to be a real cost associated with inbreeding.

INTRASPECIFIC COMPETITION HYPOTHESIS

A second cause of dispersal from the natal site may be competition with conspecifics (Dobson 1982; Moore and Ali 1984). In polygynous species such as most mammals, recall from chapter 13 that competition for mates is usually more intense among males than females. Males may be forced to disperse as they compete for

FIGURE 16-6   Abnormal sperm from inbred lion population in Africa
(a) Normal; (b) tightly coiled flagellum; (c) missing mitochondrial sheath; (d) abnormal acrosome and deranged midpiece; (e) macrocephalic with abnormal acrosome; (f) microcephalic with missing mitochondrial sheath; (g) bent flagellum; (h) bent neck with residual cytoplasmic droplet.
Source: Photo by Dave Wildt.

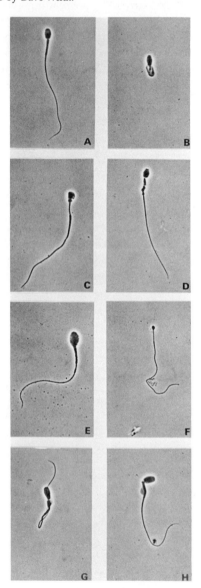

FIGURE 16-7   Migration by natal and young adult male vervet monkeys into and out of the three study groups between March 1977 and July 1982
Arrows indicate direction of movement; each line indicates one male. Letters indicate ranges of the study groups; numbers indicate ranges of regularly censused groups. Only groups with ranges adjacent to the study group are shown.

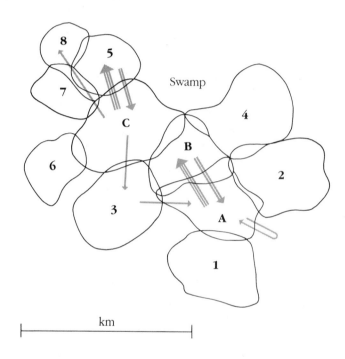

access to females. While the inbreeding avoidance hypothesis predicts that one sex should disperse, the competition hypothesis predicts that dispersal should be male-biased in polygynous species, which is certainly the case in mammals. A different type of competiton model was proposed by Hamilton and May (1977), in which animals disperse to avoid local resource competition with close relatives and thus to avoid lowering their indirect fitness. In a new habitat they are likely to be competing with nonrelatives and therefore would suffer no such cost.

FIGURE 16-8 The fate of subadult lions by 4 years of age at two African sites

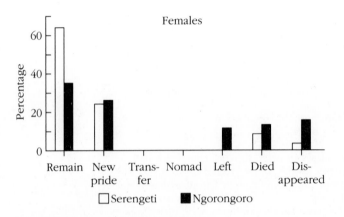

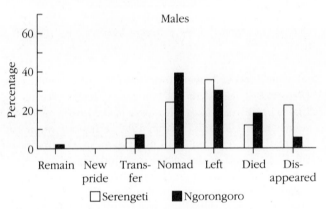

EXAMPLES OF DISPERSAL

LIONS. Dispersal patterns in lions follow the typical mammalian pattern: Females usually remain in or near their natal pride whereas males always leave, usually before four years of age, to become nomads and/or take over new prides (figure 16-8) (Pusey and Packer 1987). Competition with other males seems to be an important factor, because departures most often occur when a new coalition of males takes over the pride. However, some males appear to leave voluntarily, either in search of mating opportunities or to avoid breeding with kin.

Once a coalition takes over a new pride, it is usually ousted by another coalition within a few years, or it leaves to take over another pride. In all cases, males leave the pride before their daughters start mating. An additional factor is the fact that new coalitions of males kill the young cubs in the new pride (see chapter 12). Thus breeding males must remain in a new pride long enough to ensure the survival of their cubs. Pusey and Packer conclude that male-male competition, mate acquisition, protection of young cubs, and inbreeding avoidance all play roles in the evolution of dispersal patterns of lions.

BELDING'S GROUND SQUIRRELS. The question of why individuals disperse has been addressed at several different levels of analysis by Holecamp and Sherman (1989). Belding's ground squirrels (*Spermophilus beldingi*) follow the typical mammalian pattern: Females remain in the natal area for life, while males disperse. The proximate causes of male dispersal seem to be the effects of prenatal exposure to testosterone (organizational effects, chapter 7), and the attainment of a critical body weight. Effects of testosterone later in life (activational) seem less important, since castration of males just prior to natal dispersal did not prevent dispersal. Holecamp and Sherman were not able to test the inbreeding avoidance and competition hypotheses directly, but concluded that inbreeding avoidance was the more likely means by which dispersal increased fitness (table 16-2).

OPTIMAL INBREEDING

If inbreeding depression were the only factor involved, we might expect individuals to disperse as far as possible from relatives. However, such is not usually the case. According to Shields (1982), most species that have been adequately studied are philopatric, remaining close to the place of birth. He cites apparent cases of **outbreeding depression,** in which matings between members of different populations within a species yield less fit offspring. Members of a population may possess adaptations to local conditions that are lost with outbreeding—two areas might differ slightly in temperature, humidity, or in the types of food available. If each population is genetically adapted to these

TABLE 16-2　Why juvenile male Belding's ground squirrels disperse
Answers have been found at each of four levels of analysis.

Level of analysis	Summary of findings
Physiological mechanisms	Dispersal by juvenile males is apparently caused by organizational effects of male gonadal steroid hormones. As a result, juvenile males are more curious, less fearful, and more active than juvenile females.
Ontogenetic processes	Dispersal is triggered by attainment of a particular body mass (or amount of stored fat). Attainment of this mass or composition apparently also initiates a suite of locomotory and investigative behaviors among males.
Effects on fitness	Juvenile males probably disperse to reduce chances of nuclear family incest.
Evolutionary origins	Strong male biases in natal dispersal characterize all ground squirrel species, other ground-dwelling sciurid rodents, and mammals in general. The consistency and ubiquity of the behavior suggest that it has been selected for directly across mammalian lineages.

Source: From K. E. Holecamp and P. W. Sherman, "Why Male Ground Squirrels Disperse," in *American Scientist,* 77:232–239. Copyright © 1989. Reprinted by permission of *American Scientist,* journal of Sigma Xi, The Scientific Research Society.

conditions, they would be better off mating with individuals with those same adaptations. Perhaps that is why white-crowned sparrow (*Zonotrichia leucophrys*) females respond sexually to the songs of males from their natal area but not to males from other areas (Baker 1983); see also chapter 11.

A certain degree of inbreeding may be advantageous. Recall our discussion in chapter 13 about the costs of meiosis: In sexually reproducing organisms, the loss of genes in the offspring can be reduced by one-half if the parents are related. Furthermore, according to Shields (1982) gene complexes are less likely to be disrupted in matings between relatives. Finally, kin selection (chapter 4), which results in the evolution of cooperative behaviors, can operate only when relatives are in proximity. We predict more cooperative behavior within philopatric species and within the philopatric sex. Indeed, female Belding's ground squirrels are philopatric and engage in altruistic alarm calling, while males, who disperse away from relatives, do not alarm call (Sherman 1981); see also chapter 11.

From the preceding arguments we might predict some sort of optimal inbreeding strategy, in which matings between very close relatives (siblings, or parents and offspring) are avoided but matings with more distant relatives are favored. Bateson (1982) found that female Japanese quail (*Coturnix coturnix*) spend more time in proximity to first cousins than in proximity to siblings or more distant relatives. If these tendencies reflect later sexual preferences, an optimal level of inbreeding might be among first cousins.

Similar results were reported for white-footed mice (*Peromyscus leucopus*). In laboratory tests, estrous females preferred males who were first cousins over nonrelatives or siblings (Keane, 1990). Heavier pups and larger litters resulted from matings between first cousins than for other degrees of relationship. However, it is not yet clear how much inbreeding goes on in natural populations of this species; Wolff et al. (1988) used biochemical tests to determine maternity and paternity and concluded that little if any mating between close relatives seems to occur.

HABITAT CHOICE AND REPRODUCTIVE SUCCESS

Although it is assumed that an individual that makes a "correct" choice of habitat has higher reproductive success than one that makes an "incorrect" choice, few studies have actually measured this relationship. Witham (1980) studied the life history of the aphid *Phemphigus betae,* a plant parasite about 0.6 mm long that feeds on leaves of the cottonwood tree. In the spring, after hatching from eggs laid the previous fall in the bark of the tree, females (called *stem mothers*) move up the trunk and select a leaf on which to feed. This activity triggers the formation of a hollow gall on the leaf, within which the female produces offspring parthenogenetically.

Witham found that females settling on large leaves have higher reproductive success than females on small leaves. Not surprisingly, aphids select the largest leaves, leaving small ones vacant. However, latecomers may have to choose whether to take an already occupied large leaf or an unoccupied small one. If she takes an occupied leaf, she will have to settle further from the base, where there is less food. Stem mothers farther from the base were found to be smaller in size, and they produced fewer young than those closer to the base. Stem mothers may engage in shoving and kicking contests that last for days, with the largest aphid usually getting the basal position. Choice of habitat in these insects is nonrandom and results in higher average fitness than would random leaf selection.

Up to this point, we have assumed that habitat selection operates to increase the reproductive success of the individuals involved. Are there cases where members of one species manipulate the habitat selection of another species to their own benefit? It has long been known that parasites can modify the behavior of their intermediate hosts to increase the parasites' chances of being transmitted to their final hosts. The parasitic wasp *Aphidius nigripes* is an endoparasitoid of the potato aphid *Macrosiphum euphorbiae* (Brodeur and McNeil 1989). The wasp completes pupal development within the aphid, which becomes mummified as a result. Nonparasitized aphids are usually found on the undersurface of potato leaves, their favored feeding area, while aphids parasitized by nondiapausing larvae (those that will not overwinter as pupae) usually move to the upper surface shortly before being mummified (figure 16–9) (Brodeur and McNeil 1989). Aphids parasitized by diapausing larvae (those that will overwinter as pupae) tended to leave the host plant and mummify in concealed sites. Such pupae would thus be protected over winter and less likely to be eaten by predators or attacked by *hyperparasitoids* (i.e., parasites of the parasites!). Thus in this example, we see that an animal may select a habitat that is optimal not for itself, but for one of its parasites.

PROXIMATE FACTORS: ENVIRONMENTAL CUES

Organisms may respond directly to critical environmental factors. Speed or frequency of locomotion may be dependent on the intensity of stimulation. Isopods, such as *Porcellio*, are found in moist environments; when placed in a humidity gradient, they move faster in drier air and eventually wind up at the moist end of the gradient (Fraenkel and Gunn 1940). Likewise, the turning rates of protozoans are dependent on intensity of stimulation. (Types of orientation behavior involved in habitat selection were discussed in chapter 15.)

FIGURE 16-9 **The effects of a parasitic wasp on the behavior of the potato aphid**
(a) The effect of parasitism by wasp larvae on the distribution of aphid adults. Squares denote parasitized aphids, circles denote nonparasitized aphids. (b) The distribution of aphid mummies containing diapausing and nondiapausing individuals of the parasitoid wasp.

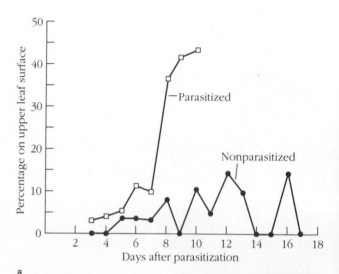

a

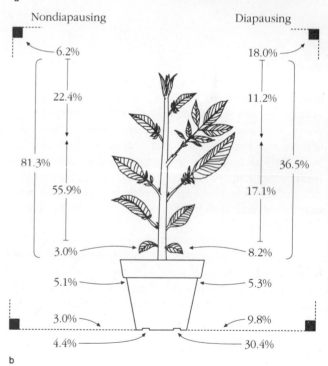

b

FIGURE 16-10   Laboratory environments for testing habitat selection in fish

(a) In all four test environments small, flat rocks were present bearing a film of algae as a food source. Variables were water level (deep or shallow) and covers under which the fish could hide (present or absent). Exploration time, as shown in (b), was least in the environment with shallow water and cover present. This habitat is the one most similar to the preferred habitat in the field.

a. TEST ENVIRONMENTS

b. EXPLORATION TIME IN TEST ENVIRONMENTS

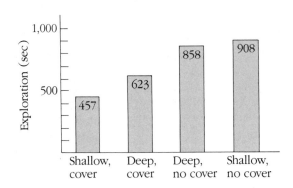

Organisms may integrate more than one environmental variable; for instance, temperature and humidity. Several species of fruit flies (*Drosophila*) prefer warm temperatures in a laboratory gradient apparatus (Prince and Parsons 1977), but only if the humidity is high. At low humidity they move to the cooler area, thereby reducing water loss and increasing the probability of survival. This adaptive pattern is a response not only to external cues but also to the individual's physiological state.

Sale (1970) hypothesized the existence of a simple mechanism of habitat selection in fish that is based on levels of exploratory behavior. Sense organs monitor specific stimuli in the environment and send a summation of pertinent stimuli back to central nervous system centers, which regulate the amount of exploration. As the constellation of cues approaches some optimum level, exploratory behavior ceases and the animal stays where it is.

An alternative hypothesis is that an animal has a cognitive map of the ideal habitat and that its behavior is goal directed. Working with a species of surgeonfish, the Hawaiian manini (*Acanthurus triostegus*), Sale (1970) tested juveniles in laboratory tanks with various water depths and bottom covers (figure 16–10). Exploration time was least in the tank with shallow water and bottom cover and highest in the tank with shallow water and no bottom cover. In choice tests and field observations, most fish preferred shallow areas with bottom cover. Thus there is no need to suggest the inheritance of complex cognitive maps and goal-directed behaviors; rather, the animal simply moves more in an unsuitable habitat and less in a suitable one.

Sale's model still does not explain how the animal "knows" what is suitable and what is not, or how stimuli from multiple cues are integrated. Nor does it explain the role of photoperiod in the response of dark-eyed juncos (*Junco hyemalis*) to photographs of their natural habitat. Wild-caught birds were presented a choice of viewing one of two 35 mm color slides showing different habitats. Birds kept in the lab under a winter photoperiod of 9L:15D (9 hours of light and 15 hours of darkness) preferred (spent more time in front of) slides of their southern winter habitat (pine and hardwood forest). After daylength was increased to 15L:9D, the birds' viewing preferences shifted to the northern summer habitat (grassland and conifer forest) (Roberts and Weigl 1984).

Social cues may also affect choice of habitat. Large sized juncos (usually males) dominate smaller individuals (usually females and juveniles) in wintering flocks. Ketterson (1979) explained the finding that females usually migrate farther south than males by hypothesizing that subordinate birds are forced to migrate farther to avoid competing with dominants. In their lab study, Roberts and Weigl (1984) found that during the short days (simulating winter) small subordinate juncos showed the strongest preference for winter scenes. Social cues may also be important among colonially nesting birds. Laughing gulls (*Larus atricilla*) were observed nesting on only one of two similar islands. When a storm destroyed the nesting island, they moved over to the other one. Thus a "suitable" habitat cannot be defined by its structural features alone; social features also play a role (Klopfer and Hailman 1965).

The proximate cues to which animals respond when selecting a habitat may not be the same as the ultimate factors that have brought about the evolution of the response (Hildén 1965). The ultimate reason a mouse chooses to live in a wooded area rather than a field might be an inability to survive the higher temperatures in the field. But rather than responding directly to temperature, it might cue on the geometric shapes of trees.

The blue tit (*Parus caeruleus*), a European relative of the chickadee, lives in oak woodlands where most of its preferred food is found (Partridge 1978). But the blue tit establishes its breeding territory each year before leaves and caterpillars (the staple food) have even appeared, so it must be using other features, such as shape of the trees, as cues to the habitat. Although birds have been studied intensively, we know little about the signals they respond to when choosing their habitat. Possible cues are stimuli from landscape; sites for nesting, singing, or feeding; food itself; or other animals. In migratory species, it is not even clear when in the life cycle a choice is made. Breeding sites may be selected in late summer or fall before migration, rather than in the spring, as is usually assumed (Brewer and Harrison 1975).

Wood frogs (*Rana sylvatica*) lay their eggs during a brief period each spring in temporary ponds that dry up in the summer. All the egg masses are deposited in one place in the pond. The physical features of this location seem less important than does the presence of an egg mass, which triggers other females to lay their eggs there (Howard 1980). By placing an egg mass in one part of the pond before any other eggs had been laid, Howard was able to induce all the females to lay their eggs there. Eggs in the center of such a mass may be protected from predators and from temperature fluctuations (Berven 1981).

Previous breeding success affects the habitat choice of some bird species. Mountain bluebirds (*Sialia currucoides*) were presented with two types of nest boxes. When breeding for the first time, most birds selected the same type of nest as the one in which they had been raised, even though they were breeding in a new territory. If they had an unsuccessful breeding attempt in one year, the next year they shifted nest sites *and nest box type* (Herlugson 1981).

In trying to develop models of habitat selection, it has become clear that rarely if ever is a single factor sufficient to explain the choice. Habitat choice may involve a stepwise sequence of decisions that is arranged hierarchically (Klopfer and Ganzhorn 1985). Thus a bird migrating to its nesting grounds may cue on gross features of the vegetation to get to the general vicinity, and shift to other more specific features, such as presence of certain species of plants, to find the exact nest site.

DETERMINANTS OF HABITAT PREFERENCE

HEREDITY

How can we sort out the roles that genetics, imprinting and learning play in habitat choice? If two animals reared from birth in identical environments are found to differ in habitat preference when they are tested as adults, we can conclude that those differences must be due to hereditary factors. When coal tits (*Parus ater*) and blue tits were reared in aviaries with no vegetation and then presented with a choice between oak and pine branches, coal tits preferred pine and blue tits preferred oak (figure 16–11). The differences correspond to the

FIGURE 16-11   Time tits spent in oak and pine habitats in the laboratory
Both wild and hand-reared blue tits preferred the oak habitat to the pine habitat. Wild and hand-reared coal tits, however, preferred the pine habitat. These preferences reflect those of each species in the wild.

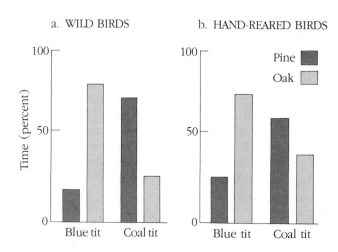

TABLE 16-3   Vegetation preference of chipping sparrows
One side of the test chamber contained oak branches; the other, pine. Findings indicate that the preference for pine could be modified to some extent by early experience.

Chipping sparrows	Percentage of time spent in pine	Percentage of time spent in oak
Wild-caught adults	71	29
Laboratory-reared, no foliage exposure	67	33
Laboratory-reared, oak foliage exposure only	46	54

Source: Data from P. H. Klopfer, "Behavioral Aspects of Habitat Selection: The Role of Early Experience," *Wilson Bulletin* 75:15–24, 1963.

distribution of these birds in nature and to the response of wild titmice in aviaries (Partridge 1974, 1978). It is perhaps not surprising that blue tits feed more efficiently in oak than do coal tits (Partridge 1976).

LEARNING AND EARLY EXPERIENCE

Some experimenters have modified the environments of young birds to test whether their genetic predisposition to respond to certain stimuli can be altered by early experience, or **habitat imprinting.** For example, when Klopfer (1963) placed wild-caught chipping sparrows (*Spizella passerina*) in a room containing both pine and oak branches, they preferred the pine, as they usually do in nature (table 16–3). Klopfer reared a group of eight nestlings in a covered cage that prevented their seeing any foliage. When tested at two months of age, these birds, like their wild-caught counterparts, preferred pine to oak. Klopfer reared a third group of ten birds in the presence of oak leaves to see if their preference could be shifted. When tested at two months, they were ambivalent, spending about equal time in each habitat. This result suggests that their inherited predisposition for pine could be affected to some extent by early experience.

After these tests, the oak-reared birds were removed to a large aviary located in a pine-hardwood forest. Within the aviary itself, however, were only some small broadleaf shrubs. When Klopfer retested the birds, their preference for pine had increased. After eight months in an aviary containing pine, the birds' preference for pine was even stronger. These results suggest that the birds retain the ability to shift to the naturally preferred vegetation type.

One of the classic studies on habitat selection in mammals is Wecker's on deermice (*Peromyscus maniculatus*) (Wecker 1963). This species, one of the more common North American rodents, is divided into many geographically variable subspecies of two general types: the long-eared, long-tailed forest form and the smaller,

short-eared, short-tailed grassland form. In the laboratory, the grassland form (*Peromyscus maniculatus bairdii*) does well in forest conditions, where its food preference and temperature tolerance are similar to those of the forest subspecies; thus experimenters assumed that the avoidance of forests in the grassland deermouse is a behavioral response (Harris 1952).

Wecker's objective was to assess the genetic basis of this behavior and to test the idea that habitat imprinting (Thorpe 1945) is important. He constructed an enclosure halfway in a forest and halfway in a grassland, released the mice in the middle, and recorded their locations (figure 16–12). The animals he tested were of the grassland subspecies (*bairdii*) and were of three basic types: (1) wild-caught in grassland; (2) offspring, reared in laboratory, of wild-caught; (3) reared in laboratory for twenty generations. Both wild-caught mice and their offspring selected the grassland half of the enclosure, regardless of previous experience. Laboratory stock and their offspring showed no preference, whether or not they had been raised in forest conditions. However, laboratory stock reared in a grassland enclosure until after weaning showed a strong preference for the grassland when tested later (table 16–4). Wecker reached the following conclusions:

1. The choice of grassland environment by grassland deermice is predetermined genetically.
2. Early grassland experience can reinforce this innate preference, but is not a necessary prerequisite for subsequent habitat selection.
3. Early experience in forest or laboratory is not sufficient to reverse the affinity of this subspecies for the grassland habitat.
4. Confinement of these deermice in the laboratory for twelve to twenty generations results in a reduction of the hereditary control over the habitat selection response.
5. Laboratory stock retain the capacity to imprint on early grassland experience but not on forest.

FIGURE 16-12

Experimental enclosure for testing habitat preference of prairie deermice is 100 feet long and 16 feet wide. Five of its 10 compartments are in a field (*top figure*) and five are in an oak-hickory woodlot. For testing, each mouse is placed in the introduction box at right of top figure. It can go from there into either the field half or the woods half of the enclosure. Each partition has a runway at one end that enables the mouse to go from one compartment to the next.

Two of the seven runways with recording treadles are labeled. Nest boxes, both of which have treadles, are in the last compartments of each figure. The instruments that make permanent records of movements of each mouse are in box at top, just to the right.

Source: From "Habitat Selection," by Stanley C. Wecker. Copyright © 1964 by Scientific American, Inc. All rights reserved. Reprinted by permission.

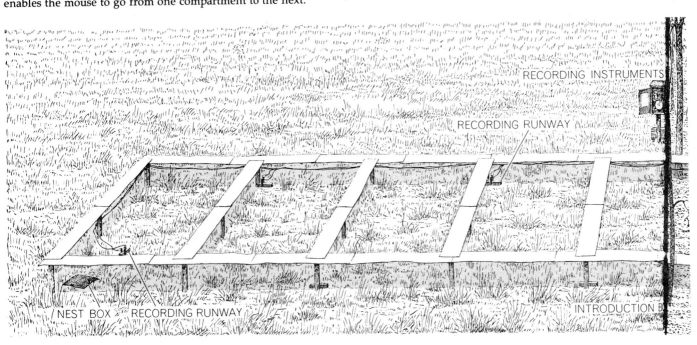

TABLE 16-4   Habitat selection by deermice (*Peromyscus maniculatus bairdii*) as a function of hereditary background and early experience

The outdoor test enclosure was forest on one side and grassland on the other. Preference was measured by the percentage of time, amount of activity, and depth of penetration by mice in each side of the enclosure.

Number of mice tested	Hereditary background	Early experience	Habitat preference
12	Grassland	Grassland	Grassland
13	Laboratory	Grassland	Grassland
12	Grassland	Laboratory	Grassland
7	Grassland	Forest	Grassland
13	Laboratory	Laboratory	None
9	Laboratory	Forest	None

Source: Data from S. C. Wecker, "The Role of Early Experience in Habitat Selection by the Prairie Deer Mouse, *Peromyscus maniculatus bairdii*," *Ecological Monographs* 33: 307–325. Copyright © by Ecological Society of America, Tempe, AZ.

Wecker also suggested that learned responses, such as habitat imprinting, are the original basis for the restriction of this subspecies to grassland environments; genetic control of this preference is secondary.

In a series of laboratory experiments designed to explore the importance of early experience on bedding preference in inbred mice (*Mus domesticus*), Anderson (1973) raised animals either on cedar shavings or on a commercial cellulose material. When he tested them later, he found that the mice preferred the bedding on which they had been raised, although females raised on cellulose drifted toward cedar shavings in subsequent tests. Naive mice preferred cedar shavings (figure 16–13).

Habitat imprinting occurs in migratory vertebrates, which commonly tend to return to the vicinity of their birth to breed. In the case of **iteroparous** organisms (those that have their young at intervals) such as birds, the adults of most species return to the same area to nest each year. **Semelparous** breeders (those that have their young all at once) such as salmon, breed only once. After feeding in the open ocean for several years, salmon return to the same upstream spawning bed where they were hatched. The salmon's olfactory system is programmed to respond to unique odors of the home stream during a critical period in the first few weeks of life (Hasler 1966).

TRADITION

Inherited tendencies and imprinting may be involved in restricting habitat choice to a small part of the potential range, but **tradition**—behavior passed from one generation to the next through the process of learning—may also be an important factor. For instance, many species of waterfowl have staging areas, where they rest and feed during migration from breeding to wintering grounds. The same areas tend to be used year after year.

Mountain sheep (*Ovis canadensis*) live in unisexual groups; females are likely to stay in the natal group but may switch to another female group when they are between one and two years of age (Geist 1971). Young rams desert the natal group after the second year of life and join all-ram bands. Mothers do not tend to chase their young away at weaning, as most other mammals do. Females follow an older, lamb-leading female, males follow the largest-horned ram in the band. When rams mature, they are followed by younger rams and pass their habitat preferences on to them.

Until the last century, mountain sheep occupied a much larger range in North America and Asia than they do today. Measures enacted to protect sheep, mainly hunting regulations, have done little to increase their numbers. A formerly inhabited part of the range that appears intact is not colonized, and transplants to suitable areas are often unsuccessful. In contrast, deer (*Odocoileus* spp.) and moose (*Alces alces*) have recolonized areas rapidly and have reached population densities higher than ever (Geist 1971).

Why have sheep failed to extend their range, while moose and deer have done so? Geist has pointed out that deer and moose, which are relatively solitary beasts, establish ranges by individual exploration after being driven out of the mother's range; sheep, in contrast, transmit home-range knowledge from generation to generation and often associate with group members for life.

FIGURE 16-13   Sleeping site selection of mice as a function of prior bedding experience and test day
Mice were born and raised on either cellulose or cedar bedding and were then given a choice of bedding types at either 30 or 60 days of age. Mice generally preferred the bedding type on which they had been raised. There was, however, a "drift" toward cedar preference across test days in females. Older females raised on cellulose initially showed no preference but by test day 3 they chose cedar.
Source: Data from L. T. Anderson, "An Analysis of Habitat Preferences in Mice as a Function of Prior Experience," *Behaviour,* 47:302–339, 1973.

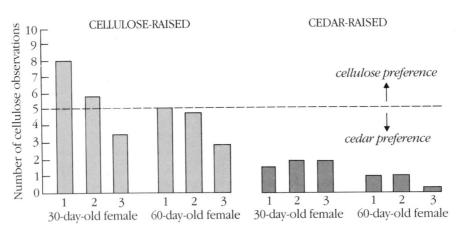

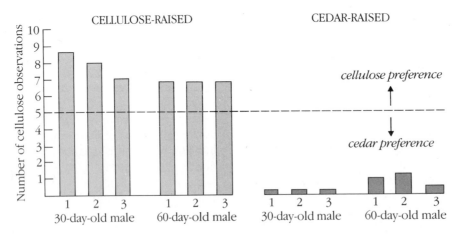

Understanding the niches of these species can help explain the sheep system, which may seem a rather poor adaptation to the environment. The moose is associated over much of its range with short-lived plant communities that follow in the wake of forest fires. Moose in these environments are *r*-selected "pioneer species" and disappear as the climax coniferous forest regenerates. Moose habitats are subject to rapid expansion after fires,

and moose must continually colonize new habitats. Each spring when her new calf is born, the cow drives away her yearling, which may wander some distance before establishing its new range.

Mountain sheep habitats are formed by stable, long-lasting climax grass communities, which exist in small patches. Geist argued that, given the distance between patches and the ease with which wolves can pick off sheep, the best strategy for the sheep is to stay on familiar ground. Sheep also have at least two and as many as seven seasonal home ranges that may be separated by twenty miles or more. These areas are visited regularly by the same sheep year after year at the same time; knowledge of the location of these ranges and the best times to visit them is transmitted from one generation to the next. Because new habitats rarely become available, there is no advantage to an individual's dispersing and attempting to colonize other areas.

THEORY OF HABITAT SELECTION

Theoretical approaches to the problem of habitat selection are in a rather early stage of development, and there is currently no single general theory (Rosenzweig 1985). At least two different approaches have been used that build on models that are developed more fully in chapters 12 and 17. If we think of habitats as patches, we can apply optimal foraging theory that enables us to predict when an animal should leave one patch and move to another in order to get the greatest benefit for the least cost. These economic models incorporate such factors as the availability of resources in various patches and the costs of getting from one patch to another. Although the resource is usually assumed to be food, nest sites or mates are other possibilities. This approach is considered more fully in the next chapter.

A second approach is the ideal free distribution, which makes predictions about the fitness of individuals as a function of population density, discussed in some detail in chapter 12. Individuals settle in habitats so that the first arrivals get the best resources. As density increases, less desirable areas are occupied and animals spread themselves out so that all have the same fitness in the absence of intraspecific competition. One obvious result of such a distribution is that rich habitats will have more individuals than will poor ones. If intraspecific competition occurs—dominance or territory—an ideal despotic distribution develops, with some individuals monopolizing the best resources (Fretwell 1972).

To conclude this chapter, we should point out that habitat selection may have an impact above the level of the individual, affecting population structure and even the formation of new species (chapter 4). Although the usual models of speciation involve the formation of geographic barriers to gene flow, reproductive isolation could evolve if members of two parts of a population came to prefer different microhabitats in the same region. If the preferences were heritable and individuals mated assortatively, a barrier to gene flow would be created that could eventually lead to the formation of new species (Rice 1987). However, such a mechanism has yet to be demonstrated.

SUMMARY

Habitat selection refers to the choice of a place to live. Factors affecting this choice can be best understood by a stepwise approach to why a species is absent from a particular place. The *transplant experiment,* in which organisms are relocated outside their natural ranges and their survival and reproduction are monitored, is an important tool. First, if the transplant is a success, we can assume that the organism's dispersal powers could not overcome *geographical barriers.* The success of introductions of foreign species by humans and data on colonization of islands suggest that this inability to disperse often explains a species' absence. Second, in a more complex case, organisms may fail to colonize an otherwise suitable area because of *behavioral responses* to specific features of the habitat. Third, the presence of such other factors as competitors, parasites, predators, or diseases may also explain a species' absence. Evidence for exclusion due to *competition* comes in large part from range expansion by one species in the absence of

another. Experimental demonstration of exclusion by *predators* requires the species to survive in the absence of the predator. Fourth, *physical and chemical factors* that are beyond the range of tolerance of organisms can restrict a species' distribution. *Temperature and moisture* are very important and may interact with other factors.

Behavior patterns may restrict organisms to a fraction of the habitat that they seem to us to be equipped to occupy. However, apparently suitable areas may in fact not be so in the long run. Behavior patterns that restrict a species' distribution may have evolved because of negative factors associated with other habitats in the evolutionary past.

Within a species' range individuals may disperse or may remain in the natal area to breed. In mammals, males typically disperse, while females are *philopatric.* In birds the opposite is generally true. By dispersing, the chances of matings between close relatives are reduced, minimizing inbreeding. Animals may also disperse to reduce competition. Although there are demonstrated costs to extreme inbreeding, outbreeding may also incur genetic costs. Several recent studies suggest that in some species there may be an optimal degree of inbreeding where first cousins are favored as mates.

A few studies have measured the impact of habitat selection on reproductive success. Among insects such as aphids, information about habitat quality and the density of competitors is used to make choices that will maximize reproductive success. However, parasites may alter the habitat selection of their hosts to maximize their own success rather than that of their host.

The proximate cues animals use in habitat selection include direct locomotive responses to such environmental variables as temperature and humidity. Little is known about the cues used by vertebrates; fish may use olfaction, and birds may respond to vegetation type. The cues may be only indirectly related to the ultimate factors that determine survival; for example, birds select breeding habitats before the leaves and staple insect foods have emerged.

Several experiments with birds and mammals have been conducted to determine the roles of genes and experience in habitat selection. Selection of the "correct" habitat is under some degree of genetic control, as has been demonstrated by studies in which animals have been reared in isolation and later tested in various habitats. However, early experience can modify later choices.

Tradition, the transmission of knowledge of habitats from one generation to the next, is thought to be important in mountain sheep. These animals live in close social groups in stable habitats, but seasonally occupy two or more ranges. Other ungulates—for example, moose—often live in unstable habitats and gain knowledge of suitable places to live through individual exploration.

Theoretical approaches to habitat selection include *optimal foraging models*, in which individuals choose *patches* (habitats) and stay in them in order to maximize the gain of some resource. The *ideal free distribution* assumes that, in the absence of competition, individuals settle in the best habitats first, with later arrivals moving into suboptimal sites.

Discussion Questions

1. Suppose that you wanted to understand why species X is present in a particular habitat and species Y, closely related to X, is absent. Outline the steps you would follow to explain the distribution.

2. Review the life history of the organism of your choice. Try to explain the dispersal patterns of each sex in light of our discussion of the costs and benefits of inbreeding and outbreeding. Include ecological as well as genetic factors in your answer.

3. Although tradition is claimed to play a role in the habitat preference of some animals, firm data are lacking. Design an experiment to demonstrate the role of tradition in habitat choice.

4. The figure to the right shows the distribution of two species of chipmunk, *Eutamias dorsalis* and *Eutamias umbrinus*, in the Great Basin area of Nevada. What conclusions about animal distribution can we draw from these data?

Source: Data from E. R. Hall, *Mammals of Nevada.* Copyright © 1948 University of California Press, Berkeley, CA.

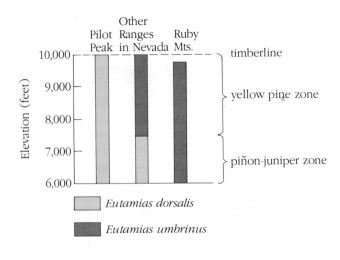

Suggested Readings

Chepko-Sade, B. D., and Z. T. Halpin, eds. 1987. *Mammalian dispersal patterns: The Effects of Social Structure on Population Genetics.* Chicago: University of Chicago Press.
The first chapter by Shields is a concise review of dispersal and mating systems. Other chapters deal with specific mammals, including deer, horses, bears, wolves, primates, and rodents.

Cody, M. L., ed. 1985. *Habitat Selection in Birds.* New York: Academic Press, Inc.
An edited volume with chapters on different species, different habitat types, and theory. Chapter 15 by Klopfer and Ganzhorn deals with behavioral aspects.

Krebs, C. J. 1985. *Ecology: The Experimental Analysis of Distribution and Abundance,* 3d ed. New York: Harper & Row.
Part 2 of this basic text treats the problem of species distribution. Chapter 5 specifically treats behavioral processes in habitat selection.

Partridge, L. 1978. Habitat selection. In *Behavioural Ecology: An Evolutionary Approach,* ed. J. R. Krebs and N. B. Davies. Oxford, England: Blackwell Scientific Publications.
One of the few attempts to treat the entire topic of habitat selection. Emphasis is on birds.

Shields, W. M. 1982. *Philopatry, Inbreeding and the Evolution of Sex.* Albany: State University of New York Press.
A provocative essay arguing that most species of plants and animals that have low fecundity are philopatric, that philopatry promotes inbreeding, and that the benefits of all but extreme inbreeding outweigh the costs.

17

FEEDING RELATIONSHIPS

The food supply—as the energy source—usually sets the upper limit to the population of a species in an area and is thus the resource for which individual organisms most often compete. In this chapter we first consider optimal foraging strategies. We then look at food-catching techniques and the relationship between social organization and food resources; and finally, conclude with a discussion of antipredator techniques. In some cases, morphological as well as behavioral adaptations are considered, since they are closely linked.

OPTIMALITY THEORY

Before we actually begin our study of foraging and foraging theory, we need to understand the basic approach that has been used by behavioral ecologists to develop models and hypotheses. Our starting point is Darwin's theory that natural selection acts as a maximizing process; individuals that deviate from the optimal design (the best possible design under the circumstances) are penalized in terms of fitness. **Optimality models** attempt to predict the particular combination of costs and benefits that will ultimately maximize the individual's inclusive fitness. More specifically, optimality models help us understand (1) the **decisions** an animal makes; e.g., Should it eat this prey type? Should it stay in this place to feed? (2) the **currency,** or what it is that is being maximized; e.g., the rate of energy intake; and (3) the **constraints,** or limits on the animal; e.g., how big a food item can it handle

(Stephens and Krebs 1986). Models will have values for, or will make assumptions about, these three components. Those values and assumptions that correctly predict an animal's behavior give us a better understanding of the way natural selection has acted to produce a behavior pattern.

Of course animals are not perfectly adapted to their environments; many forces—random mutation, rapid environmental change coupled with evolutionary lag, and so on—tend to keep organisms from ever reaching adaptive peaks (Barash 1982). However, the goal of optimality models is not to test whether or not natural selection is a maximizing process; that is generally assumed to be true. Rather, the goal is to understand the factors that cause the behavior patterns observed in nature.

Although foraging behavior has been the subject of much of the optimality research, many other behavior patterns can be analyzed in a similar manner; for example mate acquisition (chapters 13 and 14) and habitat selection (chapter 16). Two types of foraging models have been developed, called **prey models** and **patch models.** The former deals with the types of prey an individual should try to catch and eat, while the latter deals mainly with how long to stay in a food-containing patch.

FORAGING STRATEGIES

Foraging behavior research can be boiled down to several basic problems: (1) What items should an animal include in its diet? (2) If food occurs in clumps, or **patches,** what path should an animal follow within a patch to encounter the most food? (3) How long should it stay in a patch? (4) Which patch should it visit next? These are some of the (not necessarily conscious) decisions an animal must make. Whether we speak of birds as "preying" on seeds, or of lions preying on wildebeests, the problems encountered are similar. Behavioral ecologists usually assume that the **currency** (what is being maximized) is the net rate of energy intake, which is intake per unit of time. Energy **costs** are incurred in obtaining food—mainly that of the search, pursuit, handling, and eating. These costs must be subtracted from the **benefit**—the energy in the food. A rate is obtained by dividing the net energy gain by the time it takes to do all of the above:

$$
\text{net rate of energy intake} = \frac{\begin{array}{l}\text{energy from food} - \text{search energy} \\ - \text{pursuit energy} \\ - \text{handling energy} \\ - \text{eating energy}\end{array}}{\begin{array}{l}\text{search} + \text{pursuit} + \text{handling} + \text{eating} \\ \text{time} \quad\; \text{time} \quad\;\; \text{time} \quad\;\; \text{time}\end{array}}
$$

CHOICE OF FOOD ITEMS: THE OPTIMAL DIET

The barn owl (*Tyto alba*) is a highly efficient nocturnal predator of small mammals. In southwestern New Jersey, these owls roost in tree cavities or silos and forage in fields over a radius of several kilometers. Colvin (1984) found that although more than 90 percent of the available small mammals were white-footed mice (*Peromyscus leucopus*) and house mice (*Mus domesticus*), and less than 5 percent were meadow voles (*Microtus pennsylvanicus*), 70 percent of the mammals eaten were meadow voles (figure 17–1). Clearly, these owls are not simply taking prey in proportion to its abundance. What factors influence the choice of prey types?

Prey models were developed to deal with decisions an animal makes about whether to eat a prey item once encountered or to search for additional prey. Without getting into too much math, a simplified version works as follows: Once an animal has located a food item (j), it should pursue that item only if during the expected pursuit time it could not expect to locate and to catch a better food item (MacArthur 1972). If P_j is the time

needed to pursue and to capture an item just located, and P and S are the average pursuit and search times of the previous diet, item j should be added if $P_j < P + S$.

Suppose that an animal such as a bird eats small stationary insects on plant leaves. The bird must expend a lot of energy to find these prey, but they are easy to catch and eat; S is large compared to P. If the bird encounters a new insect type on a leaf, its pursuit time probably will be small, too; P_j is likely to be less than S and P, and the new item will be included. Such animals, including many species of insects, are likely to be **generalists.**

A different creature, such as a lion, may often have prey in sight, so that S is low. However, the lion has a difficult time catching anything, so P is high. The pursuit time of most new prey types the lion might sight would be greater than P and S, that is, $P_j > P + S$, and so the lion would not attempt to chase the new prey. Such "pursuers" are thus likely to be **specialists,** rarely adding newly encountered species to their diet.

A more thorough development of this model reveals several testable predictions: (1) the Zero-One Rule: prey types are either always taken on encounter or never taken; (2) prey types are ranked in terms of profitability, and added to the diet according to their ranks; (3) the inclusion of a prey type depends on its profitability, but not on its own encounter rate (Stephens and Krebs 1986).

The relationship between food choice and pursuit and search times can be applied to the productivity of the environment. In an unproductive environment where food is scarce, S is large; therefore most new items will be included, and the animal will be a generalist. In more productive habitats, S will be smaller, and the same species will be more specialized. Thus great blue herons (*Ardea herodias*) eat a much wider range of foods in the unproductive lakes of the Adirondack Mountains in New York where they breed, than they do in the more productive waters of Florida where they winter (MacArthur 1972).

Food preference has a heritable component, as Dix (1968) has demonstrated in garter snakes (*Thamnophis sirtalis*). He compared isolation-reared offspring of garter snakes collected in Florida and Massachusetts. All the snakes accepted worms and frogs, but only the Florida stock took fish readily. Florida garter snakes are more aquatic in their habits than are northern forms and frequently feed on concentrated pools of fish during the dry season in the Everglades. On the other hand, experience also affects later food preference, as discussed in chapter 9.

FIGURE 17-1   **The percentage of small mammals eaten by barn owls (a), compared to the percentage available (b); based on trapping**
In this study from southwestern New Jersey, barn owls are meadow vole specialists.

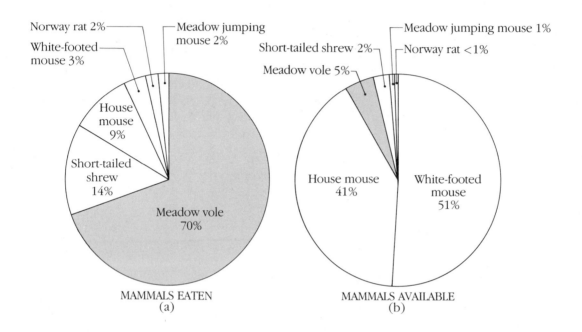

There probably is a limit to the number of different foods an animal can efficiently search for at one time. Birds sometimes overlook palatable items to zero in on one or two prey types at a time. Such foragers have formed a **search image** and will continue to hunt those prey even when they are less common than other palatable foods. This phenomenon could be due to the forager's mental limitations: of many acceptable foods, it can remember only a few during any particular search. For example, if you were asked to search a telephone book for the occurrences of twenty different first names, you might pick five and search the book once, then repeat the search with five more, and so on, rather than trying to remember all twenty names.

Just because an animal overlooks potential prey and seems to choose one prey type more frequently than its relative density would lead one to expect, it does not necessarily mean that the animal has formed a search image. Thus, hypothesizing a search image in the barn owl would be premature. We must first rule out simpler explanations, such as differences in palatability of prey types, ease of capture, or fear of a novel stimulus. Pietrewicz and Kamil (1979, 1981) tested whether blue jays (*Cyanocitta cristata*), after previous encounters with cryptic moths (*Catocala* spp.), improve their ability to detect that species and thus show a specific search image (figure 17–2). The jays viewed slides of naturally occurring prey; the pictures were of either one of the two cryptic moth species or of the substrate with no moth. When the experimenters projected a picture of a moth, the jays could get a food reward by pecking a key ten times. When no moth was present, pecking the food key not only resulted in no reward, but in a delay of the next slide as well; the correct response was to peck the advance key that started the next slide. In arranging the slides, the experimenters used either "runs," with only one species of moth shown, or "nonruns," with both species of moth shown. In both runs and nonruns, slides with no moths present were mixed in to test the jays' ability to detect the presence or absence of the cryptic moth.

The result was that in runs with only one species, the jays improved their ability to detect the moth. In nonruns when both species were mixed in, the jays did not improve. The jays thus showed an increased ability to detect prey species to which they had been recently exposed. However, the search image in this case seems to be restricted to one species of prey at a time.

FIGURE 17-2 Moth on tree trunk
Blue jays improve in their ability to detect these cryptic prey.
Source: Photo by Steve Pollick.

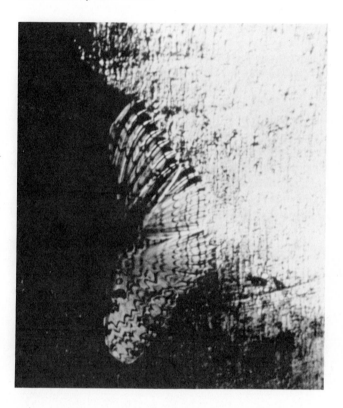

MOVEMENT RULES WITHIN A PATCH

How can we tell whether or not animals maximize the rate of energy intake? Pyke (1979b) studied a relatively simple system in which bumblebees fed on nectar from monkshood flowers and transported the food to the colony for storage (figure 17–3). After they arrived at a plant, the bees usually started foraging on the lower flowers of an inflorescence (a cluster of flowers). What rule did they follow to determine their movements to subsequent flowers? After recording thousands of actual movements in the field, Pyke came up with the following rule: The bee moves to the nearest flower not most recently visited, unless the last movement on the inflorescence was downward and was not the first movement. In the latter case, the bee moves to the closest flower not most recently visited. The bee is most likely to depart from an inflorescence after reaching the top flower, or less often, the bottom flower. Departures from other than top flowers are usually after a downward movement.

FIGURE 17-3 Bumblebee foraging on sedum flowers
Behavioral ecologists have studied these central place foragers in order to understand currencies, decisions, and evolutionary constraints in tests of optimality models.
Source: Photo by Jerome Wexler/Photo Researchers, Inc.

Pyke's next step was to try to figure out what the bees should be doing to maximize the net rate of energy intake and then to compare observed and expected behavior. He found that flowers at the bottom of the inflorescence usually had more nectar than did those at the top, so the observed tendency for bees to start on a low flower and work their way up was logical. However, a tradeoff was necessary because the cost of flying from the top of one plant to the bottom of the next was greater than the cost of going to the middle of the next plant; indeed, some bees did start in the middle. In developing expected movement rules, Pyke had to calculate (1) the time spent at each flower and the time spent flying between flowers, inflorescences, and plants; (2) the energy gain, as determined from caloric analysis of the nectar; (3) the energy cost of foraging; and finally (4) the net rate of energy intake that each possible movement rule would provide. Of a number of possible movement "rules" considered, two provided equally high rates of energy intake, and one is similar to the observed rule: Always choose the closest flower not previously visited.

Pyke's approach to the analysis of bumblebee foraging was criticized by Heinrich (1983), a long-time student of bumblebee physiology and behavior. Heinrich argued that the use of optimal foraging theory tells us little about the proximate mechanisms underlying the behavior, and may lead us in the wrong direction. In trying to explain why bees usually start at the bottom of an inflorescence and work up, he pointed out that the flowers bees feed on usually hang down, and can be most easily entered from below. When artificial inflorescences were constructed in which the top flowers had more nectar than the bottom flowers, bees continued to visit the bottom flowers first.

Other constraints may keep an animal from foraging optimally. Predators or competitors may be in the area, and may force the forager to spend time and energy defending itself in addition to the time and energy related to the optimal foraging formula. Nor can we always assume that the net rate of energy intake is the correct currency. For instance, the golden-winged sunbird (*Nectarinia reichenowi*), an African nectar-feeder that resembles a hummingbird, defends a feeding territory even in the nonbreeding season. Pyke (1979a), using data collected by Gill and Wolf (1975), tested the notion that these birds were maximizing the rate of energy intake. The birds spent far less time foraging and defending and more time just sitting than was predicted. Instead of maximizing energy intake, what they seemed

to be doing was minimizing costs, eating and defending just enough to maintain themselves. However, these data were obtained during the nonbreeding season. Presumably, foraging rates and territory defense would increase to meet the demands of reproduction in the breeding season.

HOW LONG TO STAY IN A PATCH

Leaving the problem of choice of food items, let us assume that an animal has been feeding in a patch and is gradually depleting its food supply. How long should it remain there and where should it go next? If it is maximizing its net energy gain, it should leave when its expected net gain from staying declines to its expected net gain from traveling to and foraging in a new patch (Charnov 1976). This patch model, referred to as the **marginal value theorem,** can best be explained graphically, where energy gain is plotted against time spent in patch (figure 17–4a). The travel time to the patch and the gain curve are fixed by the environment. Note that as the travel time increases, the optimum leaving time also increases (figure 17–4b). This rather simple model

---

FIGURE 17-4   Predicting when to leave a patch: The marginal value theorem

(a) As a forager stays in a patch and depletes the food resource, the rate of gain decreases. (b) Lines AB and CD are tangents to the gain curve, used to determine the optimal patch resident time (T_{opt}). For a long travel time (Line AB) the optimal patch time is greater than for a short travel time (Line CD).

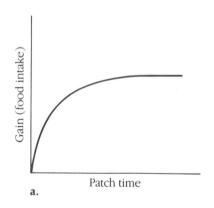

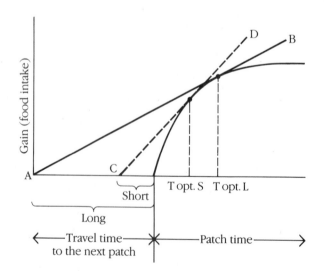

makes a number of assumptions, including that the forager "knows" the travel time between patches and can compare the current rate of intake with that of the recent past. Despite this simplicity and the assumptions, studies of a variety of insects, birds, and mammals tend to support the model (McNair 1982).

One test of the marginal value theorem was done in an aviary using great tits (*Parus major*) (Cowie 1977). Each "patch" consisted of a plastic cup filled with sawdust in which six pieces of worm were hidden. Six cups were on each of five artificial trees. Cardboard lids were placed on the cups in two ways: hard-to-remove lids fit tightly inside the cups and had to be pried off by the birds, whereas easy-to-remove lids simply lay on top of the cup and could be flipped off. Birds were tested in two situations: either all *hard* or all *easy*. Hard to remove lids corresponded to long travel times between patches, easy lids to short times. As can be seen from figure 17-5, time spent in patch increased as travel time between

patches increased, and the fit to the predicted curve was quite close. More recently, analogous models have been developed that use the rate of capture to predict the "giving up time" for foragers while actually in a patch, such as the bluegill sunfish (*Lepomis macrochirus*) that feed on midge larvae (DeVries et al. 1989).

The marginal value theorem has also been tested in **central place foragers,** animals that can carry more than one item at a time back to a central location for storage or for feeding to offspring. The problem here is not only when to leave the patch, but how many items to collect before returning to home base. Giraldeau and Kramer (1982) conducted a field experiment with chipmunks, which carry seeds back to a burrow for storage and later consumption. They placed "patches" of sunflower seeds in trays at different distances from burrows and recorded patch time, travel time, and the weight of seeds collected (load size). They found that the rate of seed collection declined as the cheek pouches filled. Also, as predicted by the model, load size and patch time increased as the travel distance increased. However, since the actual values were not as predicted, these researchers feel that more data are needed before the marginal value theorem can be accepted as a complete model.

The question of which patch to choose next has received less attention than the optimal residence time described above. Animals faced with two patches of unknown prey density might be expected to sample both, switching to the one having higher prey density (Shettleworth 1984). Without checking the other patch, they might end up in the poorer one. In a laboratory study that again used great tits, the birds were given a choice of two feeders, one at each end of an aviary. Each feeder required the bird to hop on a perch a certain number of times to get a mealworm. Initially, the birds fed at both feeders, but they gradually shifted to the one with the better reward ratio (Krebs et al. 1978).

RISK-SENSITIVE FORAGING

So far we have argued that an animal should choose the patch with the highest *average* density of prey, but we can't really expect an animal to know what an average is; its information comes from the rate at which it encounters prey. Suppose that two patches each have 100 prey per square meter, but in one the prey are uniformly distributed and in the other they are clumped. In the former, the animal will always encounter prey at the same rate, while in the latter it may encounter prey at a rate of $20/m^2$ one time, and $200/m^2$ another.

FIGURE 17-5 Test of the marginal value theorem in great tits
Time in patch as a function of travel time between patches. Line is predicted curve based on the theory, adjusted for energy expenditure.

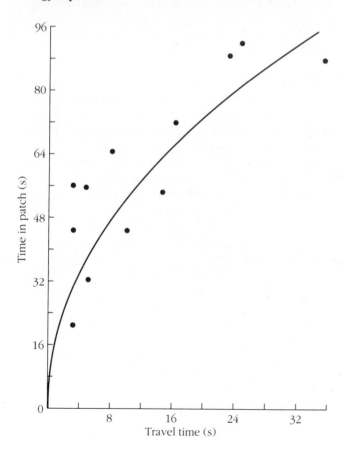

FIGURE 17-6　The standard design of risk sensitivity experiments
The forager is offered a choice between (a) a certain alternative and (b) a probabilistically determined alternative with the same mean.

Standard risk sensitivity experiment

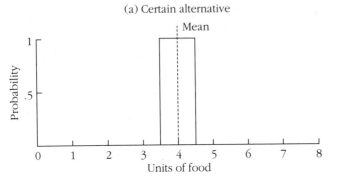

(a) Certain alternative

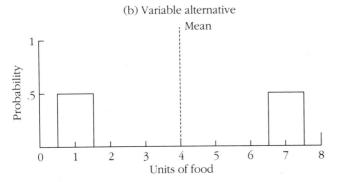

(b) Variable alternative

The word **risk** is used to refer to this variation in encounter rate. (Note that *risk* used in this sense does not mean the same thing as danger from predators!) If animals discriminate between these patches, we can say that they are **risk-sensitive** (Stephens and Krebs 1986). The prey and patch models we have discussed so far assume **risk-indifference** on the part of the predator; all that matters is maximization of energy gain.

So are real animals risk-sensitive or risk-indifferent? In most studies to date, they are definitely risk-sensitive (Stephens and Krebs 1986) (figure 17-6). In one lab study with juncos (*Junco phaenotus*), birds were given a choice between two trays of seeds, one a *certain* tray, the other a *risky* tray (Caraco et al. 1980). The *certain* tray always contained a fixed number of seeds, the mean of the seeds provided in the *risky* tray presentations. If certain trays always contained 5 seeds, risky trays might contain, for example, 0 to 10 seeds with a mean across all presentations of 5. Caraco found that when birds were getting enough food to sustain body weight they were **risk-averse,** preferring the certain tray; but when they were losing weight and facing the risk of starvation, they were **risk-prone,** preferring the risky tray. These findings have been replicated in several species of birds and mammals (Stephens and Krebs 1986).

The web-building spider *Achaearanea tepidariorum* is normally solitary and cannibalistic; when food is abundant, however, webs are aggregated. It turns out that grouped webs are more efficient at capturing prey and the capture rates are less variable than for solitary spiders. Solitary spiders in a low food area are thus risk-prone; although they don't do as well on average, occasionally they do much better than grouped spiders (Uetz 1988). Further modifications of this and other models are being tested to further our understanding of the decisions, currencies, and constraints that are important in the evolution of foraging behavior.

FEEDING TECHNIQUES

At this point we leave our rather abstract discussion of theory and deal with more practical matters, namely, the actual techniques that animals use to get their food.

TROPHIC LEVELS

As a consequence of the second law of thermodynamics, low trophic levels have more energy available, and, therefore, there is some advantage to feeding as low on the food chain as possible. One way is to feed directly on the producers, photosynthetic plants. Another way is to feed on dead organic material, as do many invertebrates, such as clams and earthworms. These animals, called **detritus feeders,** often filter food out of water or soil; they are usually sessile or slow moving. A majority of animals gather food through filter feeding and grazing, and thereby take advantage of the large supply of potential energy at low trophic levels.

Large size is often associated with grazing and filter feeding. Gorillas (*Gorilla gorilla*) and elephants (*Loxodonta africana*) feed on vegetable matter; and the largest of all animals, the blue whale (*Balaenoptera musculus*), uses its baleen mouth parts as a strainer to filter seawater for small planktonic organisms such as krill. Animals feeding at higher trophic levels are generally smaller and more active than herbivores.

FIGURE 17-7   Feeding positions within trees of Cape May warbler and yellow-rumped warbler
Shaded areas denote zones of most concentrated activity. The Cape May warbler feeds primarily in the outermost top branches of evergreen trees, the yellow-rumped warbler feeds on interior and bottom branches.

Source: Data from R. H. MacArthur, "Population Ecology of Some Warblers of Northeastern Coniferous Forests," *Ecology*, 39:599–619, 1958.

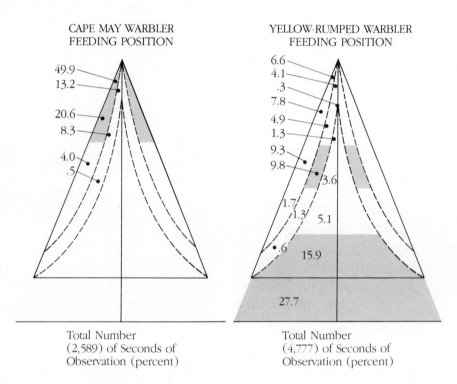

CAPE MAY WARBLER FEEDING POSITION

Total Number (2,589) of Seconds of Observation (percent)

YELLOW-RUMPED WARBLER FEEDING POSITION

Total Number (4,777) of Seconds of Observation (percent)

RESOURCE PARTITIONING

Many species divide up resources and are able to coexist only through differences in feeding behavior. MacArthur (1958) studied five species of warblers in Maine and Vermont. Each of these species of warbler—small, insect-eating birds that breed in the boreal (mainly spruce) forest—feeds primarily in a certain part of a tree; thus the Cape May warbler (*Dendroica tigrina*) feeds near the top and outermost branches, while the myrtle, or yellow-rumped, warbler (*Dendroica coronata*) feeds in all parts of the lower branches (figure 17–7). The movements of the birds differ also. The Cape May moves mostly in a vertical plane while the yellow-rumped moves mostly in a tangential plane (figure 17–8).

The feeding action of the warblers can be classified as long flights, hawking, or hovering. Thus, although the warblers are morphologically similar and can eat similar kinds of food, by feeding in different places and

in different ways, each is exposed to different kinds of food. Analysis of stomach contents has confirmed that different species of warblers do in fact eat different kinds of arthropods (MacArthur 1958).

MODIFYING FOOD SUPPLY

Some herbivores modify the food supply so as to increase it. For example, the activities of browsers and grazers in the grasslands stimulate the growth of some species of grass and prevent succession of grassland communities to other community types such as forests. Some species of intertidal limpets increase their food supply with the mucus trail they secrete during locomotion (Connor and Quinn 1984). Two species of solitary limpets (*Lottia gigantea* and *Collistella scabra*), routinely return home to a scar on a rock, depositing a mucus trail that acts as an adhesive trap for algae. This

FIGURE 17-8   Components of flight motion of Cape May and yellow-rumped warblers while feeding on insects in trees

The lengths of the lines are proportional to the total distance each species moves in radial, tangential, and vertical directions. Vertical movements are up and down parallel to the trunk of the tree; tangential movements are circular at the same elevation; and radial movements are in and out along branches, perpendicular to the trunk. Cape May move mainly vertically while yellow-rumped move mainly tangentially. Such flight differences, and the tendency to feed in different parts of the tree, expose the birds to different species of insects.

Source: Data from R. H. MacArthur, "Population Ecology of Some Warblers of Northeastern Coniferous Forests," *Ecology* 39:599–619, 1958.

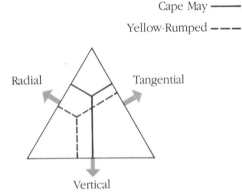

FIGURE 17-9   Fungus garden of ant

Ants of the species *Acromyrmex octospinosus* cultivate fungus as a food source. This garden, measuring approximately 31 by 11 centimeters, was found under a log in Trinidad.

Source: Photo by Neal A. Weber.

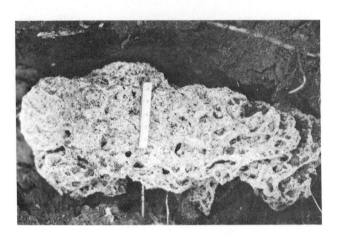

FIGURE 17-10   Leaf-cutter ants en route to underground garden in the Amazonian rain forest

Source: Photo by Stephen H. Vessey.

substance also stimulates algal growth along the trail, and the limpets feed on this algae as they retrace their path home.

The ultimate in nonhuman manipulation of producers in an ecosystem may be the "agriculture" practiced by certain fungus-growing ants widespread in the New World tropics (Weber 1966). These ants cultivate yeast or fungi on organic material that they gather and carry into their nests. The organic material ranges from caterpillar feces used by ants of the genus *Cyphomyrmex*, to the fresh leaves cut by the leaf-cutter ants of the genus *Acromyrmex*, and is placed in nest cavities as much as six meters below ground (figure 17–9).

Leaf-cutter ants form long queues as they bring cut leaf sections to the nest (figure 17–10), cut the sections into smaller pieces, and work the edges with their mandibles to make them wet and pulpy. They may deposit an anal droplet on the leaf before inserting the leaf into the garden. They place tufts of the fungal mycelium at

intervals on the substrate, and the hyphae grow out in all directions from the tuft (Weber 1966). They engage in cleaning operations to help keep foreign fungi out of the garden, but there is evidence that they also produce bacteriostatic and fungistatic factors. By converting the indigestible cellulose of the leaves into digestible sugars, the fungus makes the vast energy supply contained in the forest leaves available to the ants. Other ant species raise aphids or other plant-feeding insects in or near their nests, protect them, and feed on their "honeydew" excretions (Wilson 1971).

FIGURE 17-11 Lake-living caddisfly net

Larval insects of the genus *Polycentropus* manufacture nets with which to catch planktonic plants and animals.

Source: Drawing by Dr. Alice L. Anderson. Reprinted by permission.

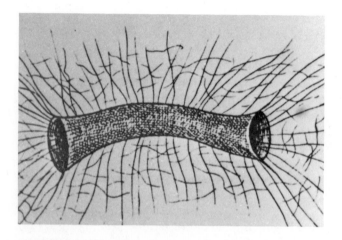

TRAPS

Some invertebrates, such as ant lion (genus *Myrmeleon*) larvae, use entrapment to get their food. They make pits in the sand, bury themselves just below the pit, and knock ants into the pit with grains of sand that they hurl at them with tosses of their head. Other traps include the webs of spiders and the underwater nets of caddis fly larvae (order Trichoptera), as seen in figure 17–11.

ELECTROMAGNETIC FIELDS

Unusual channels of communication may be used to capture food. For example, Kalmijn (1971) demonstrated that sharks can detect the weak electric fields of fish even when the latter are buried in sand (figure 17–12). Electric fish (families Gymnotidae, Mormyridae, and Gymnarchidae) generate their own electric fields (see figure 6–2) and can identify the presence of potential prey (Lissmann 1958).

AGGRESSIVE MIMICRY

A taxonomically widespread strategy for capturing prey is the use of lures, referred to as **aggressive mimicry.** In chapter 13 we saw an example of sexual enticement by fireflies, in which the female mimics the mating signal of other species and eats the male that responds.

FIGURE 17-12 Feeding responses of sharks to objects buried in sand

Solid arrows denote responses of the shark; dashed arrows denote the flow of seawater through the agar chamber. The shark responds to the magnetic field of intact fish in (a) and (b). Pieces of fish in (c) produce no magnetic field, and the shark responds to odor carried outside the agar chamber by the current. Metallic film in (d) blocks the magnetic field of the intact fish, and the shark fails to detect the fish. In (e) the shark responds to an artificially produced magnetic field, ignoring a piece of fish.

Source: Data from A. J. Kalmijn, "The Electric Sense of Sharks and Rays," *Journal of Experimental Biology*, 55:371–383, 1971.

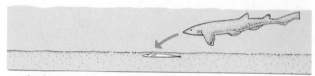

a. Shark detects fish under sand.

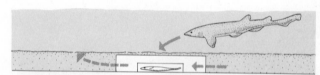

b. Shark detects fish in agar chamber. Chamber doesn't block electric field.

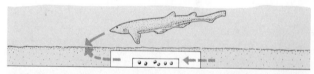

c. Shark is unable to detect pieces of fish in agar chamber.

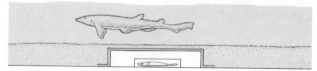

d. Fish in agar chamber is covered with metallic film.

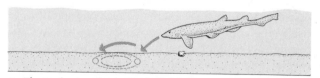

e. Electrodes produce dipole field.

FIGURE 17-13   Aggressive mimicry by fish

Angler fish (*Antennarius* spp.) and its lure are shown. In the bottom photo, a two-second time exposure shows the pattern of movement of the luring apparatus. Note the lure resembles a small fish.

Source: Photos by David B. Grobecker.

Fish of the order Lophiiformes have a modified first dorsal fin spine on the tip of the snout. At the end of the modified spine may be a fleshy appendage, a tuft of filaments, or in deep-sea forms, an organ containing light-emitting bacteria. The bait often resembles worms or crustacea, and the rest of the fish resembles an inert object, such as an algae-encrusted rock or a sponge. The fish wriggles the bait and keeps the rest of its body still. See figure 17-13 for a case in which the lure is a nearly exact replica of a small fish (Pietsch and Grobecker 1978).

Siphonophores of the phylum Coelenterata capture prey using tentacles armed with stinging cells called *nematocysts*. Some species have tentacles with branches and clustered nematocysts that look like zooplankton such as copepods or fish larvae. When the siphonophore moves its tentacles, it lures zooplankton predators into the web of nematocysts, and the predators become food for the siphonophore instead (Purcell 1980).

TOOLS

Although the use of tools was once considered an exclusively human trait, it has evolved independently in several different species of animals. In most cases the tool is an unmodified inanimate object. The sea otter (*Enhydra lutris*), for example, holds a rock on its chest and cracks shellfish, such as mussels, against it. Similarly, the Egyptian vulture (*Neophron percnopterus*) picks up rocks and drops them on ostrich eggs. The chimpanzee (*Pan troglodytes*), however, sometimes makes a modified tool by stripping leaves from a twig, which it then inserts into an ant or termite nest (figure 17-14); the insects cling to the stick, and the chimp eats those that hang on after it removes the stick. The woodpecker finch (*Cactospiza pallida*) of the Galápagos Islands uses sticks in a similar way to extract larvae from dead wood, and may modify the stick by shortening it (reviewed in Beck 1980; Lawick-Goodall 1970).

Tool use seems to have little relationship to central nervous system complexity. Rather, it enables organisms that have not evolved specialized appendages to exploit a new resource. Morphologically specialized mammals, such as the anteaters (of the order Edentata) and pangolins (of the order Pholidota), have long snouts, sticky vermiform tongues, and large claws for digging up ant nests. These features restrict their ability to eat other types of food. During the adaptive radiation of finches on the Galápagos Islands, where competition from woodpecker specialists was absent, the woodpecker finch evolved a flexible tool-using habit instead of evolving more efficient, but more restricted, morphological specializations.

Millions of years ago in Africa, tool use was important in enabling humans to compete with specialized carnivores and scavengers. Humans could retain the flexibility of the generalist while enjoying the greater efficiency of the specialist. Tools also permitted the cultivation of plants more than ten thousand years ago and were the basis of the Industrial Revolution of two hundred years ago. Each of these tool-using developments increased our energy-gathering capability and resulted in a substantial increase in population (Deevey 1960).

FIGURE 17-14 Chimpanzee using a tool to obtain ants
Chimps select sticks, modify them, and insert them into termite or ant mounds to extract the clinging insects. The infant observes the process from its mother's abdomen.
Source: Photo by James Moore from Anthro-Photo.

FEEDING AND SOCIAL BEHAVIOR

DEFENDING A TERRITORY

One way to increase net energy gain is by defending a food source against potential competitors. (For a more detailed discussion of territoriality see chapter 12.) Since an animal must spend energy advertising its presence and chasing out intruders, defending a territory will be economical only under certain circumstances.

Pied wagtails (*Motacilla alba*) are insectivorous birds; some of them defend winter feeding territories along river banks, feeding on insects that are washed up on shore. The others feed in flocks in nearby pools. Territory holders, usually males, follow a circuit up one bank and down the other, a pattern that maximizes food intake as new insects are washed ashore (Davies and Houston 1983, 1984); intruding wagtails are chased away. When food in the territories was very scarce, the owners fed elsewhere in flocks, but kept returning to the territory to evict intruders. When food was abundant, owners often shared the territory with a satellite (usually a juvenile or a female) that walked about one-half a circuit behind the owner (figure 17–15). This meant that the food available to the owner was reduced by one-half, but the owner sometimes gained because the satellite helped chase away intruders. If food declined, the owner chased the satellite away. If food became extremely abundant, owners made no effort to defend their territories. Sometimes territory-holders fed on the territory even when they could have done better in the flock. Thus they were not maximizing energy intake in the short run. This system differs from that of the nectar-feeders such as the golden-winged sunbirds (Gill and Wolf 1975) whose territory sizes vary as a function of food availability. The wagtails keep the same size area but vary their response to intruders as food supply fluctuates. Davies and Houston suggest that territory maintenance is a long-term investment to protect a more reliable food source than the pools where the flocks feed.

FIGURE 17-15   Pied wagtails feeding along a river bank
(a) Pied wagtails exploit their territories systematically. The circuit of the river bank takes, on average, 40 minutes to complete. (b) When a territory is shared between two birds, each walks, on average, half a circuit behind the other and so crops only 20 minutes worth of food renewal.

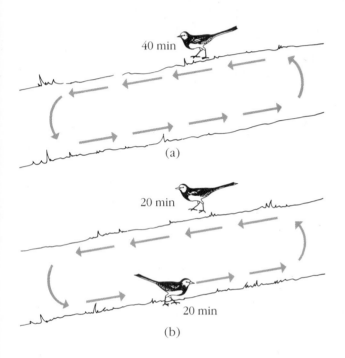

FIGURE 17-16   Communal spider web
Social spiders (*Anelosimus eximius*) from southeast Peru capturing a deerfly.
Source: Photo by Karen R. Cangialosi.

GROUP FEEDING

When food is spread widely and irregularly over the environment and cannot be defended, individuals of a species may group into flocks or herds and may associate with other species. Cody (1974) argued that by feeding in flocks, animals can recognize unexploited feeding areas more quickly. In addition, the flock can time its visits to an area so that food supplies have replenished sufficiently to make a return trip worthwhile. Assuming that the flock has exclusive use of the area, the flock could act as a "return-time regulator."

Within a foraging area, the flock's movements maximize coverage. When encountering the edge of a food patch in a feeding area, a flock is expected to move ahead half to three-quarters of the time and have a strong right- or left-turn bias, according to computer simulations. Observations of finch flocks foraging in the Mojave desert have supported these predictions (Cody 1974).

Animals feeding in groups may forage more efficiently because each individual spends less time scanning the environment for predators. For example, downy woodpeckers (*Picoides pubescens*) wintering in mixed-species flocks scanned less and fed more than did solitary individuals (Sullivan 1984).

Group living also increases the efficiency of food capture in some species of spiders. Members of a few genera build dense communal webs, which may be more than several meters across, as shown in figure 17-16. Together the spiders attack trapped prey items, drag them back to a central retreat, and feed on them communally (Buskirk 1981). Two advantages of communal webs are (1) local enhancement, where the reaction time to detect prey is reduced; and (2) the "ricochet effect," where insect prey are more likely to be caught as they are deflected from one web to another in the aggregate (Uetz 1988).

Perhaps the most spectacular food-catching enterprise is the march of an army ant colony (figure 17-17) (Franks 1989). T. C. Schneirla spent most of his life trying to understand the complex life cycle of several New World species. Each night the colony, which consists of a queen, workers, larvae, pupae, and eggs, forms a bivouac (Schneirla and Piel 1948). Workers, upwards

FIGURE 17-17 Emigrating army ants

Army ants (*Eciton burchelli*) moving along woody vine carrying white larvae. Along both sides of the column there are rows of "guard workers" including a few soldiers with large white heads. These protect the emigration column from being disturbed by many small animals.

Source: Photo by Carl Rettenmeyer/Connecticut Museum of Natural History.

of 500,000 strong, form a protective net by hooking their legs and bodies together. Each morning the bivouac dissolves and the ants begin moving outward. A column emerges along the path of least resistance and heads away from the bivouac site. There are no true leaders; workers in the lead turn back into the swarm behind them every few centimeters. The ants lay down pheromone trails that guide those that follow, and the column may branch into a fan-shaped affair, as shown in figure 17–18 and figure 17–19. Virtually all animals in the path are stung, cut into pieces, and transported to the rear. Army ants are thus able to attack and consume prey items as large as snakes, lizards, and small birds that they would not be able to handle as individuals. Arthropod life in these areas is temporarily depleted. When larvae are developing, the bivouac location changes each night (nomadic phase); during the egg-laying and egg-development phases of the reproductive cycle, the bivouac remains in the same place each night (stationary phase).

The relationship between feeding behavior and group size is evident in the great diversity of African antelope. Jarman (1974) has classified these herbivores

FIGURE 17-18 Pattern of raiding employed by army ants

In this swarm raid of army ants (*Eciton burchelli*), which can be found on Barro Colorado Island in the Panama Canal Zone, the advancing front is made up of a large mass of workers. The swarm flushes a variety of prey, mostly invertebrates, but also snakes, lizards, and birds. The queen and immature forms remain at the bivouac site.

Source: Data from C. W. Rettenmeyer, "Behavioral Studies of Army Ants," *Kansas University Science Bulletin*, 44:281–465, 1963.

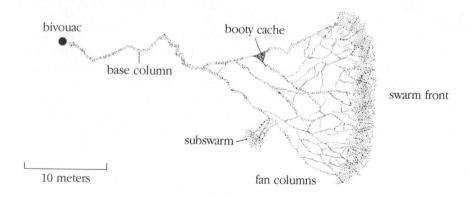

FIGURE 17-19
© 1990 by Nick Downes; from *Science*.

on the basis of five feeding styles ranging from selective species that feed on only a few highly nutritious parts of local plants, to unselective species that feed primarily on grasses and browse of low nutritive value. The selective species are small, solitary, monogamous, and monomorphic; they defend small territories. The least selective species are large, gregarious, sexually dimorphic, and polygamous; they occupy large home ranges. Jarman argued that feeding style is an important determinant of group size; both group size and the pattern of movement over the home range affect reproductive strategies and social behavior.

SOCIAL CARNIVORES

Among the mammalian carnivores, one member of the family Felidae and several members of the family Canidae have evolved complex social behavior that is related to cooperative capture of prey. All of these species have been the subjects of intensive behavioral and ecological study in their natural habitats. The evolutionary advantages of individuals who live in groups will be considered in more detail in chapter 19; here we discuss only factors that relate to diet.

FIGURE 17-20   Male lions with zebra
Male lions do relatively little killing of prey. Male on the right drags a zebra carcass that he probably scavenged from hyenas.

Source: Photo by George Schaller from Bruce Coleman.

LIONS. The lion (*Panthera leo*) lives in closed social units and is most abundant in the grasslands and open woodlands of Africa. Schaller (1972) studied the behavior and ecology of this species, its competitors, and its prey.

Lions gain from cooperation in several ways. First, by hunting in groups, they increase the resource spectrum; specifically, they add to their diet two species that an individual lion could never attack alone: buffalo (*Syncerus caffer*) and giraffe (*Giraffa camelopardalis*). Second, cooperative hunts are at least twice as successful as solitary ones. Third, a group can consume a captured prey more completely than can an individual. Fourth, the group can drive other predators and scavengers from the food. Schaller found that plains-dwelling prides kill less than half their food, relying on other predators, such as hyenas (*Crocuta crocuta*), to make kills for them (figure 17–20).

Although lions hunt in groups, the amount of cooperation is not extensive. When the lions encounter a herd of prey, such as zebras (*Equus burchelli*), the pride females fan out, sometimes encircling the herd. Lions are not endurance runners and rely on a **stalk-and-rush** tactic that depends on getting very close to prey, then surprising it with a sudden burst of speed. Although lions are more effective when hunting upwind, Schaller found no evidence that they do so more often than would be expected by chance.

Once a kill has been made, the males, which are larger than the females but which do little hunting, may drive the females from the kill. The cubs are the last to get food. Schaller believes that lion populations are regulated directly by food supply through starvation of cubs. The reproductive rate remains constant, but the mortality rate of the young can be very high, exceeding 80 percent when food is scarce. Not all cub mortality is the result of starvation; strange males that move into a group may kill cubs (Bertram 1975).

Caraco and Wolf (1975) used Schaller's data to relate ecological factors to lion foraging-group size (figure 17–21). When hunting small prey, such as Thomson's gazelles (*Gazella thomsoni*), two lions are more than twice as efficient as one, but three or more do no better than two. Since the available food for each lion per kill decreases with increasing group size, we would expect foraging groups of about 2. Caraco and Wolf went on to show that only pairs of lions can take in the minimum daily amount of food if they feed exclusively on Thomson's gazelles; Schaller reported that the mean foraging-group size ranged from 1.5 to 2, close to the expected size. In hunting larger prey, such as zebras and wildebeests (*Connochaetes taurinus*), a foraging-group size of 2 was again the most efficient, but groups of 1 to 4 could still attain their minimum daily requirement; however, Schaller found larger than expected foraging groups of about 4 to 7. Caraco and Wolf were forced to suggest that factors other than prey size, such as the higher reproductive success documented for larger prides (lionesses share in the feeding of cubs), could influence group size.

Packer (1986) recently reanalyzed the data and pointed out several erroneous assumptions in the Caraco-Wolf model. One is that hunting rate is independent of food intake. It turns out that lions resume hunting more quickly after a small meal, so less successful lions can compensate by hunting more often. In fact, Packer concluded, solitary females gain the highest food intake per hunt. Although group size is sometimes an advantage in obtaining and defending larger prey, other factors must explain the large pride size. One of those factors is better defense of cubs against infanticide by outside coalitions of males that take over the pride. Prides of 5 females seem to be optimal in protecting cubs (figure 17–22).

AFRICAN WILD DOGS. Weighing only about 18 kg, the African wild dog (*Lycaon pictus*), shown in figure 17–23, is an unlikely big game predator; however, these dogs typically catch prey weighing as much as 250 kg (Schaller 1972). Rather than using the stalk-and-rush tactic of lions, these canids are **coursers,** typically pursuing prey for many kilometers. The dogs live in mixed-sex packs that average ten adults, and when there are

FIGURE 17-21   Capture efficiency, food availability, and estimated food intake as functions of lion group size for Thomson's gazelle prey

Capture efficiency (a) increases as the lion group size increases, showing the benefits of cooperative hunting. Little increase occurs beyond hunting parties of two, however, since Thomson's gazelle is a small prey-type for lions. In (b) as the lion group size increases, the food availability declines, since there are more mouths to feed. Given the relationships in (a) and (b), the projected food intake per lion as shown in (c) is maximized at a group size of two. If Thomson's gazelle were the only prey taken, two would be the only group size in which lions would be able to obtain enough food to survive. Prey that is larger and more difficult to capture would require larger lion group sizes.

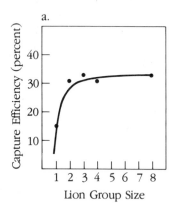
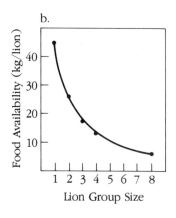
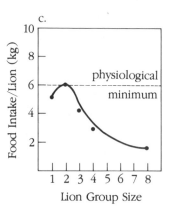

FIGURE 17-22   Costs and benefits of sociality in lions and other felids

For pride sizes of 3 to 7, the benefits of cub protection from male infanticide exceed the foraging costs, with 5 being optimal.

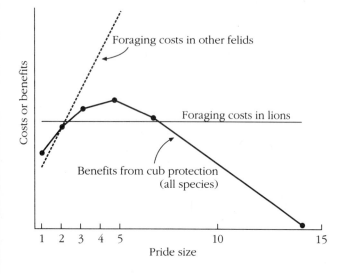

young in a den, the adults range out from it to hunt. The pack is tightly organized. Of particular interest here is the prehunt "ceremony", in which dogs draw back their lips to expose their teeth, nibble and lick each other's mouths, and run whining from pack member to pack member. The greeting seems to represent ritualized food begging. Setting out on a hunt, the dogs travel at a trot and fan out loosely over the terrain. Certain adults consistently lead some packs, but in other packs there seems to be no leader. Schaller (1972, 340) describes the hunt in his field notes:

A pack of 12 dogs becomes active at 1640; the animals greet and chase each other for 15 minutes before they trot off. About 8 km away is a large gazelle herd. The dogs race toward it at 45 km per hour. Several dogs run after one fleeing group, the rest toward another, but when a female gazelle breaks to one side, one dog pursues her immediately. She flees in a wide arc, the dog 10 m behind. The rest of the pack veers toward them and two dogs cut in front of the gazelle. She swerves to avoid them only to face two other dogs; she slows to a walk, unable to find an escape route, and a dog grabs her rump. The time is 1706.

According to Schaller, the hunting success rate for African wild dogs is extremely high: about 90 percent of all hunts result in the capture of at least one prey item.

FIGURE 17-23   African wild dogs attacking zebra
Extensive cooperation is necessary among small social
carnivores, such as African wild dogs, to enable them to
capture much larger prey, such as zebra.
Source: Photo by George Schaller from Bruce Coleman.

HYENAS. The remaining social carnivore on the African plains is the spotted hyena (*Crocuta crocuta*), which
was studied intensively by Kruuk (1972). The basic social
unit is the clan, which consists of ten to sixty hyenas of
both sexes and all ages. There is little exchange between clans; each defends a territory with a centrally
located den. Because the prey, mainly wildebeests and
zebras, migrate to the woodlands in the dry season, most
hyena clans break up at that time; some individuals
become nomadic and follow the prey. Thus food availability has a profound effect on social organization.

Although long thought to be scavengers, spotted
hyenas in the Serengeti kill more than two-thirds of
their food, a higher percentage than the lion (Schaller
1972). The method of hunting varies with the prey. In
searching for wildebeests, one or two hyenas may walk
beside a herd as close as 10 to 20 meters; suddenly one
may dash randomly toward the herd or toward a lone
wildebeest, causing the animals to run. The hyena may
run—never at full speed—after the fleeing animals, then
stop and watch, often joined by other clan members.
The function of the random chase seems to be to make
the prey run in order to spot any physically inferior individuals. Once an individual is selected, one or two

hyenas begin to chase it at speeds of 40 to 50 km/hr,
much slower than top speed for wildebeests, which can
run much faster than hyenas. Other hyenas may join in
until the wildebeest stops and turns or runs into a lake
or stream. After immobilizing the wildebeest with bites
to the hind legs or loins, they quickly disembowel it.

When hunting zebras, hyenas always operate in
packs. The group forms long before the hyenas get near
the zebras, and they ignore other prey during the search,
which clearly indicates a specific decision to hunt zebras.
For such zebra hunts, hyenas come together from all
over the range and often use a special meeting place,
or "club," for several months. Hyenas of the same clan
may hunt zebras exclusively for several days. The reproductive social unit of zebras consists of a stallion with
one to six mares and their foals. Kruuk (1972, 176) gives
the following description in his field notes:

One evening in December 1965, from 1850 to 1900 hr, hyenas
belonging to the Lakeside clan had been gathering in the rapidly approaching darkness on a small area some 300 m from
their den. They lay down there or sniffed around a bit, and
then, exactly at 1900, eight hyenas began to slowly walk off
together toward the nearest group of zebra, a family of twelve,
less than a km away in the middle of wildebeest. The hyenas
walked slowly up to the zebra, keeping very close together,
and without showing any obvious interest in them. Five minutes later, when the hyenas were very close, the zebra stopped
grazing and closed up together, their heads up. When the
hyenas were about 4 m away, the stallion turned toward them
and charged, head low and teeth bared. The hyenas scattered
out of his way and the stallion immediately turned back to his
family. The zebra bunched up and began running slowly (at
a speed of 20 to 25 km/hr) away from the hyenas. The stallion
ran just behind his family; several times he charged the hyenas
and tried to bite them, and once he kicked out with his hind
legs. The eight hyenas galloped just behind the zebra, five of
them close around the stallion and the other three just in front
of him. Several times some of the zebra barked excitedly. The
whole party still moved at the same fairly slow speed over the
plains; the zebra stayed closely bunched together and the
hyenas ran in a semicircle behind them.

When the stallion was a little farther behind than usual,
the rest of the family made a 90° turn, bunched up till they
were almost touching each other, and virtually stopped; again
the stallion attacked a hyena, chasing him right around the
zebra family. The zebras moved on again with the hyenas following. Now and then a hyena managed to get very near to
the family or even in between the zebra, biting at their flanks.
The speed still did not increase. By 1907 hr seven more hyenas
had been attracted by the commotion and there were fifteen
hyenas following the zebras, but otherwise the picture was
unchanged. Suddenly one hyena managed to grab a young
zebra while the stallion was chasing another member of the

pack. This young zebra, which was between nine and twelve months old, fell back a little, and within seconds twelve hyenas converged on it; in 30 sec they had pulled it down while the rest of the family ran slowly on. More hyenas arrived, and the little zebra was completely covered by them. At 1917, 10 min after the victim had been caught, the last hyena carried off the head and nothing remained on the spot but a dark patch on the grass and some stomach contents. Twenty-five hyenas were involved, and the whole process of dismembering took exactly 7 min.

WOLVES. The social carnivore of northern temperate and arctic regions is the wolf (*Canis lupus*), which typically preys on deer (*Odocoileus* spp.), moose (*Alces americana*), buffalo (*Bison bison*), sheep (*Ovis* spp.), caribou (*Rangifer tarandus*), and elk (*Cervus canadensis*). Pack size varies greatly (Mech 1970), but most packs have fewer than eight members. Wolves resume hunting after consuming their previous kill by traveling over their territory. Most often they use scent; the lead animals stop when they detect the odor of prey. The group may stand nose-to-nose with tails wagging for a few seconds, and then all follow the leaders directly toward the prey. They may also use chance encounter and tracking. If they detect the prey at some distance, they proceed slowly and stalk sometimes to within 10 meters of the prey. When the prey detects the wolves, it may stand its ground or flee. Once the prey is in flight, the wolves rush. Wolves must get close to their prey during the stalk-and-rush or they will quickly give up; if they cannot make an attack, they may chase the prey, usually for less than half a mile. Some earlier studies had described wolves as coursers, but Mech has concluded that they are mainly stalkers and rushers (figure 17–24).

One prey of the wolf, the musk ox (*Ovibos moschatus*), defends itself in an unusual way. Adults form a circle, facing outward and protecting the young inside. There is little evidence that wolves can break this defense; in areas where musk oxen are present, wolves eat other prey or capture an occasional postreproductive individual.

When hunting moose, which are solitary, wolves rarely attack a prime individual (between one and six years old). Most studies of wolf predation have indicated that they are highly selective of young, old, and sick prey. Although earlier reports described the main killing tactic as hamstringing, wolves actually avoid the hooves and direct bites at the rump, flanks, shoulders, neck, and nose.

While the stalk-and-rush strategy is similar to that of the social cats, wolves sometimes chase prey for long distances and may attack weakened prey at intervals over several days. In their manner of prey selection, wolves resemble the hyena and the African wild dog. However, Mech believes that wolves catch certain individuals because these prey animals are less able than others to escape or defend themselves, not because the wolves deliberately pick out certain individuals to chase. Observers of hyenas and African wild dogs have implied that these predators, in contrast, deliberately evaluate a herd and select vulnerable individuals (Kruuk 1972; Schaller 1972).

One of the biggest contributions of modern studies of social carnivores has been the debunking of much of the folklore. Although social hunters may encircle prey, ambush prey, hunt upwind, or take turns chasing prey, the amount of cooperation in hunting seems to have been overstated. Lions hunt upwind no more often than they would be expected to do by chance, and they scavenge more than half of their food. Wolves do not hamstring their prey and hyenas may kill a higher percentage of their prey than do lions.

HUMANS. Carnivory has been suggested as one of the key elements in the development of social behavior in humans. Schaller and Lowther (1969), among others, have suggested that those interested in the behavior of early humans might learn more from social carnivores than from the near relatives of humans, such as the gorilla. Humans were widely roaming scavengers and hunters for perhaps two million years. There is some doubt about how important meat was in the past; the presence of bunodont (low-cusped) molars of humans, which resemble those of the pig, suggests an omnivorous diet in the past. However, some of the selective forces favoring sociality in carnivores also could have operated on humans. Cooperation in deciding where and what to hunt, encircling of prey, relay racing in pursuit of prey, capturing large prey, protecting food from competitors, bringing back food to the rest of the group—all could have been behaviors exhibited by early humans. Compared with most mammals, humans are exceptionally good endurance runners. Even today, African Bushmen capture small game by medium pace pursuit over long distances, much as hyenas and African wild dogs do. The use of tools in capturing both large and small prey and in chasing other predators from their kills probably became important to humans more than a million years ago, as well.

FIGURE 17-24 Moose-wolf interactions
Observations at Lake Superior's Isle Royale National Park show the results of 131 separate moose-wolf interactions. Circled numbers indicate moose actually encountered by wolves. Only 6 of the 131 moose detected by wolves were killed.

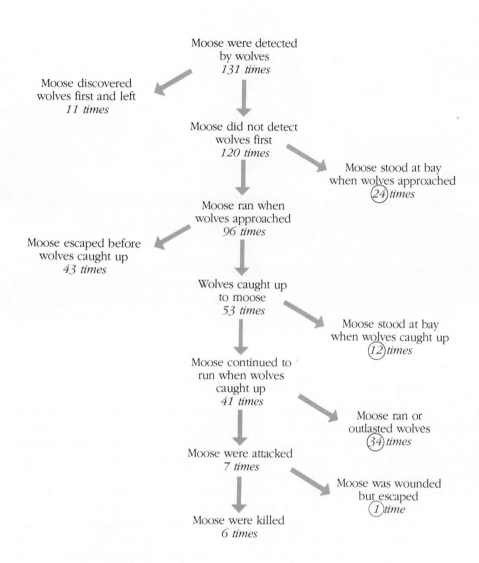

DEFENSE AGAINST PREDATORS

INDIVIDUAL STRATEGIES

Natural selection operates just as strongly on individuals with traits that reduce the likelihood of being eaten as it does on traits that increase prey-catching ability.

We have seen that feeding habits such as meat eating may favor the evolution of social behavior, but so may habits like predation avoidance. Most animals, except for a few upper trophic-level carnivores, are potential prey. Escape patterns are usually species-specific. Some social herbivores clump together when threatened (for example, zebras), while others scatter (for example, patas monkeys [*Erythrocebus patas*]).

FIGURE 17-25   **Insect defense displays**
The anterior end of an alarmed caterpillar (a) resembles the
head of a snake and frightens predators. A moth (b) exposes
large eye spots on its underwing in order to frighten away
or deflect attacks by predators.
Source: Photos (a) by Lincoln P. Brower, (b) by Thomas Eisner.

a   b

ESCAPING AND FREEZING.   Many animals keep close to
a nest, burrow, or other refuge near the center of their
home range. White-footed mice (*Peromyscus leucopus*)
usually nest in hollow trees or logs, and their use of
space typically follows what is called a bivariate normal
distribution (Vessey 1987). The probability of finding
the mouse at a particular spot declines rapidly with distance from the nest, so they are rarely in unfamiliar
space or far from shelter.

Another predator response strategy is to freeze. The
presence of protective or cryptic coloration is often associated with this behavior, as in the spotted white-
tailed deer fawn (*Odocoileus virginianus*). Some animals
carry this strategy further and feign death. For example,
the opossum (*Didelphis virginiana*), if harassed by a predator such as a dog, remains motionless on its back. The
physiological correlates of this behavior are not completely understood (Francq 1969). Hog-nosed snakes
(*Heterodon platyrhinos*) exhibit similar behavior, although they may precede it by threat behavior in which
the hog-nosed snake resembles a cobra. Freezing and
feigning death may work because many predators seem
to respond only to moving prey. For example, wolves
will not attack prey that stand motionless (Mech 1970).

Escape or defense reactions may be specific to the
type of predator. Chickens (*Gallus gallus*) head for cover
on the ground or crouch to escape aerial predators, but
they fly up into trees to escape ground predators. Each

response has its own warning call, and brain stimulation experiments have revealed that the responses have
separate neural representation in the brain (Eibl-Eibesfeldt 1975). Vervet monkeys (*Cercopithecus aethiops*) also
respond differentially to aerial and terrestrial predators
and use different alarm calls (Struhsaker 1967), as do
the Belding's ground squirrels (*Spermophilus beldingi*)
discussed earlier.

DECEPTION.   Another strategy used by vulnerable organisms is deception. The larvae of several moth species resemble snakes. When threatened, this type of
larval insect inflates its head end to form an excellent
representation of the head of a snake and it then waves
this "head" (figure 17-25a). Other larvae resemble inanimate objects in the environment, such as twigs,
leaves, bark, or even bird droppings.

Many animals have behaviors or markings that may
serve to misdirect or surprise predators. Some reef fish
have conspicuous eye spots on their tail end that are
much larger than the real eyes; these could frighten
away potential predators or misdirect their attack to less
vulnerable parts of the body. Moths (figure 17-25b) may
have similar spots. A large group of insects, the underwing moths (of the order Lepidoptera), are cryptically colored when they are at rest on tree trunks. If a
predator comes near, the moth spreads its outer wings
and reveals large, brightly colored spots on the un-
derwings that seem to surprise the predator.

FIGURE 17-26   Field experiment to test function of butterfly markings

Butterflies (*Anartia fatima*) with wing stripes intact, left, and covered with ink, right. If stripes function to reduce predation, butterflies with stripes obliterated should suffer higher mortality. Results of experiments, however, indicated no such function for the stripes.

Source: Data from R. E. Silberglied, et al., "Disruptive Coloration in Butterflies: Lack of Support in *Anartia fatima*," *Science*, 209:617–619, 1980.

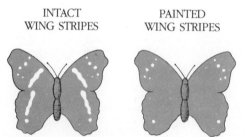

INTACT
WING STRIPES   PAINTED
WING STRIPES

FIGURE 17-27   Bombardier beetle spraying

Quinones and hydrogen peroxide are heated to the boiling point and sprayed on an attacker.

Source: Photo by Thomas Eisner.

Disruptive color patterns consist of high-contrast markings that are thought to break up the outline of an organism or create a target that directs predator attack away from vital organs. For example, many species of butterflies and reef fish have disruptive stripe patterns. Indirect evidence that they provide protection against predation comes from the finding that chemically protected or unpalatable species usually lack the markings found in palatable species; however, few experimental field tests have been attempted. Silberglied et al. (1980) obliterated the wing stripes on a species of tropical butterfly (*Anartia fatima*) with a felt-tipped pen and compared their survival with that of unaltered controls (figure 17–26). The lack of a difference between experimentals and controls in survival rate or frequency of wing damage (an indicator of predatory attack) suggests that the color patterns have some other function.

TOXICITY. Many organisms are in some way harmful to predators. Plants may contain toxins such as alkaloids; animals may have armor plates (as in the armadillo) or spines (as in the porcupine, hedgehog, and three-spined stickleback). Many invertebrates inject venom into attackers. Eisner (1966, 1970) has discovered a large number of arthropod defense mechanisms. For instance, the bombardier beetle (*Brachinus* spp.) has a plumbing system that resembles a liquid-fueled rocket. It stores quinones and hydrogen peroxide in separate reservoirs. When the beetle is disturbed, these two liquids are mixed in an outer vestibule, and in the presence of enzymes, the solution heats to the boiling point and is discharged as a noxious spray (figure 17–27).

Noxious animals tend to be conspicuously colored, presumably so that predators can easily recognize and thus avoid them. Such warning, or **aposematic,** coloration occurs, for example, in skunks (*Mephitis mephitis*) and coral snakes (*Micrurus fulvius*) (Wickler 1968).

MIMICRY. Among arthropods such as butterflies, unpalatable species tend to resemble each other and are referred to as **Müllerian mimics**—for example, the monarch (*Danaus plexippus*) and queen (*Danaus gilippus*) butterflies, both of which may be toxic because of the plants on which they feed. Müllerian mimics seem to have evolved because by looking alike and acting similarly, they provide fewer different prey types for predators to learn to avoid and thus are less likely to be eaten by mistake. Some palatable species, called **Batesian mimics,** may evolve morphologies and behaviors similar to those of unpalatable species. Batesian mimics enjoy protection because of the predators' learned avoidance of the toxic species. Brower et al. (1968) demonstrated "one trial conditioning" of blue jays (*Cyanocitta cristata*) fed monarch butterflies; consumption of one monarch butterfly often led to vomiting by the jay and avoidance of other monarchs or their mimics.

The toxicity of monarchs comes from the poisons (cardiac glycosides) they obtain from the milkweed plants they feed on as larvae. Adult females lay eggs on the toxic species of milkweed; the larvae have evolved a resistance to the poisons and a capacity to store them. How could such a system evolve if a toxic adult must be consumed before a predator can learn to avoid it?

Any butterfly with a new mutation enabling it to store the toxins through metamorphosis would probably not survive, since "uneducated" predators would be common. One possibility is that the predator need not consume the entire butterfly to reject it, and so the toxic butterfly could live to reproduce and pass on its ability to store toxins. But in many cases the vomiting is delayed, and the butterfly does not survive. Since the predator, who is now "educated" will probably never again eat another butterfly resembling that one, we seem to have a case of altruism on the part of the consumed butterfly. A second possibility is that through the process of kin selection, the toxic adult protects relatives that live in the area and share its genes—by sacrificing its life to educate the predator (Hamilton 1963). From the latter theory one predicts that sedentary populations of monarchs, in which relatives are likely to be in close proximity, should be highly toxic, but migratory populations, in which relatives are likely to be scattered, should be nontoxic, as was found to be the case by Duffey (1970).

The tendency of predators to avoid harmless prey that resemble toxic prey may be put to practical use. In a field experiment with crows (*Corvus brachyrhynchos*) chicken eggs were placed in straw nests on the ground. Eggs that were injected with a bad tasting but nonlethal toxin were painted green, while untreated eggs remained white. The crows quickly learned to avoid the green eggs, selecting the white ones even when the green eggs were not injected (Nicolaus et al. 1983). If only green eggs were present, crows tended to shift to alternate foods rather than eat eggs.

Avoidance of dangerous or unpalatable prey may occur also in the absence of previous conditioning. For example, even laboratory-reared motmots (*Eumomota superciliosa*)—lizard- and snake-eating birds—avoid models painted with the color patterns of the coral snake (Smith 1975) (figure 17–28).

SHEEP IN WOLF'S CLOTHING. An unusual case of mimicry has been reported in which a fly mimics a predaceous spider (Mather and Roitberg 1987). Snowberry flies (*Rhagoletis zephria*) are among the prey of solitary, zebra jumping spiders (*Salticus scenius*). The flies have a wing banding pattern that resembles the legs of the jumping spider (figure 17–29). Arena tests showed that spiders flee flies that display with their wings, mimicking the spider's walk, apparently mistaking them for an aggressive conspecific. Nondisplaying flies, or those with their bands covered, tended to get eaten by the spiders.

FIGURE 17-28   **Wooden snake models used to test hand-reared motmots**

Area (a) of each model was painted as indicated at left; areas (b) and (c) were plain wood. Numbers beneath each model show the percentage of pecks by the motmots. Motmots avoided those models or portions of models painted with yellow-red rings, the same colored rings that the coral snake has.

Source: Data from S. M. Smith, 1975, "Innate recognition of coral snake pattern by a possible avian predator," *Science* 187:759–60.

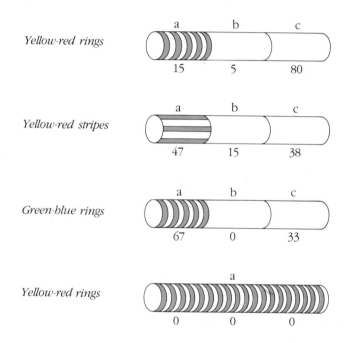

SOCIAL STRATEGIES

CONFUSION EFFECT. One of the main benefits of sociality is thought to be the protection gained, as stated in the adage, "safety in numbers." Living in flocks, herds, or schools increases the chance of spotting predators. When approached or attacked by predators, groups usually become more compact. Fish schools, however, usually form a vacuole around a predator. Tightly packed prey make a difficult target for the predator, which must select an individual toward which it can aim its attack. The predator tries to isolate an individual from the group; any individual that strays from the group or is in any way different from the rest is likely to be attacked. For example, when experimenters presented hawks with sets of ten mice, the hawks preferred the oddly colored mouse (Mueller 1971).

FIGURE 17-29 **Prey mimics predator**

(a) Front view of zebra spider and (b) posterior view of snowberry fly. Note similarity between spider legs and fly wing markings.

Source: Photo by R. Lalonde, Mather & Roitberg, *Science*, 1987, Vol. 236.

A primary means of protection is the **confusion effect,** described by Miller (1922). A variety of predators exhibit a lower prey capture rate, a longer period of hesitation, and more irrelevant behavior when they attack swarming prey than when they attack solitary prey. Factors that enhance the confusion effect are swarm number, swarm density, and the uniformity of appearance of swarm members (Milinski 1979). The mechanism producing the confusion effect is not understood, but a predator seems to be more easily frightened when it feeds in a dense swarm of prey; perhaps the predator finds it harder to detect other animals that might prey on it. In laboratory experiments with three-spined sticklebacks (*Gasterosteus aculeatus*) feeding on water fleas (*Daphnia*), fish with experience in feeding on dense swarms fed more efficiently when tested with dense swarms than did those experienced only in feeding on low-density populations (Milinski 1979).

DETECTION. We might expect to find groups of mixed species if the combination of different sensory capabilities is advantageous. Baboons and ungulates are frequently together on the African plains. The arboreal habits and keen vision of baboons and the well-developed olfactory systems of ungulates probably increase the distance at which predators can be detected. It is clear that these different species recognize and respond to each other's alarm calls.

Social living increases the likelihood that predators will be detected and allows group members to spend more time in other activities. Individuals in large, dense colonies of prairie dogs (*Cynomys* spp.) spend less time in alert postures than do those in small colonies (Hoogland 1979). Once predators are detected, alarm calls alert neighbors to danger. Marler (1955) argued that in some birds the alarm call is difficult to locate because it is a high-pitched note in a narrow-frequency range. The alarm-calling ground squirrels studied by Sherman (1977) emitted calls that were easily located, and calling squirrels were more likely to be caught than noncallers. As we saw in chapter 11, close relatives of the caller were likely to be close by and thus could benefit from this seemingly altruistic behavior.

GROUP DEFENSE: MOBBING. Some prey species are able to turn the tables and attack the predator. Mobbing of predators is common and may occur in species that usually live in widely spaced territories. If a crow (*Corvus brachyrhynchos*) flies over a field of nesting red-winged blackbirds (*Agelaius phoeniceus*), the males and sometimes the females may rise up and pursue it en masse; crows, in turn, will mob owls. Baboons (*Papio* spp.) sometimes mob lions; however, observers have disagreed about the response of baboon groups to predators. In some cases, the dominant males move out to the periphery to fend off predators while the rest of the group flees to the safety of trees (DeVore 1965). Other accounts have suggested that the males are among the first to flee, leaving behind females with young to fend for themselves (Rowell 1966).

Hoogland and Sherman (1976), when considering the advantages and disadvantages of coloniality in bank swallows (*Riparia riparia*), found that two possible advantages were increased foraging efficiency and reduced predation on adults, young, and eggs. The experimenters found that a stuffed weasel model was detected sooner and mobbed more intensely in larger colonies than in smaller colonies. When they tethered young swallows at varying distances from the colony, they found that blue jay attacks on the young swallows caused intense mobbing by adult swallows when the young were close to the colony; but when the young were farther away, the mobbing was less intense and the blue jay attacks more often successful (table 17–1).

TABLE 17-1   Field experiment to test the effectiveness of mobbing in deterring predators
A young bank swallow was tethered at varying distances from the main group of bank swallow burrows. On 17 occasions, one of several different blue jays attempted to attack the tethered young. In all cases bank swallows mobbed the jays, but the mobbing was most effective in jay deterrence close to the main group of burrows.

Behavior of blue jays	Distance from burrows to tethered young bank swallow		
	0–1 m (Bank face)	9–11 m (Bank base)	18–20 m (Center of gravel pit)
Times jays attack and kill the young bird (N)	0	2	7
Times jays attack unsuccessfully (N)	0	2	1
Times jays are deterred by adult swallows (N)	5	0	0
Successful attacks at each distance (%)	0	50	88

From J. L. Hoogland and P. W. Sherman, "Advantages and Disadvantages of Bank Swallow (*Riparia riparia*) Coloniality," *Ecological Monographs* 46:33–58. Copyright © 1976 by the Ecological Society of America, Tempe, AZ. Reprinted by permission.

Emlen and Demong (1975) reported that increased foraging is a prime advantage of coloniality and synchronous breeding; the swallows follow each other to localized, ephemeral concentrations of food.

The extent to which individuals defend the group may be a function of the degree of relatedness to those group members. Mothers can be expected to defend offspring up to the point that the benefit to the offspring exceeds the cost to themselves in reduced future reproductive output. Similarly, males will defend their offspring, and to some extent, they will defend the mates for which they have competed. However, group-living species are mostly polygynous, and confidence of paternity may be low. Thus we may expect variability in the male role, with older long-term residents being the most active in group defense. Among rhesus monkeys (*Macaca mulatta*), minor extragroup threats are handled by low-ranking peripheral males. Serious challenges, such as attacks on infants, are met by the central high-ranking males, who are more likely to be their fathers.

DISTRACTION DISPLAYS. Individuals may use distraction displays—for example, the broken-wing displays of avocets (*Recurvirostra americana*)—to attract the attention of a predator and draw it away from the nest (figure 17–30) (Sordahl 1981). Many *cursorial*, or running, mammals have white rump or tail patches; in the presence of predators the hairs are erected and the tail waved, making a conspicuous display. Observers have suggested several functions for **flagging** behavior, such as: (1) distracting the predator from other members of the group, (2) warning other group members, (3) confusing the predator when many group members are displaying, (4) signaling the predator that it has been detected, and (5) eliciting premature pursuit (figure 17–31). In the cases of (4) or (5), the predator may go off in search of less-alert prey or may be lured into an unsuccessful pursuit (Smythe 1977). Others have argued that rump patches have little to do with defense against predators and function in intraspecific social communication (Guthrie 1971).

FIGURE 17-30   American avocet performing broken-wing display
This behavior pattern presumably functions to distract predators from the nest or young. Although considered by some to be altruistic behavior, it really is most easily understood as a form of parental investment.
Source: Photo by Leonard Lee Rue III from Animals Animals.

FIGURE 17-31
THE FAR SIDE COPYRIGHT 1987 UNIVERSAL PRESS SYNDICATE. Reprinted with permission. All rights reserved.

"Forget these guys."

SUMMARY

An organism makes decisions, consciously or otherwise, about whether it should include a newly encountered item in its diet, when it should move to a new resource area (or *patch*), where it should search next, and what pattern it should use to search an area. The potential energy available from a resource must be devalued by the energy needed to search for it, to pursue it, and to handle and eat it. Behavioral ecologists have developed models to better understand decisions animals make about what and where to eat, what is being maximized by natural selection (*currency*), and what the limits are on the animal's ability to forage optimally (*constraints*).

In general, species at low trophic levels are large and relatively sedentary, often simply filtering small organisms or debris from the environment. Those at high trophic levels are likely to be few in number and active, and have high metabolic rates.

Some adaptations have expanded the range of food sources available or have enhanced the attainment of them. For example, closely related species may employ different feeding behaviors and thus reduce competition for food; herbivores may modify the environment so as to increase the food resource, as evidenced by some species of fungus-growing ants that tap a vast energy source unavailable to most animals; and animals may use a variety of aids to capture prey, such as the lures used by anglerfish or the tools used by termite-feeding chimpanzees.

Social organization and food resources are often closely related; some form of territory or group living is common. A number of studies that included spiders, insects, birds, and mammals have demonstrated that under some conditions, group feeding is more efficient than feeding alone.

Adaptations that provide defense against predation are just as elaborate as those that increase feeding efficiency. Individual tactics include *escaping*, having *cryptic*

coloration, feigning death, practicing *deception*, and having markings or behaviors that *surprise* the predator or that *misdirect* its attack. Many organisms produce or store *toxic substances* to gain protection, and nontoxic species may *mimic* the species with these toxic substances.

Living in groups may provide protection by making it difficult for a predator to single out a prey. Groups, particularly those made up of several species, may spot predators sooner or may mob a predator. Recent studies have explored the extent to which close relatives in a group warn or otherwise aid each other to escape predation.

Discussion Questions

1. Discuss the relationship between the distribution of food resources and social organization. What other factors affect social organization?

2. Discuss the similarities and differences between the evolution of alarm calls and the evolution of the ability to store substances toxic to predators.

3. Normal butterflies (*Nymphais io*) with large eyespots and altered butterflies with eyespots blotted out were presented equal numbers of times to six yellow buntings (*Emberiza citrinella*), predatory birds. Data on the buntings' escape responses are presented here (Blest 1957). Discuss these results in light of the findings of Silberglied et al. (1980) presented in the chapter.

Number of fright responses by birds (yellow buntings) when presented with butterflies

Yellow bunting	Normal butterfly (with large eyespots)	Altered butterfly (with eyespots blotted out)
1	56	16
2	11	5
3	8	4
4	18	1
5	18	3
6	17	2
Total	128	31

Source: Data from A. D. Blest, "The Function of Eyespot Patterns in the Lepidoptera," *Behaviour*, 11:209–56, 1957.

Suggested Readings

Edmunds, M. 1974. *Defence in Animals*. Harlow, Essex, England: Longman.
Review of the means by which animals avoid being killed and eaten by predators.

Kamil, A. C., J. R. Krebs, and H. R. Pulliam. 1987. *Foraging Behavior*. New York: Plenum.
Proceedings of a symposium held in 1984. Sections on theory, selectivity, patch utilization, reproductive consequences, learning, and coaching. Combines the efforts of biologists and psychologists on learning and memory.

Pyke, G. H., H. R. Pulliam, and E. L. Charnov. 1977. Optimal foraging: A selective review of theory and tests. *Quart. Rev. Biol.* 52:137–54.
One of the more intelligible treatments of a difficult subject, which is currently receiving much attention from ecologists, psychologists, and physiologists.

Stephens, D. W., and J. R. Krebs. 1986. *Foraging Theory*. Princeton: Princeton University Press.
A concise, clearly-written presentation of the basic optimal foraging models with some new variations. Can be understood without wading through all the math.

18

BEHAVIOR AND POPULATION BIOLOGY

The natural regulation of animal numbers has long fascinated ecologists, and the role of behavior in this process has been the subject of much debate. A complex of abiotic substances (minerals and organic compounds), climate (mainly temperature, moisture, and wind), and biotic factors (other organisms of the same and other species) affects the growth and reproduction of any organism. A **population** may be defined as a group of organisms of the same species in a particular place. We know that all species have a reproductive potential that is seldom realized and that no population increases without limit, as was pointed out by Darwin and by Malthus before him. The number of individuals present in a population is determined by four population forces: births, deaths, and movements out of (emigration) and into (immigration) an area. Thus the abundance of a species as well as its distribution is affected by behavior.

This chapter briefly reviews the extrinsic limits to population growth and then considers the evidence for the following possible processes of population self-regulation: (1) purely behavioral mechanisms, (2) behavioral-physiological mechanisms, and (3) behavioral-genetic mechanisms. We then discuss how such systems could have evolved by natural selection.

LIMITING FACTORS

One axiom of ecology is that only a single factor limits the growth of a population at any one time. The idea originated with Liebig (1847), a plant physiologist who studied effects of nutrients, such as nitrogen, on plant growth and is referred to as **Liebig's law of the minimum.** Extending his idea to animal populations, we can test which factor limits a population's growth if we manipulate the amount of one factor while holding others constant and record changes in the population. If food is the limiting factor, an increase in the population will result if the food supply is augmented.

For example, Holling (1959) studied the population response of some small mammals to different amounts of a staple food—cocoons of the European pine sawfly (*Neodiprion sertifer*). Shrews of the genus *Sorex* increased more or less linearly with increasing amounts of the food supply up to a population of twenty-five shrews per acre; then the population leveled off (figure 18–1). At that point, some resource other than sawflies became limiting. Likewise, deermice (*Peromyscus*) increased in numbers up to about eight per acre, then remained constant despite the food supply. The abundance of sawflies seemed to be unrelated to the population density of another shrew of the genus *Blarina*.

FIGURE 18-1   Number of mammals per acre as a function of food supply
Populations of the shrew (*Sorex*) and deermouse (*Peromyscus*) increase in size, up to a point, as food supply increases. Their numbers, therefore, are limited by food supply. Shrews of the genus *Blarina*, however, show no response to an increased cocoon supply, so their numbers must be limited by some other resource.

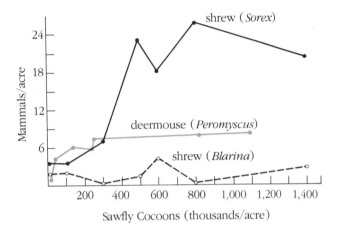

While Liebig's law of the minimum can simplify our analysis of regulatory factors, it does not help us understand the level at which these factors may operate. Ultimately, food supply usually sets the limit to the number of organisms that can live in a place—that is, the **carrying capacity** of the environment. However, some behavior patterns involved with interference competition for food, such as defense of territory, may proximally limit the population below the carrying capacity, if some individuals defend space in which to live and reproduce and thereby exclude others (see chapter 12). Thus both territory and food supply may limit the population, but operate at different levels, one proximate and the other ultimate.

The interaction of factors creates more complicated problems; for example, plants that are deficient in nitrogen are more susceptible to drought, and most terrestrial insects can tolerate higher temperatures when humidity is high than when humidity is low. In addition, the impact of a limiting factor may occur during a very short period in the history of the population and thus be difficult to identify. A few days of severe winter weather may reduce a deer herd to the point that other regulatory factors never come into play. Therefore, it should not be surprising to find little agreement among population biologists about the mechanisms of population regulation.

DENSITY-INDEPENDENT AND DENSITY-DEPENDENT FACTORS

Population regulatory factors are sometimes divided into two types: density-independent and density-dependent. **Density-independent** factors include climate, flood, fires, and so on; and their effect is not affected by population density—they kill organisms regardless of density. **Density-dependent** factors include competition, parasitism, disease, predation, and food supply, and are related directly or inversely to density, so that as population density increases, death or emigration rates increase or birth rates decrease. Actually, this division is somewhat misleading, since some of the effects of density-independent factors may be modified by population density. At high densities, organisms may gain protection from both wind and low temperature by grouping, or shelter from the elements may be limited so that surplus animals are excluded and are therefore more likely to die. Thus, although climate is usually thought of as density independent, its impact may vary with population density. If by *population regulation* we mean the maintenance of numbers within certain limits and not simply random fluctuations, we must look to density-dependent factors as the regulatory mechanisms. Only when a factor's negative effect on population growth increases as population increases can an equilibrium of numbers be maintained (Krebs 1985). Let's look closely at some of these density-dependent factors.

FOOD SUPPLY. Most ecologists agree that the supply of energy (food) most often sets the upper limit to the density of animal populations. Correlations between food abundance and population density, as shown in figure 18-1, are one type of evidence. Another approach is to supplement the natural food supply and chart the response of the population in terms of density, reproduction, mortality, and movements. In one such study, laboratory mouse-food pellets were distributed in a woodlot and the responses of the resident white-footed mice (*Peromyscus leucopus*) were noted. By the end of the study, the population was somewhat higher than was that in the control area, primarily due to higher survival of young in fall and winter (Hansen and Batzli 1978). When oats were fed to populations of deermice (*Peromyscus maniculatus*), breeding began within several weeks, even in the winter (Taitt 1981). Home ranges decreased in size and immigration tripled compared to untreated control plots, suggesting that spacing behavior of residents may have kept immigrants out prior to the addition of extra food (figure 18-2).

FIGURE 18-2   Effect of supplemental food on home range size in deermice
Home ranges were generally smaller where food was supplemented, as seen in these frequency distributions of home ranges of deermice.

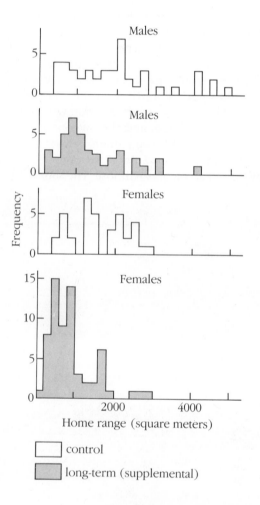

In spite of the presumed importance of food as a limiting resource for vertebrate populations, naturalists have frequently noted that starving animals are rarely found in the wild, except where humans have altered the environment. These observations have contributed to the notion that behavioral mechanisms keep numbers below the point at which the food supply is exhausted.

SHELTER. A few studies have demonstrated that the number of nesting or feeding sites limit population density. Filter feeders, such as sessile invertebrates, may compete actively for space. Barnacles (*Chthamalus stellatus* and *Balanus balanoides*) crowd each other out of attachment sites on rocks (Connell 1961).

Some species of birds nest in tree holes. In a classic study by Haartman (1956), the numbers of pied flycatchers (*Muscicapa hypoleuca*) increased significantly when nest boxes were added. The coqui (*Eleutherodactylus coqui*), a small tree frog of Puerto Rico, defends specific sites for retreats and nesting in the rain forest. When bamboo frog houses were added to some plots, they were quickly occupied by the frogs; one year later, densities and reproductive rates were significantly higher than in control plots (Stewart and Pough 1983).

PREDATORS. Although most species of animals are preyed upon by some other species, there are surprisingly few cases where predators have actually been shown to regulate the prey population. For regulation to occur, the predators must take an increasing percentage of the prey population as the number of prey increases. One approach has been to study islands, comparing those with and without predators. After censusing lizards and orb-weaving spiders on ninety-three islands in the Bahamas, Schoener and Toft (1983) found the numbers of spiders to be about ten times lower on islands with lizard predators than on those without.

INTRASPECIFIC COMPETITION. Intraspecific competition has received much attention as a regulatory factor because it is the only one that is "perfectly" density-dependent. In other words, the intensity of competition among members of the same population is likely to vary directly with the size of the population: As the population increases, so does competition for the limiting resource. Unsuccessful competitors may leave the area, die, or have lower reproduction. The other biotic regulatory factors involve other species of organisms that in turn are affected by other agents; thus they are not perfectly density-dependent. For example, predation involves a complex relationship between predator and prey. Although wolves eat caribou, they are not necessarily the regulators of the caribou population. Since the wolf population is affected by factors other than the numbers of caribou, such as diseases or the abundance of other prey species, it will not be able to track the caribou population perfectly. An outbreak of disease might reduce the size of the wolf pack and allow the caribou to escape control and to begin to destroy grazing land. Intraspecific competition among the caribou, however, will vary directly with the population density and the limiting resource.

POPULATION SELF-REGULATION

The self-regulation school of thought builds on the just mentioned reasoning about the importance of intraspecific competition and argues that certain behavioral or physiological changes take place in animals because of this competition with conspecifics. As a resource becomes scarce, the intensity of contest competition increases and causes changes within the organisms that increase the death and emigration rates and/or decrease the birth rate. The adaptive value of this system was originally thought to be the regulation of the population below the limit set by the environment. If the limiting resource is food, overexploitation could permanently lower the carrying capacity; for example, overgrazing causes soil loss from erosion by wind or water. Interference competition before such a shortage could adjust birth, death, and movement rates, and thus avoid wasted energy in reproduction.

Although proponents of self-regulation have proposed many eloquent explanations, a big stumbling block is that most mechanisms seem to require individuals to sacrifice direct fitness by not breeding or by dying for the good of the rest of the population. Thus, the evolution of self-regulatory mechanisms of population regulation appears to involve selection at the level of the group. To put the problem in human terms, let us consider the so-called tragedy of the commons (Hardin 1968). Suppose that individuals graze their flocks of sheep in a communally owned pasture. Individual sheep owners will be tempted to increase the size of their own flocks, since by adding a sheep or two, they will get more lambs and wool (figure 18-3). If the optimum number is already in the pasture, their action will only slightly reduce the total yield, and the owners will be ahead. But if all the other sheep owners behave selfishly and add a few sheep, the common pasture will soon be destroyed, and all owners will lose. The solution is for all

FIGURE 18-3   Tragedy of the commons
If these sheep were owned by several individuals and were grazed on a communally owned pasture, individual sheep owners could increase their profits by each adding a few sheep. But if each owner did that, the pasture would become overgrazed, and all the owners would suffer a loss, the tragedy of the commons. One solution would be for the owners to agree to limit the number of sheep each grazes on the pasture.
Source: Photo by Grant Heilman.

the owners to come to a common agreement, or **convention,** to limit the size of their flocks and forgo any short-term gain in favor of the long-term benefit to the whole community. (We can also take the opposite view, first expressed by economist Adam Smith (1776), that if individuals behave so as to maximize their own interests, they will in the long run promote conditions that are also optimal for the whole group. We examine this contrary view in more detail later in the chapter.)

BEHAVIORAL MECHANISMS

EPIDEICTIC DISPLAYS. Wynne-Edwards (1962, 1986) followed up on the idea of the conventions presented above and proposed that the aerial maneuvers performed by flocks of blackbirds or starlings in the evening are unrelated to feeding or predator defense. He claimed that the main purpose of this social behavior is to communicate information about the supply of potentially limiting resources. According to him, the behaviors, called **epideictic displays,** act as conventions resulting in adjustments of birth, death, and movement rates, and enable populations to track closely and not overexploit the fluctuating external environment. Wynne-Edwards included among these specifically timed communal displays the dancing of gnats and midges, the maneuvers of bats and of birds at roosting time, and the choruses of birds, bats, frogs, fish, insects, and shrimps. He believed that displays are abstract, highly symbolic conventions that produce a state of "excitation and tension closely reflecting the size and impressiveness of the display." A special time of day might be set aside for them, as well as a special place.

The mass feeding and roosting flights of birds are among the phenomena to which Wynne-Edwards applied his theory. Each night in the fall and winter, starlings (*Sturnus vulgaris*) gather by the millions (figure 18–4). Birds in a roost use a common feeding area, and Wynne-Edwards considered it a communal territory. Most birds are consistent in their choice of roost and feeding ground; the same roosts may be used for hundreds of years. Each bird tends to have its own place in the roost. As the birds go to roost in the evening, there is much communal singing and bickering over perching sites. As Wynne-Edwards (1962, 286) noted:

Moreover on fine evenings, especially early in the autumn, either a part or sometimes the whole of the noisy company not infrequently rises with a great roar of wings to engage in the most impressive aerial maneuvers over the site: the massed flock may extend in a tight formation sometimes hundreds of

yards in length, changing shape and direction like a giant amoeba silhouetted against the sky. On the grand scale these maneuvers are by no means the least of the marvels of animal adaptation, so perfect is their co-ordination and so intense the urge to excel in their performance. It seems quite irrational to dismiss what is certainly the starling's most striking social accomplishment merely as a recreation devoid of purpose or survival value, and wiser to assume that a communal exercise so highly perfected is fulfilling an important function.

Wynne-Edwards concluded that the primary function of the roost is to bring members of a population unit together for an epideictic display. The result is to stimulate the adjustment of population density through either emigration of individuals of poor quality or a reduction in the reproductive rate. The roost's function as a sleeping place is secondary to its function as the location of the morning and evening displays. In fact, the concentration in the roost makes the birds an easier target for predators than they would be if they slept individually.

FIGURE 18-4 Aggregation of starlings
Wynne-Edwards argued that the roost serves primarily as a site for the morning and evening flight displays necessary for the regulation of population size. Other researchers offer explanations based on benefits to the individual, such as group protection in the roost or the improved foraging efficiency of groups.
Source: Photo by Eric Hosking.

Such synchronous mass behavior is not limited to colonial birds. The "evening rise," characteristic of brown trout (*Salmo trutta*) among others, occurs about twenty to thirty minutes after sundown and lasts about thirty minutes. The fish cruise in small circles in a particular area. Although the rise is influenced by weather and the abundance of flies (food), the relationship is not very close, and Wynne-Edwards considered the rise to be epideictic.

In the summer, the Bogong moths (*Agrotis infusa*) of Australia form large aggregations in caves. The moths become active for a half hour or so in the morning and evening. After vibrating their wings and crawling about, the moths fly around the cave and join others outside in a dense milling flight before they return to the same cave. There is no evidence of any feeding or mating associated with this behavior. The size of the population varies during the summer, and moths with depleted fat reserves disappear. Wynne-Edwards (1962) argued that the flights are epideictic displays that communicate information about population density and that function to reduce the population through the selective dispersal of individuals in poor condition.

However appealing this hypothesis, it has been largely rejected by biologists. Wynne-Edwards himself came to realize that the evolution of these self-sacrificial behaviors for the purpose of regulating population size requires some sort of group selection. Many behavioral ecologists have used Wynne-Edwards' hypothesis as an example of group selectionist thinking at its worst and point out that most of his ideas can be explained more simply by invoking the processes of individual and kin selection, as we shall see shortly.

TERRITORY SIZE. Researchers recognized early on the possible role of territory in population regulation in birds. Because populations of all species can potentially over-reproduce, exclusive defense of a resource may force some of the surplus population into marginal habitats, where birth rates are lower and death rates are higher than in optimal areas. One can readily demonstrate the presence of surplus animals by removing residents from optimal areas and then noting whether new animals come into the area from the marginal habitats. In his analysis of more than twenty years of breeding data for the great tit (*Parus major*) from Marley Wood in England, Krebs (1970) demonstrated that population is regulated by density-dependent decreases in both clutch size and hatching success. He studied the role of territoriality in limiting the number of breeding birds as follows: First he demonstrated that defense of nest boxes causes the birds to be spaced out more than would be

FIGURE 18-5   Territorial replacement of great tits
Six pairs of great tits, which were defending territories as indicated by shaded areas on map (a), were shot. After three days, four new pairs had moved in from less suitable habitats, as indicated by shaded areas on map (b). The remaining residents also expanded their territories.

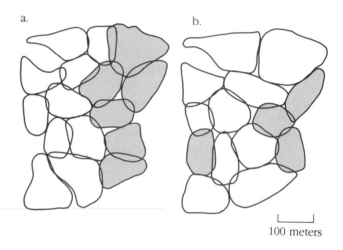

100 meters

expected if nest box occupation were random, so that breeding density is reduced locally (Krebs 1971). Next he studied the effect of removing birds. The arrival of new pairs within hours after he had removed pairs of birds suggested that surplus birds were in the area (figure 18–5). But where did these birds come from? Having color-banded most of the pairs in the area, Krebs found that nearly all the newcomers were from the hedgerows surrounding the woods. The lower reproductive success of birds that bred in hedgerows compared with those that bred in the woods demonstrates again that territories limit breeding in the optimal habitat. However, Krebs found that although territory size varied considerably from year to year, it was not related to food supply. To regulate the population, territory size would have to increase as food supply decreased. For this reason, he concluded that territory is not very important in regulating the population as a whole, and it has only local effects.

Several other lines of evidence suggest that territory size *is* related to limited resources and could therefore play a role in population regulation. The area defended by some species contains the minimum amount of resources to reproduce. For instance, in a study on red squirrels (*Tamiasciurus* spp.), Smith (1968)

A male comb-crested jacana (*Irediparra gallinacea*) escorts his offspring.

In this polyandrous species the typical sex roles are reversed, as females compete for mates and males raise the offspring (chapter 14).

Source: Photo by Terrence Mace.

PLATE 8

Communal spider web
Although the vast majority of spider species are solitary, a few to several thousand of these social spiders (*Anelosimus eximius*) live and cooperate in all daily activities (chapter 17).
Source: Photo by Karen Cangialosi.

PLATE 9

Communal nursing in lions (*Panthera leo*)
Females nurse young born to other pride members, a trait that may have evolved via kin selection since pride females are closely related (chapters 4 and 20).
Source: Photo by Anne Pusey.

PLATE 10

measured the food supply in each territory and concluded that each territory contained just enough food to provide for the energy needs of its occupant. Studies of a number of species have shown that territories are larger in marginal habitats. Nevertheless, proving that territory regulates population density is difficult.

In polygynous species such as redwings, male territory size is inversely related to habitat quality, but only other males are excluded and some feeding takes place off the territory. If the sex ratio is 50:50 and all females breed, it is not likely that male territory regulates the population as a whole. In white-footed mice (*Peromyscus leucopus*), male home ranges (which overlap to some extent and are therefore not strictly territories) decrease in size linearly with increasing density of males. It is hard to see how they could regulate the population, since the space they occupy is so compressible. Female home ranges are smaller, overlap little with those of other females, and are inelastic above a certain density; thus they might set an upper limit to the number of nesting females (Metzgar 1971).

The prevailing view of the function of territoriality is that it benefits individuals; any role in population regulation is likely to be localized. There is little experimental support for the idea that territoriality regulates entire populations, or that it has evolved as a means of population regulation per se.

DOMINANCE HIERARCHIES. Dominance hierarchies have also been implicated in population regulation. Lower-ranking individuals in groups with dominance hierarchies may have higher mortality rates and lower birth rates, as do surplus animals—those excluded from territories in prime habitats. In times of food shortage, high-ranking individuals survive, while low-ranking individuals starve or disperse. Most group-living animals are polygynous, and the effects of competition on reproductive success are severe in the male. The mating system of the elephant seal (*Mirounga angustirostris*) demonstrates this effect spectacularly: The top-ranking male does as much as 80 percent of the mating (Le Boeuf and Reiter 1988). However, a substantial turnover of males in the hierarchy takes place during the breeding season, so others do get a chance. Among redwings, nearly all females breed successfully each season; thus the male-based dominance system is ineffective in population regulation. Furthermore, for many other males in polygynous species with dominance hierarchies, the correlation between rank and reproductive success is low.

Sometimes the reproductive differential occurs in the female hierarchy. In temperate North America, paper wasp (*Polistes fuscatus*) nests are usually started each spring by a single queen, but she is joined by several other females (West Eberhard 1969). The foundress is dominant, and she suppresses reproduction in the subordinates, whose role is to regurgitate food for the foundress (trophallaxis) and to care for the young. In packs of wolves (*Canis lupus*), the alpha female inhibits mating by lower-ranking females; she is usually the only female in the pack to mate (Mech 1970). Drickamer (1974b) and Sade et al. (1976) found that among the rhesus monkeys (*Macaca mulatta*) on islands off the coast of Puerto Rico, matriarchies headed by high-ranking females are reproductively more successful than those headed by lower-ranking females. In fact, some of the latter actually lost numbers over the years. As with territoriality discussed earlier, these examples demonstrate density-dependent reproductive restraint by subordinate animals that may result in limiting population growth. However, there is no evidence that dominance hierarchies function for the purpose of regulating populations. Rather, they result from individual strategies as individuals attempt to maximize their inclusive fitness. (See chapter 12 for a further discussion of dominance.)

SOCIAL PATHOLOGY OF OVERPOPULATION. One research strategy for studying population regulation has been the creation of artificial populations in the laboratory. Such populations invariably cease to grow after following a roughly sigmoid (S-shaped) growth curve, even when food, water, and nesting space are kept in excess. Calhoun (1962, 1973) described the behavioral changes that took place in his mouse or rat "universes." Initially, males defended territories, and birth and juvenile survival rates were high. As density increased, aberrations began. One he called a **behavioral sink**: animals became conditioned to eating and drinking in the presence of others and thus restricted their activities to a few places in the cage. In mouse universes, large numbers of "grouped withdrawn" spent nearly all their time in piles. Occasionally fights broke out, and the jumping mice resembled popcorn. Those females that bore young built inadequate nests and abandoned litters.

Calhoun observed several aberrant types of behavior in male rats and mice in his universes, as well as the territorial, aggressive behavior found in most natural populations (figure 18–6). Pansexual males were

FIGURE 18-6 Researcher John Calhoun in mouse universe

Although food, water, and nesting space are present in excess of needs at all times, populations in cages such as this one stop growing and eventually die out because of the behavioral and physiological changes brought about by crowding.

Source: Photo by Nilo Olin/National Institute of Mental Health.

those that failed to discriminate between other males and females and made sexual advances to any creature that moved. Other males, called probers, were very active; although they did not compete for social rank or physical space, they showed hypersexual, homosexual, and cannibalistic behavior. Normally an estrous female rat, after being pursued by a male, returns to her burrow while the male waits outside. After the male performs a courtship dance, the female emerges and they mate. The probers, however, followed the female inside the burrow and sometimes consumed dead young that were there. Calhoun referred to another group of males as "beautiful ones." These males had flawless coats with no scars and walked about the pen with impunity, being ignored by territorial males and even nesting females.

At first, Calhoun thought these males had the highest ranking of all, but he later concluded that conspecifics did not recognize them as adult rats or mice because they were in a state of physical and social immaturity, even though of adult size. These males, when put in an uncrowded cage with receptive females, were unable to form dominance hierarchies, defend territories, or mate with females.

Females in these colonies showed delayed maturation, and few bred at all. In fact, one of Calhoun's mouse universes went extinct because the females all became too old to reproduce without ever coming into estrus. Reproductive inhibition in response to density has also been demonstrated in confined populations of other species, such as deermice (*Peromyscus maniculatus*) (Terman 1987).

Much attention has been paid to the role of agonistic behavior and interaction rates in laboratory populations of mice. Populations with particularly aggressive individuals tend to level off at lower densities than do those with less-aggressive members (Southwick 1955; Vessey 1967). When Vessey (1967), with the use of tranquilizers, lowered agonistic behavior of mice in populations that had reached the carrying capacity of the cage, litter survival improved and the population increased further (figure 18–7). These experiments suffer from several problems that make extrapolation to natural populations risky: Providing excess food, water, and nesting space creates an unrealistic environment, as does preventing emigration, which results in abnormally high densities.

Some researchers have tried to apply these findings to human populations. A number of studies have shown positive correlations between density and such variables as crime rate and mental illness, but these correlations tend to disappear when the level of income is controlled (Freedman 1980). People living in crowded, urban areas tend to have lower incomes than do those who live in low-density areas, and it is this poverty, rather than density itself, that is causally related to crime and mental illness. It also seems clear that sheer numbers per unit space are less important in social pathology than are the way space is utilized and the nature of the social interactions that take place. (See chapter 12 for more on aggression, and the book by Esser (1971) for studies on use of space.)

Although humans respond physiologically to stress like many other animals, there is no firm evidence that birth rates are reduced in response to stress. However, a number of reproductive restraints are evident in hunter-gatherer populations. Delayed onset of puberty, prolonged lactation (which delays resumption of ovulatory cycles), and preferential female infanticide are the

FIGURE 18-7 Effects of tranquilizers on fighting, litter survival, and population growth of caged wild house mice
In (a) the control population of mice reached asymptote (upper limit), then slowly declined. Four-sided figures refer to survival of litters. In (b) the population of mice reached asymptote, then increased when chlorpromazine, a tranquilizer, was added to diet (downward pointing arrow). Note the increase in litter survival and the decrease in fights when the tranquilizer was added. When the tranquilizer was removed (upward arrow), litter survival decreased, fighting increased, and population declined.

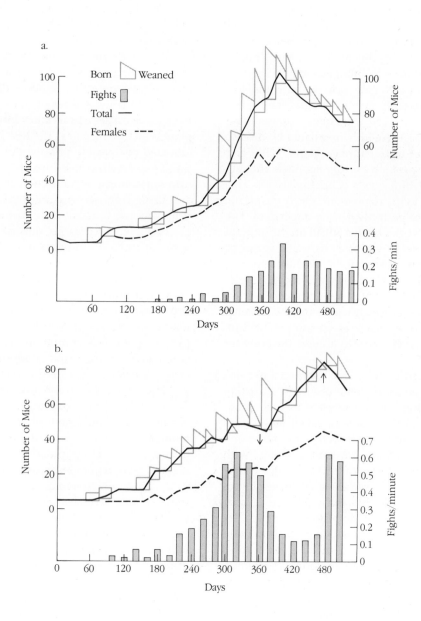

main ways family size is kept in check. Presumably these factors work in conjunction with mortality and emigration to adjust populations to the carrying capacity. Lee (1980) emphasized the importance of prolonged lactation and Dickemann (1975) argued the importance of infanticide.

BEHAVIORAL-PHYSIOLOGICAL MECHANISMS

Some animal species undergo seemingly regular nine-to-ten-year population cycles as in the case of the lynx (*Lynx canadensis*) and the snowshoe hare (*Lepus americanus*), or three- to-four-year cycles as with the vole (*Microtus* spp.) and the lemming (*Dicrostonyx* spp. and *Lemmus* spp.). For example, during cyclical peaks, hares are under practically every bush; during these population peaks a disturbance as slight as a hand clap was enough to send them into convulsions and coma, followed by death, apparently due to hypoglycemic shock (Green et al. 1939). In 1950 Christian proposed that mammalian populations could be regulated by disease caused by exhaustion of the adrenal gland, following prolonged psychological stress from agonistic interactions at high population levels (Christian 1950). This negative feedback loop, dependent on intraspecific competition, would be perfectly density-dependent. The idea grew from the work of Selye (1950) on the **general adaptation syndrome,** in which nonspecific stressors, such as heat, cold, or defeat in a fight, produce a specific physiological response. ACTH released from the anterior pituitary, under control of the hypothalamus, stimulates production of glucocorticoids by the adrenal gland. These hormones, such as cortisone, function mainly to elevate blood glucose to prepare the body for fight or flight.

The phenomenon of death due to adrenal exhaustion turned out to be an extreme case, and Christian modified his hypothesis after a series of laboratory experiments (Christian 1978). There is, in fact, an increase in adrenocortical output in response to increasing population density. These hormones, along with corticotropin-releasing factor (ACTH) and some others still being explored, seem to reduce direct fitness in numerous ways. The body's two main defense mechanisms, the immune and the inflammatory responses, are inhibited; these changes obviously increase the likelihood of morbidity or mortality. At the same time, growth and sexual maturation are inhibited, as are spermatogenesis, ovulation, and lactation (e.g., Rivier et al. 1986). Some of these effects on reproduction persist even into subsequent generations, in spite of a reduction in

population density (figure 18–8). Field data supporting these findings have come from studies of a variety of mammals, such as deer, rats, mice, and woodchucks (Christian 1978). Probably an equal number of studies have failed to find a consistent relationship between adrenocortical output and population density. Part of the problem is that the stressor is not density itself, as was seen earlier, but the agonistic behavior associated with competition for limited resources. Thus in laboratory mice, as little as two minutes per day exposure to a trained "fighter" mouse produces a pronounced stress response, lowering the body's defense against parasites (Patterson and Vessey 1973). Clearly, this negative feedback loop works under some conditions, but its generality in natural systems remains to be demonstrated.

Work with pheromones has augmented some of the above findings. Substances released in the urine produce effects on conspecifics without a physical encounter. In some species of mice, the smell of strange male urine causes a pregnancy block by preventing implantation, referred to as the *Bruce effect* (Bruce 1966); in crowded or unstable populations, the birth rate could thus be lowered. House mice (*Mus domesticus*) avoid the urine of a mouse recently defeated by another mouse, and urine from stressed mice produces an adrenocortical response in naive mice (Bronson 1971; 1979).

Grouped females void substances in urine that delay sexual maturation in other females (Drickamer 1974a). Six volatile organic compounds produced under the influence of the adrenal gland have been isolated by Novotny and colleagues (Novotny et al. 1986) and turn out to be three different classes of compounds: three are ketones, two are acetate esters, and one is 2,5-dimethylpyrazine. Various combinations of these substances, when added to the urine of adrenalectomized females, or to plain water, restore the delay effect.

The same delay in the onset of estrus has been produced in the laboratory by exposing young mice to female urine taken from high-density populations in the field (Massey and Vandenbergh 1980) In an experimental study, field populations were increased to simulate a population explosion by adding new mice (Coppola and Vandenbergh 1987). While urine from these resident females before the explosion had little effect on maturation of young mice in the lab, after the explosion these females' urine produced a pronounced delay. In contrast, odor from mature males' urine produces the opposite effect, accelerating the sexual maturation of young females (Vandenbergh 1969). Although most of these effects are best viewed as adaptations of individuals to gain reproductive advantage, they may act incidentally to regulate population size.

FIGURE 18-8   Christian's model of population regulation in small mammals (simplified version)
As population increases (upper left), social contacts increase, activating the pituitary-adrenal gland system and inhibiting the production of sex hormones. The population then declines, due to the decrease in the birth rate and the increase in the death rate. As the population declines, further social contacts are reduced, and hormone changes are reversed.

Source: Data from J. J. Christian, "Neurobehavioral Endocrine Regulation of Small Mammal Populations," in *The Pymatuning Symposia in Ecology,* 5:143–158, 1978.

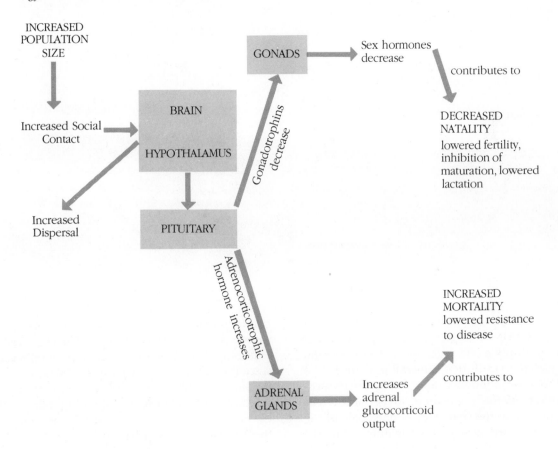

BEHAVIORAL-GENETIC MECHANISMS

The ideas examined so far are complementary; they are all based on the notion that the behavior and/or physiology of animals changes as a result of increased interaction rates associated with high population density. A competing hypothesis suggests that natural selection favors different genotypes at high and at low population levels. While working with voles (*Microtus* spp.) that have a three- to four-year population cycle, Chitty (1960) noticed that populations continued to decline even under seemingly favorable environmental conditions. Voles from declining populations were highly aggressive and intolerant, and bred poorly; voles from increasing populations were mutually tolerant and were rapid breeders. He postulated that there was a change in the quality of animals in a declining population; through natural selection, the proportion of aggressive individuals increased, and even though they could compete in crowded conditions, their reproductive rates

FIGURE 18-9   Chitty's model of population regulation in small mammals (simplified version)

In response to increased population and social contact, there is selection for aggressive individuals with low reproductive rates leading to a decline in population. Modifications of the model emphasize the role of emigration of certain phenotypes in causing the decline. This model predicts genetic changes in the population as different genotypes are favored at high versus low densities.

Source: Data from D. Chitty, "The Natural Selection of Self-Regulatory Behavior in Animal Populations," *Proceedings of the Ecological Society of Australia*, 2:51–78; and from C. Krebs, "The Lemming Cycle at Baker Lake, Northwest Territories, during 1959–1962," *Arctic Institute of North America Technical Paper No. 15, 1964.*

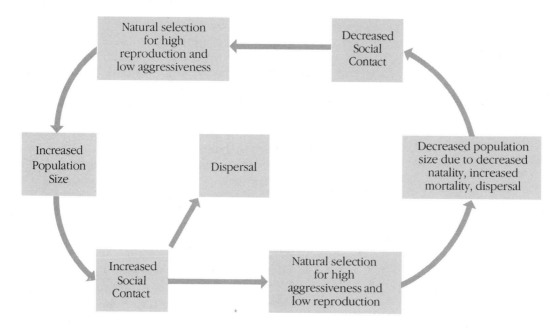

were low and the population declined (figure 18–9). It seems unlikely that a change in gene frequency could take place over only a few years, but some confirming evidence has come forth that dispersal of animals of a certain genotype can produce rapid changes in the sedentary portion of the population. Myers and Krebs (1971) reported that the frequency of one allele in the blood serum of voles was significantly different in dispersers as compared to residents. However, the behavioral changes reported in vole populations have not been linked to specific genes.

Evidence from the laboratory shows that the age of puberty can be shifted in mice (*Mus domesticus*) after only a few generations of artificial selection (Drickamer 1981). The low birth rates observed in peak populations in the wild could be due in part to selection for late-maturing individuals.

Tamarin (1980) has emphasized the role of dispersal itself as a regulating mechanism. If dispersal is blocked by a fence or is blocked naturally, as on islands,

regulation fails and high populations and depleted resources result (the *fence effect*, Krebs et al. 1973). Dispersers are likely to be at a disadvantage reproductively since they are probably in a suboptimal habitat; in addition, predation is higher on transient individuals (Metzgar 1967). However, emigrating individuals have the potential of colonizing new areas and may even be the founders of new species (Christian 1970).

Several other mechanisms have been proposed that incorporate both genetics and aggressive behavior. One starts with the fact that individuals behave less aggressively towards kin than non-kin. As a population increases, the likelihood that nonrelatives will be interacting goes up, leading to an increase in aggression, followed by population decline (Charnov and Finerty 1980). A second idea links increased heterozygosity with increased aggression. Several studies have shown that heterozygous individuals are more aggressive than homozygous individuals. As a population increases, so does dispersal and outbreeding, leading to an increase in heterozygosity. The

FIGURE 18-10   Levels of selection, individual to interdemic

Small groups of four to six lions represent families of related individuals. On the left side of a barrier is deme A. This deme shows a case of individual selection in which the individual in the upper left dies, and a case of kin selection in which the family to the right is eliminated. On the right side of the barrier, group selection occurs when the entire deme B is wiped out.

Source: Data from E. O. Wilson, "Group Selection and Its Significance for Ecology," *BioScience*, 23:631–638, 1973.

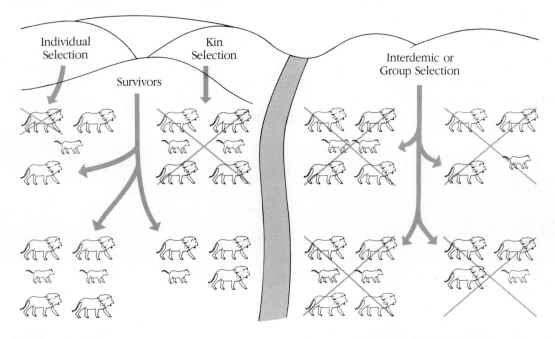

Deme A of Family Groups   *Deme B of Family Groups*

resultant increase in aggressiveness could force a decline by increasing mortality and decreasing reproduction, as descibed in the previous sections (Smith et al. 1978). So far there is little hard evidence in support of either of these hypotheses.

EVOLUTION OF POPULATION SELF-REGULATION

Proponents of self-regulation have not yet resolved two important problems. First, because most of the data are from laboratory populations where densities are often unrealistically high and where emigration is usually prevented, more testing on natural populations is needed. However, in natural populations many extrinsic factors come into play, and no single mechanism can be implicated. Second, if we accept the idea that such self-regulatory mechanisms exist, do they evolve specifically for that purpose, and if so, what is the unit of selection?

We can probably agree that self-regulation is adaptive for a population or species that has the potential for destroying or using up its resources, but many of the mechanisms we have discussed seem to be maladaptive for the individuals involved. An animal that is genetically programmed to respond to increased fighting by cutting back on reproduction will lose out, genetically speaking, to another individual that can keep on reproducing. If an animal that responds to aggression by reducing its defense mechanisms is by definition less fit, how could such a system evolve?

One answer is by group, or interdemic, selection, as Wynne-Edwards (1962, 1986) argued (figure 18–10).

Groups or populations that avoid overexploitation of the environment survive and may later colonize the habitats left vacant by imprudent groups that became extinct. Although group selection is considered by most evolutionary biologists to be theoretically possible, and although several mechanisms for it have been proposed (Lewontin and Dunn 1960; Wilson 1980), the rate of change in gene frequency would usually be too slow, when compared with the rate of change through individual selection. To maintain a trait that is adaptive to the group but not to the individual, group extinction would have to be more rapid than individual extinction—and that does not seem to be very often the case. Therefore, most population biologists favor arguments based on individual selection and reject Wynne-Edward's hypothesis.

Arguments based on kin selection offer one way around the problem. We now know that many social groups are made up of close relatives. Thus behavioral and physiological traits that reduce direct fitness but increase indirect fitness, in this case by increasing the likelihood of survival of a kinship group, could be selected for. Evolutionarily speaking, it might be worthwhile to sacrifice one's reproductive output if the survival of the group as a whole—which contains relatives who share your genes through common descent—is ensured.

A second, more widely accepted, answer to the question of self-regulation is that these so-called mechanisms are nothing more than the consequence of individuals' behaving in their own best interest. Reproductive restraint in the face of overcrowding may be the best individual strategy; later maturation or production of fewer young can save energy for a more optimal time. Ensuring the greatest genetic representation in the next generation is not synonymous with having the most offspring. Lack (1954) concluded that clutch size (the number of eggs laid) in birds evolved to an optimum number in terms of the number of offspring that will in turn survive and reproduce; laying too many eggs could result in survival of few or no young. We

can understand many of the epideictic displays described by Wynne-Edwards (1962) simply as outcomes of intraspecific competition of individuals that are behaving so as to increase direct fitness.

Similarly, at times of high density, aggressive individuals that are intolerant of others would be selected for if limiting factors were operating (Chitty's hypothesis). The most successful individuals at times of high density might be those that reduce the fitness of others through competition. We should keep in mind that fitness is a relative term that refers to increasing one's genes in the next generation relative to others'.

The inhibition of the body's defense systems as a result of increased adrenocortical output at high population is more difficult to explain (Christian's hypothesis). How could such a response be adaptive to the individual under any circumstances? The pituitary-adrenal response is part of the general adaptation syndrome; only animals that are being attacked and defeated show this response, whereas dominant animals, even in crowded situations, respond little and are like individuals at low densities. The general adaptation syndrome (GAS) enables an animal under stress to mobilize energy reserves just to survive. Other systems, such as the reproductive system, are temporarily shut down; producing offspring at such times would most likely be a waste of energy. Low rank or exclusion from a territory suitable for breeding may be a temporary situation. In fact, many territory and dominance systems are age-graded, and younger animals, usually males, attain prime territories or high rank if only they can live long enough.

The preceding arguments suggest that self-regulatory mechanisms do not evolve as such for the purpose of controlling populations. Most scientists who test hypotheses about the functions and origins of behavior patterns assume that those patterns evolved by means of natural selection's acting at the lowest level consistent with the facts (Morgan's Canon, chapter 2). This level is usually that of the individual and its offspring (Williams 1966). Arguments for selection at higher levels (group or population) are invoked only if lower level selection cannot explain the observed traits.

SUMMARY

The number of organisms present in a population equals the sum of births and immigrants minus the sum of deaths and emigrants. All populations have the capacity for rapid increase; understanding the *abiotic* and *biotic* factors in the environment that limit numbers is a central problem in ecology.

Density-independent factors, including floods, droughts, and fires, are unaffected by the individuals or the numbers of individuals in a population. *Density-dependent* factors are affected by the numbers of individuals in a population and cause increases in the death rate, decreases in the birth rate, or increases in emigration. These factors may include disease, predation, and competition.

Some evidence suggests that the behavior involved in *interference competition* for limiting resources is important in population regulation. Most of the research on the role of behavior in population regulation has focused on *self-regulatory mechanisms*, in which increases in *intraspecific competition* produce behavioral, physiological, or genetic changes. The result of these changes is that the population is regulated below the *carrying capacity*, and overexploitation of resources is avoided.

Communicating information about limiting resources might enable populations to adjust birth, death, and movement rates and to track environmental resources. Some observers argued that communal displays—such as dancing of gnats and midges, choruses of birds, as well as mass feeding and roosting flights—function to limit density. Because this idea seems to involve sacrifice of individual fitness for the good of the group, attention has shifted more recently to regulation that can be understood in terms of the action of natural selection on the individual rather than on the group.

Population growth rates can be affected by *territorial behavior*, in which some individuals are excluded from breeding or are forced to occupy marginal habitats, or by *dominance hierarchies*, in which subordinate individuals are less successful in obtaining mates or rearing offspring. In laboratory populations where emigration is usually restricted, manifestations of *social pathology* appear as the population increases. Birth rates are lowered by such things as interference in courtship or copulation, or by inadequate maternal care. Death rates also increase, and the population stabilizes or even declines. Physiological changes associated with increasing population density and agonistic behavior increase death rates and decrease birth rates in laboratory and field populations. The most important changes are in *adrenocortical hormones*, which are secreted in response to defeats suffered in social fighting. *Pheromones* also may be important in population regulation because, for example, the odor of strange male mice blocks pregnancy in females, males avoid the odor of recently defeated mice, and group-housed females produce an odor that delays sexual maturation of other females.

Another possible self-regulatory mechanism is change in gene frequencies in a population. As numbers increase, *emigration* of certain genotypes may occur. Those that remain are genetically adapted to live under crowded conditions; they are highly aggressive but have low reproductive rates. As the population declines, natural selection favors individuals that are less competitive but that have high reproductive output.

Some self-regulatory mechanisms seem to involve individual sacrifice in reproduction or in survival for the good of the population. However, it is widely believed that group selection operates too slowly to lead to the evolution of such mechanisms; selfish individuals would increase in numbers at the expense of the altruists. Reproductive restraint is probably a temporary strategy that allows resources to be devoted to surviving under intense competition by postponing reproduction until resources become more plentiful; selection here is operating at the level of the individual.

Discussion Questions

1. Understanding the factors that affect the size of populations is a central problem in ecology. Why has social behavior been implicated in so many of the theories of population regulation?

2. Natural selection acting at the level of the individual would seem to favor those with high reproductive success. How can individuals that produce fewer young ever be favored?

3. Statements suggesting that many factors contribute to regulating animal populations seem to run counter to Liebig's law of the minimum. Is there any way to resolve this apparent contradiction?

4. The data below refer to mice that were exposed two minutes per day either to trained fighter mice, to trained nonfighter mice, or to no mice for days 1 through 14. On day 8, each mouse was fed an equal number of tapeworm eggs. On day 22 the mice were killed, their adrenal and thymus glands weighed, and the intestinal tapeworms counted and weighed. Adrenal gland weight is positively related to adrenocortical hormone secretion. The thymus is involved in the immune response. Discuss these data as they may relate to population regulation and behavior.

Effect of fighter mice exposure on mean host weights, organ weights, and tapeworm number

	Groups exposed to		
	No mice	Trained non-fighter mice	Trained fighter mice
Group size	10	10	7
Adrenal weight (mg)	4.39	4.42	5.88[*]
Adrenal weight (mg)/body weight (gm) × 100	13.1	13.1	16.6[*]
Thymus weight (mg)	37.0[a]	35.0	21.9[*]
Body weight (gm)	33.7	34.0	35.8
Worm number/mouse	11.6[b]	15.4[b]	75.3[**]
Worm weight/mouse weight (mg)	6.37	4.04	82.28[**]

Source: Data from M. A. Patterson and S. H. Vessey, "Tapeworm (*Hymenolepsis nana*) Infection in Male Albino Mice: Effect of Fighting among the Hosts," *Journal of Mammalogy,* 54:784–786, 1973.
[*]Significantly different from other two groups, $p < .01$.
[**]Significantly different from other two groups, $p < .001$.
[a]Mean based on nine weights.
[b]Three of these mice had no worms.

Suggested Readings

Cohen, M. N., R. S. Malpas, and H. G. Klein, eds. 1980. *Biosocial Mechanisms of Population Regulation.* New Haven: Yale University Press.
Reviews theories of self-regulation. Several papers deal with attempts to apply Calhoun's and Christian's ideas to population regulation in humans.

Krebs, C. J. 1978. A review of the Chitty hypothesis of population regulation. *Can. J. Zool.* 56:2463–80.
Reviews hypotheses of population regulation and the types of field data needed to test them.

Krebs, C. J. 1985. *Ecology: The Experimental Analysis of Distribution and Abundance.* 3d ed. New York: Harper & Row.
Part 3 covers population growth. Chapter 16 deals specifically with theories of population regulation.

Sibly, R. M. and R. H. Smith, eds. 1985. *Behavioural Ecology: Ecological Consequences of Adaptive Behaviour.* Boston: Blackwell Scientific.
Another collection of papers from a symposium. This one explores the link between behavioral ecology and population dynamics. Sections on the relationships of feeding, spacing, breeding, and social behavior to population dynamics.

Behavior as a factor in the population dynamics of rodents: New concepts and approaches. 1987. *American Zoologist* 27:821–969.
A collection of papers reviewing current research on the role of behavior in both laboratory and field populations of mice and rats.

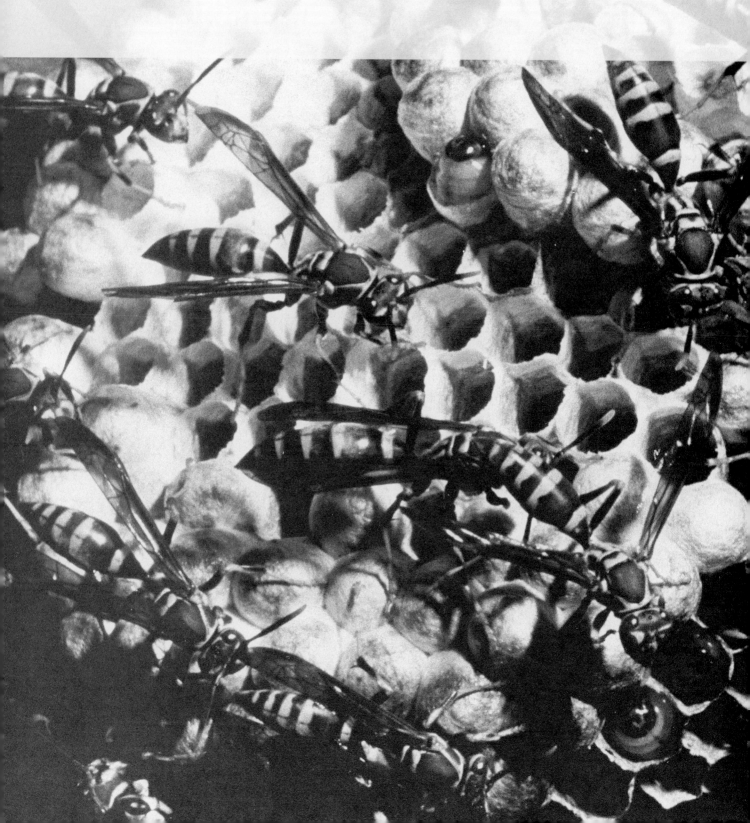

19

EVOLUTION OF BEHAVIOR PATTERNS

*T*hroughout this book we have recognized that behavior has a genetic basis, even though the specific genes involved and their exact contribution to a behavior pattern are poorly understood. We have examined the diversity of behavior patterns, the physiological mechanisms that cause them, and the biotic and abiotic environmental factors that affect them. In the last two chapters we examine how the link between genes and behavior, and the documentation of change in the frequencies of those genes, can demonstrate the evolution of behavior.

Most of this chapter deals with the evidence that behavior does in fact evolve. We consider the phenomena of phylogeny, adaptation, and the formation of new species in documenting genetic changes in behavior. At the end we describe behavioral changes that occur through nongenetic means: tradition and culture. Chapter 20 looks specifically at the evolution of social behavior.

Wilson (1975) pointed out that behavior is the part of the phenotype that is most likely to change in response to environmental change. If selection pressure shifts, behavior will usually change first, and physical structure, second. Because behavior is the evolutionary pacemaker, studying change in behavior from an evolutionary perspective is important; but the lability of behavior—that is, its distance from the genes—makes the study difficult. The events that led to a change in behavior are, of course, rooted in the past. We can never duplicate those conditions exactly, so our evidence is largely indirect.

There is a widespread tendency, particularly among nonbiologists, to talk about *progression* in evolution and to imply that some traits or species are advanced and others are primitive. The term *progression* is a holdover from orthogenetic interpretations that viewed evolution as proceeding in a predetermined direction independent of external factors. However, we should talk of progress in evolution only in reference to either conformity to phyletic trends, or to an approach to some arbitrarily designated final stage (Williams 1966): Thus forms that share traits with that arbitrary final stage could be said to be advanced, whereas those that are closest to the ancestral stage could be said to be primitive. The evidence that groups "progress" in a more general sense or that adaptations become more "perfect" can usually be countered by examples of ancient, primitive forms that are highly successful (for example, sharks) and by the fact that adaptations in one direction usually mean trade-offs in some other area. Improvements in terrestrial locomotion, for example, are usually associated with decrements in swimming ability. Nevertheless, as will become evident from the examples that follow, many researchers still speak of "improvements" in traits through evolutionary time.

FIGURE 19-1   Oblique long call in common and herring gulls

The sequence of oblique long calls reads from left to right. The difference in the call between the two species is primarily the amplitude of the movement of the (b) and (c) components. The herring gull, for example, jerks its head further down (b) while the common gull throws its head back further (c).

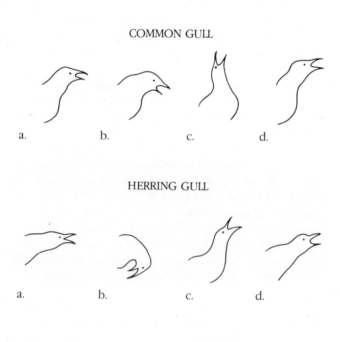

COMMON GULL

a.   b.   c.   d.

HERRING GULL

a.   b.   c.   d.

THE PACE AND MODE OF EVOLUTION

Evolutionary biologists have generally assumed that natural selection produces large-scale, qualitative change through the accumulation of small, quantitative increments. These small changes, referred to as **microevolution,** may come about through changes in thresholds in the sensory, motivational, or motor systems that control behavior patterns (Manning 1971). Changes in thresholds lead to modifications in frequency or amplitude of displays. For example, the long calls of common gulls and herring gulls (genus *Larus*), which function to advertise territory, are similar, but the two species emphasize different parts (figure 19–1). The basic pattern involves jerking the head down, then throwing it back while calling. The herring gull has increased the amplitude of the downward jerk of the head; the

common gull has increased the amplitude of the throwing back of the head (Tinbergen 1959). Such differences probably function in species recognition. Microevolution also seems important in enabling organisms to adjust to changing environments—but can it lead to the evolution of new species, as Darwin assumed? Gould and Eldredge (1977) think not, and have argued that new species arise in peripherally isolated populations through rapid change followed by periods of stasis, a process they term **punctuated equilibria,** rather than by slow, continuous transformation in the central ancestral area.

EVIDENCE FOR THE EVOLUTION OF BEHAVIOR

In this section we examine the types of evidence used to demonstrate how behavior changes in a more or less permanent way through generations. The evidence is of three general types: (1) **phylogeny,** the tracings of the patterns of evolutionary change through time; (2) **adaptation,** the way natural selection actually works; and (3) **speciation,** the formation of new species (Alexander 1978).

PHYLOGENY

A surprising amount can be learned about the history of a population or species by comparing the behavior patterns of both extinct and living species.

FOSSILS. Although the fossil record is one of the strongest lines of evidence supporting evolution, behavior patterns themselves do not leave fossil remains. However, we can infer much about behavior from bones, teeth, horns, tracks, and human and nonhuman artifacts such as tools.

Feeding behavior has been inferred from fossilized tunnels of apparently closely related worms of the Gaphoglypt family that lived 2 to 135 million years ago (figure 19–2). The lines of descent show a gradual shift from rigid behavior patterns toward an increased flexibility that enabled the worms to forage more efficiently (Seilacher 1967, 1986). There is also a trend toward specific changes in foraging patterns. The simplest foraging left tracks resembling the scribbling of children. The "program" for scribbling seems to have

---

FIGURE 19-2   Fossilized worm tunnels from millions of years ago
Worm trails in sediment provide a record of foraging behavior. Earlier types left scribbles (left), more recent types left meanders (right).
Source: Data from A. Seilacher, "Fossil Behavior," *Scientific American,* 217:72, 80, 1967.

SCRIBBLES

MEANDERS

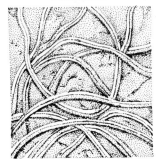
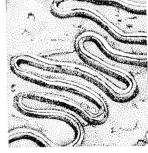

had one command: Keep heading to one side but don't stay in the same track. A more complex pattern was an outward spiral, the commands being: Keep circling in one direction and keep in touch with the spiral whorl made earlier. More complex—and more efficient in terms of covering the layer containing food—was a series of meanders. In this case the commands seem to have been: Move horizontally, keeping within a single layer of sediment; after advancing one unit of length, make a U-turn; never come closer to any other tunnel than some given distance.

Raup and Seilacher (1969) programmed a computer with these commands and generated tracks very similar to those found in the ancient sediments. These worm tunnels may have been more than deposit feeding trails or dwelling structures; they could have acted as traps for mobile microscopic organisms, much as the tunnels of moles are thought to entrap earthworms and arthropods (Ekdale 1980). Seilacher claims that scribbling appeared in Ordovician times (425 to 500 million years ago) and eventually disappeared. Complex meanders and dense double spirals did not appear until the Cretaceous period (63 to 135 million years ago).

Evidence of predation can be gained from drill holes in the fossilized shells of bivalve mollusks from the late Triassic (about 200 million years ago). These holes were made by carnivorous snails that penetrate the shells of their prey. Oddly, the habit seems to have disappeared, only to reappear some 120 million years later (Fursich and Jablonski 1984). Thus, what we might consider a

real breakthrough in predatory technique did not persist; the presumed selective advantage of such a behavior pattern may not be a good predictor of its long-term survival. In further studies of snail predation on molluskan prey during the last 100 million years, Kitchell (1986) has reached the following conclusions: Prey selection is nonrandom and is governed by the same currencies and constraints as is foraging in present-day organisms. Despite the extinction and appearance of new species of both predators and prey, there is little evidence of change and the same rules of prey and site selection have persisted for more than 100 million years.

Evidence about the behavior of extinct mammalian carnivores was obtained when Hunt et al. (1983) excavated ancient burrow systems containing the skeletons of bear dogs from the early Miocene, 20 million years ago. These animals were about the size of a wolf or hyena and apparently used the dens much as do modern carnivores.

We can also glean much information about social behavior and mating systems by studying fossils. In mammals, the amount of sexual dimorphism in body weight and length is a good predictor of the degree of polygyny (Alexander et al. 1979). Across a wide variety of species, the larger or heavier the male is relative to the female, the greater the number of females the male monopolizes. In many cases, the amount of sexual dimorphism can be estimated from fossilized remains.

ADAPTIVE RADIATION.   Isolated areas such as oceanic islands offer the possibility of tracing phylogeny. Because of small target size and large distance from a colonizing source, such places are not likely to be invaded by many species (see chapter 16). In the relative absence of interspecific competition, new forms evolve rapidly. The best studied case of such adaptive radiation is the Galápagos Islands, some 600 miles off the coast of Ecuador. Biologists believe that a single species of finch (family Fringilidae) colonized the Galápagos and radiated into fourteen species in four genera (figure 19-3). Darwin was the first to study these birds on his 1835 voyage on *H.M.S. Beagle;* Lack (1954) drew together much of the available information. Although adaptive radiation is evident in such traits as beak morphology and plumage, feeding and breeding behaviors provide evidence of a common ancestry and adaptive radiation as well. The ancestral form was probably a heavy-beaked ground finch that fed on seeds, from which evolved modern ground finches; cactus-feeding finches with long, decurved bills; insectivorous forms, including the famous woodpecker finch that uses a stick in its beak as a tool for probing underneath bark for insects; and slender-beaked warblerlike finches that feed on small insects.

FIGURE 19-3   Adaptive radiation of Darwin's finches on the Galápagos Islands

Small, isolated islands with low colonization rates serve as useful sites to study evolution and adaptation. It is likely that a single species of finch gave rise to all these types.

Source: Data from D. Lack, *Darwin's Finches: An Essay on the General Biological Theory of Evolution.* Copyright © 1961 Harper & Brothers, New York. First published in 1947 by Cambridge University Press, Cambridge, England.

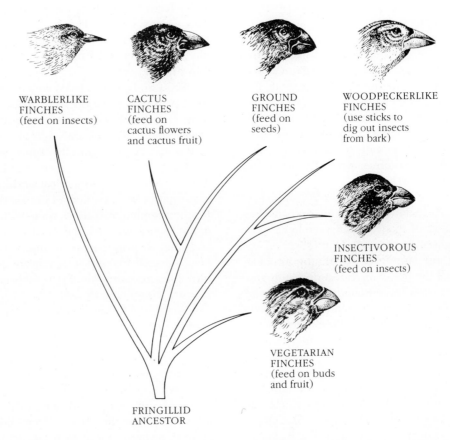

WARBLERLIKE
FINCHES
(feed on insects)

CACTUS
FINCHES
(feed on
cactus flowers
and cactus fruit)

GROUND
FINCHES
(feed on
seeds)

WOODPECKERLIKE
FINCHES
(use sticks to
dig out insects
from bark)

INSECTIVOROUS
FINCHES
(feed on insects)

VEGETARIAN
FINCHES
(feed on buds
and fruit)

FRINGILLID
ANCESTOR

EVOLUTION OF COMMUNICATIVE BEHAVIORS. In his study of gulls, Tinbergen (1959) concluded that displays are derived movements. For instance, herring gulls (*Larus argentatus*) sometimes peck the ground vigorously and pull grass—behavior that represents displacement collecting of nest material. The movement, however, is combined with attack postures; a bird pulls grass as hard as it would pull at the feathers of an opponent. Tinbergen's interpretation was that the initial movement is an attack movement redirected at the ground, while the grass-pulling behavior (displaced collecting of nest material) is superimposed. Tinbergen believed that finding the origin of this particular pattern requires no comparison with intermediate links because it is so similar to the patterns from which it is

derived. However, many displays have undergone **ritualization,** an adaptive evolutionary change away from noncommunicative functions and towards increased efficiency as a signal (see figure 11–17). In our discussion of communication in chapter 11 we found that the main sources of these derived movements are: (1) behavior patterns immediately evoked by the situation, such as attack movements triggered by the intrusion of another into one's territory, which are either performed incompletely (intention movements) or oriented at objects other than the evoking stimulus (redirected movements); or (2) movements from patterns other than those evoked by the stimulus, which are unexpected and functionally irrelevant (displacement activities).

HOMOLOGY. One of the cornerstones of research by ethologists, which began in the early 1900s, has been the idea that behavior patterns can be treated like anatomical features. They "dissected" behavior patterns to learn their "anatomy." They also demonstrated **homologies** in behavior, which are patterns shared by species through descent from a common ancestor. For instance, Van Tets (1965) studied birds of the order Pelecaniformes, scoring species for the presence or absence of behavioral traits and comparing the pattern with the phylogenetic tree as inferred from morphology (figure 19–4). Some behavior patterns, such as the prelanding call, are common to all members of the order; others, such as head-wagging seen in gannets and boobies, are restricted to one family. The general conclusion from this and other comparative studies (for example, Lorenz 1972) is that the distribution of behavioral similarities and differences in a group of species tends to be correlated with phylogenetic relationships as disclosed by morphological or other means (for example, the fossil record or protein analysis).

DOMESTICATION. Artificial selection by humans has changed the behavior of many animals we have associated with for thousands of years. Domestication involves more than just taming or socializing animals; humans also control the breeding, care, and feeding of domesticated animals. Some ethologists have argued that behavioral plasticity is greater in domestic animals than in their wild counterparts, and that **neoteny**—the persistence of juvenile characteristics in adults—is typical in domestic animals (Lorenz 1952). Others have emphasized the degeneracy of domesticants; they cite the laboratory rat. But few of these generalizations hold true when the entire spectrum of domesticated animals is considered (Price 1984; Ratner and Boice 1975). Certain species are preadapted to domestication. For instance, most carnivores maintain a nest, and urinate and defecate some distance away from it; they are thus suited to living in human habitations. Other factors facilitating the process of domestication include large social groups, the presence of males in the group year round, precocial young, and an omnivorous diet (Hale 1969).

The evolution of dog behavior provides the most striking example of selective breeding by humans (Scott and Fuller 1965). The wolf (*Canis lupus*), ancestor of the dog, and the principal carnivore of the northern hemisphere prior to the arrival of humans, has a complex, cooperative social system. As the wolf was domesticated and transformed into the dog, it spread all over the world with humans and underwent adaptive radiation. Dogs (*Canis familiaris*) were used for protection, for

hunting, and for herding other domesticated animals. The relative isolation of human tribes 8,000 to 10,000 years ago, with only occasional genetic interchange, probably encouraged the evolution of distinct dog breeds. As human contact increased with improved transportation, the different breeds of dogs mixed, and many breeds, such as those in South Africa and North and South America, lost their identity. Only in the last few hundred years have humans practiced rigid artificial selection, which has resulted in increased separation of dog breeds. The great diversity of form and behavior of current breeds illustrates the potential behavioral effects of natural and artificial selection of certain phenotypes.

Scott and Fuller (1965) concluded that the behavioral patterns of the dog and wolf are essentially the same. Selection has modified primarily the agonistic and investigatory systems. For example, the bulldog breed, in the interest of an English sport, was selected for its tendency to attack the nose of a bull and hang on, in contrast with the more typical slashing attack from the rear used by wolves. Terrier breeds have been selected for their tendency to attack prey relentlessly, regardless of any injury suffered; the usual wolf pattern is to snap and then withdraw to avoid injury. Other breeds have been selected for opposing reasons; scent hounds and bird dogs are examples of peaceable animals that can be kept in groups in a kennel.

Wolves and dogs search for prey rather than lie in wait, and therefore have a well-developed investigatory repertoire. Some dogs, such as scent hounds, are bred for their ability to follow a trail. Others, such as bird dogs, use their visual and olfactory senses much more equally: After rapid quartering of ground, they locate prey by scent only when they are a few paces away, and then they freeze. Dogs must be trained for all of these tasks, but in each case artificial selection has produced a phenotype that is predisposed to specific behavioral traits (Scott and Fuller 1965).

COMPARATIVE SERIES. Perhaps the best available evidence for the evolution of behavior has come from comparative studies of closely related forms. The presence of a series of intermediate forms between two extremes suggests what the evolutionary sequence was, but it may not represent the exact route because the living intermediates may have changed from the ancestral type. It is usually assumed that behaviors evolve from simple to complex, and such series are often referred to as progressions; but inferring goal-directed change in evolution is unwise, as we noted at the beginning of this chapter.

FIGURE 19-4 Behavioral traits and phylogeny in the Order Pelecaniformes

The top half of this figure shows the distribution of behaviors among genera. Dots indicate that the behavior is present in that genus. In the bottom half of the figure is a phylogenetic tree based on morphological traits. Note that genera that are closely related based on morphology also share the most behaviors, whereas those distantly related share the fewest.

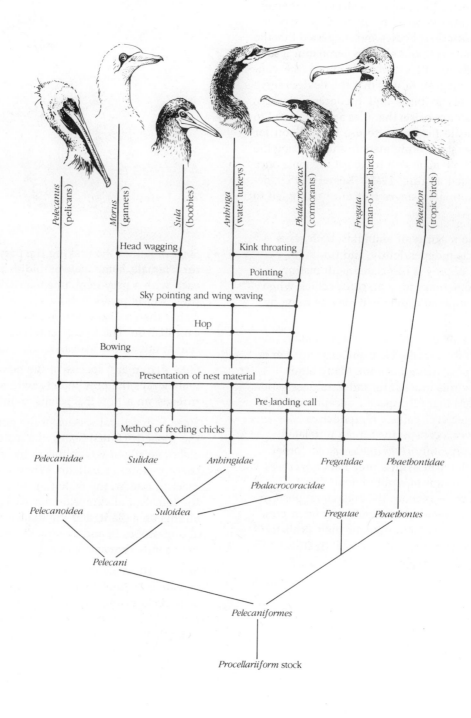

One comparative series is that of balloon flies, members of the family Empididae (order Diptera). According to Kessel (1955), the story began in 1875 when Baron Osten-Sacken was visiting in the Swiss Alps and noticed bright silvery flashes among the shadows of the fir forest. Believing that they were silvery insects he used his net to capture some, but ended up with dull-colored flies. Later he noticed that along with male flies, he had netted packets of filmy material. Osten-Sacken assumed that the males carried the material on their backs to attract females during the mating swarm. Another observer later discovered that the male flies carry the packets underneath their bodies and suggested that the somewhat flattened balloons serve as aeronautical surfboards on which the flies cavorted among the sunbeams. Another suggestion was that the devices serve as warning signals to birds and predaceous insects; however, no evidence exists that flies are distasteful to predators. The male flies seem to use the balloon for attracting females, stimulating mating, and reducing the likelihood that the female will try to eat the male once they have coupled (figure 19–5). Kessel (1955) described the evolutionary stages that most likely led to this situation:

- *Stage 1.* In the majority of empidids, both sexes capture insects independently, and no presentation of prey is associated with mating. These flies sometimes prey on conspecifics; when the male attempts to copulate, it may be eaten by the female.

- *Stage 2.* The male captures a prey item and presents it to the female. He copulates with her as she eats the prey instead of him. Many other insects follow this procedure, and natural selection theory predicts that intraspecific deception should occur if the deceiver gains a reproductive advantage over a conspecific. (In an unrelated insect, the scorpionfly *Hylobittacus apicalis* [order Mecoptera], the male with a prey item advertises to females by means of a pheromone. But sometimes a male assumes the behavioral posture of a female; he approaches a male that has a prey item, lets the male try to copulate, then steals the prey item and either eats it or uses it to attract a female [Thornhill 1979]).

FIGURE 19-5   Male balloon fly carrying balloon
Males with balloons fly in a swarm from which a female selects a mate. That male gives the balloon to the female prior to copulation. In this species (*Hilara sartor*), the silken balloon has replaced the prey item that the male offers to the female in other closely related species.

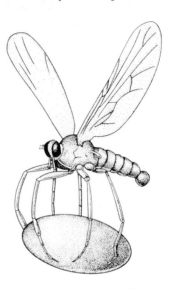

- *Stage 3.* Rather than taking the prey and searching for a female, some male empidids join other males, each with a prey item, in an aerial dance. The prey, Kessel posits, is now a stimulus for mating rather than a distraction to avoid mate cannibalism. The female enters the swarm and selects one of the males.

- *Stage 4.* In many species of the genus *Hilara*, the male wraps the prey loosely with some silken threads, an action that seems to quiet the prey.

- *Stage 5.* In several species of the genus *Empis*, from the western United States, the male applies elaborate silken wrappings to the prey, which then resembles a balloon. When male and female meet in midair, the male transfers the balloon to the female and climbs on her back. The pair alights on a plant, and the female rolls the balloon about, probes it, and eventually consumes the prey item while the male copulates with her.

- *Stage 6.* The male catches a small prey and may consume its fluid so it is no longer edible; he then constructs a complex balloon. The female accepts the balloon and plays with it during copulation, but gets no meal from it.

FIGURE 19-6   Regurgitative food offering during copulation

Food offerings during copulation occur in many kinds of animals, such as in the stilt-legged fly (family Micropezidae) shown here. This behavior may be a form of parental investment by the male. By giving the female a protein-rich meal, the male may increase the survival of their offspring. *Source:* Photo by E. S. Ross.

- *Stage 7.* The prey item is very small, of no food value to either sex, and pieces of it are plastered at the front end of the balloon. The balloon is now the sole stimulus for copulation.

- *Stage 8.* The *Hilara sartor* male gives the female a balloon that has no prey at all (figure 19–5). This behavior is thought to be the final stage in the series.

Kessel argues that the nuptial feeding of the female initially functioned to reduce the chances of the male's being consumed by the female during mating. Thornhill (1976) suggested an alternative function based on parental investment. Male insects may invest in their offspring by providing nutrition to their mates in the form of glandular secretion, prey captured by the male, regurgitative food offering (figure 19–6), or the male himself. Females may select mates on the basis of the quality of food the males offer. In the case of the balloon flies, the series seems to have evolved in the direction of deception of females by males.

It is useful, if not critical, to have an independent assessment of phylogenetic relationship with which the behavioral series can be compared. For instance, many wasps (order Hymenoptera) capture prey to provide food on which they lay their eggs and on which the larvae subsequently feed. Some species carry the prey anteriorly with the mandibles; other species carry the prey posteriorly, sometimes on a terminal abdominal segment called an "ant clamp" (Evans 1962). The series, with many intermediate forms, correlates well with other evolutionary evidence, and we can be confident of the direction of change. The selection pressure directing the change from anterior to posterior carriage is thought to be cleptoparasitism. When carrying prey with the mandibles, the wasp must put the prey down to open or dig a burrow. Other species of flies or wasps, called cleptoparasites, can then lay eggs on the prey as well and thus diminish the food resource for the wasp's offspring.

Among vertebrates, the evolution of pigmentation in the mouth-breeding cichlid fish (*Haplochromis* spp.) was studied by Wickler (1962). Presumably under the selection pressure of egg predation, some species of cichlid females pick up eggs after spawning (i.e., after the eggs have been laid and externally fertilized) and "incubate" them in the mouth. The faster the females pick up the eggs after laying, the less the chance of predation. In some species the females take the eggs before they are fertilized by the male. The male presents his anal fin, which bears yellow spots resembling eggs; the female snaps at these spots and takes in sperm, which fertilize the eggs in her mouth. Comparative studies have shown that these yellow spots evolved from less conspicuous pearly spots on the vertical fin that have no signaling function. Eibl-Eibesfeldt (1975) pointed out that the following preadaptations existed:

1. Female readiness to pick up eggs,
2. Pearly spots on the vertical fins of the male,
3. A lateral display by the male during courtship and spawning.

It is more common that the intermediate species of the series are not available; we must then draw evidence from related species that have evolved more or less individually from some common ancestor. In this situation, only a weak inference can be made by assuming that those behavior patterns common among the various living species represent truly ancestral traits.

FIGURE 19-7   Main trends in evolution of nest building in the weaverbird family (Ploceidae)

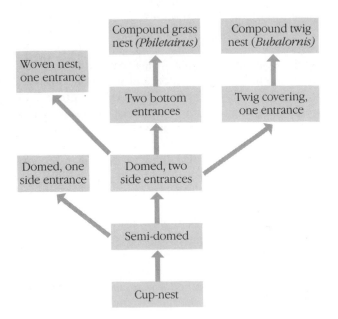

FIGURE 19-8   Weaverbird nests of the genus *Pseudonigrita* from Kenya
Note the two bottom entrance holes in each nest.
Source: Photo by Stephen H. Vessey.

Using this approach, Tinbergen (1959) compared the behaviors of different species of sea gulls (family Laridae). In particular, he examined those patterns that are relatively constant throughout a species and of the **fixed pattern** type, that is, they are developmentally stable and resistant to environmental change. The basic similarities in displays throughout the family strengthen the conclusion, previously based on morphology, that the family is **monophyletic,** that is, it originated from a single ancestral type. If a species is judged to be closely related to other species because it shares many of their characteristics, any differences can be considered to have been recently acquired.

A variety of animals construct objects for prey capture, shelter, or mate attraction—for example, spider webs, caddisfly cases and nets, bee hives, bird nests, and bowers. It is relatively easy to examine these semipermanent structures and to compare closely related species. Collias and Collias (1963) studied the nests of African weaverbirds of the family Ploceidae, tracing the evolution of the nest from the ancestral cup nest to intricate domed structures with bottom entrances (figure 19-7). Primitive members of the family begin the nest as a simple cup or platform; in some species a dome is added later (figure 19-8).

The true weavers (subfamily Ploceini) make a ring from which the roof is built, followed by the floor. Other species construct a communal thatched roof, with brood chambers of many individual pairs underneath. The researchers developed a key to the nests of the true weavers and actually suggested some taxonomic revisions in the family based on nest construction.

The techniques used to weave the nest from grasses also vary among the weavers, and the Colliases noted trends toward more regularity in the pattern, tighter weaving, and finer stitches within the subfamily (figure 19-9). In general, species making the more advanced nests used the more refined pattern of weaving. In many species, the male builds the basic nest and, along with neighboring males, advertises it to the female, who makes the final choice. Males may make many nests before one is chosen for completion by the female, implicating sexual selection in the evolution of nest construction.

BIOGENIC LAW. Can the development of behavior during an organism's life cycle tell us anything about the evolutionary history of that behavior? If changes in a characteristic tend to be added on late in an individual's development, the earlier stages could be considered primitive. Young peacocks, for example, use the primitive form of food enticement before developing the adult form (see figure 11–18). This reasoning led

FIGURE 19-9   Types of stitches used by true weaverbirds (*Ploceini*)

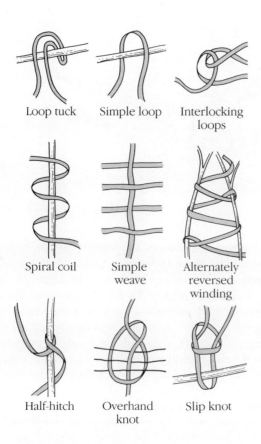

Loop tuck   Simple loop   Interlocking loops

Spiral coil   Simple weave   Alternately reversed winding

Half-hitch   Overhand knot   Slip knot

Haekel to his biogenic law: Ontogeny recapitulates phylogeny. But, as de Beer (1958), and more recently Gould (1977), have pointed out, no developing organism really goes through the actual endpoints of its ancestors. Although the mammal goes through a gill-slit stage during development, it is by no means a fish at that point. Haekel's law has been rejected by nearly all scientists, and its application to the study of the evolution of behavior is unwarranted.

STUDY OF ADAPTATION

Many studies have attempted to illustrate the adaptiveness of behavior by emphasizing the ways in which behavior patterns ensure an organism's survival and reproduction. The courtship of ring doves (chapter 7) is adaptive because it coordinates the physiological processes necessary for reproduction in the male and female. But does such evidence really tell us anything about evolution by natural selection? Natural selection leaves only the best adapted or optimal phenotypes for our inspection.

ADAPTIVE STORIES. A common practice is to study an organism's traits or behavior patterns and to propose an adaptive story for each trait considered separately. Behavior patterns or structures found to be suboptimal are considered to be the results of trade-offs among competing demands within the same organism and thus also the result of adaptation. The possibility that present traits are not adaptive to present conditions is usually not considered. Gould and Lewontin (1979), among others, have criticized this approach; they argue that present traits are constrained by phyletic heritage, developmental pathways, and general architecture and are not just the result of current selective forces. As one example, they cite adaptive stories told about the utility of the tiny front legs of the dinosaur *Tyrannosaurus*, legs so small they didn't even reach its mouth. One purported function of such legs was to help the animal rise from a prone position. While the front legs were no doubt used for something, we know them to be homologous to the conventional, full-sized limbs of dinosaur ancestors. Rather than explaining the evolution of short limbs for a specific purpose, one can propose a developmental correlation between the apparent reduction in front leg size and the relative increase in head and hind limb size. The present function of a structure or behavior is not proof that it evolved for that purpose.

We can still test hypotheses critically, however, and make predictions about the way natural selection is supposed to work. A number of researchers are currently comparing behavior patterns with theoretical optima that might be achieved in the best of all possible worlds. The models of optimal foraging discussed in chapter 17 were constructed to predict strategies based on maximizing some parameter, such as rate of energy intake. The degree to which such models fit data collected in the field permit an understanding of the workings of natural selection and other selective processes in causing changes in gene frequency. The most useful models are those that make testable and falsifiable predictions.

OBSERVING NATURAL SELECTION IN THE FIELD. The best evidence for the evolution of behavior is documentation in a natural population. Perhaps the strongest case we have so far is the work of Kettlewell (1965) on morphological change in the peppered moth, *Biston betularia*, where **melanism,** the development of dark pigmentation, was a result of the darkening of the sub-strate by soot. As the Industrial Revolution led to increased burning of coal, the bark of trees used as resting places by the moths became darker. Those moths whose darker color matched the background were assumed to be subject to less predation than were the lighter-colored moths. A behavioral change in background preference must also have occurred so that the moths remained cryptic. In recent years, the frequency of dark morphs has declined in proportion to the reduction in air pollution (Cook et al. 1986).

Intense natural selection has been demonstrated in one of the ground finches (*Geospiza fortis*) of the Galá-pagos (Boag and Grant 1981). On one of the islands, Daphne Major, regular rainfall throughout the early 1970s produced abundant seeds, and seed-eating finches thrived. In 1977, a drought occurred and the finches failed to breed, declining in numbers by 85 percent. The corresponding decline in seed supply was nonrandom, with small seeds becoming scarce, and large, hard seeds remaining relatively common. The result was strong selection for birds with large beaks that could handle the seeds, and within a short period of time, the average beak size increased dramatically. Boag (1983) went on to show that beak size is heritable and that offspring showed a strong phenotypic response to natural selection of their parents.

MATE RECOGNITION SYSTEMS AND SPECIATION

During the process of speciation, behavioral changes may take place that restrict the exchange of genes between two segments of the population, should they come in contact (see chapter 4). Of particular importance are signals associated with mate choice since they may lead to nonrandom mating within the population. Although often referred to as isolating mechanisms, there is no clear evidence that they evolve for the purpose of keeping two populations or species apart. Rather, their function may be just the opposite: to increase the likelihood that members of the same population mate with one another. Such signals become what has been called by Paterson (1980) a *Specific-Mate Recognition System*, or *SMRS*. In this view, members of a population that have adapted to local conditions will have higher fitness if they mate within the population; reproductive isolation—the inability to mate with members of other populations or species—thus becomes an incidental effect of the SMRS.

FIGURE 19-10   Spiny lizard of the genus *Sceloporus* used in head-bobbing experiments
Source: Photo courtesy of the Ohio Department of Natural Resources.

Whatever the causes of population- or species-specific signals, there are many examples of them, some of which we pointed out in chapter 11. Lizards of the genus *Sceloporous* (figure 19–10) are morphologically very similar and perform a species-specific head bob during courtship. Hunsaker (1962) developed a head-bobbing machine that moved the heads of models in various species-typical patterns. Females given a choice among models chose those that bobbed in the pattern of their own species (figure 19–11).

Smith (1966) conducted experiments on the signal function in two natural populations of sympatric but noninterbreeding species of Arctic gulls. The glaucous gull (*Larus hyperboreus*) has a yellowish eye ring; Thayer's gull (*Larus thayeri*) has a reddish purple eye ring. When the eye rings of Thayer's males were painted to look like those of glaucous males, glaucous females paired with them (figure 19–12). Once paired, however, the painted Thayer's males refused to copulate with the glaucous females until the latter were painted to look like Thayer's females. Further experiments on other sympatric gull species demonstrated the critical importance of eye-ring color in species recognition. Such manipulations, which change signals to provide the "wrong" information, are a powerful means of demonstrating signal function.

Species-recognition signals are most important when two species are sympatric and thus might interbreed. Natural selection leads to greater differences between two closely related species in areas where they overlap than in areas where they do not. Observers have

FIGURE 19-11   Head-bobbing patterns of spiny lizards
Males of each species have a distinctive head-bobbing
pattern. The height of the bob is shown on the vertical axis
and time is shown on the horizontal axis. The female is
attracted to the male that bobs in the pattern of that female's
species.

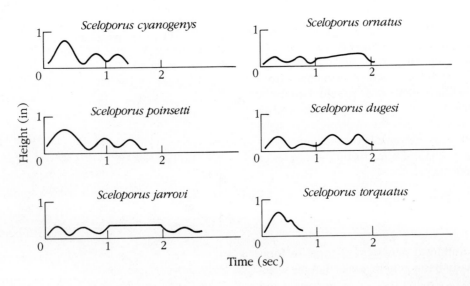

**FIGURE 19-12   Experimental study of eye-ring color in
mate recognition systems of gulls**
(a) Principal visual features thought to function in species
recognition: eye-ring, iris, mantle (top of wings and back),
and wing tips. (b) Results of altering the eye-ring color in
mated female Kumlien's gulls before copulation. A change
from purple to white resulted in the break-up of 21 of 34
pairs; other alterations and controls were relatively
ineffective.

noted this phenomenon, referred to as **character displacement,** primarily in morphological traits, particularly feeding structures, such as bill length in birds. The divergence in bill length is greatest in areas of sympatry because of intense competition between the closely related species: each specializes on a different resource. We can observe a similar effect for behavioral traits. Littlejohn (1965) recorded the songs of two species of Australian tree frogs, *Hyla ewingi* and *H. verreauxi*. In zones of sympatry, the songs appear to be quite distinct; in zones of allopatry, the songs appear to be similar (figure 19–13). In playback experiments, Littlejohn and Loftus-Hills (1968) gave gravid females a choice of two songs. They chose their own species' song when the choice was between the two species in sympatry, but showed no preference between songs of allopatric species. (See chapter 4 for further examples of character displacement.)

TRADITION

Although the evolution of persistent changes in form and behavior usually has a genetic basis, natural selection may act to produce both plasticity of expression and the tendency to transmit behavior to offspring through nongenetic means. For example, experiences of female rats can bias the behavior of offspring and even grand-offspring (Denenberg and Rosenberg 1967). The physiological mechanisms that cause these changes in behavior are poorly understood, but the effects of environment on the behavior of subsequent generations probably are more widespread and important in producing persistent changes in behavior than are generally recognized.

We have a better understanding of the role of tradition, the passing of specific forms of behavior from generation to generation by learning. Compared with genetically controlled behavior, tradition-transmitted behavior can spread through a population rapidly, sometimes in less than one generation. One common form is dialect in animal communication; birdsong has received the most attention. Males learn song during a critical period when they imitate phrases heard in their neighborhood. Male immigrants into a local area may adopt the local dialect. The adaptive significance of this phenomenon could be to enhance communication between male and female in pair bonding. It may also maintain the separateness of locally adapted populations (see chapter 11).

FIGURE 19-13   Oscillograms of songs of two Australian frog species
The amplitude of each note of the song is shown on the vertical axis, and time is shown on the horizontal axis. Under each song recording is a 50-cycles-per-second reference line. The songs of the two species are similar when comparing individuals caught in areas where the two species do not overlap (allopatry). Where the two species do overlap (sympatry), song differences are pronounced. This divergence of behavior between closely related sympatric species is referred to as character displacement.

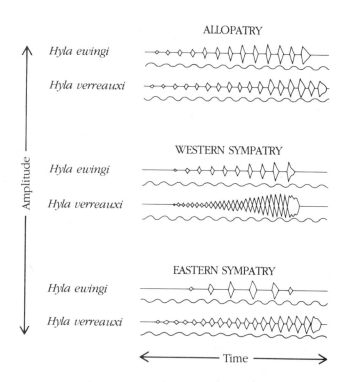

Migratory animals tend to use the same routes and return to the same places as their ancestors. Presumably the young learn the routes from their elders. Waterfowl, for example, use traditional routes and stop at the same places each year during migration, and grazing mammals use the same trails for many generations. Breeding grounds may also be used year after year; in lek species the mating arenas may persist for centuries. Tradition is particularly important in species living in closed social groups with prolonged socialization of young.

Most examples of the social transmission of acquired behavior are from insects, birds, and mammals (reviewed by Galef 1976 and Bonner 1980). Among coral reef fish, French grunts (*Haemulon flavolineatum*) exhibit social traditions of daytime schooling sites and twilight migration routes (Helfman and Schultz 1984). Individuals transplanted to new schooling sites and allowed to follow residents used the new migration routes and returned to the new sites even in the absence of resident fish. Naive, control grunts released at the same new sites failed to use the new routes or to return to the new sites.

Nonhuman primates can develop new behavior patterns through invention and incorporate them rapidly and rather permanently into their social systems. Chapter 10 discussed the techniques for washing sand from potatoes and wheat grains introduced by a Japanese macaque (*Macaca fuscata*) and copied by others in the group.

Another type of learned behavior that occasionally spreads through populations is tool use. Tools enable otherwise generalized species to specialize in particular tasks without evolving morphological adaptations. For example, animals may use sticks or similar objects, rather than specialized beaks, to extract insects from crevices in tree bark (see also chapter 17).

Although the mechanism of behavioral change involved in tradition and human culture usually is assumed to be quite different from the mechanism of genetic change brought about by natural selection and drift, learned responses can be highly stereotyped and may persist for many generations. The big difference between learned traits and genetically controlled traits is the speed with which learned traditions can spread through the population, most notably in human societies. An old controversy, which sociobiology has recently kindled anew, concerns the extent to which differences in human social behavior are influenced by genetics and by culture.

Durham (1979), among others, attempted a synthesis of human cultural and genetic evolution and argued for the coevolution of genes and culture. During the organic evolution of *Homo sapiens*, according to Durham, there was selection at the level of the individual for the capacity to modify and extend phenotypes on the basis of learning and experience. As the capacity for culture evolved via natural selection on individuals, the developing culture in each population was adaptive for the members of that population and increased its inclusive fitnesses. The process was further accelerated by group-level cultural selection, which operates analogously to natural selection and independently of it. Competition among cultures results in selection of the "more fit" cultures, those that best exploit the environment to their own advantage (Dawkins 1976). Durham emphasizes, however, that cultural fitness is dependent on the reproductive success of the individuals that make up a culture. In other words, a sort of natural selection process operates at the cultural level, but ultimately the fate of a population is determined by natural selection operating at the level of the individual.

SUMMARY

Behavior is that part of the phenotype most likely to change in response to environmental change; behavior is thus the evolutionary pacemaker. European ethologists were the first to treat behavior patterns like anatomical features, "dissecting" them to learn their structure and thus to infer their evolution. Homologous behaviors are those shared by species through descent from a common ancestor.

Evidence for the evolution of behavior comes from a variety of sources. Fossils occasionally tell us about behavior, though behavior must often be inferred from morphology. *Adaptive radiation,* the rapid appearance of new forms, as occurred on the remote islands of the Galápagos archipelago, permits comparison among closely related species and with the presumed ancestral type. Ethologists believe that communicative behavior patterns, or *displays,* have evolved from noncommunicative patterns of several types, including *intention movements, redirected movements,* and *displacement activities. Domestication* shows the potential for modifying behavior and morphology, where humans, rather than nature, are the selective agents. Through comparisons among closely related species, we can sometimes trace the probable course of evolution from one extreme to another when the intermediate forms are still extant. The adaptiveness of behavior does not, by itself, demonstrate evolution, but observers can test predictive models of the way natural selection operates. Direct observation of evolution by natural selection is strong evidence, but it is rarely documented. Two examples are industrial melanism in moths and beak size in one species of Darwin's finches.

The evolution of behavior plays a crucial role in the restriction of genetic exchange between populations in the formation of new species. We are most familiar with the species-specific mating patterns that restrict matings to members of one population. *Character displacement* occurs in areas where closely related species overlap geographically. In such areas of sympatry, the species differ more than where they are allopatric (nonoverlapping).

Natural selection may act to produce plasticity of expression and the tendency to transmit behavior to others through nongenetic means, such as learning.

Manipulating the maternal environment can produce behavioral changes in offspring and grandoffspring. Song dialects and knowledge of migration routes are examples of *tradition*, passed to others by learning. Tool use and novel feeding techniques, as seen in non-human primates, illustrate the speed with which these potentially permanent changes in behavior can spread through a population. The evolution of *culture* in humans initially involves natural selection for the ability to extend the phenotype on the basis of learning and experience. Cultures themselves may evolve in a manner analogous to natural selection.

Discussion Questions

1. Discuss some examples of homologous behavior patterns other than those already mentioned. Can you think of behaviors that are analogous—that is, behaviors that are similar between species but not because of common ancestry?

2. Discuss an example of what Gould and Lewontin (1979) have referred to as "adaptive stories" to explain the function of a behavior pattern. How could factors other than evolution by natural selection have led to such behavior?

3. The following table provides measurements (in mm) of skull length in closely related species of weasels and the culmen (the top ridge of a bird's beak) length in closely related species of nuthatches and finches. Both features reflect feeding habits. What conclusions can you draw about the role of competition in the evolution of these species?

	Sympatry			Allopatry		
	Male	*Female*	*Both sexes*	*Male*	*Female*	*Both sexes*
Weasels (skull length)						
Mustela nivalis	39.3	33.6		42.9	34.7	
Mustela ermina	50.4	45.0		46.0	41.9	
Nuthatches (culmen length)						
Sitta tephronota			29.0			25.5
Sitta neumayer			23.5			26.0
Finches (culmen length)						
Geospiza fortis			12.0			10.5
Geospiza fuliginosa			8.4			9.3

Source: Data from G. E. Hutchinson, "Homage to Santa Rosalia *or* Why Are There So Many Kinds of Animals?" in *American Naturalist* 93:145–159, 1959.

Suggested Readings

Alexander, R. D. 1962. Evolutionary change in cricket acoustical communication. *Evolution* 16:443–67.
In a classic study using the comparative method, Alexander reconstructs the evolution of cricket songs from courtship movements of the male.

Brown, J. L. 1975. *The Evolution of Behavior.* New York: Norton.
An advanced text that emphasizes the adaptedness of behavior. Draws heavily on examples from the literature.

Gould, S. J., and R. C. Lewontin. 1979. The spandrels of San Marco and the Panglossian paradigm: A critique of the adaptationist programme. *Proc. Roy. Soc. Lond.* B 205:581–98.
Argues against the tendency of biologists, especially behavioral ecologists, to explain every morphological and behavioral trait as an adaptation to current environments.

Nitecki, M. H., and J. A. Kitchell, eds. 1986. *Evolution of Animal Behavior. Paleontological and Field Approaches.* New York: Oxford Univ. Press.
Papers in the first part use paleontological evidence to deduce behavior of animals as diverse as snails and dinosaurs that lived millions of years ago; the second part contains studies of the evolution of mating systems and social behavior among recent forms.

20

EVOLUTION OF SOCIAL SYSTEMS

Two of the most rapidly advancing areas of animal behavior studies are at opposite ends of the organizational spectrum. At one end is **psychobiology,** in which the main goal is understanding the nature of the central nervous system. At the other end is **sociobiology,** study of the biological basis of social behavior.

By the 1960s, the study of animal societies increasingly emphasized field studies—particularly of insects, birds, and nonhuman primates—and focused on the relationship between social structure and ecology. Although animal behaviorists were always concerned with the apparent adaptiveness of particular social systems, they did not to any great extent investigate the mechanisms of how sociality actually evolved. Many researchers sensed a potential conflict between Darwinian evolution, in which selection acts on individuals, and the existence of sophisticated cooperative behavior. They assumed that group, population, or species benefits very likely outweighed individual sacrifice. More recently, observers consider society to be an outgrowth of the action of natural selection at the level of the individual or gene. (See chapters 4 and 18 for a further discussion of group selection.)

We can define a **society** as a group of individuals of the same species that is organized in a cooperative manner extending beyond sexual and parental behavior. In this chapter, we briefly review examples of sociality in several taxonomic groups, discuss the evolutionary advantages and disadvantages of social behavior, consider in some detail the way natural selection has brought about the evolution of social systems, and finally, mention some applications of sociobiology to our understanding of human social behavior.

EXAMPLES OF SOCIAL SYSTEMS

Sociality has evolved independently in animals ranging from invertebrates to primates. We might expect an increase in complexity of social behavior as we move from simpler to more sophisticated organisms, yet by some criteria just the opposite is true: Some of the simplest invertebrates have more complex societies than do birds or mammals. Here we briefly examine social systems in some major taxonomic groups.

COLONIAL INVERTEBRATES

Cooperation, the aggregation of individuals for mutual benefit, and interdependence are so highly developed in many invertebrate forms that deciding what constitutes an individual and what constitutes a society presents a problem. Hydrozoans of the phylum Coelenterata (Cnidaria) are a striking example of a colonial system. Individuals are physically united in colonies and specialize as reproductive or nonreproductive types. The basic unit is the polyp, which consists of a tubular body formed from two cell layers, and a central mouth surrounded by soft tentacles. Although one genus (*Hydra*) represents organisms that are single, free-living polyps, most of the genera are colonial, and the polyps (also called *zooids*) show a division of labor. In genera such as *Obelia*, some zooids feed and defend by means of stinging cells, called nematocysts, and others reproduce. Another form, which lives on the shells of

FIGURE 20-1   Portuguese man-of-war

This colonial invertebrate of the genus *Physalia* is actually a group of individuals. Four types of polyps (zooids) make up the colony, with each type specializing in either flotation, feeding, defense, or reproduction.

Source: Photo © D. P. Wilson from Eric and David Hosking.

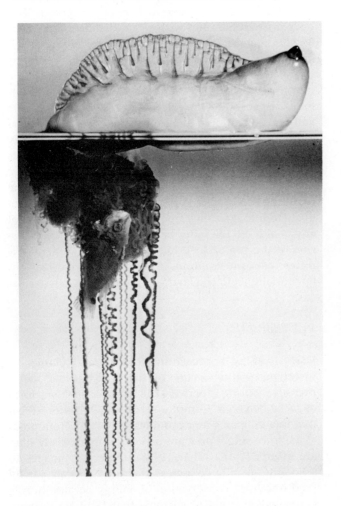

hermit crabs, has three types of zooids: feeding, reproducing, and fighting. A fourth type of zooid, found in the Portuguese man-of-war (*Physalia*) (figure 20–1), produces a gas-filled float.

In deciding whether or not an organism like *Physalia* is a society or a single individual, we should note that the entire colony comes from a single fertilized egg (zygote). The different zooids form by budding off asexually. Since all the zooids are genetically identical, the man-of-war may be considered a single individual with several organ systems. However, when comparing other Hydrozoans that show varying degrees of coloniality,

it appears that the formation of complex colonies originated from the grouping of many individual zooids; in some species the zooids forming the colony are unrelated (Wilson 1975).

From the standpoint of the individual, why have some zooids evolved into feeding or fighting structures with no reproductive potential? Marine invertebrates would seem to gain several advantages from colonial life (Wilson 1975). First, many colonial invertebrates live in shallow-water coastal areas, where sediment and wave action would devastate a single hydroid polyp. Colonies such as corals have a calcareous base that anchors and elevates the individuals above the ocean floor. Second, colonial forms such as *Physalia* are pelagic, or oceanic. Singly, the polyps are restricted largely to underwater surfaces, but by teaming up with those that form the gas-filled float, the colony can travel on the high seas. Third, the large colonial invertebrates that are sessile can smother or "shade out" competing forms. Fourth, the concentration of zooids that bear nematocysts is very high in some forms, such as *Physalia*.

SOCIAL INSECTS

Social complexity rivaling that of colonial coelenterates occurs in some groups of insects. Three traits characterize the truly social, or **eusocial** insects: (1) cooperation in the care for young; (2) reproductive castes cared for by nonreproductive castes; (3) overlap between generations such that offspring assist parents in raising siblings. Intermediate forms exist; thus quasi-social forms have only trait 1, such as orchid bees of the family Apidae, where five to twenty females each lay their eggs in a communal nest (Michner 1974). Semisocial forms have traits 1 and 2, such as the sphecoid wasp, *Microstigmus comes*, that builds a communal nest and apparently has only one egg layer (Matthews 1968). Eusociality among insects is limited to termites, ants, social wasps, and social bees (Wilson 1975).

Most of the eusocial insects are in the order Hymenoptera (wasps, bees, and ants). As an example, we look briefly at ants, which are the dominant social insects in terms of distribution and numbers. (Chapter 17 discusses army ants and fungus-growing ants.) Ant colonies (figure 20–2) range in size from a few hundred to thousands of individuals. The queens are winged when they emerge from the pupae; the workers, always female, are smaller and wingless. Often the winged males and queens engage in mass nuptial flights away from the nest; they swarm on prominent landmarks and

FIGURE 20-2   Polyrhachis ant colony
Visible are workers, larvae, and pupae. In these eusocial
insects, members of the worker caste are sterile females that
care for their mother, the queen, and help to rear their
siblings.
Source: Photo by Edward S. Ross.

the males attempt to copulate. Once the queen is inseminated, she sheds her wings, excavates a cell in the soil,
and rears the first brood of workers. In the more advanced species of ants, only one queen is usually found
in a nest. Workers typically come in several sizes: some
forage, others care for the next brood, and some—the
soldiers—defend the colony. Workers of many species
lay special alimentary, or trophic, "eggs" that nourish
other workers, the queen, and larvae. Males, which
come from unfertilized eggs, disperse, mate with
queens, and then die.

Wilson (1975) lists the following responses as typical of social insects:

1. Alarm
2. Simple attraction (multiple attraction =
   "assembly")
3. Recruitment, of individuals to a new food source
   or nest site
4. Grooming, including assisting with molting
5. Trophallaxis, the exchange of oral and anal liquid
   food
6. Exchange of solid food particles, such as trophic
   "eggs"
7. Group effect, either increasing a given activity
   (facilitation) or inhibiting it

8. Recognition, of both nestmates and members of
   particular castes
9. Caste determination, either by inhibition or by
   stimulation

Eusociality is not restricted entirely to the Hymenoptera. Termites, of the order Isoptera, are eusocial insects from a totally different ancestry, although they
have many similarities with eusocial ants (figure 20–3).
Most of the species are in the family Termitidae. Their
energy source is wood cellulose, which is digested by
flagellate protozoans or other mutualists that live in the
intestinal tracts of the termites. Winged forms of termites disperse from the colony in a mass exodus. A male
and female pair off and together begin construction of
a new nest; they do not copulate until later. In early
broods, workers of both sexes are produced; soldiers
appear in later broods; and winged reproductives are
the last to appear, sometimes many years after the
founding of the colony. Although termites do not produce trophic eggs, they frequently exchange liquid oral
and anal food, the latter being necessary for transmission of the protozoans. Like social Hymenoptera, termites use pheromones in colony organization,
territoriality, and the laying of trails.

VERTEBRATES

FISH. Nothing approaching coelenterate colonies or
insect eusociality occurs in the cold-blooded vertebrates; schooling is perhaps the most conspicuous form
of social behavior. Complex schooling behavior in fish
that live in open water provides protection from predators, since single prey are picked off more easily than
are grouped prey (Shaw 1978). Other advantages to
schooling seem to be increased ability to locate patchy
food resources, conservation of energy because heat is
generated, creation of currents that facilitate locomotion, and increased likelihood of finding mates. Other
types of social behavior in fish revolve around defense
of nest sites and usually do not involve nonparents.
Some species of fish are territorial, with the male playing
a large role in defense of nest and eggs. Cooperation by
nonparents was reported in one species of Antarctic fish,
in which a male may take over the guarding of eggs if
the female disappears (Daniels 1979). Daniels argued
that such behavior is altruistic, since there is little likelihood that that particular male is the father.

AMPHIBIANS AND REPTILES. Although amphibians and
reptiles do not show evidence of large, complex societies with division of labor, they do demonstrate many

FIGURE 20-3 Queen termite surrounded by workers
The reproductive male ("king") of the genus *Macrotermes* is in the upper left. Termites are the only eusocial insects not in the order Hymenoptera.
Source: Photo by Edward S. Ross.

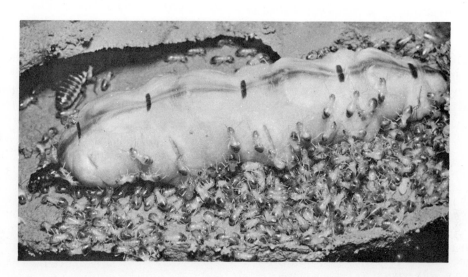

complex social interactions related to territory defense and obtaining mates; a few species engage in parental care. Large groupings are uncommon except in *hibernacula* (overwintering aggregations) and in the mating choruses of some frogs and toads. Bullfrog (*Rana catesbeiana*) males establish and defend individual territories in chorusing areas in one part of a pond. A gravid female enters a territory and is grabbed by the resident male who fertilizes the eggs externally as the female lays them within that territory (Howard 1978). A small number of the larger males do most of the mating; and eggs laid in the territories of these males have higher hatching success. Water depth, water temperature, and substrate type in these territories are optimal for egg development and are generally free of the leeches that feed on eggs laid in the territories of smaller males. Some males use alternate mating strategies. For example, a *parasitic* male does not call but remains in the territory of a large male and grabs the female before she reaches the resident male (figure 20-4). Others, referred to as *opportunistic* males, call from specific areas and act like territorial males; if challenged, however, they flee to another area and resume calling.

Many species of lizards defend territories in the breeding season. At high population densities, the territorial system may break down, and a dominance hierarchy may develop (Brattstrom 1974). In the lizard

FIGURE 20-4 Territorial male bullfrog with parasitic male
The parasitic male is the smaller of the two and is facing forward. Females are attracted to the call of the territorial male; when a female enters the territory, the parasitic male attempts to mate with her before the territorial male has a chance.
Source: Photo by Richard D. Howard.

Anolis aeneus, several females live within the territory of a male. The females, in turn, are arranged in a dominance hierarchy of at least three levels (Stamps 1973). Parental care is generally not extensive among reptilian species; exceptions are in the crocodilians: females make and then defend large nests until the young hatch (Greer 1971). Upon hatching, the young may be led or actually carried by the parents to the water's edge.

BIRDS. Many species of birds aggregate in feeding and roosting flocks, and well-organized breeding colonies are common. However, a majority of birds are monogamous territorial breeders, and even those that breed in colonies do not approach the division of labor that is characteristic of eusocial insects. Perhaps the most complex social systems are those involving **cooperative breeding,** in which nonparents share in the rearing of young. Two types of cooperative breeding have been noted: **communal nesting** and **helpers-at-the-nest.**

COMMUNAL NESTING. Members of the cuckoo family (Cuculidae) are best known for *nest parasitism*—they lay eggs in other species' nests and let the other species rear the cuckoo young—but a few species nest communally, with several females raising young in the same nest. Davis (1942), who studied some of the tropical New

---

FIGURE 20-5 Two pathways to advanced sociality in jays

According to Brown (1974) the original breeding system, as seen in California scrub jays, is one of territory defense by mated pairs. The upper route leads to colonial nesting, with mated pairs sharing a large home range and defending only a small area around the nest. The lower route leads to cooperative breeding, with nonparents helping to rear offspring.

Source: Data from J. L. Brown, "Alternate Routes to Sociality in Jays—With a Theory for the Evolution of Altruism and Communal Breeding, *American Zoologist* 14:63–80, 1974.

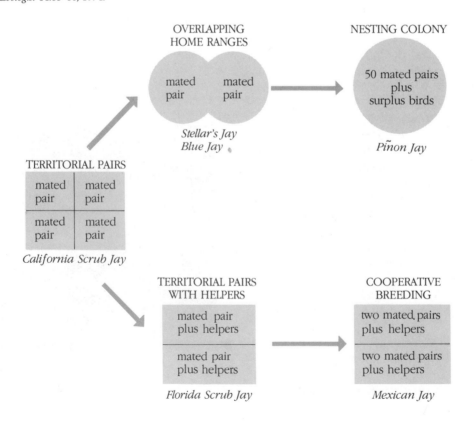

World cuckoos, reported a phylogenetic series within the subfamily. One species (*Guira guira*) nests colonially or in isolated pairs; another species, the greater ani (*Crotophaga major*), always nests in colonies. The flocks of both species are composed of mated pairs. The smooth-billed ani (*Crotophaga ani*) is promiscuous, and several females contribute to the same clutch. Territorial behavior in this species has evolved from defense of small areas by mated pairs to vigorous flock defense of the communal feeding and (scarce) nesting areas. Davis argued that flock defense evolved in a patchy environment with clumps of trees scattered as islands in grasslands. Wilson (1975) more recently pointed out that small breeding populations and genetic isolation may be involved in the evolution of this type of cooperative living.

Vehrencamp (1977) studied groove-billed anis (*Crotophaga sulcirostris*) in Costa Rica and found that considerable competition took place among adults. The most common breeding unit consisted of two pairs. Females that were paired with the dominant male tended to be the last to lay their eggs in the communal nest; they tended to throw out the eggs laid by previous females who had usually mated with the subordinate male. Most of the incubating and all of the nocturnal incubating was done by the dominant male. Although this male sired the most offspring, he paid a price because he was subject to predation as a result of sitting on the eggs all night (Vehrencamp et al. 1988).

Helpers-at-the-Nest. A more common type of cooperation in birds occurs when nonparents assist in rearing offspring of a single pair. Jays, New World members of the family Corvidae, have various grades of sociality (figure 20–5). The California scrub jay (*Aphelocoma coerulescens*) exhibits the typical system, with a pair of birds defending a territory and rearing young without help from others. Another species, the piñon jay (*Gymnorhinus cyanocephalus*), has developed colonial nesting, a system associated with a reduction of individual territory, probably in response to unpredictable food supplies (Brown 1974). As many as two hundred adult pairs nest in clusters. Each pair defends only a small space around the nest. Colony members feed in dense flocks. Possible intermediate forms are the Steller's jays (*Cyanocitta stelleri*) and blue jays (*C. cristata*), which do not nest in colonies but have widely overlapping and undefended home ranges. A second departure from the typical system is the maintenance of pair territories with cooperative breeding, where birds other than the parents assist in feeding the young and defending the territory (figure 20–6).

Florida scrub jay (*Aphelocoma coerulescens*) helpers really do help (Fitzpatrick and Woolfenden 1988; Woolfenden and Fitzpatrick 1984). They found that pairs with one or more helpers reared significantly more offspring than those without (table 20–1). The total amount of food brought to the nest did not differ between the pairs with helpers and those without. Helpers, who were older siblings of the young, brought about 30 percent of the food, thereby reducing the reproductive effort of the parents and increasing the parents' chances of surviving to produce more offspring in the future. An additional advantage of helpers is the increased vigilance and mobbing protection against predatory snakes.

FIGURE 20-6 Scrub jays in nest with helper nearby
Helpers are usually older siblings of the nestlings; they help feed young and defend the nest against predators such as snakes. (Note that three adults are present.)
Source: Photo by Glen E. Woolfenden.

TABLE 20-1   Effect of helpers on reproductive success in Florida scrub jays

Year	Number of pairs		Fledglings per pair ($\bar{x}$)		Independent young per pair ($\bar{x}$)	
	Without helper	With helper	Without helper	With helper	Without helper	With helper
1969	2	5	0	2.6	0	2.0
1970	8	8	2.0	3.5	1.3	1.8
1971	6	19	1.3	2.1	1.0	1.2
1972	13	17	0.3	1.6	0.1	1.1
1973	18	10	1.3	1.8	0.4	1.1

From G. E. Woolfenden, "Florida Scrub Jay Helpers at the Nest," in *Auk* 92:1–15. Copyright © 1975 by the American Ornithologists Union. Reprinted by permission.

Mexican jays (*Aphelocoma ultramarina*) live in flocks of eight to twenty birds, with two or more breeding pairs. Nestlings are fed by all members of the group; parents do only about half the feeding. Through long-term banding studies, Brown and Brown (1981) found that groups contained grandparents, uncles, aunts, and cousins, in addition to parents and offspring. In one unit, grandchildren helped rear their grandfather's offspring.

Helping has been reported in many other species of birds. Zahavi (1974) argued that the helpers in the groups of Arabian babblers (*Turdoides squamicaps*) he studied did more harm than good; they got in the way of the parents and attracted predators to the nest. However, in an experimental test of whether helpers really help, Brown and colleagues (1982) removed helpers from breeding groups of grey-crowned babblers, an Australian species (*Pomatostomus temporalis*). Intact control groups with a full complement of helpers produced an average of 2.4 young, while the depleted groups produced an average of 0.8 young (figure 20–7). However, evidence that helpers actually help does not explain what benefit the helpers receive and why they don't go off and have young of their own. We will deal with this problem shortly.

MAMMALS. In contrast to birds, which are mostly monogamous, mammals tend to be either solitary, where the most complex social unit is a mother with her young, or polygynous, where males monopolize several females each and harem formation is common (Eisenberg 1981). Complex social organization has evolved in some species in nearly all mammalian orders, especially in marsupials, carnivores, ungulates, and primates. Most of these systems are organized matrilineally; mothers and daughter offspring may stay together, and groups are thus composed of mothers, daughters, sisters, aunts,

FIGURE 20-7   The effect of artificial removal of helpers upon reproductive success in groups of grey-crowned babblers

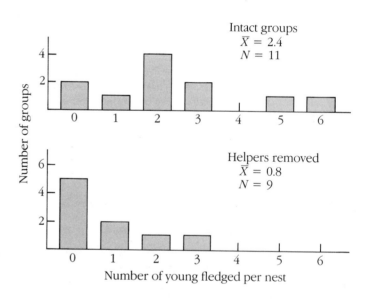

and nieces. Because of the prevalence of polygyny and the associated tendency of males to disperse as they reach sexual maturity, adult males are usually unrelated to the other adults in the group.

The most complex societies have evolved among mammals that live in relatively open habitats; grouping seems to function largely as defense against predators and protection of resources through group territory. Although cooperative rearing of young—nonmothers nursing or providing food for young—is not common, it does occur in social carnivores. For example, lionesses

FIGURE 20-8   The naked mole rat (*Heterocephalus glaber*, Bathyergidae) of East Africa
These fossorial rodents live in complex social groups approaching the eusociality of Hymenopteran insects.
Source: Photo by Graham C. Hickman, courtesy of the Mammal Slide Library.

share the nursing of cubs in the pride, and subordinate wolves regurgitate food for the alpha female and her litter.

In the banded mongoose (*Mungos mungo*) of Africa, several females breed synchronously, give birth in a communal den and then nurse each other's offspring (Rood 1986). Although the dwarf mongoose (*Helogale parvula*) is monogamous, it lives in packs of about ten individuals. Breeding is suppressed among subordinate females, as it is in wolves. These nonreproductive mongooses guard the den, bring insects to the young, and, from one report, even nurse the young of the dominant female (Rood 1980).

Rodents such as house mice and white-footed mice may form communal nests with several litters of different ages present. Jarvis (1981) demonstrated the first case of mammalian eusociality in the naked mole rat (*Heterocephalus glaber*) (figure 20-8). She captured forty members of a colony from their burrow system in Kenya and studied them for six years in an artificial burrow system in the laboratory. Only one female in the colony ever had young; mother and young were fed but not nursed by male and female adults of the worker caste; members of this caste were not seen breeding. Another caste of nonworkers assisted in keeping the young warm; males of this caste bred with the reproductive female. This species fits our criteria for eusociality: There

is cooperative care of young; there is a reproductive caste, with nonreproductives caring for reproductives; and there are offspring assisting parents in the rearing of siblings.

In the large, multimale groups found in many Old World monkeys and apes, cooperation in caring for the young does not extend to nursing or giving food to offspring other than one's own. We do not see the division of labor characteristic of colonial invertebrates or eusocial insects in any of the primates, but social roles are characteristic of some species. In rhesus monkeys (*Macaca mulatta*), the highest ranking males act as control animals (Bernstein 1966). These males protect the group against serious extragroup challenges and reduce intragroup conflict by intervening in fights between group members (Vessey 1971). Since learning and early experience play a large part in determining the social structure of primates, flexibility is perhaps the distinguishing characteristic that sets primates apart from other groups of mammals. Although kinship is important in structuring primate social systems, coalitions and friendships among nonrelatives may also be important (Smuts 1985).

WHY LIVE IN GROUPS?

It is often assumed that living in complex social groups is somehow superior to living a more solitary life, yet there are costs and benefits associated with both. Most of the **benefits** of sociality listed below can be related to two ecological factors: predation pressure and resource distribution (Alexander 1974). Keep in mind that these advantages may not, by themselves, have led to the evolution of sociality; rather they may have become secondarily advantageous once sociality had already evolved through one of the other selective pressures.

1. *Protection from physical factors.* For example, bobwhite quail (*Colinus virginianus*) survive low temperatures better when grouped than when isolated (Gerstell 1939). This benefit would result in aggregations, but not necessarily organized social groups.

2. *Protection against predators.* Detection of danger and alarm communication are faster and predator deterrence is enhanced by mobbing and group defense. Examples include prairie dogs (*Cynomys* spp.) (Hoogland 1979b), Belding's ground squirrels (*Spermophilus beldingi*) (Sherman 1977), and bank swallows (*Riparia riparia*) (Hoogland and Sherman 1976) (see also chapter 17).

3. *Assembly of sexual species for mate location.* For example, mating swarms are common in insects and in some vertebrates (see chapter 14).

4. *Location and procurement of food.* Laboratory studies demonstrate that fish in larger schools find food faster than those in smaller schools (Pitcher et al. 1982). Communal roosts in birds may serve as information centers—unsuccessful foragers learn the location of food sources by following successful foragers from the roost back to the source (Waltz 1982). It is not yet clear how the unsuccessful foragers know which birds to follow; perhaps they observe the engorged crops of the successful foragers. A related advantage is created by *tradition;* knowledge about resource location can be transmitted to subsequent generations, as it is in sheep (*Ovis canadensis*) (Geist 1971) (see also chapter 16). Increased capture rates, and capture of larger prey have often been reported (e.g., lions, chapter 17). Harris' hawks (*Parabuteo unicinctus*) form hunting parties of two to six in the nonbreeding season, and cooperate to capture jackrabbits several times larger than themselves (figure 20–9) (Bednarz 1988).

5. *Resource defense against conspecifics or competing species.* Many examples of group territoriality are included here, such as the anis studied by Davis (1942) and by Vehrencamp (1977). Among invertebrates, large colonies may have a competitive advantage over smaller groupings. The bryozoan *Bugula turrita* occurs in dense colonies on pilings and rocks in shallow water along the coasts of North America. Another species, *Schizoporella errata,* frequently overgrows the more flexible colonies of *B. turrita.* Larvae of the latter species group together when forming a colony. The resulting dense colonies are less likely to be overgrown by *S. errata* (Buss 1981).

6. *Division of labor among specialists.* Roles in complex societies (most pronounced in colonial invertebrates and in the castes of eusocial insects such as ants, wasps, bees, and termites) would seem to give members from one population a competitive advantage over those from a population lacking such divisions. Division of labor can include **mutualism,** in which two species cooperate, as when aphids provide food to their care-taker ants in return for protection from parasites and predators (Wilson 1971).

FIGURE 20-9   Cooperative hunting among hawks
Sequence of movements of a Harris' hawk implanted with a radio transmitter.

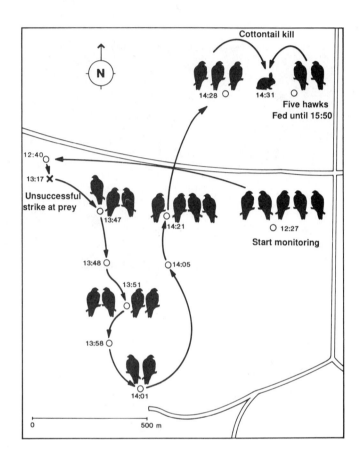

7. *Richer learning environment* for young that develop slowly. It is frequently suggested that this advantage is important for mammals in general, and primates in particular. Dependence on learning provides great plasticity but requires a long period of physiological and psychological dependence on others.

8. *Population regulation.* Wynne-Edwards (1962, 1986) hypothesized that social behavior is evolutionarily significant because organisms are able to track resources in the environment more efficiently. Intraspecific competition became ritualized into contests whose intensity was proportional to the supply of the limiting resource. Because the result of this kind of competition leads to a reduction in the reproductive success of those participating,

Wynne-Edwards believed that natural selection must be acting at the level of the group. Because of the slow pace of evolution by group selection compared with the pace by individual selection, his theory has been rejected by most behavioral ecologists. Nonetheless, social behavior may act to regulate populations incidentally to its other functions (see chapter 18).

Only a few studies have directly attempted to assess possible disadvantages of sociality. The following are several obvious **costs** of living in groups:

1. *Increased competition for resources* (food, shelter, mates) as group size increases. In prairie dog colonies (*Cynomys* spp.) the amount of agonistic behavior per individual increases as a function of group size. Also, black-tailed prairie dogs are more highly social and have higher rates of aggression than do the less social white-tails (Hoogland 1979a).

2. *Increased chance of spread of diseases and parasites.* Ectoparasites such as fleas and lice were more numerous in larger and denser prairie dog (*Cynomys* spp.) colonies than in smaller ones (Hoogland 1979a). Fleas transmit bubonic plague; plague epidemics periodically decimate prairie dog colonies, so members of dense colonies are at risk. Similarly, as cliff swallow (*Hirundo pyrrhonota*) nesting colony size increased, so did the number of blood-sucking swallow bugs increase per nest. Cliff swallow nestlings in parasitized nests lose weight and suffer higher mortality than do those in parasite-free nests (Brown and Brown 1986) (figure 20–10).

3. *Interference with reproduction,* such as cheating in parental care or killing of young by nonparents. Female white-footed mice (*Peromyscus leucopus*) with young are aggressive toward strange adults. In the absence of the mother, pups are usually killed by intruders (Wolff 1985). One of the factors influencing dispersal by male lions is defense of their cubs against infanticide by new coalitions of males (Pusey and Packer 1987). Small cubs are almost invariably killed when new males take over a pride (see chapter 16).

FIGURE 20-10 A cost of sociality
The degree of parasitism by bugs as a function of swallow colony size. As colony size increases, so does the extent of bug infestation among 10-day old nestlings.

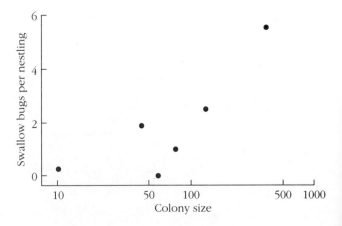

EARLY STUDIES OF SOCIAL BEHAVIOR: COOPERATION THROUGH GROUP ADVANTAGE

W. C. Allee, in *Cooperation among Animals* (1951), synthesized much of the thinking in the first half of this century about animal societies. The ideas expressed are best understood in relation to contemporary developments in the study of community ecology. Clements (1936) classified plant communities and started a school of "plant sociology," which emphasized the interrelationships of species and recognized the community as a *superorganism* with attributes transcending those of single-species populations. According to Clements, living organisms reacted to the environment and so modified it that other species could not exist there. The various assemblages of plants were so interdependent that they were distributed as single units. The idea of communities and ecosystems as superorganisms with emergent properties reached a peak with the writings of Margalef (1963), who discussed the qualities of mature and immature ecosystems. He treated the ecosystem as

FIGURE 20-11   Aggregation of brittle stars in Allee's laboratory

On the left, immediately after having been placed in a container of sea water, brittle stars are dispersed. On the right, ten minutes later, they are aggregated. If glass rods are added as substrate, the brittle stars will cling to the rods instead of to each other, thereby remaining dispersed.

Source: Data from W. C. Allee, *Cooperation Among Animals.* Copyright © 1951 University of Chicago Press, Chicago, IL.

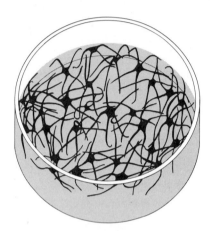

Brittle stars in
sea water

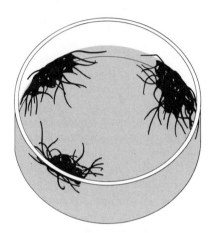

Aggregated brittle
stars ten minutes
later

a functional biological unit that organizes itself to conserve and manage information. Although not explicitly stated, such reasoning led to the notion that the whole ecosystem was the unit of selection. (For further discussion of these ideas see the ecology textbook by Krebs 1985). Animal behaviorists debated the importance of group selection, spurred on by the book *Animal Dispersion in Relation to Social Behavior*, in which Wynne-Edwards claimed that social behavior evolved mainly for the function of population regulation (Wynne-Edwards 1962). (See chapter 18 for further discussion.)

Allee recognized the importance of natural selection at the level of the individual and did much work on egoistic behaviors such as competition and dominance (Allee 1951). However, he felt that altruistic or cooperative forces were somehow stronger, and he devoted much study to what he called "natural cooperation," a force supposedly separate from natural selection. Allee believed that the phenomena he studied could be represented as several grades of social behavior, listed here in generally increasing complexity:

1. *Invertebrate coloniality.* Allee thought that the complexity and division of labor seen in such invertebrates as coelenterates was involuntary

and, therefore, somehow less advanced than the more voluntary cooperation seen in the vertebrates.

2. *Aggregation.* Organisms may be brought together by the actions of wind, tides, and other phenomena over which they have no control, such as ocean currents that bring together clusters of jellyfish.

3. *Orientation to stimuli.* Animals may be brought together by their response to environmental gradients; for example, insects aggregate around lights. Such organisms must at least have a tolerance for each other.

4. *Locomotion to favorable locations.* When resources are in patches, animals may gather, taking advantage of them; for example, birds gather in a mulberry tree when the fruit is ripe.

5. *Clumping in the absence of substrate.* Mutual attraction and clumping occur in an attempt to find missing substrate. Brittle stars (Ophiuroidea) are typically dispersed in eel grass, but in an aquarium devoid of objects they cling to each other (figure 20–11). If small glass rods are planted as artificial substrate, they disperse and attach to the rods.

6. *Sleeping group.* In many species of primates, individuals that forage in small groups during the day come together to sleep at night. Such behavior may provide superior predator protection, but may attract predators as well.

7. *Complex social life.* Allee considered highly developed social life, as is seen in some species of insects, birds, and mammals, to be an extension of sexual and family relations over a large portion of the life span. He posited that there is an "unconscious" tendency to cooperate that predisposes such species to developing complex societies.

Although he appreciated the negative effects of crowding, Allee also pointed out the unfavorable consequences of undercrowding, now often referred to as the **Allee Effect.** Typical of his experiments was the finding that goldfish (*Cyprinus* spp.) reared in a toxic colloidal suspension of silver survived longer in groups than they did alone; the slime produced by fish present in the conditioned water precipitated the silver on the bottom of the tank, thus protecting them. Other experiments showed that planarian flatworms (*Planaria* spp.) somehow gain protection from ultraviolet rays by grouping and that the effect is not due simply to the shading of one worm by another.

Allee (1951, 31), referring to the struggle for existence, best stated his philosophy about evolution in the following manner:

In the more poetic post-Darwinian days this struggle was thought of as so intense and so personal that an improved fork in a bristle or a sharper claw or an oilier feather might turn the balance toward the favored animal. Now we find that struggle for existence mainly a matter of populations, measured in the long run only, and then by slight shifts in the ratio of births to deaths.

RECENT STUDIES: COOPERATION THROUGH SELFISHNESS

Another smaller group of biologists returned to a stricter interpretation of Darwin's theory of evolution. The initiator of this movement was R. A. Fisher, who wrote *The Genetical Theory of Natural Selection* in 1930. In his 1958 revision, Fisher pointed out that his fundamental theorem referred strictly to "the progressive modification of structure or function only insofar as variations in these are of advantage to the *individual*." His theorem offered no explanation for the existence of traits that would be of use to the species to which an individual

belongs. Darwin (1859) had stated, "if it could be proved that any part of the structure of any one species had been formed for the exclusive good of another species, it would annihilate my theory." If Fisher and Darwin were correct, then we need to seek explanations for the evolution of social behavior based on natural selection at the level of the individual and its offspring, not at the group or species level that was accepted by Allee, Wynne-Edwards, and others.

The publication in 1966 of a book entitled *Adaptation and Natural Selection* by Williams led to a rapid shift in thinking among researchers working on social behavior. Williams argued that when considering any adaptation, we should assume that natural selection operates at that level necessary to explain the facts, and no higher—usually at the level of parents and their young. The argument is based on evidence that natural selection at the group or population level is so slow that it will almost always be outpaced by selection of individual phenotypes. Group selection is not impossible: If there are genes that decrease individual fitness but make it less likely that a group, population, or species will become extinct, then group selection will influence evolution (Williams 1966). Most sociobiologists today follow Williams' rules of parsimony and argue that group selection is weak and should be invoked only when lower level selection has been ruled out. The force for cooperation posited by Allee (1951), and the reproductive restraint for the good of the group of Wynne-Edwards (1986), are largely rejected. (See also chapters 4 and 18.)

Incidentally, a parallel shift in thinking took place in ecology as researchers collected new data on plant communities. Through the work of Whittaker (1975) and others, it becomes apparent that the organization of a community is best understood as an aggregation of individual species, each responding to its own set of environmental tolerances. For example, Clement's theory (1936) predicts that species along some environmental gradient—say, up the side of a mountain—come and go in groups; in fact, the species come and go almost randomly. This is not to say that species in communities do not interact, but rather that the community is not really behaving as a superorganism.

THE SELFISH HERD

One way to explain gregarious behavior in animals is to suppose that it is a form of cover-seeking in which each individual tries to reduce its chances of being caught by a predator. Hamilton (1971) suggested how

FIGURE 20-12   Frog aggregation: the selfish herd
Arrows denote the movement of frogs as they might move around the rim of a pond. The end result is an aggregation, as each frog tries to decrease its distance from its neighbor and thereby reduces the probability that it will be eaten by a predator.

Best move for one particular frog

Series of moves that can be expected

this might work in a hypothetical lily pond where some frogs and a predatory water snake live. The snake stays on the bottom of the pond most of the time and feeds at a certain time of day. Because it usually catches frogs in the water, the frogs climb out on the edge before the snake starts to hunt. They don't move inland from the rim because of even more threatening terrestrial predators. The snake surfaces at some unpredictable place and grabs the nearest frog. What should each frog do to minimize the chances of its being eaten? Hamilton showed that it will jump around the rim moving into the nearest gap between two other frogs. The end result is an aggregation (figure 20–12).

Such an example could be applied to herds of ungulates, flocks of birds, and schools of fish. When each individual behaves selfishly and minimizes its chances of being picked off, the group as a whole may be more vulnerable since the predator can go after the group rather than single out dispersed individuals. While the selfish herd might explain some aggregations, it does not explain more complex systems where individuals cooperate.

KIN SELECTION

To explain the evolution of cooperative behavior, Hamilton (1963, 1964) presented a theory incorporating both the gene and the individual as units of selection. In its simplest form, this kin selection theory suggests that if a gene that causes some kind of altruistic behavior appears, the gene's success depends ultimately not on whether it benefits the individual carrying the gene, but on the gene's benefit to itself. If the individual that is benefited by the act is a relative of the altruist—and therefore more likely than a nonrelative to be carrying that same gene—the frequency of that gene in the gene pool will increase. The more distant the relative, the less likely it will be to carry that gene, so the greater must be the ratio of recipient benefit to altruist cost. (See chapter 4.) Of course the control of an altruistic behavior pattern most likely involves more than one gene, but it is assumed that the same principles apply.

Hamilton's idea about the evolution of cooperative behavior through kin selection has stimulated considerable research. Although Allee and others had observed that most complex social groups are made up of relatives, kin selection theory makes specific predictions that can be tested. It remained to be seen if cooperative acts are distributed in some way to relatives according to the degree of relatedness. Such studies, however, require information about paternal and maternal relationships that is usually not available to the observer of a natural population. It is particularly difficult to determine paternity in species with internal fertilization, such as birds and mammals. The technique of DNA fingerprinting promises to make it possible to determine parentage in a variety of species. Another problem is the recognition factor. How do animals recognize kin in order to correctly distribute their altruistic acts? Some species can recognize relatives they have never associated with by a process called phenotype matching, where the individual uses its own kin as a template to compare with strangers who might be kin (Holmes and Sherman 1983). Perhaps a more widespread method is the formation of social attachments to those in proximity. In a natural group, proximate individuals probably will be close relatives; thus we can expect animals that are reared with nonrelatives to treat those nonrelatives as relatives. (See chapter 11 for a further discussion of kin recognition mechanisms.)

One of the most thorough tests of kin selection was Sherman's study of alarm calls in ground squirrels (Sherman 1977) (chapter 11). Individuals gave alarm calls in a way that those most likely to benefit were close kin. Males, who disperse as subadults and thus are unlikely to have relatives nearby, usually do not give alarm calls. Working with primates, Massey (1977) tested the kin selection model in rhesus monkeys. She found that within an enclosed group, monkeys aid each other in fights in proportion to their degree of relatedness. In observations of the same species in a free-ranging situation, Meikle and Vessey (1981) found that when males leave the natal group they usually join groups containing older brothers. In contrast to their behavior with nonbrothers, they associate with brothers in the new group, aid each other in fights, and avoid disrupting each other's sexual relationships.

Bertram (1976) estimated that males in a pride of lions are related on the average almost as closely as half siblings ($r = 0.22$), and females are related as closely as full cousins ($r = 0.15$). He also found that males show tolerance toward cubs and seldom compete for females in estrus, and that cubs suckle communally from the pride females. Bertram concluded that kin selection is at least partially responsible for these cooperative behaviors.

Black-backed jackals (*Canis mesomelas*) live in the brushland of Africa, where monogamous pairs defend territories, hunt cooperatively, and share food. Offspring from the previous year's litter frequently help rear their siblings by regurgitating food for the lactating mother and for the pups themselves (Moehlman 1979). The number of pups surviving correlates highly with the number of helpers (figure 20–13). An average of one-and-a-half extra pups survived for each additional helper—a yield of one pup per adult involved. Only one-half pup per adult survived when only the parents were involved. Given the fact that helpers are as closely related to their siblings as they would be to their own offspring ($r = 0.5$), yearling jackals gain by aiding their parents, genetically speaking. In addition to increasing their indirect fitness by aiding siblings, helpers may increase their direct fitness by gaining experience in rearing young, by becoming familiar with the home territory, and by possibly gaining a portion of the home territory—all of which would increase their own chances of breeding successfully at a later time. Woolfenden (1975) also cited these advantages in his study of helpers in the Florida scrub jay.

FIGURE 20-13   Black-backed jackal pup survival as a function of the number of adults in family
Young from previous years may assist their parents in rearing young. Each helper increases the number of surviving pups by a factor of about 1.5. Helpers may, therefore, increase their inclusive fitness more by rearing siblings than by attempting to breed on their own.

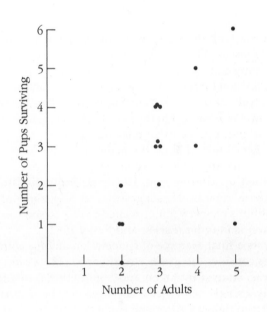

COOPERATION AMONG NON-KIN. A critical feature of a theory such as Hamilton's is that it is testable and falsifiable: While the theory cannot be proved simply by finding cases of nepotism, it can be discounted in specific cases by observing cooperation among nonrelatives in a natural population. For example, McCracken and Bradbury (1977) demonstrated that the degree of relatedness of colonial bats (*Phyllostomus hastatus*) is too low to explain their coloniality on the basis of kin selection. Another instance of cooperation among nonrelatives may be exemplified by a species of Antarctic plunder fish (*Harpagifer bispinis*) studied by Daniels (1979). Females guard their eggs, which take four to five months to hatch in the cold water. Unguarded eggs are quickly eaten. When the female is experimentally removed, males from outside the vicinity of the nest take over guard duty. Since the larval fish are planktonic,

TABLE 20-2   The coefficients of relationship (r) among close kin in hymenopteran groups

	Mother	Father	Sister	Brother	Son	Daughter	Nephew or niece
Female	0.50	0.50	0.75	0.25	0.50	0.50	0.38
Male	1.00	0.00	0.50	0.50	0.00	1.00	0.25

Source: Data from W. D. Hamilton, "The Genetic Evolution of Social Behavior I, II," *Journal of Theoretical Biology*, 7:1–52, 1964.

drifting long distances in the ocean, Daniels assumed that these replacement guards are not related to the developing eggs, and thus that kin selection is not a likely explanation for this behavior. However, to consider such acts truly altruistic, we must also know that guarding the nest is a real cost to the guard and not merely a benefit to the supposedly unrelated eggs.

Recall that in lions, the males that take over prides are frequently related; their cooperation can be explained by kin selection. However, additional studies of these same prides suggest that at least some of the coalitions are between unrelated males, so some explanation is in order (Packer and Pusey 1982).

As a final example of cooperation among nonrelatives consider the green woodhoopoe (*Phoeniculus purpureus*) studied by Ligon and Ligon (1978a) in Africa. Only a single pair breeds, but there may be as many as fourteen helpers. Although some helpers are close relatives of the breeding pair, others are nonrelatives. Clearly, factors other than kin selection must be operating. These other factors will be considered shortly.

KIN SELECTION IN SOCIAL INSECTS.   At first glance, the ultimate in unselfishness appears to be the worker honeybee. She sacrifices her life when stinging intruders in the vicinity of the colony, and she sacrifices her own reproductive success for that of the colony. She produces no eggs and raises the young of others; her direct fitness is zero. Kin selection theory suggests that it would be significant if those others were close relatives. In fact, because of a peculiarity in the genetics of hymenopteran insects, the honeybee worker is more closely related to her sisters than she would be to her own daughters! Hamilton (1964) argued that this peculiarity predisposes members of this order toward greater cooperation through kin selection. In fact, all eusocial insects, with the important exception of termites, belong to the order Hymenoptera. According to Wilson (1975), eusociality evolved independently in eleven different groups within the order Hymenoptera.

Hymenopteran males are haploid (1n), containing only one set of chromosomes that arise from an unfertilized egg; females are diploid (2n), containing two sets of chromosomes that arise from a fertilized egg. In this **haplodiploid** system, all the sperm from one male are genetically identical. If only one male mates with one queen to form the colony, the daughters will share identical genes from their father and will share half the genes from their mother. Table 20–2 shows that daughters are more closely related to each other than they would be to their own daughters. Among daughters, then, the coefficient of relationship (r) will be 0.75 instead of the usual 0.5. (See chapter 4.) Thus, any genes promoting care-giving behavior of sibs would increase faster than genes promoting investment in offspring.

From Hamilton's model, Trivers and Hare (1976) predicted that in species of eusocial insects in which the queen mates with a single male, workers should invest (i.e., provide food, defense, or other care) in female sibs over male sibs by a 3:1 margin, since they are related to sisters by 0.75 and to brothers by only 0.25. In twenty species of ants, the investment ratio was as predicted. In contrast, in slave-making ant species, the queen's brood is reared not by her daughters but by slaves, workers of other species stolen from their own nests. Since the slaves are unrelated to the brood they rear, they should not favor one sex over another. Trivers and Hare found that the investment ratio in the slave species was 1:1, as predicted.

One criticism of the haplodiploid model is that in some social Hymenoptera, more than one male may fertilize the queen. In such cases, the coefficients of relationship among daughters could be as low as 0.25, and it would be more profitable for daughters to produce their own offspring. Furthermore, in many swarm-founding wasps, there is little caste differentiation and there are multiple queens. Genetic tests demonstrate that relatedness in such colonies is low, so workers end up rearing young that are not close relatives (Queller et al. 1988). So why don't these workers produce young of their own?

Another problem for kin selection is presented by the termites, in which both sexes are diploid, but which have nevertheless evolved eusociality and sterile worker castes. Termites are dependent on symbiotic protozoans or other mutualists that digest cellulose in the termite gut. Coloniality is necessary because the protozoans are lost at each molt, and reinoculation takes place through trophallaxis (the exchange of liquid oral and anal food) among colony members (Wilson 1971). However, coloniality does not necessitate the evolution of sterile castes. One possible explanation is the recent finding that a substantial proportion of the genes in termites are linked to the X chromosome. Since the father has only one X chromosome, his daughters have identical sex-linked traits and can share more than half their genes on the average. Thus termites may have a system analogous to the haplodiploidy of Hymenoptera (Lacy 1980, 1984).

RECIPROCAL ALTRUISM

Individuals may cooperate and behave altruistically if there is a chance that they will be the recipient of such acts at a later time. Such a situation is similar to mutualism except that a time delay is involved. According to Trivers, natural selection acting at the level of the individual could produce altruistic behaviors if in the long run they benefit the organism performing them (Trivers 1971). Trivers first showed that if altruistic acts are dispensed randomly to individuals throughout a large population, genes promoting such behavior will disappear, because there is little likelihood that the recipient will pay back the altruist. However, if altruistic acts are dispensed nonrandomly among nonrelatives, genes promoting them could increase in the population if some sort of reciprocation occurs. The factors that affect that likelihood are: (1) length of lifetime—long-lived organisms will have a greater chance of meeting again to reciprocate; (2) dispersal rate—low dispersal rate will increase the chance that repeated interactions will occur; (3) mutual dependence—clumping of individuals, as occurs when avoiding predation, will increase the chances for reciprocation. In any social system nonreciprocators (cheaters) can be expected, but if cheating reaches a high enough level, altruistic acts should become infrequent.

Through the use of game theory, Axelrod and colleagues (Axelrod and Dion 1988; Axelrod and Hamilton 1981) developed a model for the evolution of cooperation between nonrelatives. The basis for their analysis was a game called Prisoner's Dilemma, in which two players have a choice of either cooperating with each

TABLE 20-3   The Prisoner's Dilemma game

The payoff to player A is shown with illustrative numerical values, where $T > R > P > S$.

Player A	Player B	
	C *Cooperation*	D *Defection*
C Cooperation	$R = 3$ Reward for mutual cooperation	$S = 0$ Sucker's payoff
D Defection	$T = 5$ Temptation to defect	$P = 1$ Punishment for mutual defection

other or defecting. The payoff matrix is shown in table 20–3. No matter what the other player does, it always pays in the short run to defect, since T, the reward for defection when the opponent cooperates, is greater than R, the reward for mutual cooperation. However, if the game continues and if both defect, they do even worse than if both cooperated; hence the dilemma. How well you do depends on the behavior of your opponent. After conducting an international computer tournament for the best solution, the highest score was obtained by the simple strategy "tit for tat," in which one player cooperates on the first move and then does whatever the other player did on the preceding move. Axelrod and Hamilton argue that this strategy is evolutionarily stable and that it shows how cooperation based on reciprocity could get started in an asocial group.

A few field studies have suggested the importance of reciprocal altruism. Working with olive baboons (*Papio anubis*) in Africa, Packer (1977) studied coalitions between males, in which two presumably unrelated males joined forces against a third male. If that third one was in consort with a female in estrus, one of the attackers might gain access to her. The pair of males tended to maintain the previously established coalition, and the next "stolen" female would be taken over by the other member of the male pair.

A different sort of reciprocity has been reported in vampire bats (*Desmodus rotundus*) by Wilkinson (1984), who studied them in Costa Rica. At night these bats feed on blood from cattle and horses, then they return to a hollow tree to roost during the day (figure 20–14). Wilkinson marked nearly two hundred bats that roosted in fourteen trees and spent 400 hours observing them

FIGURE 20-14   Vampire bat (*Desmodus rotundus*)
Blood is the sole food source for this species.
Source: Photo by William A. Wimsatt, courtesy of the Mammal Slide Library.

in their roosts. He recorded 110 cases of blood sharing, where one bat regurgitated blood that was then eaten by another bat. Not surprisingly, most of these exchanges were between mothers and offspring. In most of the other feedings, he was able to determine both the coefficient of relationship between the pair and an index of association based on how often the pair had been together in the past. After statistical analysis, Wilkinson concluded that both relatedness and association contributed significantly to the pattern of exchange (figure 20-15).

For reciprocity to persist: (1) the pairs must persist long enough to permit reciprocation, (2) the benefit to the receiver must exceed the cost to the donor, and (3) donors must recognize cheaters (those that don't reciprocate) and not feed them. Through additional studies on captive animals, Wilkinson demonstrated that vampire bats meet these conditions. It is important to note that both kin selection and reciprocal altruism are involved and that the mechanisms are not mutually exclusive.

Trivers (1971) also considered how cooperation between members of different species could have evolved through reciprocal altruism. One example he uses is that of cleaning symbiosis in fish. Many fish clean off ectoparasites from other species of fish (hosts), sometimes entering the host's mouth or gill chambers; both the cleaner and the host would seem to benefit from this mutualism (figure 20-16). Fish that are cleaners have evolved distinct colors by which they can be identified. Some unrelated species mimic cleaners, but instead of cleaning the host, they nip a piece of fin or gill. Usually cleaners and hosts meet at stations; the host signals when it is ready to be cleaned and also when it has had enough.

Why does the host show restraint and not simply eat the cleaner after it is finished instead of signaling it to go away? Ectoparasitism is a serious problem for fish, and when cleaners are removed from a coral reef, the hosts succumb to a variety of diseases (reviewed by Feder 1966). More recent attempts to replicate this experiment have failed, however (Losey 1979). Cleaners may be in short supply, and hosts may be at some risk of predation when seeking a cleaner. If cleaners and hosts live a long time and hosts use the same cleaner repeatedly, both of which seem to be true, reciprocal altruism is possible. Thus, hosts that eat cleaners after having been cleaned by them may profit in the short run since they gain an easy meal, but would later be selected against as the supply of cleaners is reduced. Gorlick et al. (1978) reviewed the evidence and the questions whether hosts really incur a cost by not eating cleaners and whether hosts even benefit from cleaning. The relationship does not seem to be the clear case of reciprocity that Trivers thought.

Lin and Michener (1972) presented a mutualistic hypothesis for the evolution of eusociality that is based on reciprocity in social insects. They proposed that groups of related or unrelated female bees had cooperated in defense against parasites and predators. This mutualism gradually has led to such a great degree of division of labor that some females gave up reproducing altogether. By cooperating, the individual gambles that the benefits it gives will be returned at a later time in a way that increases its fitness. While we can see how cooperation might evolve by means of this mechanism, kinship must be involved in the later stages when sterility appears.

FIGURE 20-15 Reciprocal food-sharing in vampire bats
Graphs A and B show the frequency of pairs observed,
based on the degree of association and relatedness. Graphs
C and D show the frequency of food-sharing, excluding
mother-young pairs. Both degree of association and
relationship independently predict food-sharing,
implicating both reciprocal altruism and kin selection in the
evolution of this behavior.

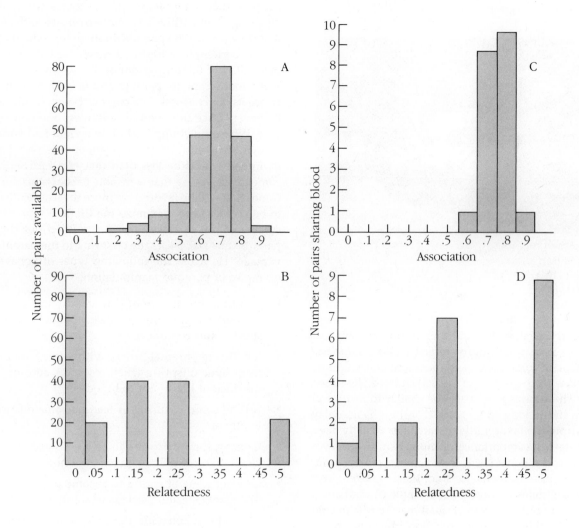

FIGURE 20-16   Cleaner fish removing ectoparasites from around mouth of host fish
In this case of mutualism, both the cleaner and the host seem to benefit. The relationship may be an example of reciprocal altruism. If the host refrains from eating the cleaner after being cleaned, the cleaner can perform its services for the host on another occasion. However, recent evidence suggests that there is little, if any, cost to not eating cleaners and that the host may not even benefit from being cleaned.
Source: Photo from Douglas Faulkner/Sally Faulkner Collection.

MUTUALISM

What about the situation where both individuals benefit from the relationship and there is no apparent cost to either? Such interactions, referred to as *mutualism,* could explain some cases of cooperation, and might be included with the notion of the selfish herd, described earlier. These cases would also be similar to reciprocal altruism in the long run, except that there is no delay in returning the favor. Little research has actually been done to test the importance of mutualism in the evolution of social behavior; it is difficult to separate mutualism from selfish behavior since there is never a cost to the individuals involved. An example of mutualism might be the resting flocks of quail that huddle in cold weather. In this simple aggregation, there is no apparent cost to individuals and there is an immediate benefit: protection from freezing. Usually, however, behavioral ecologists look for examples of apparent altruism, where costs to the actor seem to outweigh the benefits, and ignore cases of mutualism.

PARENTAL MANIPULATION OF OFFSPRING

In the discussion of parent-offspring conflict in chapter 14, we observed that offspring with only half of each parent's genes demand a greater investment from each parent than the parent should be willing to give. Our first reaction might be that offspring should win such conflicts since they are the ones that must survive and pass the genes on to future generations. But Alexander (1974) argued that parents should win in the long run. Suppose that some offspring have genes that give the offspring a competitive advantage over sibs to the extent that they reduce the parents' lifetime reproductive success. For example, a highly competitive baby bird that pushed all of its siblings out of the nest would receive more food from its parents and would probably increase its direct fitness. The parents' fitness would suffer, however, since they would only raise one young that year. When the young bird grew up and had young of its own, it would pass those competitive traits on and its fitness would be less than that of a bird with less competitive young that were able to coexist in the nest. Genes that will be favored are those that cause offspring to behave so as to maximize the lifetime reproductive success of the *parent.* Alexander pointed out that the parents thus *manipulate* the offspring to the parents' advantage. He listed the following types of behavior as examples of parental manipulation:

- Limiting the amount of parental care given to each offspring so that all have an equal chance to survive and reproduce

- Restricting parental care or withholding it entirely from some offspring when resources become insufficient for an entire brood

- Killing some offspring or feeding some offspring to others

- Causing some offspring to be temporarily or facultatively sterile helpers at the nest

- Causing some offspring to become permanent (obligately sterile) workers or soldiers

The last type of behavior, the extreme form of parental manipulation, is a third hypothesis—in addition to haplodiploidy and reciprocity—to explain the evolution of eusociality in insects. The queen uses some of her offspring to help rear others, and thereby forces them to sacrifice their reproductive success in order to rear males and potential queens that have a competitive edge over offspring of other queens.

TABLE 20-4   Ecological constraints that severely limit personal reproduction

Type of constraint	*Cause of constraint*
1. Breeding openings are nonexistent	Species has specialized ecological requirements; suitable habitat is saturated and marginal habitat is rare (stable environments)
2. Cost of rearing young is prohibitive	Unpredictable season of extreme environmental harshness (fluctuating, erratic environments)

Result: Grown offspring postpone dispersal and are retained in the parental unit. The population becomes subdivided into stable, social, kin groups.

From S. T. Emlen, "The Evolution of Helping, I. An Ecological Constraints Model," in *American Naturalist*, 119:29–39. Copyright © 1982 by the University of Chicago Press, Chicago, IL.

Alexander's reasoning was criticized by Dawkins (1976) because the argument can just as easily be turned around so that the offspring always win, as parents who are manipulated by offspring will leave more offspring themselves. Although parental manipulation is not inevitable, it is at least possible as an explanation of some types of apparent cooperation.

ECOLOGICAL FACTORS IN COOPERATION

When trying to understand the mechanisms involved in the evolution of cooperation, it is necessary to consider the type of environment in which the population lives. Emlen (1982, 1984) developed an *ecological constraints model* to explain the evolution of cooperative breeding in birds and mammals. He considers environmental factors that restrict the chances for individuals to breed independently (table 20-4).

One condition that might favor staying at home and helping the parents or others rear offspring is a stable, predictable environment. In these cases, unoccupied territories are absent or rare; young have little chance of dispersing and breeding on their own. Thus Stacey (1979) found that in communally breeding acorn woodpeckers (*Melanerpes formicivorous*), the fewer territories vacant, the more yearlings retained as helpers. Presumably, the young remain at home on familiar ground to gain experience and social status, and await an opportunity to obtain and defend a territory of their own. In the meantime, they can increase their indirect fitness by aiding relatives. There is also a chance, as is the case in the Florida scrub jay, that all or part of their parents' territory will eventually become vacant.

Having explained helping on the basis of habitat saturation in stable environments, it is ironic to find that cooperative breeding is most common in arid regions of Africa and Australia, where rainfall is highly variable and unpredictable. In this case, Emlen (1982, 1984) argues that the same behavior results for different reasons. Working with the white-fronted bee-eater (*Merops bullockoides*) Emlen found that in harsh years when rain was low and insects were scarce, the number of helpers was high. In good years, the birds were more likely to breed independently, needing little or no help to get enough food for their young.

While the ecological constraints model may explain why some animals stay home and refrain from breeding, why should they bother to help? We have already mentioned the kin selection argument, whereby indirect fitness is enhanced. But often nonrelatives or very distant relatives are aided, as in the green woodhoopoe and the bee-eater. A number of purely selfish possibilities have been suggested, such as gaining experience and social status that might help later in an independent breeding effort. Another possibility is reciprocal altruism, as proposed for the woodhoopoes by Ligon and Ligon (1978b). There is some evidence that the birds who join breeding groups and help nonrelatives "inherit" that flock and are helped themselves when breeding in a later effort. Larger flocks have a better chance of getting a breeding territory and providing for young. A similar argument was used by Rood (1983) to explain why unrelated dwarf mongooses guard and feed young of the dominant pair. He observed one case where a female was later assisted by the very young she had helped rear.

SOCIOBIOLOGY AND HUMAN SOCIAL BEHAVIOR

Does any of the preceding discussion have relevance to the evolution of human social systems? The usual arguments made against such attempts are that human behavior is too heavily shaped by culture and that our behavior is so far removed from genetic control that natural selection can have little to do with our present societies. In addition, both social and natural scientists resist the idea that Darwinian evolution, with its emphasis on competition among individuals, is largely responsible for present-day human morphology and behavior. They emphasize the importance of random processes, such as genetic drift, in changing gene frequencies. Also, cultural evolution, although possibly influenced by natural selection at the group level, is not thought by them to be a result of selection of individuals. Unfortunately, much of the argument boils down to a new version of the old nature-nurture controversy. Social scientists generally attribute an overwhelming input to environment, and natural scientists tend to stress the underlying importance of genetic heritage.

In the last two decades, a few anthropologists, biologists, and psychologists have tested the ability of the theories of Darwin, Fisher, Hamilton, Trivers, and others to explain some aspects of human social behavior. A few examples are considered here.

HUMAN MATING SYSTEMS

Polygyny is prevalent in mammals, and in many species, sexual dimorphism in body size and mating systems are related in such a way that the average number of females a male can reproductively monopolize is positively correlated with the ratio of male to female size. The data on primates are shown in figure 20–17. Human males are 5 to 12 percent taller than females, which should indicate that they are mildly polygynous. In fact, many societies do have harem polygyny (figure 20–18) or promiscuity, which leads to greater variance in reproductive success in males than in females (Alexander et al. 1979). According to Murdock (1967) 708, or 83 percent, of 849 human societies sampled had at least occasional polygyny. Alexander and his colleagues further demonstrated that in societies in which monogamy is practiced (excluding those in which monogamy is institutionally imposed), males are less

FIGURE 20-17   Relationship between harem size and degree of sexual dimorphism in primates
Each point represents a different species of primate. Males of species with the largest difference in male/female size have the largest harems.

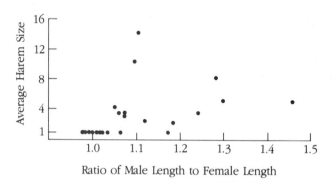

FIGURE 20-18   Polygyny in humans
A Bedouin man with his two wives (right), niece (left), and child. Polygyny is practiced in a large number of human societies.
Source: Photo by Lila Abu-Lughod from Anthro-Photo.

dimorphic than in societies in which polygyny is practiced. We might also note that like males of other polygynous species, human males take longer than females to mature, suffer a higher rate of mortality, and senesce more rapidly.

Trivers and Willard (1973) hypothesized that high-ranking females in good condition should invest more heavily in sons than in daughters (see chapter 13). As predicted by their model, human males receive more parental care than females, and sons are favored in high-ranking families among stratified polygynous societies such as caste systems (Dickemann 1979). In another study of differential investment in humans, Smith et al. (1984) analyzed one thousand probated wills. Wealthy decedents significantly favored male kin, and poor decedents favored female kin. Because in polygynous species reproductive success is more variable in males than in females (Trivers and Willard 1973), parental investment in males is more risky. However, wealthy families can increase the reproductive competitiveness of their sons through heavy material investment. Poor families take the safer option of investing in females. This argument assumes, as has been shown in some societies, that there is a positive correlation between reproductive success and material wealth.

KIN SELECTION IN HUMANS

Testing the theory of kin selection in humans offers potential rewards because of the relative ease with which we can obtain data on genealogical relationships. Hamilton (1964, 19) stated the hypothesis as follows:

The social behavior of a species evolves in such a way that in each distinct behavior-evoking situation the individual will seem to value his neighbor's fitness against his own according to the coefficients of relationship appropriate to that situation.

The Yanomamo, an Amerindian tribe in southern Venezuela, live in villages of several hundred individuals. Chagnon and Bugos (1979) recorded a crisis situation—an axe fight—precipitated when members of a recent splinter group visited their original village. One of the male visitors (Mohesiwa) was insulted by a resident woman, and he beat her. When she told her story, her kinsmen were angered, and the dispute escalated into an axe fight with several injuries. Chagnon and Bugos analyzed the fight in detail and compared the coefficients of relatedness. The supporters of Mohesiwa, one of the principals in the fight, were related to him on the average 7.8 times more than they were to the principal opponent, who was a half-brother of the beaten woman. Chagnon and Bugos argued that kinship behavior in this society is consistent with pre-

dictions based on Hamilton's theory. More generally, the Yanomamo engage in raids to avenge attacks on kinsmen. Chagnon reported that 44 percent of all men in the groups he studied had participated in killing someone in another group (Chagnon 1988). These "unokais," as they are called, seemed to have elevated status in their groups, as they had more wives and offspring than did men who had not killed.

Closer to home, Daly and Wilson (1982) studied homicides in Detroit among cohabitants. Out of 98 murders, 76 (77 percent) were by nonrelatives living in the home, usually the spouse. When corrected for their representation in the family, relatives were ten times less likely to kill each other than were nonrelatives. Kinship is also important in predicting patterns of child abuse. Parents are far more likely to abuse their stepchildren than their genetic offspring (Daly and Wilson 1981). An even stronger relationship is evident when comparing cases of child homicide, where stepparents are greatly over-represented (figure 20–19) (Daly and Wilson 1988).

Kin selection may also explain other aspects of parental behavior. Compared to other mammals, human males are capable of great investment in offspring, but confidence of paternity must be high to make the investment profitable. In polygynous societies, a male may not be sure that his wife's children are his own. In that case, he might do better to invest in his sister's children, since they are likely to share at least one-eighth of his genes. As predicted, in highly polygynous societies the mother's brother often cares for his sister's children (Alexander 1974).

Returning to the analysis of wills by Smith et al. (1984), we see that close relatives received more wealth than did distant relatives; younger relatives, with greater reproductive potential, received more wealth than did older relatives. In these cases, humans could be acting to maximize their inclusive fitness.

Although some social systems can be understood as partly the result of kin selection, humans engage in reciprocity, which transcends close kinship. Wilson (1978) argued that reciprocity between distantly related or unrelated individuals is the key to human society. The constraints of rigid kin selection have been broken by our use of written and spoken language to fashion long-remembered agreements upon which civilizations can be built.

FIGURE 20-19   Homicides of children by parents in Canada, 1974–1983

Age-specific rates of homicide victimization by (a) genetic parents (*N* = 341 victims) or (b) stepparents (*n* = 67), Canada, 1974 to 1983.

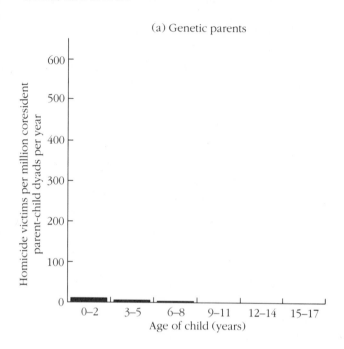

(a) Genetic parents

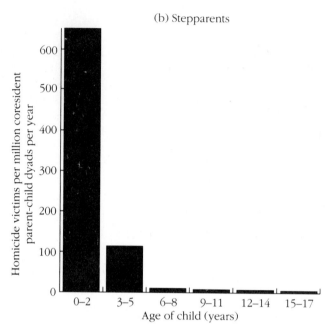

(b) Stepparents

SUMMARY

A *society* may be defined as a group of individuals of the same species, organized in a cooperative manner that extends beyond sexual behavior. An example of a complex social system is that fashioned by a colonial invertebrate such as the Portuguese man-of-war, in which individuals specialize in the functions of feeding, defense, locomotion, and reproduction. Other examples are the social systems of eusocial insects, cooperatively nesting birds, and primates.

The potential *benefits* of social behavior include protection from physical elements, predator detection and defense, group defense of resources, and division of labor. The *costs* are increased competition, spread of contagious diseases, and interference in reproduction.

Most early researchers of animal social systems assumed that traits favoring cooperation could be selected for even when they were detrimental to individuals. Allee studied cooperation in a variety of animals and argued that *altruistic* forces are stronger than

those forces that promote selfishness. He believed that natural cooperation was a force separate from natural selection, and he implied in his arguments that behavior and structure could evolve for the benefit of the group or species.

Research on the evolution of social behavior in the last two decades has invoked *Williams's rule of parsimony*: When considering any adaptation, assume that natural selection operates at that level necessary to explain the facts, and no higher. Usually that is the level of parents and their young; group selection is invoked only as a last resort. Organisms cooperate for genetically *selfish* reasons rather than for the good of the group.

Aggregations may occur as individuals attempt to reduce their own chances of being picked off by a predator. *Kin selection* may explain cooperation among individuals that share genes by common descent. Individuals may behave so as to lower their own fitness but increase their inclusive fitness. For example, in many eusocial insects, *haplodiploidy* results in female workers being more closely related to their sisters than they would be to their daughters; this relationship may have led to the evolution of sterile workers.

Reciprocal altruism may be important in long-lived organisms that do not disperse extensively. Thus an individual stands a good chance of being "paid back" if he or she cooperates with others. Reciprocity is possibly involved in the evolution of *mutualistic* relationships between different species.

Parents may *manipulate* their own offspring to behave in ways beneficial to the parents instead of to themselves, the offspring. Genes giving some offspring a competitive edge against sibs could be selected against if the reproductive success of the parent is lowered. In addition to kin selection, both reciprocal altruism and parental manipulation of offspring could explain the evolution of eusociality in insects.

The application of kin selection and reciprocity to the evolution of human social behavior has met with much resistance among social scientists, who downplay the role of natural selection of favored genotypes in determining human behavior. Nonetheless, some anthropologists, psychologists, and biologists are testing these ideas against the anthropological evidence and are coming up with suggestive results.

Discussion Questions

1. How might you argue that colonial invertebrates such as the Portuguese man-of-war have a more "perfect" society than does a group of chimpanzees?

2. What are the differences between kin selection and group selection? In what ways can they be considered similar?

3. Contrast Allee's and Wynne-Edwards's attitudes toward the evolution of social behavior with those of Hamilton, Trivers, and Alexander.

4. Distinguishing characteristics of human sexuality are the concealment of ovulation in women and the lack of pronounced monthly change in female sexual receptivity. Speculate on the possible evolutionary importance of these characteristics. See Alexander and Noonan (1979) and Burley (1979) for alternative explanations.

5. The data below represent lifetime reproductive success for the !Kung San, an undisturbed population of hunter-gatherers in Africa's Kalahari Desert. What conclusions can you reach about the social system in this population?

From *SEX, EVOLUTION, AND BEHAVIOR*, Second Edition, by Martin Daly and Margo Wilson. Copyright © 1983 by PWS Publishers. Reprinted by permission of Wadsworth, Inc.

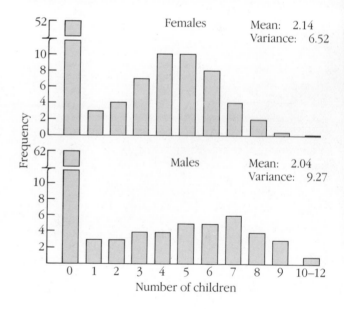

Females Mean: 2.14 Variance: 6.52

Males Mean: 2.04 Variance: 9.27

Frequency

Number of children

Suggested Readings

Alexander, R. D. 1974. The evolution of social behavior. *Ann. Rev. Ecol. Syst.* 5:325–83.
One of the first comprehensive attempts to understand the evolution of sociality through Darwinian evolution. Thoroughly develops the idea of parental manipulation of offspring.

Alexander, R. D., and D. W. Tinkle, eds. 1981. *Natural Selection and Social Behavior.* New York: Chiron Press.
From a symposium held in 1978. Pioneering articles on theory and on field studies of organisms ranging from insects to humans.

Barash, D. P. 1982. *Sociobiology and Behavior.* 2d ed. New York: Elsevier.
Clearly written summary of the discipline for the general reader. Lots of detailed examples and simplified explanations of theory.

Campbell, B., ed. 1972. *Sexual Selection and the Descent of Man 1871–1971.* Chicago: Aldine.
Assembled on the one hundredth anniversary of Darwin's treatment of the subject, this volume contains a number of important papers on sexual selection and human evolution.

Rubenstein, D. I., and R. W. Wrangham, eds. 1986. *Ecological Aspects of Social Evolution. Birds and Mammals.* Princeton: Princeton University Press.
A collection of reports of long-term field studies on both monogamous and polygynous species. A summary chapter by the editors provides general principles for understanding the connection between ecology and sociality.

Wilson, E. O. 1975. *Sociobiology: The New Synthesis.* Cambridge: Harvard University Press.
Reviews the basic principles of social organization, then gives examples from different taxa. Clearly written and still the most comprehensive treatment available.

REFERENCES

Able, K. P. 1974. Environmental influences on the orientation of nocturnal bird migrants. *Anim. Behav.* 22:225–39.

———. 1978. Field studies of the orientation cue hierarchy of nocturnal songbird migrants. In *Animal Migration, Navigation, and Homing,* ed. Schmidt-Koenig and W. T. Keeton. New York: Springer-Verlag.

———. 1980. Mechanisms of orientation, navigation and homing. In *Animal Migration, Orientation, and Navigation,* ed. S. A. Gauthreaux. New York: Academic Press.

———. 1982. Field studies of avian nocturnal migration orientation. I. Interaction of sun, wind and stars as directional cues. *Anim. Behav.* 30:761–67.

———. 1989. Skylight polarization patterns and the orientation of migratory birds. *J. Exp. Biol.* 141:241–56.

Able, K. P., and M. A. Able. 1990a. Ontogeny of migratory orientation in the Savannah sparrow (*Passerculus sandwichensis*): Mechanisms of sunset orientation. *Anim. Behav.* in press.

———. 1990b. Ontogeny of migratory orientation in the Savannah sparrow (*Passerculus sandwichensis*): Calibration of the magnetic compass. *Anim. Behav.* in press.

Able, K. P., V. P. Bingman, P. Kerlinger, and W. Gergits. 1982. Field studies of avian nocturnal migratory orientation. II. Experimental manipulation in white-throated sparrows (*Zonotrichia albicollis*) released aloft. *Anim. Behav.* 30:768–73.

Abramson, C. I. 1986. Aversive conditioning in honeybees (*Apis mellifera*). *J. Comp. Psychol.* 100:108–16.

Abu-Gideiri, Y. B. 1966. The behaviour and neuroanatomy of some developing teleost fishes. *J. Zool.* 149:215–41.

Ader, R., and P. M. Conklin. 1963. Handling of pregnant rats: Effects on emotionality of their offspring. *Science* 142:411–12.

Adkins-Regan, E. 1981. Early organizational effects of hormones. In *Neuroendocrinology of Reproduction,* ed. N. Adler. New York: Plenum.

———. 1987. Sexual differentiation in birds. *Trends. Neurosci.* 10:517–22.

Adkins-Regan, E., and M. Ascenzi. 1987. Social and sexual behaviour of male and female zebra finches treated with oestradiol during the nestling period. *Anim. Behav.* 35:1100–1112.

Adler, K., and D. H. Taylor. 1973. Extraocular perception of polarized light by orienting salamanders. *J. Comp. Physiol.* 87:203–12.

Albrecht, H. 1966. Zur Stammesgeschichte einiger Bewegungsweisen bei Fischen untersucht am Verhalten von *Haplochromis* (Pisces, Cichlidae). *Zeit. Tierpsychol.* 23:270–301.

Alexander, B. K., and J. Hughes. 1971. Canine teeth and rank in Japanese monkeys (*Macaca fuscata*). *Primates* 12:91–93.

Alexander, R. D. 1961. Aggressiveness, territoriality, and sexual behavior in field crickets (Orthoptera: Gryllidae). *Behaviour* 17:130–223.

———. 1974. The evolution of social behavior. *Ann. Rev. Ecol. Syst.* 5:325–83.

Alexander, R. D., J. L. Hoogland, R. D. Howard, K. M. Noonan, and P. W. Sherman. 1979. Sexual dimorphisms and breeding systems in pinnipeds, ungulates, primates and humans. In *Evolutionary Biology and Human Social Behavior: An Anthropological Perspective,* ed. N. A. Chagnon and W. Irons. N. Scituate, MA: Duxbury.

Alexander, R. D., and K. M. Noonan. 1979. Concealment of ovulation, parental care, and human social evolution. In *Evolutionary Biology and Human Social Behavior: An Anthropological Perspective,* ed. N. A. Chagnon and W. Irons. N. Scituate, MA: Duxbury.

Alkon, D. L. 1980. Cellular analysis of a gastropod (*Hermissenda crassicornis*) model of associative learning. *Biol. Bull.* 159:505–60.

———. 1983. Learning in a marine snail. *Sci. Amer.* 249(July):70–85.

Allee, W. C. 1951. *Cooperation among Animals.* Chicago: University of Chicago Press.

Altmann, J. 1974. Observational study of behavior: Sampling methods. *Behaviour* 49:227–67.

Altmann, J., G. Hausfater, and S. A. Altmann. 1988. Determinants of reproductive success in savannah baboons, *Papio cynocephalus.* In *Reproductive Success,* ed. T. H. Clutton-Brock. Chicago: University of Chicago Press.

American Ornithologists' Union. 1988. Report of the committee on use of wild birds in research. *Auk,* 105(Suppl.): 1A–41A

American Society of Mammalogists. 1987. Acceptable field methods in mammalogy: preliminary guidelines approved by the American Society of Mammalogists. *J. Mammal.* 68(Suppl.): 1–18.

Anand, B. K., and J. R. Brobeck. 1951. Hypothalamic control of food intake in rats and cats. *Yale J. Biol. Med.* 24:123–40.

Anderson, B., and S. M. McCann. 1955. A further study of polydipsia evoked by hypothalamic stimulation in the goat. *Acta Physiol. Scand.* 33:333–46.

———. 1956. The effect of hypothalamic lesions on water intake of the dog. *Acta Physiol. Scand.* 35:312–20.

Anderson, L. T. 1973. An analysis of habitat preference in mice as a function of prior experience. *Behaviour* 47:302–39.

Andersson, M. 1982. Female choice selects for extreme tail length in a widowbird. *Nature* 299:818–20.

Animal Behavior Society and Association for the Study of Animal Behaviour. 1986. Guides for the use of animals in research. *Anim. Behav.* 34:315–18.

Arbib, M. A. 1972. *The Metaphorical Brain.* New York: Wiley-Interscience.

Archer, J. 1973. Tests for emotionality in rats and mice: A review. *Anim. Behav.* 21:205–35.

———. 1988. *The Behavioural Biology of Aggression.* New York: Cambridge University Press.

Ardrey, R. 1966. *The Territorial Imperative.* New York: Atheneum.

Arendash, G. W., and R. A. Gorski. 1982. Enhancement of sexual behavior in female rats by neonatal transplantation of brain tissue from males. *Science* 217:1276–78.

Arendt, J., M. Aldhous, and V. Marks. 1986. Alleviation of jet lag by melatonin: Preliminary results of controlled double blind trial. *Brit. Med. J.* 292:1170.

Arey, L. B. 1954. *Developmental Anatomy.* Philadelphia: W. B. Saunders.

Arling, G. L., and H. F. Harlow. 1967. Effects of social deprivation on maternal behavior of rhesus monkeys. *J. Comp. Physiol. Psychol.* 64:371–77.

Arnold, A. P., and R. A. Gorski. 1984. Gonadal steroid induction of structural sex differences in the central nervous system. *Ann. Rev. of Neurosci.* 7:413–42.

Aronson, L. R., and M. L. Cooper. 1966. Seasonal variation in mating behavior in cats after desensitization of the glans penis. *Science* 152:226–30.

Aschoff, J. 1960. Exogenous and endogenous components in circadian rhythms. *Cold Spr. Harb. Symp. Quant. Biol.* 25:11–28.

———. 1963. Comparative physiology: Diurnal rhythms. *Ann. Rev. Physiol.* 25:581–600.

———. 1979. Circadian rhythms: Influences of internal and external factors on the period measured in constant conditions. *Zeit. Tierpsychol.* 49:225–49.

———. 1981. *Handbook of Behavioral Neurobiology.* Vol. 4: *Biological Rhythms.* New York: Plenum.

Askenmo, C. E. H. 1984. Polygyny and nest selection in the pied flycatcher. *Anim. Behav.* 32:972–80.

Attneave, F. 1959. *Applications of Information Theory to Psychology.* New York: Holt, Rinehart and Winston.

Austad, S. N. 1982. First male sperm priority in the bowl and doily spider, *Frontinella pyramitela* (Walckenaer). *Evolution* 36:777–85.

———. 1983. A game theoretical interpretation of male combat in the bowl and doily spider (*Frontinella pyramitela*). *Anim. Behav.* 31:59–73.

Axelrod, R., and D. Dion. 1988. The further evolution of cooperation. *Science* 242:1385–90.

Axelrod, R., and W. D. Hamilton. 1981. The evolution of cooperation. *Science* 211:1390–96.

Bachus, S. E., and E. S. Valenstein. 1979. Individual behavioral responses to hypothalamic stimulation persist despite destruction of tissue surrounding electrode tip. *Physiol. Behav.* 23:421–26.

Bailey, E. D. 1966. Social interaction as a population regulating mechanism in mice. *Can. J. Zool.* 44:1007–12.

Baker, M. C. 1983. The behavioral response of female Nuttall's white-crowned sparrows to male song of natal and alien dialects. *Behav. Ecol. Sociobiol.* 12:309–15.

Baker, R. 1980. Goal orientation by blindfolded humans after long-distance displacement: Possible involvement of a magnetic sense. *Science* 210:555–57.

Baker, R. R., and M. A. Bellis. 1988. 'Kamikaze' sperm in mammals? *Anim. Behav.* 36:936–38.

Bakker, T. C. M. 1986. Aggressiveness in sticklebacks (*Gasterosteus aculeatus* L.): A behaviour-genetic study. *Behaviour* 56:1–144.

Balda, R. P. 1980. Recovery of cached seeds by a captive *Nucifraga caryocatactes. Z. Tierpsychol.* 52:331–46.

———. 1987. Avian impacts on pinyon-juniper woodlands. 525–533, In *Proceedings: Pinyon-Juniper Conference,* ed. R. L. Everett. U.S. Forest Service Technical Report INT-215.

Balda, R. P., and A. C. Kamil. 1988. The spatial memory of Clark's nutcrackers (*Nucifraga columbiana*) in an analogue of the radial arm maze. *Anim. Learn. Behav.* 16:116–22.

———. 1989. A comparative study of cache recovery by three corvid species. *Anim. Behav.* 38: 486–95.

Bandler, R. J., Jr., and J. P. Flynn. 1974. Nerve pathways from the thalamus associated with regulation of aggressive behavior. *Science* 183:96–99.

Banfield, A. W. 1954. Preliminary investigation of the barren ground caribou. II. Life history, ecology and utilization. *Can. Wild. Serv. Wild. Mgmt. Bull,* No. 10B.

Barash, D. P. 1973a. The social biology of the Olympic marmot. *Anim. Behav. Monogr.* 6:171–249.

———. 1973b. Social variety in the yellow-bellied marmot (*Marmota flaviventris*). *Anim. Behav.* 21:579–84.

———. 1974a. The evolution of marmot societies: A general theory. *Science* 185:415–20.

———. 1974b. Social behavior of the hoary marmot (*Marmota caligata*). *Anim. Behav.* 22:257–62.

———. 1982. *Sociobiology and Behavior.* 2d ed. New York: Elsevier.

Barbour, R. W., W. H. Davis, and M. D. Hassell. 1966. The need of vision in homing by *Myotis sodalis. J. Mammal.* 47:356–57.

Barfield, R. J. 1971. Activation of sexual and aggressive behavior activated by androgen implanted into the male ring dove brain. *Endocrinology* 89:1470–76.

Barfield, R. J., D. E. Busch, and K. Wallen. 1972. Gonadal influence on agonistic behavior in the male domestic rat. *Horm. Behav.* 3:247–59.

Barlow, G. W. 1968. Ethological units of behavior. In *The Central Nervous System and Fish Behavior*, ed. D. Ingle. Chicago: University of Chicago Press.

Barnes, B. M. 1989. Freeze avoidance in a mammal: Body temperatures below 0° C in an arctic hibernator. *Science* 244:1593–95.

Barnett, S. A. 1956. Behavior components in the feeding of wild and laboratory rats. *Behaviour* 9:24–43.

———. 1963. *The Rat: A Study in Behavior.* Chicago: Aldine.

Barnwell, F. H. 1966. Daily and tidal patterns of activity in individual fiddler crabs (genus *Uca*) from the Woods Hole region. *Biol. Bull.* 130:1–17.

Barraclough, C. A. 1967. Modifications in reproductive function after exposure to hormones during the prenatal and postnatal period. In *Neuroendocrinology*, vol. 2, ed. L. Martini and W. F. Ganing. New York: Academic Press.

Barron, D. H. 1941. The functional development of some mammalian neuromuscular mechanisms. *Biol. Rev.* 16:1–33.

Barth, F. G. 1982. Spiders and vibratory signals: Sensory reception and behavioral significance. In *Spider Communication: Mechanisms and Ecological Significance*, eds. P. N. Witt and J. S. Rovner. Princeton, NJ: Princeton Univ. Press.

Barth, R. H. 1965. Insect mating behavior: Endocrine control of a chemical communication system. *Science* 149:882–83.

———. 1968. The comparative physiology of reproductive processes in cockroaches. I. Mating behavior and its endocrine control. *Adv. Reprod. Physiol.* 3:167–201.

Bastock, M. 1956. A gene mutation which changes a behavior pattern. *Evolution* 10:421–39.

Bastock, M., and A. Manning. 1955. The courtship of *Drosophila melanogaster*. *Behaviour* 8:85–111.

Bateman, A. J. 1948. Intra-sexual selection in *Drosophila*. *Heredity* 2:349–68.

Bateson, P. P. G. 1978. Early experience and sexual preferences. In *Biological Determinants of Sexual Behavior*, ed. J. B. Hutchinson. New York: Wiley.

———. 1982. Preference for cousins in Japanese quail. *Nature* 295:236–37.

Beach, F. A. 1950. The snark was a boojum. *Amer. Psychol.* 5:115–24.

———. 1960. Experimental investigations of species-specific behavior. *Amer. Psychol.* 15:1–18.

———. 1976. Sexual attractivity, proceptivity and receptivity in female mammals. *Horm. Behav.* 7:105–38.

Beach, F. A., and A. M. Holz-Tucker. 1949. Effects of different concentrations of androgen upon sexual behavior in castrated male rats. *J. Comp. Physiol. Psychol.* 42:433–53.

Beck, B. B. 1980. *Animal Tool Behavior: The Use and Manufacture of Tools by Animals.* New York: Garland STPM.

Beck, S. D. 1968. *Insect Photoperiodism.* New York: Academic Press.

Bednarz, J. C. 1988. Cooperative hunting in Harris' hawks (*Parabuteo unicinctus*). *Science* 239:1525–27.

Beecher, M. D. 1982. Signature systems and kin recognition. *Amer. Zoologist* 22:477–90.

Beecher, M. D., M. B. Medvin, P. K. Stoddard, and P. Loesch. 1986. Acoustic adaptations for parent-offspring recognition in swallows. *Exp. Biol.* 45:179–93.

Bekoff, A. 1976. Ontogeny of leg motor output in the chick embryo: A neural analysis. *Brain Res.* 106:271–91.

———. 1978. A neuroethological approach to the study of the ontogeny of coordinated behavior. In *Development of Behavior*, eds. G. Burghardt and M. Bekoff. New York: Garland STPM Press.

Bekoff, A., P. S. G. Stein, and V. Hamburger. 1975. Coordinated motor output in the hindlimb of the 7-day chick embryo. *Proc. Nat. Acad. Sci. U.S.A.* 72:1245–48.

Bekoff, M. 1974a. Social play in coyotes, wolves and dogs. *Bioscience* 24:225–30.

———. 1974b. Social play in mammals. *Am. Zool.* 14:265–436.

———. 1974c. Social play and play-soliciting by infant canids. *Am. Zool.* 14:323–40.

———. 1977. Social communication in canids: Evidence for the evolution of a stereotyped mammalian display. *Science* 197:1097–99.

———. 1978. Social play: Structure, function and the evolution of a cooperative social behavior. In *Development of Behavior*, ed. G. Burghardt and M. Bekoff. New York: Garland STPM Press.

Bekoff, M., and J. A. Byers. 1981. A critical reanalysis of the ontogeny and phylogeny of mammalian social and locomotor play: An ethological hornet's nest. In *Behavioral Development*, ed. K. Immelmann et al. New York: Cambridge University Press.

Bell, G. 1982. *The Masterpiece of Nature: The Evolution and Genetics of Sexuality.* Berkeley: University of California Press.

Bellrose, F. C. 1964. Radar studies of waterfowl migration. *Trans. N. Amer. Wild. Nat. Conf.* 29:128–143.

———. 1967. Orientation in waterfowl migration. *Proc. Ann. Biol. Colloq.*, Oreg. State Univ. 27:73–99.

———. 1971. The distribution of nocturnal migrants in the air space. *Auk* 88:397–424.

Bennett, M. A. 1940. The social hierarchy in ring doves. II. The effect of treatment with testosterone propionate. *Ecology* 21:148–65.

Bentley, D. 1969. Intracellular activity in cricket neurosis during the generation of song patterns. *Zeit. Vergleich. Physiol.* 62:267–83.

Bentley, D., and R. R. Hoy. 1970. Postembryonic development of adult motor patterns in crickets: A neural analysis. *Science* 170:1409–11.

———. 1974. The neurobiology of cricket song. *Science* 231:34–44.

Benzer, S. 1973. Genetic dissection of behavior. *Sci. Amer.* 229:24–37.

Bercovitch, F. B. 1986. Male rank and reproductive activity in savanna baboons. *Int. J. Primatol.* 7:533–50.

Berkowitz, L. 1969. *The Roots of Aggression.* New York: Atherton.

Berkowitz, L., and A. LePage. 1967. Weapons as aggression-eliciting stimuli. *J. Per. Soc. Psychol.* 7:202–7.

Berman, C. M. 1980. Mother-infant relationships among free-ranging rhesus monkeys on Cayo Santiago: a comparison with captive pairs. *Anim. Behav.* 28:860–73.

Bernstein, I. S. 1966. Analysis of a key role in a capuchin (*Cebus albifrons*) group. *Tulane Stud. Zool.* 13:49–54.

Beroza, M., and E. F. Knipling. 1972. Gypsy moth control with the sex attractant pheromone. *Science* 177:19–27.

Bertram, B. C. R. 1976. Kin selection in lions and in evolution. In *Growing Points in Ethology*, ed. P. P. G. Bateson and R. A. Hinde. New York: Cambridge University Press.

Bertram, B. P. 1975. The social system of lions. *Sci. Amer.* 232:54–65.

Berven, K. A. 1981. Mate choice in the wood frog, *Rana sylvatica*. *Evolution* 35:707–22.

Beugnon, G. 1986. Learned orientation in landward swimming in the cricket *Pteronemobius lineolatus*. *Behav. Proc.* 12:215–26.

Birch, H. C. 1945. The relation of previous experience to insightful problem-solving. *J. Comp. Physiol. Psychol.* 38:367–83.

Birdsall, D. A., and D. Nash. 1973. Occurrence of successful multiple insemination of females in natural populations of deer mice (*Peromyscus maniculatus*). *Evolution* 27:106–10.

Bitterman, M. E. 1960. Toward a comparative psychology of learning. *Amer. Psychol.* 15:704–12.

———. 1965a. The evolution of intelligence. *Sci. Amer.* 212:92–100.

———. 1965b. Phyletic differences in learning. *Amer. Psychol.* 20:396–410.

———. 1975. The comparative analysis of learning. *Science* 188:699–709.

Blaich, C. F., and D. B. Miller. 1986. Call responsivity of mallard ducklings (*Anas platyrhynchos*). IV. Effects of social experience. *J. Comp. Psychol.* 100:401–5.

Blair, W. F. 1958. Mating call in the speciation of anuran amphibians. *Amer. Nat.* 92:27–51.

———. 1974. Character displacement in frogs. *Amer. Zool.* 14:1119–25.

Blest, A. D. 1957. The function of eyespot patterns in the Lepidoptera. *Behaviour* 11:209–56.

Block, G. D., and T. L. Page. 1978. Circadian pacemakers in the nervous system. *Ann. Rev. Neurosci.* 1:19–34.

Block, R. A., and J. V. McConnell. 1967. Classically conditioned discrimination in the planarian, *Dugesia dorotocephala*. *Nature* 215:1465–66.

———. 1975. The comparative analysis of learning. *Science* 188:699–709.

Boag, P. T. 1983. The heritability of external morphology in Darwin's ground finches (Geospiza) on Isla Daphne Major, Galapagos. *Evolution* 37:377–84.

Boag, P. T., and P. R. Grant. 1981. Intense natural selection in a population of Darwin's finches (Geospizinae) in the Galapagos. *Science* 214:82–85.

Boice, R., 1972. Some behavioral tests of domestication in Norway rats. *Behaviour* 42:198–231.

Bolles, R. 1970. Species-specific defense reactions and avoidance learning. *Psychol. Rev.* 77:32–48.

Bonner, J. T. 1980. *Evolution of Culture in Animals*. Princeton: Princeton University Press.

Borgia, G. 1985. Bower quality, number of decorations and mating success of male satin bowerbirds (*Ptilonorhynchus violaceus*): An experimental analysis. *Anim. Behav.* 33:266–71.

Bouissou, M. F. 1978. Effect of injections of testosterone propionate on dominance relationships in a group of cows. *Horm. Behav.* 11:388–400.

Bouissou, M. F., and V. Gaudioso. 1982. Effect of early androgen treatment on subsequent social behavior in heifers. *Horm. Behav.* 16:132–46.

Boulos, Z., A. M. Rosenwasser, and M. Terman. 1980. Feeding schedules and the circadian organization of behavior in the rat. *Behav. Brain Res.* 1:39–65.

Bovjberg, R. V. 1956. Some factors affecting aggressive behavior in crayfish. *Physiol. Zool.* 29:127–36.

Bowen, B. W., A. B. Meylan, and J. C. Avise. 1989. An odyssey of the green sea turtle: Ascension Island revisited. *Proc. Nat. Acad. Sci. U.S.A.* 86:573–576.

Boyan, G. S. 1981. Two-tone suppression of an identified auditory neuron in the brain of the cricket, *Gryllus bimaculatus* (DeGreer). *J. Comp. Physiol.* 144:117–25.

Boycott, B. B. 1965. Learning in the octopus. *Sci. Amer.* 212(3):42–50.

Brace, R. C., J. Pavey, and D. L. Quicke. 1979. Intraspecific aggression in the colour morphs of the anemone *Actinia equina*: the 'convention' governing dominance ranking. *Anim. Behav.* 27:553–61.

Brady, J. 1967a. Control of the circadian rhythm of activity in the cockroach. I. The role of the corpora cardiaca, brain and stress. *J. Exp. Biol.* 47:153–63.

———. 1967b. Control of the circadian rhythm of activity in the cockroach. II. The role of the subesophageal ganglion and ventral nerve cord. *J. Exp. Biol.* 47:165–78.

———. 1969. How are insect circadian rhythms controlled? *Nature* 223:781–84.

———. 1979. *Biological Clocks*. Baltimore: University Press.

———. 1982. Circadian rhythms in animal physiology. In J. Brady (ed.) *Circadian Time Keeping*. London: Cambridge Univ. Press.

———. 1988. The circadian organization of behavior: Timekeeping in the tsetse fly, a model system. *Adv. Stud. Behav.* 18:153–91.

Brattstrom, B. H. 1974. The evolution of reptilian social behavior. *Amer. Zool.* 14:35–49.

Brazier, M. A. B. 1968. *The Electrical Activity of the Nervous System*. Baltimore: Williams & Wilkins.

Breed, M. D. 1983. Nestmate recognition in honey bees. *Anim. Behav.* 31:86–91.

Breed, M. D., and W. J. Bell. 1983. Hormonal influences on invertebrate aggressive behavior. In *Hormones and Aggressive Behavior*, ed. B. B. Svare. 577–90. New York: Plenum.

Brewer, R., and K. G. Harrison. 1975. The time of habitat selection in birds. *Ibis* 117:521–22.

Broadhurst, P. L. 1963. The Choice of Animal for Behaviour Studies. *Laboratory Animals Centre Collected Papers* 12:65–80.

———. 1965. The inheritance of behavior. *Science J.* 24:39–43.

Brodeur, J., and J. N. McNeil. 1989. Seasonal microhabitat selection by an endoparasitoid through adaptive modification of host behavior. *Science* 244:226–28.

Bronson, F. 1971. Rodent pheromones. *Biol. Reprod.* 4:344–57.

Bronson, F. H. 1979. The reproductive ecology of the house mouse. *Quart. Rev. Biol.* 54:265–99.

Brookman, M. A. 1978. Sensitivity of the homing pigeon to an earth-strength magnetic field. In *Animal Migration, Navigation, and Homing*, ed. K. Schmidt-Koenig and W. T. Keeton. New York: Springer-Verlag.

Brookshire, K. H. 1970. Comparative psychology of learning. In *Learning: Interactions,* ed. M. H. Marx. New York: Macmillan.

Brower, J. V. 1958. Experimental studies of mimicry in some North American butterflies. I. The monarch, *Danaus plexippus,* and viceroy, *Limenitis archippus archippus. Evolution* 12:32–47.

Brower, J. V., and L. P. Brower. 1964. Birds, butterflies and plant poisons: A study in ecological chemistry. *Zoologica* 49:137–59.

Brower, L. P., W. H. Calvert, L. E. Hedrick, and J. Christian. 1977. Biological observations on an overwintering colony of monarch butterflies (*Danaus plexippus Danaidae*) in Mexico. *J. Lepid. Soc.* 31:232–42.

Brower, L. P., W. N. Ryerson, L. L. Coppinger, and S. C. Glazier. 1968. Ecological chemistry and the palatability spectrum. *Science* 161:1349–51.

Brown, C. H., and P. M. Waser. 1984. Hearing and communication in blue monkeys (*Cercopithecus mitis*). *Anim. Behav.* 32:66–75.

Brown, C. R., and M. B. Brown. 1986. Ectoparasitism as a cost of coloniality in cliff swallows (*Hirundo pyrrhonota*). *Ecology* 67:1206–18.

Brown, J. L. 1964. The evolution of diversity in avian territorial systems. *Wilson Bull.* 76:160–69.

———. 1974. Alternate routes to sociality in jays—with a theory for the evolution of altruism and communal breeding. *Amer. Zool.* 14:63–80.

Brown, J. L., and E. R. Brown. 1981. Extended family system in a communal bird. *Science* 211:959–60.

Brown, J. L., E. R. Brown, S. D. Brown, and D. D. Dow. 1982. Helpers: Effects of experimental removal on reproductive success. *Science* 215:421–22.

Brown, R. E. 1986. Social and hormonal factors influencing infanticide and its suppression in adult male Long-Evans rats (*Rattus norvegicus*). *J. Comp. Psychol.* 100:155–61.

Brown, S. D., R. J. Dooling, and K. O'Grady. 1988. Peripheral organization of acoustic stimuli by budgerigars (*Melopsittacus undulatus*). III. Contact calls. *J. Comp. Psychol.* 102:236–47.

Brown, W. L., and E. O. Wilson. 1956. Character displacement. *Syst. Zool.* 5:49–64.

Bruce, H. M. 1961. Observations on the suckling stimulus and lactation in the rat. *J. Reprod. Fertil.* 2:17–34.

———. 1966. Smell as an exteroceptive factor. *J. Anim. Sci.,* Suppl. 25:83–89.

Bruderer, B., and P. Steidinger. 1972. Methods of quantitative and qualitative analysis of bird migration with a tracking radar. *NASA Spec. Publ.* NASA SP-262:151–67.

Bruner, J. S., A. Jolly, and K. Sylva, eds. 1976. *Play: Its Role in Development and Evolution.* New York: Basic Books.

Buchsbaum, R. 1938. *Animals without Backbones.* Chicago: University of Chicago Press.

Buckle, G. R., and L. Greenberg. 1981. Nestmate recognition in sweat bees (*Lasioglossum zephyrum*): Does an individual recognize its own odour or only odours of its nestmates? *Anim. Behav.* 29:802–9.

Bull, J. J. 1980. Sex determination in reptiles. *Quart. Rev. Biol.* 55:3–21.

Bull, J. J., W. H. N. Gutzke, and D. Crews. 1988. Sex reversal by estradiol in three reptilian orders. *Gen. Comp. Endocrinol.* 70:425–28.

Bullock, T. H. 1973. Seeing the world through a new sense: Electroreception in fish. *Amer. Scientist* 61:316–25.

Burghardt, G. M. 1966. Stimulus control of the prey attack response in naive garter snakes. *Psychnom. Sci.* 4:37–38.

———. 1967. The primacy of the first feeding experience in the snapping turtle. *Psychonomic Science* 7:383–84.

———. 1969. Comparative prey attack studies in newborn snakes of the genus *Thamnophis. Behaviour* 33:77–144.

Burghardt, G. M., and E. H. Hess. 1966. Food imprinting in the snapping turtle, *Chelydra serpentia. Science* 151:108–9.

Burghardt, G. M., and C. H. Pruitt. 1975. Role of the tongue and senses in feeding of naive and experienced garter snakes. *Physiol. Behav.* 14:185–94.

Burke, T., N. B. Davies, M. W. Bruford, and B. J. Hatchwell. 1989. Parental care and mating behaviour of polyandrous dunnocks *Prunella modularis* related to paternity by DNA fingerprinting. *Nature* 338:249–50.

Burley, N. 1979. The evolution of concealed ovulation. *Amer. Nat.* 114:835–58.

Burley, N., G. Krantzberg, and P. Radman. 1982. Influence of colour-banding on the conspecific preferences of zebra finches. *Anim. Behav.* 27:686–98.

Burnet, B., and K. Connolly. 1974. Activity and sexual behavior in *Drosophila melanogaster.* In *Genetics of Behaviour,* ed. J. H. F. van Abeelen. Amsterdam: North-Holland.

Burnet, B., K. Connolly, M. Kearney, and R. Cook. 1973. Effects of male paragonial gland secretion on sexual receptivity and courtship behaviour of female *Drosophila melanogaster. J. Insect. Physiol.* 19:2421–31.

Burtt, E. T. 1974. *The Senses of Animals.* London: Wykeham.

Buskirk, R. E. 1981. Sociality in Arachnida. In *Social Insects,* ed. H. R. Hermann. New York: Academic Press.

Buss, L. W. 1981. Group living, competition, and the evolution of cooperation in a sessile invertebrate. *Science* 213:1012–14.

Calhoun, J. B. 1962. Population density and social pathology. *Sci. Amer.* 206:139–48.

———. 1973. Death squared: The explosive growth and demise of a mouse population. *Proc. Roy. Soc. Med.* 66:80–88.

Calvert, W. H., and L. P. Brower. 1981. The importance of forest cover for the survival of overwintering monarch butterflies (*Danaus plexippus Danaidae*). *J. Lepid. Soc.* 35:216–25.

Capranica, R. R. 1976a. Morphology and physiology of the auditory system. In *Handbook of Frog Neurobiology,* ed. R. Llinas and W. Precht, 551–75. Berlin: Springer-Verlag.

———. 1976b. The auditory system of anurans. In *Physiology of Amphibia,* Vol. 3, ed. B. Lofts, 443–66. New York: Academic Press.

Capranica, R. R., and A. J. M. Moffat. 1977. Place mechanism underlying frequency analysis in the toad's inner ear. *J. Acoust. Soc. Amer.* 62 (Suppl.):S36.

———. 1980. Nonlinear properties of the peripheral auditory system of anurans. In *Comparative Studies of Auditory Processing in Vertebrates,* ed. A. Popper and R. Fay, 139–65. Berlin: Springer-Verlag.

———. 1983. Neurobehavioral correlates of sound communication in anurans. In *Advances in Vertebrate Neuroethology,* ed. J. P. Ewert, R. R. Capranica and D. J. Ingle, 701–30. New York: Plenum Publishing.

Caraco, T., and L. L. Wolf. 1975. Ecological determinants of group sizes of foraging lions. *Amer. Nat.* 109:343–52.

Caraco, T., S. Martindale, and T. S. Whitham. 1980. An empirical demonstration of risk-sensitive foraging preferences. *Anim. Behav.* 28:820–30.

Carlisle, D. B., and P. E. Ellis. 1959. La persistance des glandes ventrales céphaliques chez les criquets solitaires. *Comptes Rendu* 249:1059–60.

Carlson, A. D., and J. Copeland. 1978. Behavioral plasticity in the flash communication systems of fireflies. *Amer. Scientist* 66:340–46.

Carpenter, F. L., and R. E. MacMillen. 1976. Threshold model of feeding territoriality and test with a Hawaiian honeycreeper. *Science* 194:639–42.

Carr, A. 1965. The navigation of the green turtle. *Sci. Amer.* 212:78–86.

———. 1967. Adaptive aspects of the scheduled travel of Chelonia. *Proc. Ann. Biol. Colloq.*, Oreg. State Univ. 27:35–36.

Carr, A. J., and P. J. Coleman. 1974. Seafloor spreading theory and the odyssey of the green turtle. *Nature* 249:128–30.

Cavalli-Sforza, L. L., and W. F. Bodmer. 1971. *Genetics of Human Populations*. San Francisco: W. H. Freeman.

Chadab, R., and C. W. Rettenmeyer. 1975. Mass recruitment by army ants. *Science* 188:1124–25.

Chagnon, N. A. 1988. Life histories, blood revenge, and warfare in a tribal population. *Science* 239:985–92.

Chagnon, N. A., and P. E. Bugos, Jr. 1979. Kin selection and conflict: An analysis of a Yanomamo ax fight. In *Evolutionary Biology and Human Social Behavior: An Anthropological Perspective*, ed. N. A. Chagnon and W. Irons. N. Scituate, MA: Duxbury Press.

Charniaux-Cotton, H., and L. H. Kleinholz. 1964. Hormones in invertebrates other than insects. In *The Hormones*, vol. 4, ed. G. Pincus, K. V. Thimann, and E. B. Astwood. New York: Academic Press.

Charnov, E., and J. Finerty. 1980. Vole population cycles: A case for kin-selection? *Oecologia* 45:1–2.

Charnov, E. L. 1976. Optimal foraging: The marginal value theorem. *Theor. Pop. Biol.* 9:129–36.

Chase, J., and R. A. Suthers. 1969. Visual obstacle avoidance by echolocating bats. *Anim. Behav.* 17:201–7.

Cheney, D. L., and R. M. Seyfarth. 1983. Nonrandom dispersal in free-ranging vervet monkeys: Social and genetic consequences. *Am. Nat.* 122:392–412.

Cheng, M. F. 1979. Progress and prospects in ring dove research: A personal view. *Adv. Stud. Behav.* 9:97–130.

Chevalier-Skolnikoff, S. 1974. Male-female, female-female, and male-male sexual behavior in the stumptail monkey, with special attention to the female orgasm. *Arch. Sex. Behav.* 3:95–116.

Chitty, D. 1960. Population processes in the vole and their relevance to general theory. *Can. J. Zool.* 38:99–113.

———. 1967. The natural selection of self-regulatory behavior in animal populations. *Proceedings of the Ecological Society of Australia* 2:51–78.

Chitty, D., and H. Chitty. 1962. Population trends among the voles at Lake Vyrnwy, 1932–60. *Symp. Theriologicum, Brno.* 1960:67–76.

Christenson, T. E. 1984. Behaviour of colonial and solitary spiders of the Theridid species *Anelosimus eximus. Anim. Behav.* 32:725–34.

Christian, J. J. 1950. The adreno-pituitary system and population cycles in mammals. *J. Mammal.* 31:247–59.

———. 1970. Social subordination, population density and mammalian evolution. *Science* 168:84–90.

———. 1978. Neurobehavioral endocrine regulation of small mammal populations. In *Populations of Small Mammals under Natural Conditions.* ed. D. P. Snyder. The Pymatuning Symposia in Ecology 5:143–58.

Christian, J. J., and D. E. Davis. 1964. Endocrines, behavior and population. *Science* 146:1550–60.

Clark, A. B. 1978. Sex ratio and local resource competition in a prosimian primate. *Science* 201:163–65.

Clarke, C. A., and P. M. Sheppard. 1960. Supergenes and mimicry. *Heredity* 14:175–85.

Clemens, L. G. 1974. Neurohormonal control of male sexual behavior. In *Reproductive Behavior*, ed. W. Montagna and S. Sadler. New York: Plenum.

Clemens, L. G., B. A. Gladue, and L. P. Coniglio. 1978. Prenatal endogenous androgenic influences on masculine sexual behavior and genital morphology in male and female rats. *Horm. Behav.* 10:40–53.

Clements, F. E. 1936. Nature and structure of the climax. *J. Ecol.* 24:252–84.

Cloudsley-Thompson, J. L. 1952. Studies in diurnal rhythms. II. Changes in the physiological responses of the woodlouse *Oniscus aspellus* L to environmental stimuli. *J. Exp. Biol.* 29:295–303.

———. 1960. Adaptive functions of circadian rhythms. *Cold. Spr. Harb. Symp. Quant. Biol.* 24:361–67.

Clutton-Brock, T. H., ed. 1988. *Reproductive Success.* Chicago: University of Chicago Press.

Clutton-Brock, T. H., S. D. Albon, and F. E. Guinness. 1984. Maternal dominance, breeding success and birth sex ratios in red deer. *Nature* 308:358–60.

Clutton-Brock, T. H., S. D. Albon, R. M. Gibson, and F. E. Guiness. 1979. The logical stag: Adaptive aspects of fighting in red deer. *Anim. Behav.* 27:211–25.

Clutton-Brock, T. H., F. E. Guinness, and S. D. Albon. 1982. *Red Deer: Behavior and Ecology of Two Sexes.* Chicago: University of Chicago Press.

Cody, M. L. 1974. Optimization in ecology. *Science* 183:1156–64.

Colby, D. R., and J. G. Vandenbergh. 1974. Regulatory effects of urinary pheromones on puberty in the mouse. *Biol. Reprod.* 11:268–79.

Cole, J. E., and J. A. Ward. 1970. An analysis of parental recognition by the young of the cichlid fish, *Etroplus maculatus* (Bloch). *Z. Tierpsychol.* 27:156–276.

Cole, L. J. 1916. Twinning in cattle with special reference to the free-martin. *Science* 43:177.

Colgan, P. W. 1989. *Animal Motivation.* New York: Chapman and Hall.

Collias, N. E., and E. C. Collias. 1963. Evolutionary trends in nest building by the weaverbirds (Ploceidae). *Proc. XIII Intern. Ornith. Cong., Amer. Ornith. Union* 518–30.

Collier, G., and C. R. Rovee-Collier. 1982. A comparative analysis of optimal foraging behavior: laboratory simulations. In *Foraging Behavior: Ecological, Ethological, and Psychological Approaches*, ed. A. C. Kamil and T. D. Sargent, 39–76. New York: Garland STPM Press.

Colvin, B. A. 1984. Barn owl foraging behavior and secondary poisoning hazard from rodenticide use on farms. Ph.D. dissertation, Bowling Green State University.

Comer, C., and P. Grobstein. 1981. Tactually elicited prey acquisition behavior in the frog, *Rana pipiens*, and a comparison with visually elicited behavior. *J. Comp. Physiol.* 142:141–50.

Connell, J. H. 1961. The influence of interspecific competition and other factors on the distribution of the barnacle *Chthamalus stellatus*. *Ecology* 42:710–23.

Connor, V. M., and J. F. Quinn. 1984. Stimulation of food species growth by limpet mucus. *Science* 225:843–44.

Cook, A. 1971. Habituation in a freshwater snail (*Limnaea stagnalis*). *Anim. Behav.* 17:679–82.

Cook, L. M., G. S. Mani, and M. E. Varley. 1986. Postindustrial melanism in the peppered moth. *Science* 231:611–13.

Cooper, R., and J. Zubek. 1958. Effects of enriched and restricted early environments on the learning ability of bright and dull rats. *Can. J. Psychol.* 12:159–64.

Copeland, J. 1983. Male firefly mimicry. *Science* 221:484–85.

Coppola, D. M., and J. G. Vandenbergh. 1987. Induction of a puberty-regulating chemosignal in wild mouse populations. *J. Mammal.* 68:86–91.

Corbet, P. S. 1957. The life history of the emperor dragonfly, *Anax imperator* Leach. *J. Anim. Ecol.* 26:1–69.

———. 1960. Patterns of circadian rhythms in insects. *Cold Spr. Harb. Symp. Quant. Biol.* 25:357–60.

Corning, W. C., and S. Kelly. 1973. Platyhelminthes: The turbellarians. In *Invertebrate Learning*, Vol. 1, ed. W. C. Corning, J. A. Dyal, and A. O. D. Willows. New York: Plenum.

Corning, W. C. and von Burg, R. 1973. Protozoa. In *Invertebrate Learning*, Vol. 1. ed. W. C. Corning, J. A. Dyal, and A. O. D. Willows. New York: Plenum.

Coulson, J. C. 1966. The influence of the pair bond and age on the breeding biology of the kittiwake gull, *Rissa tridactyla*. *J. Animal. Ecol.* 35:269–79.

Cowie, R. J. 1977. Optimal foraging in great tits (*Parus major*). *Nature* 268:137–39.

Cox, C. R., and B. J. L. Boeuf. 1977. Female incitation of male competition: A mechanism in sexual selection. *Amer. Nat.* 111:317–35.

Craig, W. 1914. Male doves reared in isolation. *J. Anim. Behav.* 4:121–33.

Crawford, J. D. 1984. Orientation in a vertical plane: The use of light cues by an orb-weaving spider, *Araneus diadematus* Clerk. *Anim. Behav.* 32:162–71.

Crews, D. 1975. Psychobiology of reptilian reproduction. *Science* 189:1059–65.

———. 1977. The annotated *Anole*: Studies on the control of lizard reproduction. *Amer. Scientist* 65:428–34.

———. 1979. Neuroendocrinology of lizard reproduction. *Biol. Reprod.* 20:51–73.

———. 1980. Interrelationships among ecological, behavioral and neuroendocrine processes in the reproductive cycle of *Anolis carolinensis* and other reptiles. *Adv. Stud. Behav.* 11:1–75.

———. 1983. Alternative reproductive tactics in reptiles. *BioScience* 33:562–66.

Crews, D., and N. Greenberg. 1981. Function and causation of social signals in lizards. *Amer. Zool.* 21:273–94.

Crews, D., J. S. Rosenblatt, and D. S. Lehrman. 1974. Effects of unseasonal environmental regime, group presence, group composition and males' physiological state on ovarian recrudescence in the lizard *Anolis carolinensis*. *Endocrinology* 95:102–6.

Cross, B. A., and R. G. Dyer. 1972. Ovarian modulation of unit activity in the anterior hypothalamus of the cyclic rat. *J. Physiol.* (Lond.) 222:25P.

Crow, J. F., and M. Kimura. 1970. *Introduction to Population Genetics Theory*. New York: Harper & Row.

Crowcroft, P. 1973. *Mice All Over*. Brookfield, IL: Chicago Zoological Park.

Crowcroft, P., and F. Rowe. 1957. Social organization and territorial behaviour in the wild house mouse (*Mus musculus* L.). *Proc. Zool. Soc. Lond.* 140:517–31.

Crump, A. J., and J. Brady. 1979. Circadian activity patterns in three species of tsetse fly: *Glossina palpalis, austeni* and *morsitans*. *Physiol. Entomol.* 4:311–18.

Cullen, E. 1957. Adaptations in the kittiwake to cliff nesting. *Ibis* 99:275–302.

———. 1960. Experiment on the effect of social isolation on reproductive behavior in the three-spined stickleback. *Anim. Behav.* 8:235.

Czeisler, C. A., R. E. Kronauer, J. S. Allan, J. F. Duffy, M. E. Jewett, E. N. Brown, and J. M. Ronda. 1989. Bright light induction of strong (Type 0) resetting of the human circadian pacemaker. *Science* 244:1328–33.

Daan, S., and J. Aschoff. 1982. Circadian contributions to survival. In *Vertebrate Circadian Systems*, ed. J. Aschoff, S. Daan, and G. A. Groos. Berlin: Springer-Verlag.

Daan, S., and A. J. Lewy. 1984. Scheduled exposure to daylight: A potential strategy to reduce jet-lag following transmeridian flight. *Psychopharmacol. Bull.* 20:566–68.

Daly, M. 1978. The cost of mating. *Amer. Nat.* 112:771–74.

Daly, M., and M. I. Wilson. 1981. Abuse and neglect of children in evolutionary perspective. In *Natural Selection and Social Behavior*, ed. R. D. Alexander and D. W. Tinkle. New York: Chiron Press.

———. 1982. Homicide and kinship. *Amer. Anthrop.* 84:372–78.

———. 1988. Evolutionary social psychology and family homicide. *Science* 242:519–24.

Daniels, R. A. 1979. Nest guard replacement in the antarctic fish *Harpigifer bispinus*: Possible altruistic behavior. *Science* 205:831–33.

Darchen, R., and B. Delage. 1970. Facteur déterminant les castes chez les Trigones. *Compt. Ren. Acad. Sci.* (Paris) 270:1372–73.

Darwin, C. 1871. *The Descent of Man, and Selection in Relation to Sex*. New York: D. Appleton.

———. 1873. *On the Expression of the Emotions in Man and Animals*. New York: D. Appleton.

Darwin, C. R. 1845. *The Voyage of the Beagle*. London: Dent.

———. 1859. *The Origin of Species*. London: Dent.

Datta, L. G., S. Milstein, and M. E. Bitterman. 1960. Habitat reversal in the crab. *J. Comp. Physiol. Psychol.* 53:275–78.

Davidson, J. M. 1966a. Characteristics of sex behavior in male rats following castration. *Anim. Behav.* 14:266–72.

————. 1966b. Activation of the male rat's sexual behavior by intracerebral implantation of androgen. *Endocrinology* 79:783–94.

Davies, N. B. 1978. Territorial defence in the speckled wood butterfly (*Pararge aegeria*): the resident always wins. *Anim. Behav.* 26:138–47.

————. 1983. Polyandry, cloaca-pecking and sperm competition in dunnocks. *Nature* 302:334–36.

Davies, N. B., and A. I. Houston. 1983. Time allocation between territories and flocks and owner-satellite conflict in foraging pied wagtails, *Motacilla alba. J. Anim. Ecol.* 52:621–34.

————. 1984. Territory economics. In *Behavioural Ecology: An Evolutionary Approach*, 2d ed., ed. J. R. Krebs and N. B. Davies. Oxford, England: Blackwell Scientific Publications, Ltd.

Davies, N. B., and A. Lundberg. 1984. Food distribution and a variable mating system in the dunnock, *Prunella modularis. J. Anim. Ecol.* 53:895–912.

Davis, D. E. 1942. The phylogeny of social nesting habits in the Crotophaginae. *Quart. Rev. Biol.* 17:115–34.

————. 1951. The relation between level of population and pregnancy of Norway rats. *Ecology* 32:459–61.

————. 1963. The physiological analysis of aggressive behavior. In *Social Behavior and Organization among Vertebrates*, ed. W. Etkin. Chicago: University of Chicago Press.

Davis, W. H., and H. B. Hitchcock. 1965. Biology and migration of the bat, *Myotis lucifungus*, in New England. *J. Mammal.* 46:296–313.

Dawe, A. R., and W. A. Spurrier. 1972. The blood-borne "trigger" for natural mammalian hibernation in the 13-lined ground squirrel and the woodchuck. *Cryobiology* 9:163–72.

Dawkins, R. 1976. *The Selfish Gene.* New York: Oxford University Press.

————. 1982. *The Extended Phenotype.* Oxford: Oxford University Press.

Dawkins, R., and T. R. Carlisle. 1976. Parental investment, mate desertion and a fallacy. *Nature* 262:131–33.

Dawkins, R., and J. R. Krebs. 1978. Animal signals: Information or manipulation? In *Behavioural Ecology: An Evolutionary Approach*, eds. J. R. Krebs and N. B. Davies. Oxford, England: Blackwell Scientific Publications, Ltd.

de Beer, G. 1958. *Embryos and Ancestors.* 3d ed. London: Oxford University Press.

DeCoursey, P. J. 1960. Phase control of activity in a rodent. *Cold. Spring Harb. Symp. Quant. Biol.* 25:49–56.

————. 1961. Effect of light on the circadian activity rhythm of the flying squirrel, *Glaucomys volans. Zeit. Physiol.* 44:331–54.

————. 1983. Biological timekeeping. *Biology of Crustacea* 7:107–62.

————. 1986. Circadian photoentrainment: Parameters of phase delaying. *J. Biol. Rhythms* 1:171–86.

Deevey, E. 1960. The human population. *Sci. Amer.* 203:195–204.

Delgado, J. M. R. 1963. Cerebral heterostimulation in a monkey colony. *Science* 141:161–63.

————. 1966. Aggressive behavior evoked by radio-stimulation in monkey colonies. *Amer. Zool.* 6:669–81.

————. 1967. Social rank and radio stimulated aggressiveness in monkeys. *J. Nerv. Ment. Disease* 144:383–90.

DeLong, K. T. 1967. Population ecology of feral house mice. *Ecology* 48:611–34.

Denenberg, V. H., and K. M. Rosenberg. 1967. Nongenetic transmission of information. *Nature* 216:549–50.

Denenberg, V. H., and A. E. Whimbey. 1963. Behavior of adult rats is modified by the experiences their mothers had as infants. *Science* 142:1192–93.

de Roth, S. 1974. Communication in the crayfish (*Orconectes rusticus*). Master's thesis, Bowling Green State University.

Derscheid, J. M. 1947. Strange parrots. *Avic. Mag.* 53:44–49.

de Souza, H. M. L., A. B. da Cunha, and E. P. dos Santos. 1970. Adaptive polymorphism of behavior evolved in laboratory populations of *Drosophila willistoni. Am. Nat.* 104:175–89.

Dethier, V. G. 1962. *To Know a Fly.* Englewood Cliffs, NJ: Prentice-Hall.

————. 1976. *The Hungry Fly.* Cambridge: Harvard University Press.

Dethier, V. G., and D. Bodenstein. 1958. Hunger in the blowfly. *Zeit. Tierpsychol.* 15:129–40.

Dethier, V. G., and A. Gelperin. 1967. Hyperphagia in the blowfly. *J. Exp. Biol.* 47:191–200.

Dethier, V. G., and E. Stellar. 1961. *Animal Behavior: Its Evolutionary and Neurological Basis.* Englewood Cliffs, NJ: Prentice-Hall.

Deutsch, J. A. 1960. *Structural Basis of Behavior.* Chicago: University of Chicago Press.

————. 1973. *Physiological Basis of Memory.* New York: Academic Press.

————. 1978. *Comparative Animal Behavior.* New York: McGraw-Hill.

DeVoogd, T., and F. Nottebohm. 1981. Gonadal hormones influence dendritic growth in the adult avian brain. *Science* 214:202–4.

DeVore, I., ed. 1965. *Primate Behavior: Field Studies of Monkeys and Apes.* New York: Holt, Rinehart and Winston.

DeVries, D. R., R. A. Stein, and P. L. Chesson. 1989. Sunfish foraging among patches: The patch-departure decision. *Anim. Behav.* 37:455–64.

deWilde, J. 1975. An endocrine view of metamorphosis, polymorphism and diapause in insects. *Amer. Zool.* (sup. 1):13–28.

Dewsbury, D. A. 1972. Patterns of copulatory behavior in male mammals. *Quart. Rev. Biol.* 47:1–33.

————. 1975. Diversity and adaptation in rodent copulatory behavior. *Science* 190:947–54.

————. 1978. *Comparative Animal Behavior.* New York: McGraw-Hill.

————. 1982. Ejaculate cost and male choice. *Amer. Nat.* 119:601–10.

————. 1984. *Comparative Psychology in the Twentieth Century.* Stroudsburg, PA: Hutchinson Ross.

————. 1985. *Leaders in the Study of Animal Behavior.* Lewisburg, PA: Bucknell University Press.

Dewsbury, D. A., D. L. Lanier, and A. Miglietta. 1980. A laboratory study of climbing behavior in 11 species of muroid rodents. *Am. Mid. Nat.* 103:66–72.

Diamond, J. M. 1974. Colonization of exploded volcanic islands by birds: The supertramp strategy. *Science* 184:803–6.

————. 1978. Niche shifts and the rediscovery of interspecific competition. *Amer. Scientist* 66:322–31.

Dickemann, M. 1975. Demographic consequences of infanticide in man. *Ann. Rev. Ecol. Syst.* 6:107–37.

———. 1979. Female infanticide, reproductive strategies, and social stratification: A preliminary model. In *Evolutionary Biology and Human Social Behavior: An Anthropological Perspective*, ed. N. A. Chagnon and W. Irons. N. Scituate, MA: Duxbury Press.

Dill, P. A. 1977. Development of behaviour in alevins of Atlantic salmon, *Salmo salar*, and rainbow trout, *S. gairdneri*. *Anim. Behav.* 25:116–21.

Dillon, L. S. 1978. *Evolution: Concepts and Consequences*. 2d ed. St. Louis: Mosby.

Dingle, H. 1980. Ecology and evolution of migration. In *Animal Migration, Orientation and Navigation*, ed. S. A. Gauthreaux. New York: Academic Press.

Dix, M. W. 1968. Snake food preference: Innate intraspecific geographic variation. *Science* 159:1478–79.

Dobson, F. S. 1982. Competition for mates and predominant juvenile male dispersal in mammals. *Anim. Behav.* 30:1183–92.

Domjan, M. 1980. Ingestional aversion learning: Unique and general processes. *Adv. Stud. Behav.* 11:276–337.

Dörner, G., and M. Kawakami, eds. 1978. *Hormones and Brain Development*. New York: Elsevier North-Holland.

Doty, R. L. 1981. Olfactory communication in humans. *Chemical Senses* 6:351–76.

Dowling, J. E. 1976. Physiology and morphology of the retina. In *Frog Neurobiology*, ed. R. Llinas and L. Precht. Berlin: Springer-Verlag.

Drickamer, L. C. 1970. Seed preferences in wild caught *Peromyscus maniculatus bairdi* and *P. leucopus noveboracensis*. *J. Mammal.* 51:191–94.

———. 1972. Experience and selection behavior as factors in the food habits of *Peromyscus*: Use of olfaction. *Behaviour* 41:269–87.

———. 1974a. Sexual maturation of female house mice: Social inhibition. *Dev. Psychobiol.* 7:257–65.

———. 1974b. A ten-year summary of population and reproduction data for free-ranging *Macaca mulatta* at La Parguera, Puerto Rico. *Folia Primatologica* 21:61–80.

———. 1975. Daylength and sexual maturation of female house mice. *Develop. Psychobiol.* 8:561–70.

———. 1979. Acceleration and delay of first vaginal estrus in wild *Mus musculus*. *J. Mammal.* 60:215–16.

———. 1981. Selection for age of sexual maturation in mice and the consequences for population regulation. *Behav. Neur. Biol.* 31:82–89.

———. 1982a. Delay and acceleration of puberty in female mice by urinary chemosignals from other females. *Develop. Psychobiol.* 15:433–42.

———. 1982b. Acceleration and delay of sexual maturation in female house mice by urinary cues: Dose levels and mixing urine from different sources. *Anim. Behav.* 30:456–60.

———. 1983. Male acceleration of puberty in female mice. *J. Comp. Psychol.* 97:191–200.

———. 1989. Pheromones: Behavioral and biochemical aspects. In *Advances in Comparative and Environmental Physiology*, ed. J. Balthazart, 269–348. Berlin: Springer-Verlag.

Drickamer, L. C., and J. E. Hoover. 1979. Effects of urine from pregnant and lactating female house mice on sexual maturation of juvenile females. *Develop. Psychobiol.* 12:545–51.

Drickamer, L. C., and R. X. Murphy. 1978. Female mouse maturation: Effects of excreted and bladder urine from juvenile and adult males. *Develop. Psychobiol.* 11:63–72.

———. 1986. Puberty-influencing chemosignals in mice: ecological and evolutionary considerations. In *Chemical Signals in Vertebrates*, vol. 4, ed. D. Duvall, D. Muller-Schwarze, and R. M. Silverstein, 441–55. New York: Plenum.

———. 1988. Acceleration and delay of sexual maturation in female house mice (*Mus domesticus*) by urinary chemosignals: Mixing urine sources in unequal proportions. *J. Comp. Psychol.* 102:215–21.

Drickamer, L. C., and J. Stuart. 1984. Peromyscus: Snow tracking and possible cues used for navigation. *Amer. Mid. Nat.* 111:202–4.

Drickamer, L. C., and J. G. Vandenbergh. 1973. Predictors of dominance in the female golden hamster (*Mesocricetus auratus*). *Anim. Behav.* 21:564–70.

Drickamer, L. C., J. G. Vandenbergh, and D. R. Colby. 1973. Predictors of dominance in the male golden hamster (*Mesocricetus auratus*). *Anim. Behav.* 21:557–63.

Drickamer, L. C., and S. H. Vessey. 1973. Group changing in male free-ranging rhesus monkeys. *Primates* 14:359–68.

Drummond, H. 1981. The nature and description of behavior patterns. In *Perspectives in Ethology*, vol. 4, eds. P. P. G. Bateson and P. H. Klopfer. New York: Plenum.

Drury, W. H., and J. A. Keith. 1962. Radar studies of songbird migration in coastal New England. *Ibis* 104:449–89.

Drury, W. H., and I. C. T. Nisbet. 1964. Radar studies of orientation of songbird migrants in southeastern New England. *Bird Banding* 35:69–119.

D'Udine, D., and E. Alleva. 1983. Early experience and sexual preferences in rodents. In *Mate Choice*, ed. P. P. G. Bateson, 311–27. New York: Cambridge University Press.

Duffey, S. S. 1970. Cardiac glycosides and distastefulness: Some observations on the palatability spectrum of butterflies. *Science* 169:78–79.

Dunbar, R. I. M. 1976. Some aspects of research design and their implications in the observational study of behavior. *Behaviour* 58:78–98.

Duncan, J. R. and D. M. Bird. 1989. The influence of relatedness and display effort on the mate choice of captive female American kestrels. *Anim. Behav.* 37:112–17.

Durham, W. H. 1979. Toward a coevolutionary theory of human biology and culture. In *Evolutionary Biology and Human Social Behavior: An Anthropological Perspective*, ed. N. A. Chagnon and W. Irons. N. Scituate, MA: Duxbury.

Dyal, J. A., and W. C. Corning. 1973. Invertebrates learning and behavior taxonomies. In *Invertebrate Learning*. Vol. 1, ed. W. C. Corning, J. A. Dyal, and A. O. D. Willows. New York: Plenum.

Ealey, E. H. M. 1963. The ecological significance of delayed implantation in a population of the hill kangaroo (*Macropus robustus*). In *Delayed Implantation*, ed. A. C. Enders. Chicago: University of Chicago Press.

Eberhard, W. G. 1988. Memory of distances and directions moved as cues during temporary spiral construction in the spider *Leucauge mariana* (Araneae: Araneidae). *J. Insect. Behav.* 1:51–66.

Ebert, P. D., and J. S. Hyde. 1976. Selection for agonistic behavior in wild female *Mus musculus*. *Behav. Genet* 6:291–304.

Edwards, D. A. 1968. Mice: Fighting by neonatally androgenized females. *Science* 161:1027–28.

Ehrenfeld, D. W., and A. Carr. 1967. The role of vision in the sea-finding orientation of the green turtle (*Chelonia mydas*). *Anim. Behav.* 15:25–36.

Ehrman, L., and P. A. Parsons. 1981. *Behavior Genetics and Evolution.* New York: McGraw-Hill.

Eibl-Eibesfeldt, I. 1975. *Ethology: The Biology of Behavior.* 2d ed. New York: Holt, Rinehart and Winston.

Eisenberg, J. 1981. *The Mammalian Radiations.* Chicago: University of Chicago Press.

Eisner, T. E. 1966. Beetle spray discourages predators. *Natur. Hist.* 75:42–47.

———. 1970. Chemical defenses against predators in Arthropods. In *Chemical Ecology,* ed. E. Sondheimer and J. B. Simeone. New York: Academic Press.

Ekdale, A. A. 1980. Graphoglyptid burrows in modern deep-sea sediment. *Science* 207:304–6.

Elmes, G. W. 1971. An experimental study on the distribution of heathland ants. *J. Anim. Ecol.* 40:495–99.

Elner, R. W., and R. N. Hughes. 1978. Energy maximization in the diet of the shore crab, *Carcinus maenas. J. Anim. Ecol.* 47:103–16.

Elsner, N., and A. V. Popov. 1978. Neuroethology of acoustic communication. *Adv. Insect Physio.* 13:229–335.

Eltringham, S. K. 1978. Methods of capturing wild animals for marking purposes. In *Animal Marking.* ed. B. Stonehouse, 13–23. Baltimore: University Park Press.

Emlen, J. T. 1952a. Social behavior in nesting cliff swallows. *Condor* 54:177–99.

———. 1952b. Flocking behavior in birds. *Auk* 69:160–70.

Emlen, S. T. 1967a. Migratory orientation in the indigo bunting. *Passerina cyanea. Auk* 84:463–89.

———. 1967b. Migratory orientation in the indigo bunting. I. Evidence for use of celestial cues. *Auk* 84:309–42.

———. 1970. Celestial rotation: Its importance in the development of migratory orientation. *Science* 170:1198–1201.

———. 1972. An experimental analysis of the parameters of bird song eliciting species recognition. *Behaviour* 41:130–71.

———. 1975a. Migration, orientation and navigation. In *Avian Biology,* vol. 5, ed. D. S. Farner and J. R. King. New York: Academic Press.

———. 1975b. The stellar-orientation system of a migratory bird. *Sci. Amer.* 233(2):102–11.

———. 1978. The evolution of cooperative breeding in birds. In *Behavioural Ecology: An Evolutionary Approach,* eds. J. R. Krebs and N. B. Davies. Oxford, England: Blackwell Scientific Publications, Ltd.

———. 1982. The evolution of helping. I. An ecological constraints model. *Am. Nat.* 119:29–39.

———. 1984. Cooperative breeding in birds and mammals. In *Behavioural Ecology: An Evolutionary Approach,* 2d ed., ed. J. R. Krebs and N. B. Davies. Oxford, England: Blackwell Scientific Publications, Ltd.

Emlen, S. T., and N. J. Demong. 1975. Adaptive significance of synchronized breeding in a colonial bird: A new hypothesis. *Science* 188:1029–31.

Emlen, S. T., and L. W. Oring. 1977. Ecology, sexual selection, and the evolution of mating systems. *Science* 197:215–23.

Emlen, S. T., and P. H. Wrege. 1988. The role of kinship in helping decisions among white-fronted bee-eaters. *Behav. Ecol. Sociobiol.* 23:305–15.

Enright, J. T. 1970. Ecological aspects of endogenous rhythmicity. *Ann. Rev. Ecol. Syst.* 1:221–38.

Erickson, C. J., and D. S. Lehrman. 1964. Effect of castration of male ring doves on ovarian activity of females. *J. Comp. Physiol. Psychol.* 58:164–66.

Erlenmeyer-Kimling, L., and L. F. Jarvik. 1963. Genetics and intelligence: A review. *Science* 142:1477–78.

Erwin, J., G. Mitchell, and T. Maple. 1973. Abnormal behavior in non-isolate-reared rhesus monkeys. *Psychol. Reports* 33:515–23.

Eskin, A., and J. S. Takahashi. 1983. Adenylate cyclase activation shifts the phase of a circadian pacemaker. *Science* 220:82–84.

Esser, A. H. 1971. *Behavior and Environment: The Use of Space by Animals and Men.* New York: Plenum.

Evans, H. E. 1962. The evolution of prey-carrying mechanisms in wasps. *Evolution* 16:468–83.

Ewert, J. P. 1980. *Neuroethology.* Berlin: Springer-Verlag.

———. 1984. Tectal functions that underlie prey-catching and predator avoidance behaviors in toads. In *Comparative Neurology of the Optic Tectum,* ed. H. Vanegas, 247–416. New York: Plenum Publishing.

———. 1985. Concepts in vertebrate neuroethology. *Anim. Behav.* 33:1–29.

Fabricius, E. 1964. Crucial periods in the development of the following response in young nidifugous birds. *Zeit. Tierpsychol.* 21:326–37.

Fagen, R. 1981. *Animal Play Behavior.* New York: Oxford University Press.

Falconer, D. S. 1981. *Introduction to Quantitative Genetics.* 2d ed. Essex England: Longman Group.

Falls, J. B., and R. J. Brooks. 1975. Individual recognition by song in white-throated sparrows. II. Effects of location. *Can. J. Zool.* 53:1412–20.

Farner, D. S. 1955. The annual stimulus for migration: Experimental and physiologic aspects. In *Recent Studies in Avian Biology,* ed. A. Wolfson. Urbana: University of Illinois Press.

———. 1964. The photoperiodic control of reproductive cycles in birds. *Amer. Scientist* 52:137–56.

Feder, H. 1966. Cleaning symbioses in the marine environment. In *Symbiosis.* vol. 1, ed. S. M. Henry. New York: Academic Press.

———. 1981. Experimental analysis of hormone actions on the hypothalamus, anterior pituitary and ovary. In *Neuroendocrinology of Reproduction,* ed. N. Adler. New York: Plenum.

Feng, A. S., P. M. Narins, and R. R. Capranica. 1975. Three populations of primary auditory fibers in the bullfrog (*Rana catesbiana*): Their peripheral origins and frequency sensitivities. *J. Comp. Physiol.* 100:221–29.

Ferguson, D. E., H. F. Landreth, and J. P. McKeown. 1967. Sun compass orientation of the northern cricket frog, *Acris crepitans. Anim. Behav.* 15:45–53.

Ferguson, D. E., H. F. Landreth, and M. R. Turnipseed. 1965. Astronomical orientation of the southern cricket frog, *Acris gryllus. Copeia* (1965):58–66.

Ferreira, A. J. 1965. Emotional factors in the prenatal environment. *J. Nerv. Ment. Disorders* 141:108–18.

Feshbach, S., and R. D. Singer. 1971. *Television and Aggression.* San Francisco: Jossey-Bass.

Finkenstadt, T., N. T. Adler, T. O. Allen, S. O. E. Ebbeson, and J. P. Ewert. 1985. Mapping of brain activity in mesencephalic and diencephalic structures of toads during presentation of visual key stimuli: A computer assisted analysis of (^{14}C)2DG autoradiographs. *J. Comp. Physiol.* 156:433–45.

Fisher, R. A. 1958. *Genetical Theory of Natural Selection*. New York: Dover.

Fitzpatrick, J. W., and G. E. Woolfenden. 1988. Components of lifetime reproductive success in the Florida scrub jay. In *Reproductive Success*, ed. T. H. Clutton-Brock. Chicago: University of Chicago Press.

Fletcher, H. J. 1965. The delayed-response problem. In *Behavior of Nonhuman Primates*. Vol. 1. ed. A. M. Schrier, H. F. Harlow, and F. Stollnitz. New York: Academic Press.

Fletcher, T. J. 1975. The environmental and hormonal control of reproduction in male and female red deer (*Cervus elaphus*). Ph.D. Thesis. University of Cambridge, England.

———. 1978. The induction of male sexual behavior in red deer (*Cervus elaphus*) by the administration of testosterone to hinds and estradiol 17B to stags. *Horm. Behav.* 11:74–88.

Flynn, J. P. 1967. The neural basis of aggression in cats. In *Neurophysiology and Emotion*, ed. D. C. Glass. New York: Rockefeller University Press and Russell Sage Foundation.

Ford, E. B. 1964. *Ecological Genetics*. New York: Wiley.

Foster, J. B. 1966. The giraffe of Nairobi National Park: home range, sex ratios, the herd, and food. *E. Afr. Wildl. J.* 4: 139–48.

Fouts, R. 1973. Acquisition and testing of gestural signs in four young chimpanzees. *Science* 180:978–80.

Fox, M. W. 1975. The behavior of cats. In *Behavior of Domestic Animals*, ed. E. S. E. Hafez. Baltimore: Williams & Wilkins.

Fox, M. W., and A. L. Clark. 1971. The development and temporal sequencing of agonistic behavior in the coyote (*Canis latrans*). *Z. Tierpsychol.* 28:262–78.

Fraenkel, G. 1975. Interactions between ecdysone, bursicon and other endocrines during puparium formation and adult emergence in flies. *Amer. Zool.* 15 (sup. 1):29–41.

Fraenkel, G., and D. L. Gunn. 1940. *Orientation of Animals: Kineses, Taxes, and Compass Reactions*. New York: Oxford University Press.

Francq, E. N. 1969. Behavioral aspects of feigned death in the opossum *Didelphis marsupialis*. *Amer. Mid. Nat.* 81:556–68.

Franks, N. R. 1989. Army ants: A collective intelligence. *Amer. Scientist* 77:138–45.

Fraser, D. F. 1976. Coexistence of salamanders in the genus *Plethodon*: A variation on the Santa Rosalia theme. *Ecology* 57:238–51.

Free, J. B. 1965. The allocation of duties among worker honeybees. *Symp. Zool. Soc. Lond.* 14:39–59.

Freedle, R., and M. Lewis. 1977. Prelinguistic conversations. In *Interaction, Conversation, and the Development of Language*, ed. M. Lewis and L. A. Rosenblum. New York: Wiley.

Freedman, J. L. 1980. Human reactions to population density. In *Biosocial Mechanisms of Population Regulation*, ed. M. N. Cohen, R. S. Malpass, and H. G. Klein. New Haven: Yale University Press.

Fretwell, S. D. 1972. *Populations in a Seasonal Environment*. Princeton, NJ: Princeton University Press.

Fretwell, S. D., and H. L. Lucas. 1970. On territorial behaviour and other factors influencing habitat distribution in birds. I. Theoretical development. *Acta Biotheoretica* 19:16–36.

Fuller, C. A., R. Lydic, F. M. Sulzman, H. E. Albers, B. Tepper, and M. C. Moore-Ede. 1981. Circadian rhythm of body temperature persists after suprachiasmatic lesions in the squirrel monkey. *Am. J. Physiol.* 241:R385–91.

Fuller, J. L. 1967. Experimental deprivation and later behavior. *Science* 158:1645–52.

Fuller, J. L., and W. R. Thompson. 1978. *Foundations of Behavior Genetics*. St. Louis, MO: Mosby.

Fursich, F. T., and D. Jablonski. 1984. Late Triassic naticid drill holes: Carnivorous gastropods gain a major adaptation but fail to radiate. *Science* 224:78–80.

Futuyma, D. J. 1987. *Evolutionary Biology*. 2d ed. Sunderland, Mass: Sinauer Assoc.

Galef, B. G. 1971. Social factors in the poison avoidance and feeding behavior of wild and domesticated rat pups. *J. Comp. Physiol. Psychol.* 75:341–57.

———. 1976. Social transmission of acquired behavior: A discussion of tradition and social learning in vertebrates. *Adv. Stud. Behav.* 6:77–100.

———. 1977. Mechanisms for the social transmission of food preferences from adult to weanling rats. In *Learning and Mechanisms in Food Selection*, ed. L. M. Barker, M. Best, and M. Domjan. Waco: Baylor University Press.

———. 1981. Development of olfactory control of feeding-site selection in rat pups. *J. Comp. Physiol. Psychol.* 95:615–22.

Galef, B. G., and M. M. Clark. 1971. Social factors in the poison avoidance and feeding behavior of wild and domesticated rat pups. *J. Comp. Physiol. Psych.*, 75:341–57.

Galef, B. G., and L. Heiber. 1976. The role of residual olfactory cues in the determination of feeding site selection and exploration patterns of domestic rats. *J. Comp. Physiol. Psychol.* 90:727–39.

Galef, B. G., and P. W. Henderson. 1972. Mother's milk: A determinant of the feeding preferences of weaning rat pups. *J. Comp. Physiol. Psychol.* 78:213–19.

Galef, B. G., D. J. Kennett, and M. Stein. 1985. Demonstrator influence on observer diet preference: Effects of simple exposure and the presence of a demonstrator. *Anim. Learn. Behav.* 13:25–30.

Galef, B. G., D. J. Kennett, and S. W. Wigmore. 1984. Transfer of information concerning distant foods in rats: A robust phenomenon. *Anim. Learn. Behav.* 12:292–96.

Galef, B. G., J. R. Mason, G. Preti, and N. J. Bean. 1988. Carbon disulfide: A semiochemical mediating socially-induced diet choice in rats. *Physiol. Behav.* 42:119–24.

Galef, B. G., A. Mischinger, and S. A. Malenfant. 1987. Hungry rats' following of conspecifics to food depends on the diets eaten by potential leaders. *Anim. Behav.* 35:1234–39.

Gallistel, C. R., C. T. Piner, T. O. Allen, N. T. Adler, E. Yadin, and M. Negin. 1982. Computer assisted analysis of 2-DG autoradiographs. *Neurosci. Biobehav. Rev.* 6:409–20.

Gandelman, R., F. S. vom Saal, and J. M. Reinisch. 1977. Contiguity to male fetuses affects morphology and behavior of female mice. *Nature* 266:722–24.

Ganzhorn, J. U. 1986. Feeding behavior of *Lemur catta* and *Lemur fulvus*. *Int. J. Primatol.* 7:17–30.

Garcia, J., W. G. Hankins, and K. W. Rusiniak. 1976. Flavor aversion studies. *Science* 192:265–66.

Gardner, B. T., and R. A. Gardner. 1971. Two-way communication with an infant chimpanzee. In *Behavior of Nonhuman Primates*, eds. A. M. Schrier and F. Stollnitz. New York: Academic Press.

Gardner, R. A., and B. T. Gardner. 1969. Teaching sign language to a chimpanzee. *Science* 165:664–72.

Gardner, R. A., B. T. Gardner, and T. E. Van Cantfort. 1989. *Teaching sign language to chimpanzees.* Albany, NY: SUNY Press.

Gaston, S. 1971. The influence of the pineal organ on the circadian activity rhythm in birds. In *Biochronometry*, ed. M. Menaker. Washington, D.C.: National Academy of Sciences.

Gauthreaux, S. A., Jr. 1971. A radar and direct visual study of passerine spring migration in southern Louisiana. *Auk* 88:343–65.

Geist, V. 1971. *Mountain Sheep: A Study in Behavior and Evolution.* Chicago: University of Chicago Press.

Gelber, B. 1965. Studies of the behavior of *Paramecium aurelia*. *Anim. Behav. Suppl.* 1:21–29.

Gerhardt, C. 1979. Vocalizations of some hybrid treefrogs: Acoustic and behavioral analyses. *Behaviour* 49:130–51.

Gerhardt, H. C. 1974a. Significance of some spectral features in mating call recognition in the green treefrog (*Hyla cinerea*). *J. Exp. Biol.* 61:229–41.

———. 1974b. Vocalizations of some hybrid treefrogs: Acoustic and behavioral analyses. *Behaviour* 49:130–51.

———. 1975. Sound pressures levels and radiation patterns of the vocalizations of some North American tree frogs and toads. *J. Comp. Physiol.* 102:1–12.

———. 1982. Sound pattern recognition in some North American treefrogs (Anura: Hylidae): Implications for mate choice. *Amer. Zool.* 2:581–95.

———. 1987. Evolutionary and neurobiological implications of selective phonotaxis in the green treefrog, *Hyla cinerea*. *Anim. Behav.* 35:1479–89.

Gerstell, R. 1939. Certain mechanisms of quail losses revealed by laboratory experimentation. Trans. *Fourth N. Amer. Wildl. Inst.* 462–67.

Giantonio, G. W., N. L. Lund, and A. A. Gerall. 1970. Effects of diencephalic and rhinencephalic lesions on the male rat's sexual behavior. *J. Comp. Physiol. Psychol.* 73:38–46.

Gibson, R. M., and J. W. Bradbury. 1985. Sexual selection in lekking sage grouse: phenotypic correlates of male mating success. *Behav. Ecol. Sociobiol.* 18:117–23.

Gilbert, L. I. 1974. Endocrine action during insect growth. *Rec. Prog. Horm. Res.* 30:347–84.

Gilhousen, H. C. 1927. The use of the vision and of the antennae in the learning of crayfish. *Univ. Calif. Publ. Physiol.* 7:73–89.

Gill, F. B., and L. L. Wolf. 1975. Economics of feeding territoriality in the golden-winged sunbird. *Ecology* 56:333–45.

Gilliard, E. T. 1963. The evolution of bower birds. *Sci. Amer.* 209:38–46.

———. 1969. *Birds of Paradise and Bower Birds.* Garden City, NY: Natural History Press.

Gilraldeau, L. A., and D. L. Kramer. 1982. The marginal value theorem: A quantitative test using load size variation in a central place forager, the eastern chipmunk *Tamias striatus*. *Anim. Behav.* 30:1036–42.

Girgis, M. 1971. The role of the thalamus in the regulation of aggressive behavior. *Int. J. Neurol.* 8:327–51.

Gochee, C., W. Rasband, and L. Sokoloff. 1980. Computerized densitometry and color coding of [^{14}C]-deoxyglucose autoradiographs. *Ann. Neurol.* 7:359–70.

Goldfoot, D. A., S. M. Essock-Vitale, C. S. Asa, J. E. Thornton, and A. I. Leshner. 1978. Anosmia in male rhesus monkeys does not alter copulatory activity with cycling females. *Science* 199:1095–96.

Goldfoot, D. A., H. Loon, W. Groeneveld, and A. K. Slob. 1980. Behavioral and physiological evidence of sexual climax in the female stumptailed macaque (*Macaca arctoides*). *Science* 208:1477–79.

Golding, D. W. 1972. Studies in the comparative neuroendocrinology of polychaete reproduction. *Gen. Comp. Endocr.* (sup. 3):580–90.

Gorlick, D. L., P. D. Atkins, and G. S. Losey. 1978. Cleaning stations as water holes, garbage dumps, and sites for the evolution of reciprocal altruism? *Amer. Nat.* 112:314–53.

Gottlieb, G. 1963. A naturalistic study of imprinting in wood ducklings (*Aix sponsa*). *J. Comp. Physiol. Psychol.* 56:86–91.

———. 1968. Prenatal behavior of birds. *Quart. Rev. Biol.* 43:148–74.

———. 1971. *Development of Species Identification in Birds.* Chicago: University of Chicago Press.

Gottlieb, G., and J. G. Vandenbergh. 1968. Ontogeny of vocalization in duck and chick embryos. *J. Exp. Zool.* 168:307–26.

Gould, J. L. 1975. Honeybee recruitment: The dance-language controversy. *Science* 189:685–93.

———. 1976. The dance-language controversy. *Quart. Rev. Biol.* 51:211–44.

Gould, S. J. 1977. *Ontogeny and Phylogeny.* Cambridge, MA: Harvard University Press.

Gould, S. J., and N. Eldredge. 1977. Punctuated equilibria: The tempo and mode of evolution reconsidered. *Paleobiology* 3:115–51.

Gould, S. J., and R. C. Lewontin. 1979. The spandrels of San Marco and the Panglossian paradigm: A critique of the adaptionist programme. *Proc. Roy. Soc. Lond.* B 205:581–98.

Gowaty, P. A., and S. J. Wagner. 1988. Breeding season aggression of female and male eastern bluebirds (*Sialia sialis*) to models of potential conspecific and interspecific egg dumpers. *Ethology* 78:238–50.

Goy, R. W. 1970. Experimental control of psycho-sexuality. *Phil. Trans. Roy. Soc. Lond.*, ser. B. 259:149–62.

Goy, R. W., F. B. Bercovitch, and M. C. McBrair. 1988. Behavioral masculinization independent of genital masculinization in prenatally androgenized female rhesus macaques. *Horm. Behav.* 22:552–70.

Graber, R. R. 1968. Nocturnal migration in Illinois: Different points of view. *Wilson Bull.* 80:36–71.

Grant, E. C., and J. H. MacKintosh. 1963. A comparison of the social postures of some common laboratory rodents. *Behaviour* 21:246–59.

Grant, P. R. 1981. Speciation and the adaptive radiation of Darwin's finches. *Amer. Sci.* 69:653–63.

Grant, P. R., and I. Abbott. 1980. Interspecific competition, null hypotheses and island biogeography. *Evolution* 34:332–41.

Grant, P. R., and B. R. Grant. 1980. The breeding and feeding characteristics of Darwin's finches on Isla Genovesa, Galapagos. *Ecol. Monogr.* 50:381–410.

Green, R. G., C. L. Larson, and J. F. Bell. 1939. Shock disease as the cause of the periodic decimation of the snowshoe hare. *Amer. J. Hyg.* 30:83–102.

Greenberg, N., and D. Crews. 1983. Physiological ethology of aggression in amphibians and reptiles. In *Hormones and Aggressive Behavior*, ed. B. B. Svare, 469–506. New York: Plenum.

Greenwood, P. J. 1980. Mating systems, philopatry, and dispersal in birds and mammals. *Anim. Behav.* 28:1140–62.

Greer, A. E. Jr. 1971. Crocodilian nesting habits and evolution. *Fauna* 2:20–28.

Greer, M. A. 1955. Suggestive evidence of a primary "drinking center" in the hypothalamus of the rat. *Proc. Soc. Exp. Biol.* 89:59–62.

Griffin, D. A. 1955. Bird navigation. In *Recent Studies in Avian Biology*, ed. A. Wolfson. Urbana: University of Illinois Press.

———. 1970. Migrations and homing of bats. In *Biology of Bats*, vol. 1, ed. W. A. Wimsatt. New York: Academic Press.

———. 1974. *Bird Migration*. New York: Dover.

Griffin, D. A., and R. Galambos. 1941. The sensory basis of obstacle avoidance by flying bats. *J. Exp. Zool.* 86:481–506.

Griffin, D. R. 1958. *Listening in the Dark*. New Haven: Yale University Press.

Gromko, M. H., D. G. Gilbert, and R. C. Richmond. 1984. Sperm transfer and use in the multiple mating system of Drosophila. In *Sperm Competition and the Evolution of Animal Mating Systems*, ed. R. L. Smith. New York: Academic Press.

Gross, M. R., and R. Shine. 1981. Parental care and mode of fertilization in ectothermic vertebrates. *Evolution* 35:775–93.

Grossfield, J. and B. Sakri. 1972. Divergence in the neural control of oviposition in *Drosophila. J. Insect Physiol.* 18:237–41.

Grota, L. J., and K. B. Eik-Nes. 1967. Plasma progesterone concentrations during pregnancy and lactation in the rat. *J. Reprod. Fertil.* 13:83–91.

Guhl, A. M. 1961. Gonadal hormones and social behavior. In *Sex and Internal Secretions*, ed. W. C. Young. Baltimore: Williams & Wilkins.

Gulick, W. L., and H. Zwick. 1966. Auditory sensitivity of the turtle. *Psychol. Rec.* 16:47–53.

Guthrie, R. D. 1971. A new theory of mammalian rump patch evolution. *Behaviour* 38:132–45.

Gwinner, E. 1986a. *Circannual Rhythms*. Berlin: Springer-Verlag.

———. 1986b. Circannual rhythms in the control of avian migration. *Adv. Stud. Anim. Behav.* 16:191–228.

Gwynne, D. T. 1981. Sexual difference theory: Mormon crickets show role reversal in mate choice. *Science* 213:779–80.

———. 1984. Courtship feeding increases female reproductive success in bushcrickets. *Nature* 307:361–63.

Haartman, L. V. 1956. Territory in the pied flycatcher *Muscicapa hypoleuca. Ibis* 98:460–75.

Hailman, J. P. 1967. The ontogeny of an instinct: The pecking response in chicks of the laughing gull (*Larus atricilla* L.) and related species. *Behavior*, Suppl. No. 15:1–159.

———. 1969. How an instinct is learned. *Sci. Amer.* 221:98–106.

———. 1973. Anatomical and physiological basis of embryonic motility in birds and mammals. In *Studies on the Development of Behavior and the Nervous System*, ed. G. Gottlieb. New York: Academic Press.

Hale, E. B. 1969. Domestication and the evolution of behavior. In *Behavior of Domestic Animals*, 2d ed, ed. E. S. E. Hafez. London: Balliere, Tindall, and Cassell.

Hale, E. B., W. M. Schleidt, and M. W. Schein. 1969. The behavior of turkeys. In *Behavior of Domestic Animals*, ed. E. S. E. Hafez. Baltimore: Williams & Wilkins.

Hall, E. T. 1966. *The Hidden Dimension*. Garden City, NY: Doubleday.

Halliday, T. R. 1983. Motivation. 1. Causes and Effects. In *Animal Behaviour*, ed. T. R. Halliday and P. J. B. Slater, 100–33. New York: W. H. Freeman.

Halpern, M., and N. Frumin. 1979. Roles of the vomeronasal and olfactory systems in prey attack and feeding in adult garter snakes. *Physiol. Behav.* 22:1183–89.

Hamburger, V. 1963. Some aspects of the embryology of behavior. *Quart. Rev. Biol.* 38:342–65.

Hamilton, W. D. 1963. The evolution of altruistic behavior. *Amer. Nat.* 97:354–56.

———. 1964. The genetical evolution of social behaviour I, II. *J. Theoret. Biol.* 7:1–52.

———. 1967. Extraordinary sex ratios. *Science* 156:477–88.

———. 1971. Geometry for the selfish herd. *J. Theoret. Biol.* 31:295–311.

Hamilton, W. D., and R. M. May. 1977. Dispersal in stable habitats. *Nature* 269:578–81.

Hamilton, W. D., and M. Zuk. 1984. Heritable true fitness and bright birds: a role for parasites? *Science* 218:384–87.

Hansen, E. W. 1966. The development of maternal and infant behavior in the rhesus monkey. *Behaviour* 27:107–49.

Hansen, L. P., and G. O. Batzli. 1978. The influence of food availability on the white-footed mouse: Populations in isolated woodlots. *Can. J. Zool.* 56:2530–41.

Harcourt, A. H. 1989. Deformed sperm are probably not adaptive. *Anim. Behav.* 37:863–64.

Hardin, G. 1968. The tragedy of the commons. *Science* 162:1243–48.

Harding, C. F. 1983. Hormonal influences on avian aggressive behavior. In *Hormones and Aggressive Behavior*, ed. B. B. Svare. New York: Plenum.

———. 1965. Sexual behavior in the rhesus monkey. In *Sex and Behavior*, ed. F. Beach. New York: Wiley.

Harker, J. E. 1960. Endocrine and nervous factors in insect circadian rhythms. *Cold Spr. Harb. Symp. Quant. Biol.* 25:279–87.

———. 1964. The physiology of diurnal rhythms. *Camb. Monogr. Exp. Biol.* 13:1–114.

Harlow, H. F. 1949. The formation of learning sets. *Psychol. Rev.* 56:51–65.

———. 1951. Primate learning. In *Comparative Psychology*. ed. C. P. Stone. Englewood Cliffs, NJ: Prentice-Hall.

———. 1962. Development of affection in primates. In *Roots of Behavior*, ed. E. L. Bliss. New York: Harper & Row.

Harlow, H. F., and M. K. Harlow. 1962a. Social deprivation in monkeys. *Sci. Amer.* 207:136–46.

Harlow, H. F., and M. K. Harlow. 1962b. The effect of rearing conditions on behavior. *Bull. Menninger Clinic* 26:213–24.

Harlow, H. F., and M. K. Harlow. 1969. Age-mate or peer affectional system. *Adv. Study Behav.* 2:333–83.

Harlow, H. F., M. K. Harlow, and S. J. Suomi. 1971. From thought to therapy: Lessons from a primate laboratory. *Amer. Scientist* 59:538–49.

Harlow, H. F., and S. J. Suomi. 1971. Social recovery by isolate-reared monkeys. *Proc. Nat. Acad. Sci. USA* 68:1534–38.

Harlow, H. F., and R. R. Zimmerman. 1959. Affectional responses in the infant monkey. *Science* 130:421–32.

Harnly, M. H. 1941. Flight capacity in relation to phenotypic and genotypic variations in the wings of *Drosophila melanogaster. J. Exp. Zool.* 88:263–73.

Harper, D. G. C. 1982. Competitive foraging in mallards: 'ideal free' ducks. *Anim. Behav.* 30:575–84.

Harris, G., and S. Levine. 1965. Sexual differentiation of the brain and its experimental control. *J. Physiol.* 181:379–400.

Harris, V. T. 1952. An experimental study of habitat selection by prairie and forest races of the deer mouse, *Peromyscus maniculatus. Contrib. Lab. Vert. Biol., Univ. Mich.* 56:1–53.

Harrison, J. L. 1952. Moonlight and pregnancy of Malayan forest rats. *Nature* 170:73–74.

Hart, B. L. 1974. Medical preoptic-anterior hypothalamic area and sexual behavior of male dogs. *J. Comp. Physiol. Psychol.* 86:328–49.

Hartl, D. L. 1988. *A Primer of Population Genetics.* Sunderland, MA: Sinauer Assoc. 305 pp.

Hartwick, P., J. Kiepenheuer, and K. Schmidt-Koenig. 1978. Further experiments on the olfactory hypothesis of pigeon homing. In *Animal Migration, Navigation, and Homing,* ed. K. Schmidt-Koenig and W. T. Keeton. New York: Springer-Verlag.

Hasler, A. D. 1960. Guideposts of migrating fishes. *Science* 132:785–92.

———. 1966. *Underwater Guideposts: Homing of Salmon.* Madison: University of Wisconsin Press.

Hausfater, G., and S. B. Hrdy, eds. 1984. *Infanticide: Comparative and Evolutionary Perspectives.* New York: Aldine.

Hawkins, R. D., T. W. Abrams, T. J. Carew, and E. R. Kandel. 1983. A cellular mechanism of classical conditioning in *Aplysia*: Activity-dependent amplification of presynaptic facilitation. *Science* 219:400–5.

Hayes, J. H. 1951. *The Ape in Our House.* New York: Harper Brothers.

Hazlett, B. A., and W. H. Bossert. 1965. A statistical analysis of the aggressive communications systems of some hermit crabs. *Anim. Behav.* 13:357–73.

Hebb, D. O. 1949. *The Organization of Behavior.* New York: Wiley.

———. 1958. *Textbook of Psychology.* Philadelphia: Saunders.

Hegner, R. E., and J. C. Wingfield. 1986. Social modulation of gonadal development and circulating hormone levels during autumn and winter. In *Behavioural Rhythms,* ed. Yvon Queinec and N. Delvolve, 109–17. Toulouse, France: Privat.

———. 1987a. Effects of experimental manipulation of testosterone levels on parental investment and breeding success in male house sparrows. *Auk* 104:462–69.

———. 1987b. Effects of brood-size manipulation on parental investment, breeding success, and reproductive endocrinology of house sparrows. *Auk* 104:470–80.

Heiligenberg, W. 1966. The stimulation of territorial singing in house crickets (*Acheta domesticus*). *Zeit. Vergleich. Physiol.* 49:459–64.

Heinrich, B. 1983. Do bumblebees forage optimally, and does it matter? *Amer. Zool.* 23:273–81.

Helfman, G. S., and E. T. Schultz. 1984. Social transmission of behavioral traditions in a coral reef fish. *Anim. Behav.* 32:379–84.

Hendrickson, A. E., N. Wagoner, and W. M. Cowan. 1972. Autoradiographic and electron microscopic study of retino-hypothalamic connections. *Zeit. Zellforsch.* 125:1–26.

Herlugson, C. J. 1981. Nest site selection in mountain bluebirds. *Condor* 83:252–55.

Herndon, J. G., A. A. Perachio, and M. McCoy. 1979. Orthogonal relationship between electrically elicited social aggression and self-stimulation from the same brain sites. *Brain Research* 171:374–80.

Hess, E. 1959. Imprinting. *Science* 30:133–41.

Hiebert, S. M., P. K. Stoddart, and P. Arcese. 1989. Repertoire size, territory acquisition and reproductive success in the song sparrow. *Anim. Behav.* 37:266–73.

Hildén, O. 1965. Habitat selection in birds. A review. *Ann. Zool. Fenn.* 2:43–75.

Hilgard, E. R., and G. H. Bower. 1975. *Theories of Learning,* 4th ed. Englewood Cliffs, NJ: Prentice-Hall.

Hinde, R. A. 1965. Interaction of internal and external factors in integration of canary reproduction. In *Sex and Behavior,* ed. F. A. Beach. New York: Wiley.

———. 1970. *Animal Behaviour,* 2d ed. New York: McGraw-Hill.

———. 1973. Constraints on learning—An introduction to the problems. In *Constraints on Learning: Limitations and Predispositions,* ed. R. A. Hinde and J. Stevenson-Hinde. New York: Academic Press.

———. 1973. On the design of check-sheets. *Primates* 14:393–406.

Hinde, R. A., and L. Davies. 1972. Removing infant rhesus from mother for 13 days compared with removing mother from infant. *J. Child. Psychol. Psychiat.* 19:199–211.

Hinde, R. A., M. E. Leighton-Shapiro, and L. McGinnis. 1978. Effects of various types of separation experience on rhesus monkeys five months later. *J. Child. Psychol. Psychiat.* 19:199–211.

Hinde, R. A., and L. McGinnis. 1977. Some factors influencing the effects of temporary mother-infant separation—Some experiments with rhesus monkeys. *Psychol. Med.* 7:197–212.

Hinde, R. A., and Y. Spencer-Booth. 1967. The behaviour of socially living rhesus monkeys in their first two-and-a-half years. *Anim. Behav.* 15:169–96.

———. 1970. Individual differences in the responses of rhesus monkeys to a period of separation from their mothers. *J. Child. Psychol. Psychiat.* 11: 159–76.

———. 1971. Effect of brief separation from mother on rhesus monkeys. *Science* 173:111–18.

Hinde, R. A., and J. Stevenson-Hinde, eds. 1973. *Constraints on Learning: Limitations and Predispositions.* New York: Academic Press.

Hirsch, J. 1967. Behavior-genetic analysis. In *Behavior-Genetic Analysis,* ed. J. Hirsch. New York: McGraw-Hill.

———. ed. 1967. *Behavior-Genetic Analysis.* New York: McGraw-Hill.

———. 1975. Jensenism: The bankruptcy of "science" without scholarship. *Educ. Theory* 25:3–27.

———. 1981. To "unfrock the charlatans." *Sage Race Rel. Abstr.* 6:1–67.

Hirsch, J., and J. C. Boudreau. 1958. Studies in experimental behavior genetics. I. The heritability of phototaxis in a population of *Drosophila melanogaster. J. Comp. Physiol. Psychol.* 51:647–51.

Hirsch, J., T. R. McGuire, and A. Vetta. 1980. Concepts of behavior genetics and misapplications to humans. In *Evolution of Human Social Behavior*, ed. J. S. Lockard. New York: Elsevier.

Hodgkin, A. L. 1971. *Conduction of the Nervous Impulse*. Springfield, Ill.: Charles C. Thomas.

Hodos, W., and C. B. G. Campbell. 1969. Scala naturae: Why is there no theory in comparative psychology? *Psychol. Rev.* 76:337–50.

Hoffman, K. 1959. Die Aktivitätsperiodik von im 18- und 36-Stunden-tag erbrüteten Eideschsen. *Zeit. Vergleich. Physiol.* 42:422–32.

Holecamp, K. E., and P. W. Sherman. 1989. Why male ground squirrels disperse. *Amer. Scient.* 77:232–39.

Hölldobler, B., and C. D. Michener. 1980. Mechanisms of identification and discrimination in social hymenoptera. In *Evolution of Social Behavior: Hypotheses and Empirical Tests*, ed. H. Markl. Deerfield Beach, FL: Verlag Chemie.

Holling, C. S. 1959. The components of predation as revealed by a study of small mammal predation of the European pine sawfly. *Can. Entomol.* 91:293–320.

Holmes, W. G. 1988. Kinship and development of social preferences. In *Developmental Psychobiology and Behavioral Ecology*, ed. E. M. Blass, 389–413. New York: Plenum Press.

Holmes, W. G., and P. W. Sherman. 1983. Kin recognition in animals. *Amer. Scientist* 71:46–55.

Honegger, H. W. 1976. Locomotor activity in *Uca crenulata* and the response to two zeitgebers, Light-dark and tides. In *Biological Rhythms in the Marine Environment*, ed. P. J. DeCoursey, 93–102. Columbia, SC: University of South Carolina Press.

Hood, K. E., and R. B. Cairns. 1988. A developmental-genetic analysis of aggressive behavior in mice. II. Cross-sex inheritance. *Behav. Genet.* 18:605–20.

Hoogland, J. L. 1979a. Aggression, ectoparasitism, and other possible costs of prairie dog (Sciuridae: *Cynomys* spp.) coloniality. *Behaviour* 69:1–35.

———. 1979b. The effect of colony size on individual alertness of prairie dogs (Sciuridae: *Cynomys* spp.). *Anim. Behav.* 27:394–407.

———. 1982. Prairie dogs avoid extreme inbreeding. *Science* 215:1639–41.

Hoogland, J. L., and P. W. Sherman. 1976. Advantages and disadvantages of bank swallow (*Riparia riparia*) coloniality. *Ecol. Monogr.* 46:33–58.

Hopkins, C. D. 1974. Electric communication in fish. *Amer. Scientist* 62:426–37.

Hopkins, C. D., and A. H. Bass. 1981. Temporal coding of species recognition signals in an electric fish. *Science* 212:85–87.

Hotta, Y., and S. Benzer. 1972. The mapping of behavior in *Drosophila* mosaics. *Nature* 240:527–35.

———. 1973. Courtship in *Drosophila* mosaics: Sex-specific foci for sequential action patterns. *Proc. Nat. Acad. Sci. USA* 73:4154–58.

Howard, H. E. 1920. *Territory in Bird Life*. New York: Atheneum.

Howard, R. D. 1978. The evolution of mating strategies in bullfrogs, *Rana catesbeiana*. *Evolution* 32:850–71.

———. 1980. Mating behaviour and mating success in wood frogs. *Anim. Behav.* 28:705–16.

———. 1988. Reproductive success in two species of anurans. In *Reproductive success*, ed. T. H. Clutton-Brock. Chicago: University of Chicago Press.

Hoy, R. R., A. Hoikkala, and K. Kaneshiro. 1988. Hawaiian courtship songs: evolutionary innovation in communication signals of *Drosophila*. *Science* 240:217–19.

Hoy, R. R., and R. C. Paul. 1972. Genetic control of song specificity in crickets. *Science* 180:82–83.

Hoy, R. R., G. S. Pollack, and A. Moiseff. 1982. Species-recognition in the field cricket, *Teleogryllus oceanicus*: Behavioral and neural mechanisms. *Amer. Zool.* 22:597–607.

Hrdy, S. B. 1977a. Infanticide as a primate reproductive strategy. *Amer. Scientist* 65:40–49.

———. 1977b. *Langurs of Abu: Female and Male Strategies of Reproduction*. Cambridge: Harvard University Press.

Hubel, D. H., and T. N. Wiesel. 1962. Receptive fields, binocular interaction and functional architecture in the cat's visual cortex. *J. Physiol.* 160:106–54.

———. 1965. Receptive fields and functional architecture in two non-striate visual areas of the cat. *J. Neurophysiol.* 28:229–89.

———. 1976. Spatial vision in anurans. In *The Amphibian Visual System: A Multidisciplinary Approach*, ed. K. V. Fite. New York: Academic Press.

———. 1977. Functional architecture of macaque monkey visual cortex. *Proc. Royal Soc. London* Series B 198:1–59.

———. 1979. Brain mechanisms of vision. *Sci. Amer.* 241(Sept.):150–62.

Huber, F. 1978. The insect nervous system and insect behaviour. *Anim. Behav.* 26:969–81.

———. 1983a. Implications of insect neuroethology for studies on vertebrates. In *Advances in Neuroethology*, ed. J. P. Ewert, R. R. Capranica, and D. J. Ingle, 91–138. New York: Plenum Press.

———. 1983b. Neural correlates of orthopteran and cicada phonotaxis. In *Neuroethology and Behavioral Physiology*, ed. F. Huber and H. Markl, 108–35. Berlin: Springer-Verlag.

Huck, U. W., and E. M. Banks. 1980. The effects of cross-fostering on the behaviour of two species of North American lemmings, *Dicrostonyx groenlandicus* and *Lemmus trimucronatus*. *Anim. Behav.* 28:1046–52.

Hudson, D. J., and M. E. Lickey. 1977. Weak negative coupling between the circadian pacemakers of the eyes of *Aplysia*. *Neurosci. Abst.* 3:179.

Hulse, F. 1971. *The Human Species*. New York: Random House.

Hunsaker, D. 1962. Ethological isolating mechanisms in the *Sceloporus torquatus* group of lizards. *Evolution* 16:62–74.

Hunt, R. M., X. Xiang-Xu, and J. Kaufman. 1983. Miocene burrows of extinct bear dogs: Indication of early denning behavior of large mammalian carnivores. *Science* 221:364–66.

Huntingford, F. A. 1986. Development of behaviour in fish. In *The Behavior of Teleost Fishes*, ed. T. J. Pitcher, 47–68. Baltimore: Johns Hopkins University Press.

Hutchinson, G. E. 1957. Concluding remarks. *Cold Spr. Harb. Symp. Quant. Biol.* 22:415–27.

Hutchinson, J. B. 1969. Changes in hypothalamic responsiveness to testosterone in male Barbary dove (*Streptopelia risoria*). *Nature* 222:176.

———. 1971. Effects of hypothalamic implants of gonadal steroids on courtship behavior in Barbary doves (*Streptopelia risoria*). *J. Endocr.* 50:97–113.

———. 1978. Hypothalamic regulation of male sexual responsiveness to androgen. In *Biological Determinants of Sexual Behavior*, ed. J. B. Hutchinson. New York: Wiley.

Hutt, C. 1966. Exploration and play in children. *Symp. Zool. Soc. Lond.* 18:23–44.

———. 1967a. Temporal effects on response decrement and stimulus satiation in exploration. *Brit. J. Psychol.* 58:365–73.

———. 1967b. Effects of stimulus novelty on manipulatory exploration in an infant. *J. Child Psychol. Psychiat.* 8:241–47.

———. 1970a. Specific and diversive exploration. *Adv. Child Dev. Behav.* 5:119–80.

———. 1970b. Curiosity in young children. *Science J.* 6:68–71.

Hutt, C., and R. Bhanvani. 1972. Predictions from play. *Nature* 237:171–72.

Huxley, J. 1974. *Evolution: The Modern Synthesis*. London: Allen and Unwin.

Huxley, J. S. 1914. The courtship-habits of the great-crested grebe (*Podiceps cristatus*); with an addition to the theory of sexual selection. *Proc. Zool. Soc. Lond.* 35:491–562.

Immelmann, K., and C. Beer, 1989. A *Dictionary of Ethology*. Cambridge, MA: Harvard University Press.

Immelmann, K. 1965. Objektfixierung geschlechtlicher Triebhandlung bei Prachtfinken. *Naturwissenschaften* 52:169–70.

———. 1972. Sexual and other long-term aspects of imprinting in birds and other species. *Adv. Stud. Behav.* 4:147–74.

Ingle, D. 1973. Two visual systems in the frog. *Science* 181:1053–55.

Inglis, L. R., and A. J. Isaacson. 1978. The responses of dark-bellied brent geese to models of geese in various postures. *Anim. Behav.* 26:953–58.

Ingram, J. C. 1978. Primate markings. In *Animal Marking*, ed. B. Stonehouse, 169–74. Baltimore: University Park Press.

Irwin, R. E. 1988. The evolutionary importance of behavioural development: The ontogeny and phylogeny of bird song. *Anim. Behav.* 36:814–24.

Itani, J. 1958. On the acquisition and propagation of a new food habit in the troop of Japanese monkeys at Takasaki-yama. *Primates* 1:131–48.

Jacklet, J. W. 1969. Circadian rhythm of optic nerve impulses recorded in darkness from isolated eye of *Aplysia*. *Science* 164:562–63.

———. 1973. Neuronal population interactions in a circadian rhythm in *Aplysia*. In *Neurobiology of Invertebrates*, ed. J. Salanki. Budapest: Akademiai Kiado.

———. 1976. Circadian rhythms in the nervous system of a marine gastropod, Aplysia. In *Biological Rhythms in the Marine Environment*, ed. P. J. DeCoursey, 17–31. Columbia, SC: University of South Carolina Press.

Jacklet, J. W., and J. Geronimo. 1971. Circadian rhythm: Populations of interacting neurons. *Science* 174:299–302.

Jackson, J. R. 1963. Nesting of keas. *Notornis* 10:319–26.

Jacobson, M. 1974. A plentitude of neurons. In *Aspects of Neurogenesis*, vol. 2, ed. G. Gottlieb. New York: Academic Press.

———. 1978. *Developmental Neurobiology*, 2d ed. New York: Plenum.

Jacobssen, S., and T. Jarvik. 1976. Anti-predator behaviour of two-year-old hatchery-reared Atlantic salmon (*Salmo salar*) and a description of the predatory behaviour of burbot (*Lota lota*). *Zool. Rev.* 38:57–70.

Jallon, J. M., C. Antony, and O. Benamar. 1981. Un anti-aphrodisiaque produit par les males de *Drosophila melanogaster* et transfere aux femelles los de la copulation. *C. R. Acad. Sci. Paris* 292:1147–49.

Janowitz, H. D., and M. I. Grossman. 1949. Some factors affecting the food intake of normal dogs and dogs with esophagotomy and gastric fistules. *Amer. J. Physiol.* 159:143–48.

Jarman, P. J. 1974. The social organization of antelope in relation to their ecology. *Behaviour* 48:215–67.

Jarvis, J. U. M. 1981. Eusociality in a mammal: cooperative breeding in naked mole rat colonies. *Science* 212:571–73.

Jenni, D. A. 1974. Evolution of polyandry in birds. *Amer. Zool.* 14:129–44.

Jensen, A. R. 1967. Estimation of the limits of heritability of traits by comparison of monozygotic and dizygotic twins. *Proc. Nat. Acad. Sci.* (USA) 58:149–56.

———. 1973. *Educability and Group Differences*. New York: Harper & Row.

Jensen, D. D. 1965. Paramecia, planaria and pseudo-learning. *Anim. Behav. Suppl.* 1:9–20.

Jerison, H. J. 1973. *Evolution of the Brain and Intelligence*. New York: Academic Press.

Joerman, G., U. Schmidt, and C. Schmidt. 1988. The mode of orientation during flight and approach to landing in two phyllostomid bats. *Ethology* 78:332–40.

Johansson, B. 1959. Brown fat: A review. *Metabolism* 8:221–40.

John, E. R. 1967. *Mechanisms of Memory*. New York: Academic Press.

———. 1972. Switchboard versus statistical theories of learning and memory. *Science* 177:850–64.

Johnson, C. G. 1969. *Migration and Dispersal of Insects by Flight*. London: Methuen.

Johnson, D. F., and G. Collier. 1987. Caloric regulation and patterns of food choice in a patchy environment: The value and cost of alternative foods. *Physiol. Behav.* 39:351–59.

———. 1989. Patch choice and meal size of foraging rats as a function of the profitability of food. *Anim. Behav.* 38: 285–97.

Johnson, R. N. 1972. *Aggression in Man and Animals*. Philadelphia: Saunders.

Joslin, J. K. 1977a. Rodent long distance orientation ("homing"). *Adv. Ecol. Res.* 10:63–90.

———. 1977b. Visual cues used in orientation in white-footed mice, *Peromyscus leucopus*: A laboratory study. *Amer. Mid. Nat.* 98:303–18.

Jung-Hoffman, I. 1966. Die Determination von Königin und Arbeiterin der Honigbee. *Z Bienenforschung* 8:296–322.

Jussiaux, M., and C. Trillaud. 1979. Comparaison entres des techniques de vasectomie et d'androgenisation pour la détection des chaleurs. *Cereopa, Journee d'Etude*, March 1979.

Kalmijn, A. J. 1971. The electric sense of sharks and rays. *J. Exp. Biol.* 55:371–83.

Kamil, A. C., and R. P. Balda. 1985. Cache recovery and spatial memory in Clark's nutcrackers (*Nucifraga columbiana*). *J. Exp. Psychol.: Anim. Behav. Proc.* 11:95–111.

Kamil, A. C., R. P. Balda, and K. Grim. 1986. Revisits to emptied cache sites by Clark's nutcrackers (*Nucifraga columbiana*). *Anim. Behav.* 34:1289–98.

Kamil, A. C., and T. D. Sargent, eds. 1981. *Behavior: Ecological, Ethological, and Psychological Approaches.* New York: Garland Publishing.

Kaufman, G. A. 1989. Use of fluorescent pigments to study social interactions in a small nocturnal rodent, *Peromyscus maniculatus. J. Mammal.* 70:171–74.

Kawai, M. 1965. Newly acquired precultural behavior of the natural troop of Japanese monkeys on Koshima Island. *Primates* 6:1–30.

Kawakami, M., E. Terasawa, and T. Ibuki. 1970. Changes in multiple unit activity of the brain during the estrous cycle. *Neuroendocrinology* 6:30–48.

Kayser, C. 1965. Hibernation. In *Physiological Mammalogy*, vol. 3, ed. W. Mayer and R. W. van Gelder. New York: Academic Press.

Keane, B. 1990. The effect of relatedness on reproductive success and mate choice in the white-footed mouse, *Peromyscus leucopus. Anim. Behav.* 39:264–73.

Keeton, W. T. 1969. Orientation by pigeons: Is the sun necessary? *Science* 165:922–28.

———. 1970. Orientation by pigeons. *Science* 168:153.

———. 1971. Magnets interfere with pigeon homing. *Proc. Nat. Acad. Sci. USA* 68:102–6.

———. 1974. The orientation and navigational basis of homing in birds. *Adv. Stud. Behav.* 5:47–132.

Keeton, W. T., and A. I. Brown. 1976. Homing behavior of pigeons not disturbed by application of an olfactory stimulus. *J. Comp. Physiol.* 105:259–66.

Keeton, W. T., M. L. Kreithen, and K. L. Hermayer. 1976. Orientation by pigeons deprived of olfaction by nasal tubes. *J. Comp. Physiol.* 114:289–99.

Keller, R. 1975. Das Spielverhalten der Keas (*Nestor notabilis* Gould) des Zürcher Zoos. *Z. Tierpsychol.* 38:393–408.

———. 1976. Beitrag zur Biologie und Ethologie der Keas (*Nestor notabilis*) des Zürcher Zoos. *Zool. Beitr.* 22:111–56.

Kelley, D. B., J. I. Morrell, and D. W. Pfaff. 1975. Autoradiographic localization of hormone-concentrating cells in the brain of an amphibian, *Xenopus laevis.* I. Testosterone. *J. Comp. Neurol.* 164:47–62.

Kelly, R., J. W. Deutsch, S. S. Carlson, and J. A. Wagner. 1979. Biochemistry of neurotransmitter release. *Ann. Rev. Neurosci.* 2:399–446.

Kennleyside, M. H., and F. T. Yamamoto. 1962. Territorial behaviour of juvenile Atlantic salmon (*Salmo salar*). *Behaviour* 19:139–69.

Kessel, E. L. 1955. The mating activities of balloon flies. *Syst. Zool.* 4:97–104.

Ketterson, E. D. 1979. Aggressive behavior in wintering dark-eyed juncos: Determinants of dominance and their possible relation to geographic variation in sex ratio. *Wilson Bull.* 91:371–83.

Kettlewell, H. B. D. 1965. Insect survival and selection for pattern. *Science* 148:1290–96.

Kety, S. S. 1982. The evolution of concepts of memory: An overview. In *Neural Basis of Behavior*, ed. A. L. Beckman. New York: SP Medical.

King, J. A. 1955. Social behavior, social organization, and population dynamics in a black-tailed prairie dog town in the Black Hills of South Dakota. *Contributions to the Laboratory of Vertebrate Biology, University of Michigan.* 67:1–123.

———. 1957. Parameters relevant to determining the effect of early experience upon the adult behavior of animals. *Psychol. Bull.* 55:46–48.

———. 1967. Behavioral modification of the gene pool. In *Behavior-Genetic Analysis*, ed. J. Hirsch. New York: McGraw-Hill.

———. 1968. Species specificity and early experience. In *Early Experience and Behavior*, ed. G. Newton and S. Levine. Springfield, IL: Thomas.

———. 1969. A comparison of longitudinal and cross-sectional groups in the development of behavior of deer mice. *An. N.Y. Acad. Sci.* 159:696–709.

King, J. A., and R. G. Weisman. 1964. Sand-digging contingent upon bar-pressing in deer mice (*Peromyscus*). *Anim. Behav.* 12:446–50.

Kitchell, J. A. 1986. The evolution of predator-prey behavior: Naticid gastropods and their molluskan prey. In *Evolution of Animal Behavior. Paleontological and Field Approaches*, ed. M. H. Nitecki and J. A. Kitchell. New York: Oxford Univ. Press.

Kitching, J. A., and F. J. Ebling. 1967. Ecological studies at Lough Ine. *Adv. Ecol. Res.* 4:197–291.

Kleiman, D. G., and J. F. Eisenberg. 1973. Comparisons of canid and felid social systems from an evolutionary perspective. *Anim. Behav.* 21:637–59.

Kleindeinst, H. U. 1980. Biophysikalische Untersuchungen am Gehörsystem von Feldgrillen. Dissertation, University of Bonn.

Kleinholz, L. H. 1970. A progress report on the separation and purification of crustacean neurosecretory pigmentary-effector hormones. *Gen. Comp. Endocr.* 14:578–88.

Klopfer, P. 1957. An experiment with empathic learning in ducks. *Amer. Nat.* 91:61–63.

Klopfer, P. H. 1963. Behavioral aspects of habitat selection: The role of early experience. *Wilson Bull.* 75:15–22.

Klopfer, P. H., D. K. Adams, and M. S. Klopfer. 1964. Maternal imprinting in goats. *Proc. Nat. Acad. Sci. USA* 52:911–14.

Klopfer, P. H., and J. U. Ganzhorn. 1985. Habitat selection: Behavioral aspects. In *Habitat selection in birds*, ed. M. L. Cody. New York: Academic Press.

Klopfer, P. H., and J. P. Hailman. 1965. Habitat selection in birds. *Adv. Stud. Behav.* 1:279–303.

Klopfer, P. H., and M. S. Klopfer. 1968. Maternal "imprinting" in goats: Fostering of alien young. *Zeit. Tierpsychol.* 25:862–66.

Kodric-Brown, A., and J. H. Brown. 1984. Truth in advertising: The kinds of traits favored by sexual selection. *Am. Nat.* 124:309–23.

Koenig, W. D., R. L. Mumme, and F. A. Pitelka. 1984. The breeding system of the acorn woodpecker in central coastal California. *Z. Tierpsychol.* 65:289–308.

Kogure, M. 1933. The influence of light and temperature on certain characters of the silkworm, *Bombyx mori. J. Dept. Agr. Kyushu Univ.* 4:1–93.

Köhler, W. 1925. *The Mentality of Apes*. New York: Harcourt, Brace.

Kohsaka, S., K. Takamatsu, E. Aoki, and Y. Tsukada. 1979. Metabolic mapping of chick brain after imprinting using [^{14}C] 2-deoxyglucose technique. *Brain Res.* 172:539–44.

Komisaruk, B. I. 1978. The nature of the neural substrate of female sexual behaviour in mammals and its hormonal sensitivity: Review and speculations. In *Biological Determinants of Sexual Behaviour*, ed. J. B. Hutchison. Chichester, England: Wiley.

Konopka, R. J. 1979. Genetic dissection of the *Drosophila* circadian system. *Fed. Proc.* 38:2602–5.

Kramer, G. 1949. Über Richtungstendenzen bei der nächtlichen Zugunruhe gekäfigten Vögel. In *Ornithologie als biologische Wissenschaft*, ed. E. Mayr and E. Schüz. Heidelberg: Winter.

———. 1950. Orientierte Zugaktivitätgekäfigter Sing-vögel. *Naturwissenschaften* 37:188.

———. 1951. Eine neue Methode zur Erforschung der Zugorientierung und die bisher damit erzielten Ergebnisse. *Proc. Xth Inter. Ornithol. Congr.*, Uppsala. 271–80.

Krasne, F. B. 1973. Learning in Crustacea. In *Invertebrate Learning*, Vol. 2., ed. W. C. Corning, J. A. Dyal, and A. O. D. Willows. New York: Plenum.

Kravitz, E. A. 1988. Hormonal control of behavior: Amines and the biasing of behavioral output in lobsters. *Science* 241:1775–81.

Krebs, C. J. 1964. The lemming cycle at Baker Lake, Northwest Territories, during 1959–1962. Arctic Institute of North America Technical Paper No. 15.

———. 1966. Demographic changes in fluctuating populations of *Microtus californicus*. *Ecol. Monogr.* 36:239–73.

———. 1985. *Ecology: The Experimental Analysis of Distribution and Abundance*. 3d ed. New York: Harper & Row.

Krebs, C. J., M. S. Gaines, B. L. Keller, J. H. Myers, and R. H. Tamarin. 1973. Population cycles in small rodents. *Science* 179:35–41.

Krebs, J. R. 1970. Regulation of numbers in the great tit (Aves: Passeriformes). *J. Zool. Lond.* 162:317–33.

———. 1971. Territory and breeding density in the great tit, *Parus major* L. *Ecology* 52:2–22.

———. 1978. Optimal foraging: Decision rules for predators. In *Behavioral Ecology*, ed. J. R. Krebs and N. B. Davies. Sunderland, MA: Sinauer.

Krebs, J. R., and N. B. Davies. 1987. *An Introduction to Behavioral Ecology*. Sunderland, MA: Sinauer Assoc.

Krebs, J. R., and R. Dawkins. 1984. Animal signals: Mind-reading and manipulation. In *Behavioural Ecology: An Evolutionary Approach*, 2d ed., eds. J. R. Krebs and N. B. Davies. Oxford, England: Blackwell Scientific Publications, Ltd.

Krebs, J. R., A. Kacelnik, and P. Taylor. 1978. Test of optimal sampling by foraging great tits. *Nature* 275:27–31.

Kreithen, M. L., and W. T. Keeton. 1974a. Detection of changes in atmospheric pressure by the homing pigeon, *Columba livia*. *J. Comp. Physiol.* 89:73–82.

———. 1974b. Attempts to condition homing pigeons to magnetic stimuli. *J. Comp. Physiol.* 91:355–62.

Krieger, D. T. 1980. Ventromedial hypothalamic lesions abolish food-shifted circadian adrenal and temperature rhythmicity. *Endocrinology* 106:649–54.

Kroodsma, D. E. 1978. Aspects of learning in the ontogeny of bird song: Where, from whom, when, how many, which and how accurately? In *Development of Behavior*, ed. G. Burghardt and M. Bekoff. New York: Garland STPM Press.

———. 1981. Ontogeny of bird song. In *Behavioral Development*, ed. K. Immelmann et al. New York: Cambridge University Press.

Kroodsma, D. E., and R. Pickert. 1984. Repertoire size, auditory templates and selective vocal learning in songbirds. *Anim. Behav.* 32:395–99.

Kruuk, H. 1972. *The Spotted Hyena: A Study of Predation and Social Behavior*. Chicago: University of Chicago Press.

Kummer, H. 1971. *Primate Societies: Group Techniques of Ecological Adaptation*. Chicago: Aldine-Atherton.

Kuo, Z. Y. 1967. *Dynamics of Behavior Development*. New York: Random House.

Lack, D. 1933. Habitat selection in birds with special reference to the effects of afforestation on the Breckland avifauna. *J. Anim. Ecol.* 2:239–62.

———. 1954. *The Natural Regulation of Animal Numbers*. London: Oxford University Press.

———. 1961. *Darwin's Finches: An Essay on the General Biological Theory of Evolution* (New York: Harper & Bros.) First published in 1947 by Cambridge University Press, Cambridge, Eng.

Lacy, R. C. 1980. The evolution of eusociality in termites: A haplodiploid analogy? *Amer. Nat.* 116:449–51.

———. 1984. The evolution of termite eusociality: Reply to Leinaas. *Am. Nat.* 123:876–78.

Lagerspetz, K., and T. Heino. 1970. Changes in social reactions resulting from early experience with another species. *Psychol. Rep.* 27:255–62.

Laird, L. M. 1978. Marking fish. In *Animal Marking*, ed. B. Stonehouse, 95–101. Baltimore: University Park Press.

Landau, I. T., and W. G. Holmes. 1988. Mating of captive thirteen-lined ground squirrels and the annual timing of estrus. *Horm. Behav.* 22:474–87.

Lane-Petter, W. 1978. Identification of laboratory animals. In *Animal Marking*, ed. B. Stonehouse, 35–40. Baltimore: University Park Press.

Larkin, T. S., and W. T. Keeton. 1976. Bar magnets mask the effect of normal magnetic disturbances on pigeon orientation. *J. Comp. Physiol.* 110:227–31.

Lashley, K. S. 1929. *Brain Mechanisms and Intelligence*. Chicago: University of Chicago Press.

Le Boeuf, B. J. 1974. Male-male competition and reproductive success in elephant seals. *Amer. Zool.* 14:163–76.

Le Boeuf, B. J., and J. Reiter. 1988. Lifetime reproductive success in northern elephant seals. In *Reproductive Success*, ed. T. H. Clutton-Brock. Chicago: University of Chicago Press.

Lee, R. B. 1980. Lactation, ovulation, infanticide, and women's work: A study of hunter-gatherer population regulation. In *Biosocial Mechanisms of Population Regulation*, ed. M. N. Cohen, R. S. Malpass, and H. G. Klein. New Haven: Yale University Press.

Lehner, P. N. 1979. *Handbook of Ethological Methods*. New York: Garland STPM Press.

Lehrman, D. S. 1953. A critique of Konrad Lorenz's theory of instinctive behavior. *Quart. Rev. Biol.* 28:337–69.

———. 1955. The physiological basis of parental feeding behavior in ring doves (*Streptopelia risoria*). *Behaviour* 7:241–86.

————. 1958a. Effect of female sex hormones on incubation behavior in the ring dove (*Streptopelia risoria*) *J. Comp. Physiol. Psychol.* 51:142–45.

————. 1958b. Induction of broodiness by participation in courtship and nest-building in the ring dove (*Streptopelia risoria*). *J. Comp. Physiol. Psychol.* 51:32–36.

————. 1959. Hormonal responses to external stimuli in birds. *Ibis* 101:478–96.

————. 1961. Hormonal regulation of parental behavior in birds and infrahuman mammals. In *Sex and Internal Secretions*, ed. W. C. Young. Baltimore: Williams & Wilkins.

————. 1964. The reproductive behavior of ring doves. *Sci. Amer.* 211:48–54.

————. 1965. Interaction between internal and external environments in the regulation of the reproductive cycle of the ring dove. In *Sex and Behavior*, ed. F. A. Beach. New York: Wiley.

————. 1970. Semantic and conceptual issues in the nature-nurture problem. In *Development and Evolution of Behavior*, ed. L. R. Aronson et al. San Francisco: Freeman.

Lehrman, D. S., P. N. Brody, and R. P. Wortis. 1961. The presence of the mate and of nesting material as stimuli for the development of incubation behavior and for gonadotrophin secretion in the ring dove (*Streptopelia risoria*). *Endocrinology* 68:507–16.

Lent, P. C. 1966. Calving and related social behavior in the barren-ground caribou. *Zeit. Tierpsychol.* 23:701–56.

Leshner, A. I. 1975. A model of hormones and agonistic behavior. *Physiol. Behav.* 15:225–35.

Lettvin, J. Y., Maturana, H. R., McCulloch, W. S., and W. H. Pitts. 1959. What the frog's eye tells the frog's brain. *Proc. Inst. Radio Engineers* 47:1940–51.

Levine, S. 1967. Maternal and environmental influences on the adrenocortical responses to stress in weanling rats. *Science* 156:258–60.

————. 1968. Hormones and conditioning. *Nebr. Symp. Motiv.* 16:85–101.

Lewis, E. R., and P. M. Narins. 1985. Do frogs communicate with seismic signals? *Science* 227:187–89.

Lewontin, R. C., and L. C. Dunn. 1960. The evolutionary dynamics of a polymorphism in the house mouse. *Genetics* 45:705–22.

Ley, W. 1968. *Dawn of Zoology*. Englewood Cliffs, NJ: Prentice-Hall.

Licht, P. 1973. Influence of temperature and photoperiod on the annual ovarian cycle in the lizard *Anolis carolinensis*. *Copeia* 1973:465–72.

Liebig, J. 1847. *Chemistry Application to Agriculture and Physiology*. 4th ed. London: Taylor and Walton.

Ligon, J. D., and S. H. Ligon. 1978a. Communal breeding in green woodhoopoes as a case for reciprocity. *Nature* 276:496–98.

————. 1978b. The communal social system of the green woodhoopoe in Kenya. *Living Bird* 17:159–98.

Lillie, F. R. 1916. The theory of the free-martin. *Science* 43:611–13.

Lin, N. A., and C. D. Michener. 1972. Evolution of sociality in insects. *Quart. Rev. Biol.* 47:131–59.

Lincoln, F. C. 1950. Migration of birds. *U.S. Fish and Wild. Serv. Washington, D.C. Circ.* 16:1–102.

Lindauer, M. 1961. *Communication among Social Bees*. Cambridge: Harvard University Press.

Lindzey, G., H. D. Winston, and M. Manosevitz. 1963. Early experience, genotype, and temperament in *Mus musculus*. *J. Comp. Physiol. Psychol.* 56:622–29.

Lindzey, J., and D. Crews. 1988. Psychobiology of sexual behavior in a whiptail lizard, *Cnemidophorus inornatus*. *Horm. Behav.* 22:279–93.

Linn, I. J. 1978. Radioactive techniques for small mammal marking. In *Animal Marking*, ed. B. Stonehouse, 177–91. Baltimore: University Park Press.

Lissmann, H. W. 1958. On the function and evolution of electric organs in fish. *J. Exp. Biol.* 35:156–91.

Littlejohn, M. J. 1965. Premating isolation in the *Hyla ewingi* complex (Anura: Hylidae). *Evolution* 19:234–43.

Littlejohn, M. J., and J. J. Loftus-Hills. 1968. An experimental evaluation of premating isolation in the *Hyla ewingi* complex. *Evolution* 22:659–63.

Lively, C. M. 1987. Evidence from a New Zealand snail for the maintenance of sex by parasitism. *Nature* 328:519–21.

Lloyd, J. E. 1965. Aggressive mimicry in *Photuris*: Firefly *femmes fatales*. *Science* 149:653–54.

————. 1980. Male *Photuris* fireflies mimic sexual signals of their females' prey. *Science* 210:669–71.

Lockard, R. B. 1971. Reflections on the fall of comparative psychology: Is there a message for us all? *Amer. Psychol.* 26:168–79.

Lockyer, C. H., and S. G. Brown. 1981. The migration of whales. In *Animal Migration*, ed. D. J. Aidley, 105–37. New York: Cambridge University Press.

Loeb, J. 1918. *Forced Movements, Tropisms and Animal Conduct*. Philadelphia: Lippincott.

Lohrer, W. 1961. The chemical acceleration of the maturation process and its hormonal control in the male of the desert locust. *Proc. Roy. Soc. Lond.* (ser. B) 153:380–97.

Lohrer, W., and F. Huber. 1966. Nervous and endocrine control of sexual behavior in a grasshopper (*Comphocerus rufus* L., Acridinae). *Soc. Exp. Biol. Symp.* 20:381–400.

Loizos, C. 1967. Play behavior in high primates: A review. In *Primate Ethology*, ed. D. Morris. Chicago: Aldine.

Lombard, R. E., and I. R. Straughan. 1974. Functional aspects of anuran middle ear structures. *J. Exp. Biol.* 61:71–93.

Lombardi, J. R., and J. G. Vandenbergh. 1977. Pheromonally induced sexual maturation in females: Regulation by the social environment of the male. *Science* 196:545–46.

Longhurst, A. R. 1976. Vertical migration. In *Ecology of the Seas*, ed. D. H. Cushing and J. J. Walsh. Philadelphia: Saunders.

Lorden, J. F., and G. A. Oltmans. 1978. Alteration of the characteristics of learned taste aversion by manipulation of serotonin levels in the rat. *Pharmacol. Biochem. Behav.* 8:13–18.

Lorenz, K. 1935. Der Kumpan in der Umwelt des Vogels. *J. Ornith.* 83:137–213, 289–413.

————. 1937. Über die Bildung des Instinktbegriffes. *Die Naturwissenschaften* 19:289–300.

————. 1950. The comparative method in studying innate behavior patterns. *Symp. Soc. Exp. Biol.* 4:221–68.

————. 1952. *King Solomon's Ring*. New York: Thomas Y. Crowell.

————. 1966. *On Aggression*. New York: Harcourt, Brace & World.

Lorenz, K. Z. 1972. Comparative studies on the behavior of Anatinae. In *Function and Evolution of Behavior: An Historical Sample from the Pens of Ethologists*, ed. P. H. Klopfer and J. P. Hailman. Reading, MA: Addison-Wesley.

Losey, G. S. 1979. The fish cleaning symbiosis: Proximate causes of host behaviour. *Anim. Behav.* 27:669–85.

Lott, D. F. 1962. The role of progesterone in the maternal behavior of rodents. *J. Comp. Physiol. Psychol.* 55:610–13.

Lott, D. F., and J. S. Rosenblatt. 1969. Development of maternal responsiveness during pregnancy in the rat. In *Determinants of Infant Behavior*, vol. 4, ed. B. M. Foss. London: Methuen.

Low, B. S. 1978. Environmental uncertainty and the parental strategies of marsupials and placentals. *Am. Nat.* 112:197–213.

Lowery, G. H. 1946. Evidence of trans-Gulf migration. *Auk* 63:175–211.

Lubin, M., M. Leon, H. Moltz, and M. Numan. 1972. Hormones and maternal behavior in the male rat. *Horm. Behav.* 3:369–74.

Luttge, W. G., and R. E. Whalen. 1970. Dihydrotestosterone, androstenodione, testosterone: Comparative effectiveness in masculinizing and defeminizing reproductive systems in male and female rats. *Horm. Behav.* 1:265–81.

Lutz, P. E., and C. E. Jenner. 1964. Life-history and photoperiodic responses of nymphs of *Tetragoneura cynosura* (Say). *Biol. Bull.* 127:304–16.

Lyman, C. P., J. S. Willis, A. Malan, and L. C. H. Wang. 1982. *Hibernation and Torpor in Mammals and Birds*. New York: Academic Press.

Lynch, C. B., and J. P. Hegmann. 1972. Genetic differences influencing behavioral temperature regulation in small mammals. I. Nesting by *Mus musculus. Behav. Gen.* 2:43–54.

MacArthur, R. H. 1958. Population ecology of some warblers of northeastern coniferous forests. *Ecology* 39:599–619.

————. 1965. Patterns of species diversity. *Biol. Rev.* 40:410–533.

————. 1972. *Geographical Ecology: Pattern in the Distribution of Species*. New York: Harper & Row.

MacArthur, R. H., and E. O. Wilson. 1967. *Theory of Island Biogeography*. Princeton: Princeton University Press.

Machin, K. E., and H. W. Lissman. 1960. The mode of operation of the electric receptors in *Gymnarchus niloticus. J. Exp. Biol.* 37:801–11.

Madison, D. M. 1985. Activity rhythms and spacing. In *Biology of the New World Microtus*, ed. R. H. Tamarin, 373–419. Special Publication of the American Society of Mammalogists #8.

Manning, A. 1966. Corpus allatum and sexual receptivity in female *Drosophila melanogaster. Nature* 211:1321–22.

————. 1967. The control of sexual receptivity in female *Drosophilia. Anim. Behav.* 15:239–50.

————. 1971. Evolution of behavior. In *Psychobiology: Behavior from a Biological Perspective*, ed. J. L. McGaugh. New York: Academic Press.

Manning, A., and T. E. McGill. 1974. Early androgen and sexual behavior in female house mice. *Horm. Behav.* 5:19–31.

Margalef, R. 1963. On certain unifying principles in ecology. *Amer. Nat.* 97:357–74.

Margoliash, D. 1983. Acoustic parameters underlying the responses of song-specific neurons in the white-crowned sparrow. *J. Neurosci.* 3:1039–57.

————. 1986. Preference for autogenous song by auditory neurons in a song system nucleus of the white-crowned sparrow. *J. Neurosci.* 6:1643–61.

————. 1987. Neural plasticity in birdsong learning. In *Imprinting and Cortical Plasticity*, ed. J. P. Rauschecker and P. Marler, 23–54. New York: Wiley.

Marler, P. 1955. Characteristics of some animal calls. *Nature* 176:6–8.

————. 1967. Animal communication signals. *Science* 157:769–74.

————. 1968. Aggregation and dispersal: Two functions in primate communication. In *Primates: Studies in Adaptation and Variability*, ed. P. C. Jay. New York: Holt, Rinehart and Winston.

————. 1970. A comparative approach to vocal learning: Song development in white-crowned sparrows. *J. Comp. Physiol. Psychol.* 7:1–25.

————. 1973. A comparison of vocalizations of red tailed monkeys and blue monkeys, *Cercopithecus ascanius* and *C. mitis*, in Uganda. *Zeit. Tierpsychol.* 33:223–47.

————. 1976. On animal aggression: The roles of strangeness and familiarity. *Amer. Psychol.* 31:239–46.

Marler, P., and P. Mundinger. 1971. Vocal learning in birds. In *Ontogeny of Vertebrae Behavior*, ed. H. Moltz. New York: Academic Press.

Marler, P., and S. Peters. 1977. Selective vocal learning in a sparrow. *Science* 198:519–21.

————. 1981. Sparrows learn adult song and more from memory. *Science* 213:780–82.

————. 1982. Structural changes in song ontogeny in the swamp sparrow *Melospiza georgiana. Auk* 99:446–58.

Marler, P., and M. Tamura. 1962. Song variation in three populations of white-crowned sparrows. *Condor* 64:368–77.

Marsden, H. M. 1968. Agonistic behavior of young rhesus monkeys after changes induced in social rank of their mothers. *Anim. Behav.* 16:38–44.

Marshall, A. J. 1960. Annual periodicity in the migration and reproduction of birds. *Cold Spr. Harb. Symp. Quant. Biol.* 25:499–505.

Martin, N. L., E. O. Price, S. J. R. Wallach, and M. R. Dally. 1987. Fostering lambs by odor transfer: The add-on experiment. *J. Anim. Sci.* 64:1378–83.

Marx, J. L. 1980. Ape-language controversy flares up. *Science* 207:1330–33.

Mason, W. A. 1960. The effects of social restriction on the behavior of rhesus monkeys. I. Free social behavior. *J. Comp. Physiol. Psychol.* 53:582–89.

————. 1961a. The effects of social restriction on the behavior of rhesus monkeys. II. Tests of gregariousness. *J. Comp. Physiol. Psychol.* 54:287–90.

————. 1961b. The effects of social restriction on the behavior of rhesus monkeys. III. Dominance tests. *J. Comp. Physiol. Psychol.* 54:694–99.

Massey, A. 1977. Agonistic aids and kinship in a group of pigtail macaques. *Behav. Ecol. Sociobiol.* 2:31–40.

Massey, A., and J. G. Vandenbergh. 1980. Puberty delay by a urinary cue from female house mice in feral populations. *Science* 209:821–22.

———. 1981. Puberty acceleration by a urinary cue from male mice in feral populations. *Biol. Reprod.* 24:523–27.

Masson, G. M. 1948. Effects of estradiol and progesterone on lactation. *Anat. Rec.* 102:513–21.

Mast, S. O. 1938. Factors involved in the process of orientation of lower organisms in light. *Biol. Rev.* 17:68–90.

Masterton, R. B., M. E. Bitterman, C. B. G. Campbell, and N. Hotton. 1976. *Evolution of Brain and Behavior in Vertebrates.* Hillsdale, NJ: Erlbaum.

Mather, M. H., and B. D. Roitberg. 1987. A sheep in wolf's clothing: Tephritid flies mimic spider predators. *Science* 236:308–10.

Mathews, G. V. T. 1968. *Bird Navigation*, 2d ed. Cambridge: Cambridge University Press.

Matthews, R. W. 1968. *Microstigmus comes:* sociality in a specid wasp. *Science* 160:787–88.

Maturana, H. R., et al. 1960. Anatomy and physiology of vision in the frog (*Rana pipiens*). *J. Gen. Physiol. Suppl.* 43:129–75.

Maxim, P. E. 1977. Self-stimulation of a hypothalamic site in response to tension or fear. *Physiol. Behav.* 18:197–201.

Maynard Smith, J. 1974. The theory of games and the evolution of animal conflict. *J. Theoret. Biol.* 47:209–21.

———. 1976. Evolution and the theory of games. *Amer. Sci.* 64:41–45.

Maynard Smith, J., and G. R. Price. 1973. The logic of animal conflict. *Nature* 246:15–18.

Mayr, E. 1963. *Animal Species and Evolution.* Cambridge: Harvard University Press.

———. 1970. *Populations, Species and Evolution.* Cambridge: Harvard University Press.

———. 1977. Darwin and natural selection. *Amer. Sci.* 65:321–27.

Mazur, A. 1983. Hormones, aggression, and dominance in humans. In *Hormones and Aggressive Behavior*, ed. B. B. Svare. New York: Plenum.

McCarty, R., and C. H. Southwick. 1977. Cross-species fostering: Effects on the olfactory preference of *Onychomys torridus* and *Peromyscus leucopus*. *Behav. Biol.* 19:255–60.

McConnell, J. V., A. L. Jacobson, and D. P. Kimble. 1959. The effects of regeneration upon retention of a conditioned response in the planarian. *J. Comp. Physiol. Psychol.* 52:1–5.

McCracken, G. F., and J. W. Bradbury. 1977. Paternity and genetic heterogeneity in the polygynous bat, *Phyllostomus hastatus*. *Science* 198:303–6.

McDonald, D. L., and L. G. Forslund. 1978. The development of social preferences in the voles, *Microtus montanus* and *Microtus canicaudus:* Effects of cross-fostering. *Behav. Biol.* 22:457–508.

McEwen, B. S., P. G. Davis, B. Parsons, and D. W. Pfaff. 1979. The brain as a target for steroid hormone action. *Ann. Rev. Neurosci.* 2:65–112.

McFarland, D. 1981. *The Oxford Companion to Animal Behaviour.* Oxford, England: Oxford University Press.

McGuire, T. R., and T. Tully. 1986. Food-search behavior and its relation to the central excitatory state in the genetic analysis of the blowfly (*Phormia regina*). *J. Comp. Psychol.* 100:52–58.

McIntosh, T. K., and R. J. Barfield. 1984a. Brain monoaminergic control of male reproductive behavior. I. Serotonin and the post-ejaculatory refractory period. *Behav. Brain Res.* 12:255–65.

———. 1984b. Brain monoaminergic control of male reproductive behavior. II. Dopamine and the post-ejaculatory refractory period. *Behav. Brain Res.* 12:267–73.

———. 1984c. Brain monoaminergic control of male reproductive behavior. III. Norepinephrine and the post-ejaculatory refractory period. *Behav. Brain Res.* 12:275–81.

McKenna, J. J. 1981. Primate infant caregiving behavior. In *Parental Care in Mammals*, ed. D. J. Gubernick and P. H. Klopfer, 389–416. New York: Plenum.

McLaren, I. A. 1963. Effects of temperature on growth of zooplankton and the adaptive value of vertical migration. *J. Fish. Res. Bd. Can.* 20:685–727.

McNab, B. K. 1963. Bioenergetics and the determination of home range size. *Amer. Nat.* 97:133–40.

McNair, J. N. 1982. Optimal giving-up times and the marginal value theorem. *Am. Nat.* 119:511–29.

Meany, M. J., J. Stewart, and W. W. Beatty. 1982. The influence of glucocorticoids during the neonatal period on the development of play-fighting in Norway rat pups. *Horm. Behav.* 16:475–91.

Mech, L. D. 1970. *The Wolf: The Ecology and Behavior of an Endangered Species.* Garden City, NY: Natural History Press.

Mech, L. D. 1970. *The Wolf.* Garden City, N.Y.: Doubleday.

Meier, A. H. 1973. Daily hormone rhythms in the white-throated sparrow. *Amer. Sci.* 61:184–87.

Meier, G. W. 1965. Other data on the effects of social isolation during rearing upon adult reproductive behavior in the rhesus monkey (*Macaca mulatta*). *Anim. Behav.* 13:228–31.

Meikle, D. B., B. L. Tilford, and S. H. Vessey. 1984. Dominance rank, secondary sex ratio and reproduction of offspring in polygynous primates. *Amer. Nat.* 124:173–88.

Meikle, D. B., and S. H. Vessey. 1981. Nepotism among rhesus monkey brothers. *Nature* 294:160–61.

Melnick, D. J., M. C. Pearl, and A. F. Richard. 1984. Male migration and inbreeding avoidance in wild rhesus monkeys. *Am. J. Primatol.* 7:229–43.

Menaker, M., J. S. Takahashi, and A. Eskin. 1978. The physiology of circadian pacemakers. *Ann. Rev. Physiol.* 40:501–26.

Menaker, M., and N. Zimmerman. 1976. Role of the pineal in the circadian system of birds. *Amer. Zool.* 16:45–55.

Menzel, E. W. 1971. Communication about the environment in a group of young chimpanzees. *Folia Primat.* 15:220–32.

Merle, J. 1969. Fonctionnement ovarien et réceptivité sexuelle de *Drosophila melanogaster* après implantation de fragments de l'appareil génitale mâle. *J. Insect Physiol.* 14:1159–68.

Merrell, D. J. 1949. Selective mating in *Drosophila melanogaster*. *Genetics* 34:370–89.

———. 1953. Selective mating as a cause of gene frequency changes in laboratory populations of *Drosophila melanogaster*. *Evolution* 7:287–96.

Metzgar, L. H. 1967. An experimental comparison of screech owl predation on resident and transient white-footed mice (*Peromyscus leucopus*). *J. Mammal.* 48:387–91.

———. 1971. Behavioral population regulation in the woodmouse, *Peromyscus leucopus*. *Amer. Mid. Nat.* 86:434–48.

Meyer, D. R., and H. F. Harlow. 1952. Effects of multiple variables on delayed response performance by monkeys. *J. Genet. Psycol.* 81:53–61.

Meyer, D. R., F. R. Treichler, and P. M. Meyer. 1965. Discrete-trial training techniques and stimulus variables. In *Behavior of Nonhuman Primates.* Vol. 1., ed. A. M. Schrier, H. F. Harlow, and F. Stollnitz. New York: Academic Press.

Michael, R. P., and R. W. Bonsall. 1977. Chemical signals and primate behavior. In *Chemical Signals in Vertebrates,* ed. D. Muller-Schwartze and M. M. Mozell. New York: Plenum.

Michel, R. 1972. Étude expérimentale de l'influence des glandes prothoraciques sur l'activité de vol du Criquet Pèlerin *Schistocerca gregaria. Gen. Comp. Endocr.* 19:96–101.

Michener, C. D. 1974. *The Social Behavior of the Bees: A Comparative Study.* Cambridge, MA: Belknap Press.

Michener, C. D. 1974. *Social Behavior of the Bees.* Cambridge, MA: Harvard University Press.

Milinski, M. 1979. Can an experienced predator overcome the confusion of swarming prey more easily? *Anim. Behav.* 27:1122–26.

Miller, D. B. 1982. Alarm call responsivity of mallard ducklings: I. The acoustical boundary between behavioral inhibition and excitation. *Develop. Psychobiol.* 16:185–94.

———. 1985. Methodological issues in the ecological study of learning. In *Issues in the Ecological Study of Learning,* ed. T. D. Johnston and A. T. Pietrewicz, 73–95. Hillsdale, NJ: Lawrence Erlbaum Associates.

———. 1988. Development of instinctive behavior. In *Handbook of Behavioral Neurobiology,* Vol. 9, *Developmental Psychobiology and Behavioral Ecology,* ed. E. Blass, 415–44. New York: Plenum.

Miller, D. B., and C. F. Blaich. 1987. Alarm call responsivity of mallard ducklings: V. Age-related changes in repetition rate specificity and behavioral inhibition. *Devel. Psychobiol.* 20:571–86.

Miller, D. B., and G. Gottlieb. 1978. Maternal vocalizations of mallard ducks (*Anas platyrhynchos*). *Anim. Behav.* 26:1178–94.

Miller, F. L., F. W. Anderka, C. Vithayasai, and R. L. McClure. 1975. Distribution, movements, and socialization of barrenground caribou radio-tracked on their calving and post-calving areas. In *First International Reindeer and Caribou Symposium,* eds. J. R. Luick, C. Lent, D. R. Klein and P. G. White, 423–35. Fairbanks, AK: University of Alaska.

Miller, J. W. 1975. Much ado about starlings. *Natur. Hist.* 84 (7):38–45.

Miller, L. A., and J. T. Emlen, Jr. 1975. Individual chick recognition and family integrity in the ring-billed gull. *Behaviour* 52:124–44.

Miller, N. E. 1957. Experiments on motivation studies combining psychological, physiological, and pharmacological techniques. *Science* 126:1271–78.

Miller, R. C. 1922. The significance of the gregarious habit. *Ecology* 3:122–26.

Mitchell, G. 1970. Abnormal behavior in primates. In *Primate Behavior: Developments in Field and Laboratory Research,* vol. 1, ed. L. Rosenblum. New York: Academic Press.

Mitchell, G., and E. M. Brandt. 1970. Behavioral differences related to experience of mother and sex of infant in the rhesus monkey. *Dev. Psychol.* 3:149.

Mock, D. W. 1984. Siblicidal aggression and resource monopolization in birds. *Science* 225:731–33.

Mock, D. W., and B. J. Ploger. 1987. Parental manipulation of optimal hatch asynchrony in cattle egrets: an experimental study. *Anim. Behav.* 35:150–60.

Moehlman, P. D. 1979. Jackal helpers and pup survival. *Nature* 277:382–83.

Moffat, A. J. M., and R. R. Capranica. 1978. Middle ear sensitivity in anurans measured by light scattering spectroscopy. *J. Comp. Physiol.* 127:97–107.

Moller, A. P. 1988. Ejaculate quality, testes size and sperm competition in primates. *J. Hum. Evol.* 17:479–88.

Moller, P. 1976. Electric signals and schooling behavior in a weakly electric fish, *Marcusenius cyprinoides* L. (Mormyriformes). *Science* 193:697–99.

Moltz, H. 1965. Contemporary instinct theory and the fixed action pattern. *Psychol. Rev.* 72:27–47.

Moltz, H., D. Robbins, and M. Parks. 1966. Caesarian delivery and the maternal behavior of primiparous and multiparous rats. *J. Comp. Physiol. Psychol.* 61:455–60.

Money, J. 1977. Human hermaphroditism. In *Human Sexuality in Four Perspectives,* ed. F. A. Beach. Baltimore: Johns Hopkins University Press.

Money, J., and A. A. Ehrhardt. 1972. *Man and Woman, Boy and Girl: Differentiation and Dimmorphism of Gender Identity from Conception to Maturity.* Baltimore: Johns Hopkins University Press.

Moore, F., and P. Kerlinger. 1987. Stopover and fat deposition by North American wood warblers (Parulinae) following spring migration over the Gulf of Mexico. *Oecologia* 74:47–54.

Moore, J., and R. Ali. 1984. Are dispersal and inbreeding avoidance related? *Anim. Behav.* 32:94–112.

Moore, J. A. 1984. Science as a way of knowing—evolutionary biology. *Amer. Zool.* 24:467–534.

Moore, R. Y. 1979. The anatomy of the central neural mechanisms regulating endocrine rhythms. In *Endocrine Rhythms,* ed. D. Krieger. New York: Raven Press.

———. 1980. Suprachiasmatic nucleus, secondary synchronizing stimuli and the central neural control of circadian rhythms. *Brain Res.* 183:13–28.

Moore, R. Y., and V. B. Eichler. 1972. Loss of a circadian adrenal corticosterone rhythm following suprachiasmatic lesions in the rat. *Brain Res.* 42:201–6.

Moore, R. Y., and N. J. Lenn. 1972. A retinohypothalamic projection in the rat. *J. Comp. Neurol.* 146:1–14.

Moore-Ede, M. C., F. M. Sulzman, and C. A. Fuller. 1982. *The Clocks That Time Us.* Cambridge: Harvard University Press.

Morgan, C. L. 1896. *Introduction to Comparative Psychology.* London: Scott.

Morrell, J. I., D. B. Kelley, and D. W. Pfaff. 1975. Sex steroid binding in the brains of vertebrates: Studies with light microscopic autoradiography. In *The Ventricular System in Neuroendocrine Mechanisms,* ed. K. M. Knigge et al. Basel, Switzerland: Karger.

Morrell, J. I., and D. W. Pfaff. 1981. Autoradiographic technique for steroid hormone localization. In *Neuroendocrinology of Reproduction,* ed. N. Adler. New York: Plenum.

Morris, D. 1956. The feather postures of birds and the problem of the origin of social signals. *Behaviour* 9:75–113.

Morris, R. G. M. 1981. Spatial localization does not require the presence of local cues. *Learn. Motivat.* 12:239–60.

Morse, D. H. 1980. *Behavioral Mechanisms in Ecology*. Cambridge, MA: Harvard University Press.

Morton, M. L., and L. R. Mewaldt. 1962. Some effects of castration on migratory sparrows (*Zonotrichia atricapilla*). *Physiol. Zool.* 35:237–47.

Moyer, K. E. 1976. *Psychobiology of Aggression*. New York: Harper & Row.

Moynihan, M. 1956. Notes on the behaviour of some North American gulls. 1. Aerial hostile behaviour. *Behaviour* 10:126–78.

Mudry, K. M. 1978. A Comparative Study of the Response Properties of Higher Auditory Nuclei in Anurans. Ph.D. Dissertation, Cornell University, Ithaca, NY.

Mueller, H. 1966. Homing and distance-orientation in bats. *Zeit. Tierpsychol.* 23:403–21.

Mueller, H., and J. T. Emlen. 1957. Homing in bats. *Science* 126:307–8.

Mueller, H. C. 1971. Oddity and specific searching image more important than conspicuousness in prey selection. *Nature* 233:345–46.

Muller, H. J. 1964. The relation of recombination to mutational advance. *Mutation Res.* 1:2–9.

Murdock, G. P. 1967. *Ethnographic Atlas*. Pittsburgh: University of Pittsburgh Press.

Murie, O. J. 1935. Alaska-Yukon caribou. *N. Amer. Fauna* (U.S. Dept. Agric.) 54:1–93.

Murphy, M. R. 1980. Sexual preferences of male hamsters: Importance of preweaning and adult experience, vaginal secretion and olfactory or vomeronasal sensation. *Behav. Neural Biol.* 30:323–40.

Myers, J. H., and C. J. Krebs. 1971. Genetic, behavioral and reproductive attributes of dispersing field voles *Microtus pennsylvanicus* and *Microtus ochrogaster*. *Ecol. Monogr.* 41: 53–78.

National Institutes of Health. 1985. *Guide for the Care and Use of Laboratory Animals*. Washington, DC: U.S. Department of Health and Human Services.

Neumann, D. 1966. Die lunare and tägliche Schlüpfperiodik der Mücke *Clunio*-Steuerung und Abstimmung auf die Gezeitenperiodik. *Zeit. Vergleich. Physiol.* 53:1–61.

Newman, H. H., F. N. Freeman, and K. J. Holzinger. 1937. *Twins: A Study of Heredity and Environment*. Chicago: University of Chicago Press.

Nice, M. M. 1941. The role of territory in bird life. *Amer. Mid. Nat.* 26:441–87.

Nicolaus, L. K., J. F. Cassel, R. B. Carlson, and C. B. Gustavson. 1983. Taste-aversion conditioning of crows to control predation on eggs. *Science* 220:212–14.

Nicoll, C. S., and J. Meites. 1959. Prolongation of lactation in the rat by litter replacement. *Proc. Soc. Exp. Biol. Med.* 101:81–82.

Noakes, D. L. G. 1978. Ontogeny of behavior in fishes: A survey and suggestions. In *Development of Behavior*, ed. G. Burghardt and M. Bekoff. New York: Garland STPM Press.

Noirot, E. 1972. Ultrasounds and maternal behavior in small mammals. *Dev. Psychobiol.* 5:371–87.

Nordenskiöld, E. 1928. *History of Biology*. New York: Knopf.

Nottebohm, F. 1975. Vocal behavior in birds. In *Avian Biology*, Vol. 5, ed. J. R. King and D. S. Farner. New York: Academic Press.

———. 1980. Brain pathways for vocal learning in birds: A review of the first 10 years. In *Progress in Psychobiology and Physiological Psychology*, ed. J. M. Sprague and A. N. Epstein. 9:85–124.

———. 1981. A brain for all seasons: Cyclical anatomical changes in song control nuclei in the canary brain. *Science* 214:1368–70.

Nottebohm, F., and M. Nottebohm. 1976. Left hypoglossal dominance in the control of canary and white-crowned sparrow song. *J. Comp. Physiol.* 108:171–92.

Nottebohm, F., M. E. Nottebohm, and L. Crane. 1986. Developmental and seasonal changes in canary song and their relation to changes in the anatomy of song-control nuclei. *Behav. Neural Biol.* 46:445–71.

Novotny, M., B. Jemiolo, S. Harvey, D. Wiesler, and A. Marchlewska-Koj. 1986. Adrenal-mediated endogenous metabolites inhibit puberty in female mice. *Science* 231:722–25.

Numan, M. 1974. Medial preoptic area and maternal behavior in the female rat. *J. Comp. Physiol. Psychol.* 87:746–59.

Numan, M., M. Leon, and H. Moltz. 1972. Interference with prolactin release and maternal behavior of female rats. *Horm. Behav.* 3:29–38.

O'Connell, M. E., C. Reboulleau, H. H. Feder, and P. Silver. 1981a. Social interactions and androgen levels in birds. I. Female characteristics associated with increased plasma androgen levels in the male ring dove (*Streptopelia risoria*). *Gen. Comp. Endocrinol.* 44:454–63.

O'Connell, M. E., R. Silver, H. H. Feder, and C. Reboulleau. 1981b. Social interactions and androgen levels in birds. II. Social factors associated with a decline in plasma androgen levels in male ring doves (*Streptopelia risoria*). *Gen. Comp. Endocrinol.* 44:464–69.

Oldham, R. S., and H. C. Gerhardt. 1975. Behavioral isolation of the treefrogs *Hyla cinerea* and *Hyla gratiosa*. *Copeia* 1975:223–31.

Olds, J. 1956. A preliminary mapping of electrical reinforcing effects in the rat brain. *J. Comp. Physiol. Psychol.* 49:281–85.

Olgivie, D. W., and R. H. Stinson. 1966. Temperature selection in *Peromyscus* and laboratory mice. *J. Mammal.* 47:655–60.

Oliverio, A. 1983. Genes and behavior: An evolutionary perspective. *Adv. Study Behav.* 13:191–217.

Oppenheim, R. W. 1970. Some aspects of embryonic behavior in the duck (*Anas platyrhynchos*). *Anim. Behav.* 18:335–52.

———. 1972. Pre-hatching and hatching behavior in birds: A comparative study of altrical and precocial species. *Anim. Behav.* 20:644–55.

———. 1982. Preformation and epigenesis in the origins of the nervous system and behavior: Issues, concepts and their history. *Perspectives in Ethology* 5:1–100.

Oppenheim, R. W., W. Chu-Wang, and J. L. Maderut. 1978. Cell death of motorneurons in the chick embryo and spinal cord. III. *J. Comp. Neurol.* 177:87–112.

Orians, G. H. 1960. Autumnal breeding in the tricolored blackbird. *Auk* 77:379–98.

———. 1961. The ecology of blackbird (*Agelaius*) social systems. *Ecol. Monogr.* 31:285–312.

———. 1969. On the evolution of mating systems in birds and mammals. *Amer. Nat.* 103:589–603.

Orians, G. H., and G. Collier. 1963. Competition and blackbird social systems. *Evolution* 17:449–59.

Oring, L. W., and D. B. Lank. 1982. Sexual selection, arrival times, philopatry and site fidelity in the polyandrous spotted sandpiper. *Behav. Ecol. Sociobiol.* 10:185–91.

Orr, R. T. 1970. *Animals in Migration.* New York: Macmillan.

Packer, C. 1977. Reciprocal altruism in *Papio anubis. Nature* 265:441–43.

———. 1979. Inter-troop transfer and inbreeding avoidance in *Papio anubis. Anim. Behav.* 27:1–36.

———. 1986. The ecology of sociality in felids. In *Ecological Aspects of Social Evolution,* ed. D. I. Rubenstein and R. W. Wrangham. Princeton: Princeton University Press.

Packer, C., and A. E. Pusey. 1982. Cooperation and competition within coalitions of male lions: Kin selection or game theory? *Nature* 296:740–42.

Page, T. L. 1982. Transplantation of the cockroach circadian pacemaker. *Science* 216:73–75.

Page, T. L., P. C. Caldarola, and C. S. Pittendrigh. 1977. Mutual entrainment of bilaterally distributed circadian pacemakers. *Proc. Nat. Acad. Sci. USA* 74:1277–81.

Panksepp, J. 1971. Aggression elicited by electrical stimulation of the hypothalamus in albino rats. *Physiol. Behav.* 6:321–29.

Papi, F. 1982. Olfaction and homing in pigeons: Ten years of experiments. In *Avian Navigation,* ed. F. Papi and H. G. Walraff, 149–59. Berlin: Springer-Verlag.

Papi, F., V. Fiaschi, S. Benvenuti, and N. E. Baldaccini. 1972. Olfaction and homing in pigeons. *Monit. Zool. Ital.* (n.s.) 6:85–95.

Papi, F., and L. Pardi. 1953. Ricerche sull'orientamento di *Talitrus saltator* Montagu (Crustacea-Amphipoda). *Z. Vergl. Physiol.* 35:490–518.

Parker, G. A. 1970a. The reproductive behaviour and the nature of sexual selection in *Scatophaga stercoraria* L. (Diptera: Scatophagidae) IV. Epigamic recognition and competition between males for the possession of females. *Behaviour* 37:113–39.

———. 1970b. Sperm competition and its evolutionary consequences in the insects. *Biol. Rev.* 45:525–68.

———. 1974. The reproductive behavior and the nature of sexual selection in *Scatophaga stercoraria* L. IX. Spatial distribution of fertilization rates and evolution of male search strategy within the reproductive area. *Evolution* 28:93–108.

———. 1978. Searching for mates. In *Behavioural Ecology: An Evolutionary Approach,* ed. J. R. Krebs and N. B. Davies. Oxford, England: Blackwell Scientific Publications, Ltd.

———. 1984. Evolutionarily stable strategies. In *Behavioural Ecology: An Evolutionary Approach,* 2d ed., eds. J. R. Krebs and N. B. Davies. Oxford, England: Blackwell Scientific Publications, Ltd.

Parker, G. A., R. R. Baker, and V. G. F. Smith. 1972. The origin and evolution of gamete dimorphism and the male-female phenomenon. *J. Theoret. Biol.* 36:529–53.

Partridge, L. 1974. Habitat selection in titmice. *Nature* 247:573–74.

———. 1976. Field and laboratory observations on the foraging and feeding techniques of blue tits (*Parus caeruleus*) and coal tits (*Parus ater*) in relation to their habitats. *Anim. Behav.* 24:534–44.

———. 1978. Habitat selection. In *Behavioural Ecology: An Evolutionary Approach,* ed. J. R. Krebs and N. B. Davies. Oxford, England: Blackwell Scientific Publications, Ltd.

Paterson, H. E. H. 1980. A comment on "mate recognition systems". *Evolution* 34:330–31.

Paton, J. A., and F. Nottebohm. 1984. Neurons generated in the adult brain are recruited into functional circuits. *Science* 225:1046–48.

Patterson, D. J. 1973. Habituation in a protozoan *Vorticella convallaria. Behaviour* 45:304–11.

Patterson, I. J. 1978. Tags and other distant-recognition markers for birds. In *Animal Marking,* ed. B. Stonehouse, 54–62. Baltimore: University Park Press.

Patterson, M. A., and S. H. Vessey. 1973. Tapeworm (*Hymenolepis nana*) infection in male albino mice: Effect of fighting among the hosts. *J. Mammal.* 54:784–86.

Payne, K. B., W. R. Langbauer, Jr., and E. M. Thomas. 1986. Infrasonic calls of the Asian elephant *(Elephas maximus). Behav. Ecol. Sociobiol.* 18:297–301.

Payne, R. S. and S. McVay. 1971. Songs of humpback whales. *Science* 173:585–97.

Pener, M. P. 1965. On the influence of corpora allata on maturation and sexual behavior of *Schistocerca gregaria. J. Zool. Lond.* 147:119–36.

Pengelley, E. T., and S. J. Asmundson. 1974. Circannural rhythmicity in hibernating animals. In *Circannual Clocks,* ed. E. T. Pengelley. New York: Academic Press.

Pennycuik, C. J. 1978. Identification using natural markings. In *Animal Marking,* ed. B. Stonehouse, 147–58. Baltimore: University Press.

Pennycuik, C. J., and J. Rudnai. 1970. A method of identifying individual lions *Panther leo* with an analysis of the reliability of identification. *J. Zool. Lond.* 160:497–508.

Pepperberg, I. M. 1987a. Acquisition of the same/different concept by an African grey parrot (*Psittacus erithacus*): Learning with respect to categories of color, shape, and material. *Animal Learning & Behavior* 15:423–32.

———. 1987b. Evidence for conceptual quantitative abilities in the African grey parrot: Labeling of cardinal sets. *Ethology* 75:37–61.

Peters, S., W. A. Searcy, and P. Marler. 1980. Species song discrimination in choice experiments with territorial male swamp and song sparrows. *Anim. Behav.* 28:393–404.

Petersen, J. C. B. 1972. An identification system for zebra (*Equus burchelli,* Gray). *E. Afr. Wildl. J.* 10:59–63.

Pfaff, D. W. 1981. Electrophysiological effects of steroid hormones in brain tissue. In *Neuroendocrinology of Reproduction,* ed. N. Adler. New York: Plenum.

Pfaff, D. W., and M. Keiner. 1973. Atlas of estradiol-concentrating cells in the central nervous system of the female rat. *J. Comp. Neurol.* 151:121–30.

Pfaffenberger, C. J., J. L. Fuller, B. E. Ginsberg, and S. W. Bielfelt. 1976. *Guide Dogs for the Blind: Their Selection, Development and Training.* Amsterdam: Elsevier.

Phoenix, C. H. 1974. Prenatal testosterone in the nonhuman primate and its consequences for behavior. In *Sex Differences in Behavior,* eds. R. C. Friedman, R. N. Richart and R. L. Van de Wiele. New York: Wiley.

———. 1977. Induction of sexual behavior in ovariectomized rhesus females with 19-hydroxytestosterone. *Horm. Behav.* 8:356–62.

Phoenix, C. H., R. W. Goy, A. A. Gerrall, and W. C. Young. 1959. Organizing action of prenatally administered testosterone propionate on the tissues mediating mating behavior in the female guinea pig. *Endocrinology* 65:369–82.

Pietrewicz, A. T., and A. C. Kamil. 1979. Search image formation in the blue jay (*Cyanocitta cristata*). *Science* 204:1332–33.

———. 1981. Search images and the detection of cryptic prey: An operant approach. In *Foraging Behavior: Ecological, Ethological, and Psychological Approaches,* ed. A. C. Kamil and T. D. Sargent. New York: Garland STPM Press.

Pietsch, T. W., and D. B. Grobecker. 1978. The compleat angler: Aggressive mimicry in an Antennariid anglerfish. *Science* 201:369–70.

Pinsker, H., I. Kupfermann, V. Castellucci, and E. R. Kandel. 1970. Habituation and dishabituation of the gill-withdrawal reflex in *Aplysia*. *Science* 167:1740–42.

Pitcher, T. J., A. E. Magurran, and I. J. Winfield. 1982. Fish in larger shoals find food faster. *Behav. Ecol. Sociobiol.* 10:149–51.

Pittendrigh, C. S. 1954. On temperature independence in the clock system controlling emergence time in *Drosophila*. *Proc. Nat. Acad. Sci. USA* 40:1018–29.

———. 1960. Circadian rhythms and the circadian organization of living systems. *Cold Spr. Harb. Symp. Quant. Biol.* 25:159–84.

Pittman, R., and R. W. Oppenheim. 1979. Cell death of motorneurons in the chick embryo spinal cord. IV. *J. Comp. Neurol.* 187:425–46.

Pleim, E. T., and R. J. Barfield. 1988. Progesterone versus estrogen facilitation of female sexual behavior in intracranial administration to female rats. *Horm. Behav.* 22:150–59.

Plomin, R., J. C. DeFries, and J. C. Loehlin. 1977. Genotype-environment interaction and correlation in the analysis of human behavior. *Psych. Bull.* 84:309–22.

Pollack, G. S., and R. R. Hoy. 1979. Temporal pattern as a cue for species-specific calling song recognition in crickets. *Science* 204:429–32.

Poole, J. H., K. Payne, W. R. Langbauer, Jr., and C. J. Moss. 1988. The social contexts of some very low frequency calls of African elephants. *Behav. Ecol. Sociobiol.* 22:385–92.

Poulsen, H. R., and D. Chiszar. 1975. Interaction of predation and intraspecific aggression in bluegill sunfish *Lepomis macrochirus*. *Behaviour* 55:268–86.

Powley, T. L., and R. E. Keesey. 1970. Relationship of body weight to the lateral hypothalamic feeding syndrome. *J. Comp. Physiol. Psychol.* 70:25–36.

Pratt, C. L., and G. P. Sackett. 1967. Selection of social partners as a function of peer contact during rearing. *Science* 155:1133–35.

Premak, D. 1971. Language in chimpanzee? *Science* 172:808–22.

Price, E. O. 1972. Domestication and early experience effects on escape conditioning in the Norway rat. *J. Comp. Physiol. Psychol.* 79:51–55.

———. 1984. Behavioral aspects of animal domestication. *Quart. Rev. Biol.* 59:1–32.

Price, E. O., and P. L. Belanger. 1977. Maternal behavior of wild and domestic Norway rats. *Behav. Biol.* 20:60–69.

Price, E. O., G. C. Dunn, J. A. Talbot, and M. R. Dally. 1984. Fostering lambs by odor transfer: The substitution experiment. *J. Anim. Sci.* 59:301–7.

Price, E. O., and U. W. Huck. 1976. Open-field behavior of wild and domestic Norway rats. *Anim. Learn. Behav.* 4:125–30.

Prince, G. J., and P. A. Parsons. 1977. Adaptive behaviour of *Drosophila* adults in relation to temperature and humidity. *Aust. J. Zool.* 25:285–90.

Pugesek, B. 1981. Increased reproductive effort with age in the California gull (*Larus californicus*). *Science* 212:822–23.

———. 1983. The relationship between parental age and reproductive effort in the California gull (*Larus californicus*). *Behav. Ecol. Sociobiol.* 13:161–71.

Purcell, J. E. 1980. Influence of siphonophore behavior upon their natural diets: Evidence for aggressive mimicry. *Science* 209:1045–47.

Pusey, A. E., and C. Packer. 1987. The evolution of sex-biased dispersal in lions. *Behaviour* 101:275–310.

Pyke, G. H. 1979a. The economics of territory size and time budget in the golden-winged sunbird. *Am. Nat.* 114:131–45.

———. 1979b. Optimal foraging in bumblebees: Rule of movement between flowers within inflorescences. *Anim. Behav.* 27:1167–81.

Quadagno, D. M., and E. M. Banks. 1970. The effect of reciprocal cross-fostering on the behavior of two species of rodents, *Mus musculus* and *Baiomys taylori ater*. *Anim. Behav.* 18:379–90.

Quartermus, C., and J. A. Ward. 1969. Development and significance of two motor patterns used in contacting parents by young orange chromides (*Etroplus maculatus*). *Anim. Behav.* 17:624–35.

Queller, D. C., J. E. Strassmann, and C. R. Hughes. 1988. Genetic relatedness in colonies of tropical wasps with multiple queens. *Science* 242:1155–57.

Quiatt, D. 1979. Aunts and mothers: Adaptive implications of allomaternal behavior in nonhuman primates. *Amer. Anthropol.* 81:310–19.

Rainey, R. C. 1959. Some new methods for the study of flight and migration. *Proc. 15th Inter. Congr. Zool.*, London. 866–70.

———. 1962. The mechanisms of desert locust swarm movements and the migration of insects. *Proc. 11th Inter. Congr. Ent.* 3:47–49.

Ralls, K. 1971. Mammalian scent marking. *Science* 171: 443–49.

Ralph, M. R., and M. Menaker. 1988. A mutation of the circadian system in golden hamsters. *Science* 241:1225–27.

Randall, J. A. 1984. Territorial defense and advertisement by footdrumming in bannertail kangaroo rats (*Dipodomys spectabilis*) at high and low population densities. *Behav. Ecol. Sociobiol.* 16:11–20.

Ratner, S. C., and R. Boice. 1975. Effects of domestication on behaviour. In *Behaviour of Domestic Animals*. 3d ed., ed. E. S. E. Hafez. Baltimore, MD: Williams & Wilkins.

Ratner, S. C., and A. R. Gilpin. 1974. Habituation and retention of habituation of responses to air puff of normal and decerebrate earthworms. *J. Comp. Physiol. Psychol.* 86:911–18.

Ratner, S. C., and K. R. Miller. 1959. Classical conditioning in earthworms, *Lumbricus terrestris*. *J. Comp. Physiol. Psychol.* 52:102–5.

Raup, D. M., and A. Seilacher. 1969. Fossil foraging behavior: Computer simulation. *Science* 166:994–95.

Rawson, K. S. 1960. Effects of tissue temperature on mammalian activity rhythms. *Cold Spr. Harb. Symp. Quant. Biol.* 25:105–14.

Reinisch, J. M. 1981. Prenatal exposure to synthetic progestins increases potential for aggression in humans. *Science* 211:1171–73.

Reiter, R. J. 1974a. Pineal regulation of the hypothalamic-pituitary axis: Gonadotrophins. In *Handbook of Physiology, Endocrinology*, vol. 4, part 2, ed. E. Knobil and W. H. Sawyer. Washington, D.C.: American Physiological Society.

———. 1974b. Circannual reproductive rhythms in mammals related to photoperiod and pineal function: A review. *Chronobiology* 1:365–95.

———. 1980. The pineal gland: A regulator of regulators. *Prog. in Psychobiol. and Physiol. Psych.* 9:323–56.

Renner, M. 1957. Neue Versuche über den Zeitsinn der Honigbiene. *Zeit. Vergleich. Physiol.* 40:85–118.

———. 1959. Über ein weiteres Versetzungsexperiment zur Analyse des Zeitsinnes und der Sonnenorientierung der Honigbiene. *Zeit. Vergleich. Physiol.* 42:449–83.

———. 1960. The contribution of the honey bees to the study of time-sense and astronomical orientation. *Cold Spr. Harb. Symp. Quant. Biol.* 25:361–67.

Renner, M. J. 1987. Experience-dependent changes in exploratory behavior in the adult rat *(Rattus norvegicus)*: overall activity level and interactions with objects. *J. Comp. Psychol.* 101:94–100.

Reppert, S. M., D. R. Weaver, S. A. Rivkees, and E. G. Stopa. 1988. Putative melatonin receptors in a human biological clock. *Science* 242:78–81.

Ressler, R. H., R. B. Cialdini, M. L. Ghoca, and S. M. Kleist. 1968. Alarm pheromone in the earthworm *Lumbricus terrestris*. *Science* 161:597–99.

Rettenmeyer, C. W. 1963. Behavioral studies of army ants. *Kansas University Science Bulletin* 44:281–465.

Rheinlander, J., H. C. Gerhardt, D. D. Yager, and R. R. Capranica. 1979. Accuracy of phonotaxis by the green treefrog *(Hyla cinerea)*. *J. Comp. Physiol.* 133:247–55.

Ribbands, C. R. 1953. *Behaviour and Social Life of Honey Bees.* London: Bee Research Assoc.

Rice, W. R. 1987. Speciation via habitat specialization: the evolution of reproductive isolation as a correlated character. *Evolutionary Ecology* 1:301–14.

Richards, M. P. 1967. Maternal behavior in rodents and lagomorphs. In *Advances in Reproductive Physiology*, vol.2, ed. A. McLaren. New York: Academic Press.

Richardson, W. J. 1971. Spring migration and weather in eastern Canada: A radar study. *Amer. Birds* 25:684–90.

———. 1972. Autumn migration and weather in eastern Canada. *Amer. Birds* 26:10–17.

Richter, C. P. 1922. A behavioristic study of the activity of the rat. *Comp. Psychol. Monogr.* 1:1–55.

Riechert, S. E. 1984. Games spiders play III: Cues underlying context-associated changes in agonistic behaviour. *Anim. Behav.* 32:1–15.

Ringo, J. M. 1976. A communal display in Hawaiian *Drosophila* (Diptera: Drosophilidae). *An. Entomol. Soc. Am.* 69:209–14.

———. 1978. The development of behavior in *Drosophila*. In *Development of Behavior*, ed. G. Burghardt and M. Bekoff. New York: Garland STPM Press.

Rivier, C., J. Rivier, and W. Vale. 1986. Stress-induced inhibition of reproductive functions: Role of endogenous corticotropin-releasing factor. *Science* 231:607–09.

Roberts, E. P., Jr., and P. D. Weigl. 1984. Habitat preference in the dark-eyed junco *(Junco hyemalis)*: The role of photoperiod and dominance. *Anim. Behav.* 32:709–14.

Roberts, R. C. 1967. Some concepts and methods in quantitative genetics. In *Behavior-Genetic Analysis*, ed. J. Hirsch. New York: McGraw-Hill.

Roberts, S. K. 1966. Circadian activity rhythms in cockroaches. III. The role of endocrine and neural factors. *J. Cell Physiol.* 67:473–86.

———. 1974. Circadian rhythms in cockroaches. Effects of the optic lobe lesions. *J. Comp. Physiol.* 88:21–30.

Robinson, B. W., M. Alexander, and G. Bowne. 1969. Dominance reversal resulting from aggressive responses evoked by brain telestimulation. *Physiol. Behav.* 4:749–52.

Roeder, K. D. 1970. Episodes in insect brains. *Amer. Scientist* 58:378–89.

Roeder, K. D., and A. E. Treat. 1961. The detection and evasion of bats by moths. *Amer. Scientist* 49:135–48.

Rood, J. P. 1980. Mating relationships and breeding suppression in the dwarf mongoose. *Anim. Behav.* 28:143–50.

———. 1983. The social system of the dwarf mongoose. In *Recent Advances in the Study of Mammalian Behavior*, ed. J. F. Eisenberg and D. G. Kleiman. Special Pub. No. 7, Amer. Soc. Mammal.

———. 1986. Ecology and social evolution in the mongooses. In *Ecological Aspects of Social Evolution. Birds and Mammals*, ed. D. I. Rubenstein and R. W. Wrangham. Princeton: Princeton University Press.

Rose, G. J., R. Zelick, and S. Rand. 1988. Auditory processing of temporal information in a neotropical frog is independent of signal intensity. *Ethology* 77:330–36.

Rose, R. N., I. S. Bernstein, and T. P. Gordon. 1975. Consequences of social conflict on plasma testosterone levels in rhesus monkeys. *Psychosomatic Med.* 37:50–61.

Roseler, P. F., I. Roseler, and A. Strambi. 1986. Role of ovaries and ecdysteroid in dominance hierarchy establishment among foundresses of the primitively social wasp, *Polistes gallicus. Behav. Ecol. Sociobiol.* 18:9–13.

Rosenblatt, J. S. 1967. Nonhormonal basis for maternal behavior in the rat. *Science* 156:1512–14.

———. 1970. Views on the onset and maintenance of maternal behavior in the rat. In *Development and Evolution of Behavior*, ed. L. R. Aronson et al. San Francisco: Freeman.

Rosenblatt, J. S., and L. R. Aronson. 1958. The decline in sexual behavior of male cats after castration with special reference to the role of prior sexual experience. *Behaviour* 12:285–338.

Rosenblatt, J. S., and D. S. Lehrman. 1963. Maternal behavior of the laboratory rat. In *Maternal Behavior in Mammals*, ed. H. L. Rheingold. New York: Wiley.

Rosenblatt, J. S., H. I. Siegel, and A. D. Mayer. 1979. Progress in the study of maternal behavior in the rat: Hormonal, nonhormonal, sensory, and developmental aspects. *Adv. Stud. Behav.* 10:226–311.

Rosenblum, L. A., and I. C. Kaufman. 1967. Laboratory observations of early mother-infant relations in pigtail and bonnet macaques. In *Social Communication among Primates*, ed. S. A. Altmann. Chicago: University of Chicago Press.

———. 1968. Variations in infant development and response to maternal loss in monkeys. *Amer. J. Orthopsych.* 38:418–26.

Rosenzweig, M. L. 1985. Some theoretical aspects of habitat selection. In *Habitat Selection in Birds*, ed. M. L. Cody. New York: Academic Press.

Ross, D. M. 1965. Complex and modifiable behavior patterns in *Calliactis* and *Stomphia*. *Amer. Zool.* 5:573–80.

Ross, R. M., G. S. Losey, and M. Diamond. 1983. Sex change in a coral reef fish: dependence of stimulation and inhibition on relative size. *Science* 221:574–75.

Rottman, S. J., and C. T. Snowden. 1972. Demonstration and analysis of an alarm pheromone in mice. *J. Comp. Physiol. Psychol.* 81:483–90.

Rowell, T. E. 1966. Forest living baboons in Uganda. *J. Zool.* 149:344–64.

———. 1974. The concept of social dominance. *Behav. Biol.* 11:131–54.

Rozin, P. 1968. Specific aversions and neophobia resulting from vitamin deficiency or poisoning in half wild and domestic rats. *J. Comp. Physiol. Psychol.* 66:82–88.

Rumbaugh, D. M. 1968. The learning and sensory capacities of the squirrel monkey in phylogenetic perspective. In *The Squirrel Monkey*, ed. L. A. Rosenblum and R. W. Cooper. New York: Academic Press.

Rumbaugh, D. M., and T. V. Gill. 1976. The mastery of language-type skills by the chimpanzee (Pan). In *Origins and Evolution of Language and Speech*, eds. S. R. Harnad, H. D. Steklis, and J. Lancaster. Annals of the New York Academy of Sciences. 280:562–78.

Rusak, B., and I. Zucker. 1979. Neural regulation of circadian rhythms. *Physiol. Rev.* 59:449–526.

Rushforth, N. D. 1973. Behavioral modifications in coelenterates. In *Invertebrate Learning*. Vol. 1., ed. W. C. Corning, J. A. Dyal, and A. O. D. Willows. New York: Plenum.

Russell, P. F., and T. R. Rao. 1942. On relation of mechanical obstruction and shade to ovipositing of *Anopheles culifacies*. *J. Exp. Zool.* 91:303–29.

Rust, C. C. 1965. Hormonal control of pelage cycles in the short-tailed weasel (*Mustela erminea bangsi*). *Gen. Comp. Endocr.* 5:222–31.

Rust, C. C., and R. K. Meyer. 1969. Coat color, molt and testis size in male short-tailed weasels treated with melatonin. *Science* 165:921–22.

Ryan, M. J., and W. Wilczynski. 1988. Coevolution of sender and receiver: Effect on local mate preference in cricket frogs. *Science* 240:1786–88.

Sade, D. S. 1965. Some aspects of parent-offspring and sibling relations in a group of rhesus monkeys, with a discussion of grooming. *Amer. J. Phys. Anthrop.* 23:1–17.

———. 1967. Determinants of dominance in a group of free-ranging rhesus monkeys. In *Social Communication among Primates*, ed. S. Altmann. Chicago: University of Chicago Press.

Sade, D. S., K. Cushing, P. Cushing, J. Dunaif, A. Figueroa, J. R. Kaplan, C. Laver, D. Rhodes, and J. Schneider. 1976. Population dynamics in relation to social structure on Cayo Santiago. *Yearbook Phys. Anthrop.* 20:253–62.

Sale, P. F. 1970. A suggested mechanism for habitat selection by the juvenile manini *Acanthurus triostegus sandvicensis* Streets. *Behaviour* 35:27–44.

Salzen, E. A., and C. C. Meyer. 1967. Imprinting: Reversal of a preference established during the critical period. *Nature* 215:785–86.

Salzen, E. A., and W. Sluckin. 1959. The incidence of the following response and the duration of responsiveness in domestic fowl. *Anim. Behav.* 7:172–79.

Sanders, G. D. 1973. The cephalopods. In *Invertebrate Learning*. Vol. 3., ed. W. C. Corning, J. A. Dyal, and A. O. D. Willows. New York: Plenum.

Sarnat, H. B., and M. G. Netsky. 1974. *Evolution of the Nervous System*. New York: Oxford University Press.

Sauer, E. G. F. 1957. Die Sternenorientierung nächtlich ziehender Grasmücken (*Sylvia atricapilla*, *borin* und *curruca*). *Zeit. Tierpsychol.* 14:29–70.

———. 1963. Migration habits of golden plovers. *Proc. Inter. Ornithol. Congr., 13th.* 454–67.

Saunders, D. S. 1978. Internal and external coincidence and the apparent diversity of photoperiodic clocks in the insects. *J. Comp. Physiol.* A 127:197–207.

Savage-Rumbaugh, E. S. 1986. *Ape Language: From Conditioned Response to Symbol*. New York: Columbia University Press.

Savage-Rumbaugh, E. S., D. M. Rumbaugh, and S. Boysem. 1980. Do apes use language? *Amer. Scientist* 68:49–61.

Scapini, F. 1986. Inheritance of direction finding in sandhoppers. In *Orientation in Space*, ed. G. Beugnon, 111–19. Toulouse, France: Editions Pivat.

Schaller, G. B. 1972. *The Serengeti Lion: A Study of Predator-Prey Relations*. Chicago: University of Chicago Press.

Schaller, G. B., and G. Lowther. 1969. The relevance of carnivore behavior to the study of early hominids. *Southwest. J. Anthrop.* 25:307–41.

Schein, M. W., and E. B. Hale. 1959. The effect of early social experience on male sexual behavior of androgen injected turkeys. *Anim. Behav.* 7:189–200.

Schenkel, R. 1956. Zur Deutung der Balzleisturgen einiger Phasianiden und Tetraoniden. *Ornithologische Beobachter* 53:182–201.

Schjelderup-Ebbe, T. 1922. Beitrage zur Sozialpsychologie des Haushuhns. *Zeit. Psychol.* 88:225–52.

Schleidt, W. 1974. "How fixed is the fixed action pattern?" *Zeit. Tierpsychol.* 36:184–211.

Schmidt-Koenig, K., and J. B. Phillips. 1978. Local anesthesia of the olfactory membrane and homing in pigeons. In *Animal Migration, Navigation, and Homing*, ed. K. Schmidt-Koenig and W. T. Keeton. New York: Springer-Verlag.

Schmidt-Koenig, K., and H. J. Schlichte. 1972. Homing in pigeons with reduced vision. *Proc. Nat. Acad. Sci. USA* 69:2446–47.

Schmidt-Koenig, K., and C. Walcott. 1973. Flugwege und Verbleib von Brieftauben mit getrübten Haftschalen. *Naturwissenschaften* 60:108–9.

Schneider, D. 1974. The sex-attractant receptor of moths. *Sci. Amer.* 231:28–35.

Schneiderman, H. A. 1972. Insect hormones and insect control. In *Insect Juvenile Hormones*, ed. J. Menn and M. Beroza. New York: Academic Press.

Schnierla, T. C. 1950. The relationship between observation and experimentation in the field study of behavior. *Ann. N.Y. Acad. Sci.* 51:1022–44.

Schnierla, T. C., and G. Piel. 1948. The army ant. *Sci. Amer.* 178:16–23.

Schoener, T. W., and C. A. Toft. 1983. Spider populations: Extraordinarily high densities on islands without top predators. *Science* 219:1353–55.

Schröder, J. H., and M. Sund. 1984. Inheritance of water-escape performance and water-escape learning in mice. *Behav. Genet.* 14:221–34.

Schultz, F. 1965. Sexuelle Prägung bei Anatiden. *Zeit. Tierpsychol.* 22:50–103.

Schumacher, M., J. C. Hendrick, and J. Balthazart. 1989. Sexual differentiation in quail: Critical period and hormonal specificity. *Horm. Behav.* 23:130–49.

Schwagmeyer, P. L. 1988. Scramble-competition polygyny in an asocial mammal: male mobility and mating success. *Am. Nat.* 131:885–92.

Scott, D. 1978. Identification of individual Bewick's swans by bill patterns. In *Animal Marking,* ed. B. Stonehouse, 160–68. Baltimore: University Press.

Scott, J. P. 1966. Agonistic behavior of mice and rats: A review. *Amer. Zool.* 6:683–701.

———. 1972. *Animal Behavior.* 2d ed. Chicago: University of Chicago Press.

———. 1975. Violence and the disaggregated society. *Aggressive Behav.* 1:235–60.

———. 1976. The control of violence: Human and nonhuman societies compared. In *Violence in Animal and Human Societies,* ed. A. Neal. Chicago: Nelson-Hall.

Scott, J. P., and E. Fredericson. 1951. The causes of fighting in mice and rats. *Physiol. Zool.* 24:273–309.

Scott, J. P., and J. L. Fuller. 1965. *Genetics and the Social Behavior of the Dog.* Chicago: University of Chicago Press.

Seilacher, A. 1967. Fossil behavior. *Sci. Amer.* 217 (2): 72–80.

———. 1986. Evolution of behavior as expressed in marine trace fossils. In *Evolution of Animal Behavior. Paleontological and Field Approaches,* ed. M. H. Nitecki and J. A. Kitchell. New York: Oxford University Press.

Seligman, M. E. P. 1970. On the generality of the laws of learning. *Psychol. Rev.* 77:406–18.

Selye, H. 1950. *Stress.* Montreal: Acta.

Seyfarth, R. M., D. L. Cheyney, and P. Marler. 1980. Monkey responses to three different alarm calls: Evidence of predator classification and semantic communication. *Science* 210:801–3.

Shannon, C. E., and W. Weaver. 1949. *Mathematical Theory of Communication.* Urbana: University of Illinois Press.

Shapiro, D. Y. 1979. Social behavior, group structure, and the control of sex reversal in hermaphroditic fish. In *Advances in the Study of Behavior,* ed. J. S. Rosenblatt, R. A. Hinde, and C. Beer. New York: Academic Press.

Sharp, F. R., J. S. Kauer, and G. M. Shepherd. 1975. Local sites of activity-related glucose metabolism in rat olfactory bulb during olfactory stimulation. *Brain Res.* 98:596–600.

Shaw, E. 1978. Schooling fishes. *Amer. Scientist* 66:166–75.

Shell, W. F., and A. J. Riopelle. 1957. Multiple discrimination learning in raccoons. *J. Comp. Physiol. Psychol.* 50:585–87.

Sheridan, M., and R. H. Tamarin. 1988. Space use, longevity, and reproductive success in meadow voles. *Behav. Ecol. Sociobiol.* 22:85–90.

Sherman, P. W. 1977. Nepotism and the evolution of alarm calls. *Science* 197:1246–53.

———. 1981. Kinship, demography, and Belding's ground squirrel nepotism. *Behav. Ecol. Sociobiol.* 8:251–59.

Shettlesworth, S. J. 1972. Constraints on learning. *Adv. Stud. Behav.* 4:1–68.

———. 1984. Learning and behavioural ecology. In *Behavioural Ecology: An Evolutionary Approach,* 2d ed., ed. J. R. Krebs and N. B. Davies. Oxford: Blackwell Scientific Publications, Ltd.

Shields, J. 1962. *Monozygotic Twins Brought Up Apart and Brought Up Together.* London: Oxford University Press.

Shields, W. M. 1982. *Philopatry, Inbreeding, and the Evolution of Sex.* Albany: State University of New York Press.

Shockley, W. B. 1969. Human quality problems and research taboos. In *New Concepts and Directions in Education,* ed. J. A. Pintus. Greenwich, Conn.: Educational Records Bureau.

Silberglied, R. E., A. Aiello, and D. M. Windsor. 1980. Disruptive coloration in butterflies: Lack of support in *Anartia fatima. Science* 209:617–19.

Silberglied, R. E., J. G. Shepherd, and J. L. Dickinson. 1984. Eunuchs: The role of apyrene sperm in Lepidoptera? *Am. Nat.* 123:255–65.

Silver, R. 1978. The parental behavior of ring doves. *Amer. Sci.* 66:209–15.

Simonds, P. E. 1965. The bonnet macaque in South India. In *Primate Behavior: Field Studies of Monkeys and Apes,* ed. I. DeVore. New York: Holt, Rinehart and Winston.

Simons, E. L. 1972. *Primate Evolution.* New York: Macmillan.

Simpson, J., I. B. M. Riedel, and N. Wilding. 1968. Invertase in the hypopharyngeal glands of the honeybee. *J. Apicult. Res.* 7:29–36.

Skinner, B. F. 1938. *Behavior of Organisms: An Experimental Analysis.* New York: Appleton-Century-Crofts.

———. 1953. *Science and Human Behavior.* New York: Macmillan.

Slater, P. J. B. 1983. The study of communication. In *Animal Behaviour,* ed. T. R. Halliday and J. B. Slater. New York: Freeman.

———. 1989. Bird song learning: Causes and consequences. *Ethol. Ecol. Evol.* 1:19–46.

Smetzer, B. 1969. Night of the palolo. *Nat. Hist.* 78:64–71.

Smith, A. 1776. *Wealth of Nations.* New York: Modern Library. 1937.

Smith, C. C. 1968. The adaptive nature of social organization in the genus of tree squirrels *Tamiasciurus. Ecol. Monogr.* 38:31–63.

Smith, D. G. 1976. An experimental analysis of the function of red-winged blackbird song. *Behaviour* 56:136–56.

———. 1981. The association between rank and reproductive success of male rhesus monkeys. *Amer. J. Primat.* 1:83–90.

Smith, J. E. 1950. Some observations on the nervous mechanisms underlying the behavior of starfish. *Symp. Soc. Exp. Biol. Med.* 4:196–220.

Smith, M., M. Manlove, and J. Joule. 1978. Spatial and temporal dynamics of the genetic organization of small mammal populations. In *Populations of Small Mammals under Natural Conditions,* ed. D. Snyder. Pittsburgh: University of Pittsburgh.

Smith, M. S., B. J. Kish, and C. B. Crawford. 1984. Inheritance of wealth as human kin investment. Paper presented to the Animal Behavior Society, Cheney, Washington.

Smith, N. G. 1966. Evolution of some arctic gulls (*Larus*): An experimental study of isolating mechanisms. *Ornith. Monogr.* 4:1–99.

Smith, R. E., and R. J. Hock. 1963. Brown fat: Thermogenic effector of arousal in hibernators. *Science* 149:199–200.

Smith, R. T. 1965. A comparison of socioenvironmental factors in monozygotic and dizygotic twins, testing an assumption. In *Methods and Goals in Human Behavior Genetics*, ed. S. G. Vandenberg. New York: Academic Press.

Smith, S. M. 1975. Innate recognition of coral snake pattern by a possible avian predator. *Science* 187:759–60.

Smith, W. J. 1984. *Behavior of Communicating*. 2d ed. Cambridge: Harvard University Press.

Smuts, B. B. 1985. *Sex and Friendship in Baboons*. New York: Aldine.

Smythe, N. 1977. The function of mammalian alarm advertising: Social signals or pursuit invitation? *Amer. Nat.* 111:191–94.

Snodgrass, R. E. 1956. *Anatomy of the Honeybee*. Ithaca, NY: Cornell University Press.

Snyder, N., and H. Snyder. 1970. Alarm response of *Diadema antillarium*. *Science* 168:276–78.

Sokoloff, L., M. Reivich, C. Kennedy, M. Des Rosiers, C. Patlak, K. Pettigrew, O. Sakurada, and M. Shinohara. 1972. The [^{14}C]-deoxyglucose method for the measurement of local cerebral glucose utilization: Theory, procedure and normal values in the conscious and anesthetized rat. *J. Neurochem.* 28:897–916.

Sokolove, P. G. 1975. Localization of the cockroach optic lobe circadian pacemaker with microlesions. *Brain Res.* 87:13–21.

Sontag, L. W., and R. F. Wallace. 1935. The movement response of the human fetus to sound stimuli. *Child Dev.* 6:253–58.

Sordahl, T. A. 1981. Sleight of wing. *Nat. Hist.* 90(8): 42–49.

Southern, W. E. 1969. Orientation behavior of ring-billed gull chicks and fledglings. *Condor* 71:418–25.

———. 1972. Magnets disrupt the orientation of juvenile ring-billed gulls. *BioScience* 22:476–79.

Southwick, C. H. 1955. The population dynamics of confined house mice supplied with unlimited food. *Ecology* 36:212–25.

———. 1967. An experimental study of intragroup agonistic behavior in rhesus monkeys, *Macaca mulatta*. *Behaviour* 28:182–209.

Southwick, C. H., and L. H. Clark. 1968. Interstrain differences in aggressive and exploratory activity of inbred mice, Part A. *Comm. Behav. Biol.* 1:49–59.

Southwick, C. H., M. F. Siddiqi, M. Y. Farooqui, and B. C. Pal. 1974. Xenophobia among freeranging rhesus groups in India. In *Primate Aggression Territoriality and Xenophobia*, ed. R. L. Halloway. New York: Academic Press.

Southwood, T. R. E. 1978. Marking invertebrates. In *Animal Marking*, ed. B. Stonehouse, 102–5. Baltimore: University Park Press.

Spalding, D. A., 1873. Instinct with original observations on young animals. *Macmillan's* 27:282–93. Reprinted in *Brit. J. Anim. Behav.* 2:2–11.

Spencer, R. 1978. Ringing and related durable methods of marking birds. In *Animal Marking*, ed. B. Stonehouse, 43–53. Baltimore: University Park Press.

Sperry, R. A., J. S. Stamm, and N. Miner. 1956. Relearning tests for interocular transfer following division of optic chiasma and corpus callosum in rats. *J. Comp. Physiol. Psychol.* 49:529–33.

Sperry, R. W. 1961. Cerebral organization and behavior. *Science* 133:1749–57.

———. 1974. Lateral specialization in the surgically separated hemispheres. In *The Neurosciences: Third Study Program*, ed. F. O. Schmitt and F. G. Worden. Cambridge, MA: MIT Press, pp. 5–20.

Spieth, H. T. 1966. Courtship behavior of endemic Hawaiian *Drosophila*. *Univ. Texas Publ.* 6615:245–313.

Stacey, P. B. 1979. Habitat saturation and communal breeding in the acorn woodpecker. *Anim. Behav.* 27:1153–66.

Staddon, J. E. R. 1983. *Adaptive Behavior and Learning*. New York: Cambridge University Press.

Stamm, J. S. 1954. Genetics of hoarding. I. Hoarding differences between homozygous strains of rats. *J. Comp. Psychol.* 47:157–61.

Stamps, J. 1983. Territoriality and the defence of predator-refuges in juvenile lizards. *Anim. Behav.* 31:857–70.

Stamps, J. A. 1973. Displays and social organization in female *Anolis aeneus*. *Copeia* 1973 (2):264–72.

Stanley, S. M. 1981. *The New Evolutionary Timetable*. New York: Basic Books.

Stebbins, G. L. 1966. *Processes of Organic Evolution*. Englewood Cliffs, N.J.: Prentice-Hall.

Stephan, F. K., J. M. Swann, and C. L. Sisk. 1979a. Anticipation of 24-hour feeding schedules in rats with lesions of the suprachiasmatic nucleus. *Behav. Neural Biol.* 25:346–63.

———. 1979b. Entrainment of circadian rhythms by feeding schedules in rats with suprachiasmatic lesions. *Behav. Neural Biol.* 25:545–54.

Stephan, F. K., and I. Zucker. 1972. Circadian rhythms in drinking behavior and locomotor activity of rats are eliminated by hypothalamic lesions. *Proc. Nat. Acad. Sci. USA* 69:1583–86.

Stephens, D. W., and J. R. Krebs. 1986. *Foraging Theory*. Princeton: Princeton University Press.

Stern, J. M., and L. Rogers. 1988. Experience with younger siblings facilitates maternal responsiveness in pubertal Norway rats. *Dev. Psychobiol.* 21:575–90.

Stewart, M. M., and F. H. Pough. 1983. Population density of tropical forest frogs: Relation to retreat sites. *Science* 221:570–72.

Stonehouse, B., ed. 1978. *Animal Marking*. Baltimore: University Park Press.

Struhsaker, T. T. 1967. Auditory communication among vervet monkeys (*Cercopithecus aethiops*). In *Social Communication among Primates*, ed. S. A. Altmann. Chicago: University of Chicago Press.

Sullivan, K. A. 1984. The advantages of social foraging in downy woodpeckers. *Anim. Behav.* 32:16–22.

Suomi, S. J. 1973. Surrogate rehabilitation of monkeys reared in total social isolation. *J. Child. Psychol. Psychiat.* 14:71–77.

Suomi, S. J., and H. F. Harlow. 1977. Early separation and behavioral maturation. In *Genetics, Environment and Intelligence*, ed. A. Oliverio. Elsevier: North-Holland.

Suomi, S. J., H. F. Harlow, and M. A. Novak. 1974. Reversal of social deficits produced by isolation rearing in monkeys. *J. Hum. Evol.* 3:527–34.

Suthers, R. A. 1966. Optomotor responses by echolocating bats. *Science* 152:1102–4.

Sweeney, B., and J. W. Hastings. 1960. Effects of temperature upon diurnal rhythms. *Cold Spr. Harb. Symp. Quant. Biol.* 25:87–104.

Taitt, M. J. 1981. The effect of extra food on small rodent populations: I. Deermice (*Peromyscus maniculatus*). *J. Anim. Ecol.* 50:111–24.

Takahashi, J. S., and M. Zatz. 1982. Regulation of circadian rhythmicity. *Science* 217:1104–11.

Tamarin, R. H. 1980. Dispersal and population regulation in rodents. In *Biosocial Mechanisms of Population Regulation*, ed. M. N. Cohen, R. S. Malpass, and H. G. Klein. New Haven: Yale University Press.

Tamarkin, L., S. Brown, and B. Goldman. 1975. Neuroendocrine regulation of seasonal reproductive cycles in the hamster. *Abstr. 5th Ann. Mtg. Soc. Neurosci.*, p. 458.

Terasawa, E., and C. H. Sawyer. 1969. Changes in electrical activity in the rat hypothalamus related to electrochemical stimulation of adenohypophyseal function. *Endocrinology* 85:143–51.

Terkel, J., and J. S. Rosenblatt. 1968. Maternal behavior induced by maternal blood plasma injection into virgin rats. *J. Comp. Physiol. Psychol.* 65:479–82.

———. 1971. Aspects of nonhormonal maternal behavior in the rat. *Horm. Behav.* 2:161–71.

———. 1972. Humoral factors underlying maternal behavior at parturition: Cross transfusion between freely moving rats. *J. Comp. Physiol. Psychol.* 80:365–71.

Terman, C. R. 1987. Intrinsic behavioral and physiological differences among laboratory populations of prairie deermice. *Amer. Zool.* 27:853–66.

Terrace, H. S., L. A. Petitto, R. J. Sanders, and T. G. Bever. 1979. Can an ape create a sentence? *Science* 206:891–902.

Thiessen, D. D., and P. Yahr. 1970. Central control of territorial marking in the Mongolian gerbil (*Meriones unguiculatus*). *Physiol. Behav.* 5:275–78.

Thiessen, D. D., P. Yahr, and K. Owen. 1973. Regulatory mechanisms of territorial marking in the Mongolian gerbil. *J. Comp. Physiol. Psychol.* 82:382–93.

Thompson, R., and J. V. McConnell. 1955. Classical conditioning in the planarian, *Dugesia dorotocephala*. *J. Comp. Physiol. Psychol.* 48:65–68.

Thompson, W. R. 1957. Influence of prenatal maternal anxiety on emotionality in young rats. *Science* 125:698–99.

Thornhill, R. 1976. Sexual selection and paternal investment in insects. *Amer. Nat.* 110:153–63.

———. 1979. Adaptive female-mimicking behavior in a scorpionfly. *Science* 205:412–14.

Thorpe, W. H. 1945. The evolutionary significance of habitat selection. *J. Anim. Ecol.* 14:67–70.

———. 1956. *Learning and Instinct in Animals.* Cambridge: Harvard University Press.

———. 1963. *Learning and Instinct in Animals*, 2d ed. London: Methuen.

Thorpe, W. H., and F. G. W. Jones. 1937. Olfactory conditioning in a parasitic insect and its relation to the problem of host selection. *Proc. Roy. Soc. London*, (Series B) 124:56–81.

Tinbergen, N. 1948. Social releasers and the experimental method required for their study. *Wilson Bull.* 60:6–52.

———. 1951. *The Study of Instinct.* Oxford: Clarenden Press.

———. 1953. *The Herring Gull's World.* London: Collins.

———. 1958. *Curious Naturalists.* New York: American Museum of Natural History.

———. 1959. Comparative studies of the behavior of gulls (Laridae): A progress report. *Behaviour* 15:1–70.

———. 1960. *The Herring Gull's World.* Garden City, NY: Doubleday.

———. 1963a. On aims and methods of ethology. *Zeitschrift für Tierpsychologie.* 20:410–33.

———. 1963b. The shell menace. *Nat. Hist.* 72:28–35.

Tinbergen, N., G. J. Broekhuysen, F. Feekes, J. C. W. Houghton, H. Kruuk, and E. Szulc. 1962. Egg shell removal by the black-headed gull *Larus ridibundus* L.: A behavior component of camouflage. *Behaviour* 19:74–118.

Tinbergen, N., and W. Kruyt. 1938. Uber die Orientierung des Bienenwolfes (*Philanthus triangulum* Fabr.): III Die Bevorzugung bestimmter Wegmarken. *Zeit. Vergl. Physiol.* 25:292–334.

Tinbergen, N., and A. C. Perdeck. 1950. On the stimulus situation releasing the begging response in the newly hatched herring gull chick (*Larus a argentatus* Ponstopp). *Behaviour* 3:1–38.

Toates, F. M. 1986. *Motivational Systems.* Cambridge, England: Cambridge University Press.

Tollman, J., and J. A. King. 1956. The effects of testosterone propionate on aggression in male and female C57BL/10 mice. *Anim. Behav.* 4:147–49.

Tomback, D. 1980. How nutcrackers find their seed stores. *Condor* 82:10–19.

Toran-Allerand, C. D. 1978. Gonadal hormones and brain development: Cellular aspects of sexual differentiation. *Amer. Zool.* 18:553–65.

Travis-Neideffer, M. N., J. D. Niedeffer, and S. F. Davis. 1982. Free operant single and double alternation in the albino rat: A demonstration. *Bull. Psychonom. Soc.* 19:287–90.

Trivers, R. L. 1971. The evolution of reciprocal altruism. *Quart. Rev. Biol.* 46:35–57.

———. 1972. Parental investment and sexual selection. In *Sexual Selection and the Descent of Man 1871–1971*, ed. B. Campbell. Chicago: Aldine.

———. 1974. Parent-offspring conflict. *Amer. Zool.* 14:249–64.

———. 1985. *Social Evolution.* Reading, MA: Benjamin Cummings.

Trivers, R. L., and H. Hare. 1976. Haplodiploidy and the evolution of the social insects. *Science* 191:249–63.

Trivers, R. L., and D. E. Willard. 1973. Natural selection of parental ability to vary the sex ratio of offspring. *Science* 179:90–92.

Truman, J. W. 1971. Physiology of insect ecdysis. I. The eclosion behavior of saturniid moths and its hormonal release. *J. Exp. Biol.* 54:805–14.

Truman, J. W., and O. S. Dominick. 1983. Endocrine mechanisms organizing invertebrate behavior. *BioScience* 33:546–51.

Truman, J. W., A. M. Fallon, and G. R. Wyatt. 1976. Hormonal release of programmed behavior in silk moths. *Science* 194:1432–33.

Truman, J. W., and L. M. Riddiford. 1970. Neuroendocrine control of ecdysis in silkmoths. *Science* 167:1624–26.

Turner, C. D., and J. T. Bagnara. 1976. *General Endocrinology.* Philadelphia: Saunders.

Twiggs, D. G., H. B. Popolow, and A. A. Gerall. 1978. Medial preoptic lesions and male sexual behavior: Age and environmental interactions. *Science* 200:1414–15.

Uetz, G. W. 1988. Group foraging in colonial web-building spiders: Evidence for risk sensitivity. *Behav. Ecol. Sociobiol.* 22:265–70.

Ukegbu, A. A., and F. A. Huntingford. 1988. Brood value and life expectancy as determinants of parental investment in male three-spined sticklebacks (*Gasterosteus aculeatus*). *Ethology* 78:72–86.

Ulrich, R. E., R. R. Hutchinson, and N. H. Azrin. 1965. Pain-elicited aggression. *Psych. Record* 15:111–26.

Upchurch, M., and J. M. Wehner. 1988. Differences between inbred strains of mice in Morris water maze performance. *Behav. Genet.* 18:55–68.

Uvarov, B. 1966. *Grasshoppers and Locusts.* New York: Cambridge University Press.

Valenstein, E. S. 1969. Behavior elicited by hypothalamic stimulation: A preoptency hypothesis. *Brain Behav. Evol.* 2:295–316.

Valenstein, E. S., V. C. Cox, and J. Kakolewski. 1970. A reexamination of the role of the hypothalamus in motivation. *Psychol. Rev.* 77:16–31.

Valenstein, P., and D. Crews. 1977. Mating-induced termination of behavioral estrus in the female lizard *Anolis carolinensis. Horm. Behav.* 9:362–70.

Van Hemel, P. E., and J. S. Myer. 1970. Satiation of mouse killing by rats in an operant situation. *Psychon. Sci.* 21:129–30.

van Lawick, H., and J. van Lawick-Goodall. 1970. *Innocent Killers.* Boston: Houghton Mifflin.

van Lawick-Goodall, J. 1968. A preliminary report on expressive movements and communication in the Gombe Stream chimpanzees. In *Primates: Studies in Adaptation and Variability,* ed. P. Jay. New York: Holt, Rinehart and Winston.

van Lawick-Goodall, J. 1970. Tool-using in primates and other vertebrates. *Adv. Study Behav.* 3:195–249.

Van Tets, G. F. 1965. A comparative study of some social communication patterns in the Pelecaniformes. *Ornith. Monogr.* 2:1–88.

van Tyne, J., and A. J. Berger, 1959. *Fundamentals of Ornithology.* New York: Wiley.

Van Valen, L. 1973. A new evolutionary law. *Evol. Theory* 1:1–30.

Vandenbergh, J. G. 1965. Hormonal basis of sex skin in male rhesus monkeys. *Gen. Comp. Endocr.* 5:31–34.

———. 1967. Effect of the presence of a male on the sexual maturation of female mice. *Endocrinology* 81:345–48.

———. 1969a. Endocrine coordination in monkeys: Male sexual responses to the female. *Physiol. Behav.* 4:261–64.

———. 1969b. Male odor accelerates female sexual maturation in mice. *Endocrinology* 84:658.

———. 1971. The effects of gonadal hormones on the aggressive behavior of adult golden hamsters. *Anim. Behav.* 19:589–94.

———. 1973. Effects of gonadal hormones on the flank gland of the golden hamster. *Horm. Res.* 4:28–33.

Vandenbergh, J. G. and D. M. Coppola. 1986. The physiology and ecology of puberty modulation by primer pheromones. *Adv. Stud. Behav.* 16:71–108.

Vandenbergh, J. G., and L. C. Drickamer. 1974. Reproductive coordination among free-ranging rhesus monkeys. *Physiol. Behav.* 13:373–76.

Vandenbergh, J. G., L. C. Drickamer, and D. R. Colby. 1972. Social and dietary factors in the sexual maturation of female mice. *J. Reprod. Fertil.* 28:397–405.

Vandenbergh, J. G., and S. Vessey. 1968. Seasonal breeding of free-ranging rhesus monkeys and related ecological factors. *J. Reprod. Fert.* 15:71–79.

Vander Wall, S. B., and Balda. 1981. Ecology and evolution of food storage behavior in conifer-seed-caching corvids. *Z. Tierpsychol.* 56:217–42.

Vaughan, T. A., and R. M. Hansen. 1964. Experiments on interspecific competition between two species of pocket gophers. *Amer. Mid. Nat.* 72:444–52.

Vehrencamp, S. L. 1977. Relative fecundity and parental effort in communally nesting anis, *Crotophaga sulcirostris. Science* 197:403–5.

Vehrencamp, S. L., R. R. Koford, and B. S. Bowen. 1988. The effect of breeding-unit size on fitness components in groove-billed anis. In *Reproductive Success,* ed. T. H. Clutton-Brock. Chicago: University of Chicago Press.

Vernikos-Danellis, J., and C. D. Winget. 1979. The importance of light, postural and social cues in the regulation of the plasma cortical rhythms in man. In *Chronopharmacology,* eds. A. Reinberg and F. Halbert. New York: Pergamon.

Vessey, S. H. 1967. Effects of chlorpromazine on aggression in laboratory populations of wild house mice. *Ecology* 48:367–76.

———. 1971. Free-ranging rhesus monkeys: Behavioural effects of removal, separation and reintroduction of group members. *Behaviour* 40:216–27.

———. 1973. Night observations of free ranging rhesus monkeys. *Amer. J. Phys. Anthrop.* 38:613–20.

———. 1987. Long-term population trends in white-footed mice and the impact of supplemental food and shelter. *Amer. Zool.* 27:879–90.

Vessey, S. H., and H. M. Marsden. 1975. Oviduct ligation in rhesus monkeys causes maladaptive epimeletic (care-giving) behavior. In *Contemporary Primatology,* ed. S. Kondo, M. Kawai, and A. Ehara. Basel: S. Karger.

Vessey, S. H., and D. B. Meikle. 1984. Free-ranging rhesus monkeys: Adult male interactions with infants and juveniles. In *Paternal Behavior,* ed. D. Taub. New York: Van Nostrand Reinhold.

Villars, T. A. 1983. Hormones and aggressive behavior in teleost fish. In *Hormones and Aggressive Behavior,* ed. B. B. Svare. New York: Plenum.

Vince, M. A. 1964. Social facilitation of hatching in bobwhite quail. *Anim. Behav.* 12:531–34.

———. 1966. Artificial acceleration of hatching in quail embryos. *Anim. Behav.* 14:389–94.

———. 1969. Embryonic communication, respiration and the synchronization of hatching. In *Bird Vocalizations: Their Relations to Current Problems in Biology and Psychology: Essays Presented to W. H. Thorpe,* ed. R. A. Hinde. Cambridge: Cambridge University Press.

Vince, M. A. 1973. Effects of external stimulation on the onset of lung ventilation and the time of hatching in the fowl, duck and goose. *Brit. J. Poult. Sci.* 14:389–401.

Vinogradova, Y. B. 1965. An experimental study of the factors regulating induction of imaginal diapause in the mosquito *Aedes togoi* Theob. *Entomol. Rev.* 44:309–15.

Vivien-Roels, B., J. Arendt, and J. Bradtke. 1979. Circadian and circannual fluctuations of pineal indoleamines (serotonin and melatonin) in *Testudo hermanni* Gmelin (Reptilia, Chelonia). I. Under natural conditions of photoperiod and temperature. *Gen. Comp. Endocr.* 37:197–210.

vom Saal, F. S. 1983. Models of early hormonal effects on intrasex aggression in mice. In *Hormones and Aggressive Behavior*, ed. B. B. Svare. New York: Plenum.

———. 1989. Sexual differentiation in litter-bearing mammals: Influence of sex of adjacent fetuses in utero. *J. Anim. Sci.* 67:1824–40.

vom Saal, F. S., and F. H. Bronson. 1978. In utero proximity of female mouse fetuses to males: Effect on reproductive performance during later life. *Biol. Reprod.* 19:842–53.

———. 1980a. Variation in length of the estrous cycle in mice due to former intrauterine proximity to male fetuses. *Biol. Reprod.* 22:777–80.

———. 1980b. Sexual characteristics of adult female mice are correlated with their blood testosterone levels during prenatal development. *Science* 208:597–99.

von Frisch, K. 1967a. *Bees, Their Vision, Chemical Senses and Language*. Ithaca, NY: Cornell University Press.

———. 1967b. *Dance Language and Orientation of Bees*. Cambridge: Harvard University Press.

Waage, J. K. 1979. Dual function of the damselfly penis: Sperm removal and transfer. *Science* 203:916–18.

Walcott, C. 1969. A spider's vibration receptor: Its anatomy and physiology. *Amer. Zool.* 9:133–44.

Walcott, C., J. L. Gould, and J. L. Kirschvink. 1979. Pigeons have magnets. *Science* 205:1027–29.

Walcott, C. H. 1972. Bird navigation. *Natural History* 81(June): 32–43.

———. 1977. Magnetic fields and orientation of homing pigeons under sun. *J. Exp. Biol.* 70:105–23.

Waldman, B. 1982. Sibling association among schooling toad tadpoles: Field evidence and implications. *Anim. Behav.* 30:700–13.

Walker, B. W. 1949. Periodicity of spawning of the grunion, *Leuresthes tenuis*. Ph.D. thesis, University of California, Los Angeles.

Waloff, Z. 1958. The behaviour of locusts in migrating swarms. *Proc. 10th Inter. Congr. Ent.*, Montreal 2:567–70.

Walraff, H. G. 1986. Relevance of olfaction and atmospheric odours to pigeon homing. In *Orientation in Space*, ed. G. Beugnon, 71–80. Toulouse, France: Editions Privat.

Walsh, J. 1983. Wide world of reports. *Science* 220:804–5.

Waltz, E. C. 1982. Resource characteristics and the evolution of information centers. *Am. Nat.* 119:73–90.

Ward, P. 1971. The migration patterns of *Quelea quelea* in Africa. *Ibis* 113:275–97.

Warren, J. M. 1973. Learning in vertebrates. In *Comparative Psychology: A Modern Survey*, ed. D. A. Dewsbury and D. A. Rethlingshafer. New York: McGraw-Hill.

Warren, J. M., and A. Barron. 1956. The formation of learning sets by cats. *J. Comp. Physiol. Psychol.* 49:227–31.

Waterman, T. H. 1966. Systems analysis and the visual orientation of animals. *Amer. Sci.* 54:15–45.

Waterman, T. H., and H. Hashimoto. 1974. E-vector discrimination by the goldfish optic tectum. *J. Comp. Physiol.* 95:1–12.

Watkins, L. R., and D. J. Mayer. 1982. Organization of endogenous opiate and nonopiate pain control systems. *Science* 216:1185–92.

Watson, J. B. 1930. *Behaviorism*. New York: Norton.

Watt, W. B., P. A. Carter, and K. Donohue. 1986. Females' choice of "good genotypes" as mates is promoted by an insect mating system. *Science* 233:1187–90.

Weaver, N. 1966. Physiology of caste determination. *Ann. Rev. Entomol.* 11:79–102.

Weber, N. A. 1966. Fungus-growing ants. *Science* 153:587–604.

Webster, D. G., M. H. Williams, R. D. Owens, V. B. Geiger, and D. A. Dewsbury. 1981. Digging behavior in 12 taxa of muroid rodents. *Anim. Learn. Behav.* 9:173–77.

Wecker, S. C. 1963. The role of early experience in habitat selection by the prairie deer mouse, *Peromyscus maniculatus bairdi*. *Ecol. Monogr.* 33:307–25.

Wells, M. J., and J. Wells. 1957. The function of the brain of *Octopus* in tactile discrimination. *J. Exp. Biol.* 34:131–42.

———. 1959. Hormonal control of sexual maturity in *Octopus*. *J. Exp. Biol.* 36:1–33.

Wells, P. H. 1973. Honey bees. In *Invertebrate Learning*. Vol. 2., ed. W. C. Corning, J. A. Dyal, and A. O. D. Willows. New York: Plenum.

Wenner, A. M. 1967. Honeybees: Do they use the distance information contained in their dance maneuver? *Science* 155:847–49.

———. 1971. *Bee Language Controversy*. Boulder, CO: Educational Programs Improvement Corp.

———. 1974. Information transfer in honeybees: A population approach. In *Nonverbal Communication*, vol. 1, ed. L. Krames, P. Pliner, and T. Alloway. New York: Plenum.

Werblin, F. S., and J. E. Dowling. 1969. Organization of the retina of the mud puppy, *Necturus maculosus*. II. Intracellular recording. *J. Neurophysiol.* 32:339–55.

Werren, J. 1980. Sex ratio adaptations to local mate competition in a parasitic wasp. *Science* 208:1157–59.

Werren, J. H. 1983. Sex ratio evolution under local mate competition. *Evolution* 37:116–24.

West Eberhard, M. J. 1969. The social biology of Polistine wasps. *Misc. Pub. Mus. Zool.*, (University of Michigan, Ann Arbor) 140:1–101.

Whittaker, R. H. 1975. *Communities and Ecosystems*. 2d. ed. New York: Macmillan.

Wickler, W. 1962. Ei-Attrapen und Maulbrüten bei afrikanischen Cichliden. *Zeit. Tierpsychol.* 19:129–64.

———. 1968. *Mimicry in Plants and Animals*. London: World University Library.

———. 1969. Zur Sociologie des Brabantbuntbarshes, *Tropheus moorei* (Pisces, cichlidae) *Zeit. für Tierpsychol.* 26:967–87.

———. 1972. *The Sexual Code: The Social Behavior of Animals and Men*. Garden City, N.Y.: Doubleday.

Wilcox, R. S. 1972. Communication by surface waves: Mating behavior of a water strider (Gerridae). *J. Comp. Physiol.* 80:255–66.

———. 1979. Sex discrimination in *Gerris remigis*: Role of a surface wave signal. *Science* 206:1325–27.

Wildt, D. E., M. Bush, K. L. Goodrowe, C. Packer, A. E. Pusey, J. L. Brown, P. Joslin, and S. J. O'Brien. 1987. Reproductive and genetic consequences of founding isolated lion populations. *Nature* 329:328–31.

Wiley, R. H. 1973. Territoriality and nonrandom mating in sage grouse, *Centrocercus urophasianus*. *Anim. Behav. Monogr.* 6:85–169.

Wilkinson, G. S. 1984. Reciprocal food sharing in the vampire bat. *Nature* 308:181–84.

Wille, A., and E. Orozco. 1970. The life cycle and behavior of the social bee *Lasioglossum* (*Dialictus*) *umbripenne*. *Rev. Biol. Trop.* (Costa Rica): 17:199–245.

Williams, G. C. 1966. *Adaptation and Natural Selection: A Critique of Some Current Evolutionary Thought*. Princeton: Princeton University Press.

———. 1975. *Sex and Evolution*. Princeton: Princeton University Press.

———. 1979. The question of adaptive sex ratio in outcrossed vertebrates. *Proc. Roy. Soc. Lond.* B 205:567–80.

Williams, T. C., and J. M. Williams. 1967. Radio tracking of homing bats. *Science* 155:1435–36.

———. 1970. Radio tracking of homing and feeding flights of a neotropical bat, *Phyllostomus hastatus*. *Anim. Behav.* 18:302–9.

Williams, T. C., J. M. Williams, and D. R. Griffin. 1966. The homing ability of the neotropical bats, *Phyllostomus hastatus*, with evidence for visual orientation. *Anim. Behav.* 14:468–73.

Willows, A. O. D. 1967. Behavioral acts elicited by stimulation of single, identifiable brain cells. *Science* 157:570–74.

Wilson, D. S. 1980. *The Natural Selection of Populations and Communities*. Menlo Park, CA: Benjamin/Cummings.

Wilson, E. O. 1971. *Insect Societies*. Cambridge: Harvard University Press.

———. 1973. Group selection and its significance for ecology *BioScience* 23:631–38.

———. 1975. *Sociobiology: The New Synthesis*. Cambridge: Harvard University Press.

———. 1978. *On Human Nature*. Cambridge: Harvard University Press.

Wilson, E. O., and W. H. Bossert. 1971. *A Primer of Population Biology*. Sunderland, Mass. Sinauer.

Wiltschko, W. 1972. The influence of magnetic total intensity and inclination on directions preferred by migrating European robins (*Erithacus rubecula*). *NASA Spec. Publ.* NASA SP-262:569–78.

Wiltschko, W., and R. Wiltschko. 1972. The magnetic compass of European robins, *Erithacus rubecula*. *Science* 176:62–64.

Witham, T. G. 1980. The theory of habitat selection: Examined and extended using *Pemphigus* aphids. *Am. Nat.* 115:449–66.

Wittenberger, J. F. 1981. *Animal Social Behavior*. Boston, MA: Duxbury Press.

Wodinsky, J. 1977. Hormonal inhibition of feeding and death in *Octopus*: Control by optic gland secretion. *Science* 198:948–51.

Wolff, J. O. 1985. Maternal aggression as a deterrent to infanticide in *Peromyscus leucopus* and *P. maniculatus*. *Anim. Behav.* 33:117–23.

Wolff, J. O., K. I. Lundy, and R. Baccus. 1988. Dispersal, inbreeding avoidance, and reproductive success in white-footed mice. *Anim. Behav.* 36:456–65.

Wolff, J. R. 1981. Some morphogenetic aspects of the development of the central nervous system. In *Behavioral Development*, ed. K. Immelmann et al. New York: Cambridge University Press.

Wolfson, A. 1948. Bird migration and the concept of continental drift. *Science* 108:23–30.

Woody, C. D. 1982. *Memory, Learning and Higher Function*. New York: Springer-Verlag.

Woolfenden, G. E. 1975. Florida scrub jay helpers at the nest. *Auk* 92:1–15.

Woolfenden, G. E., and J. W. Fitzpatrick. 1984. *The Florida Scrub Jay: Demography of a Cooperative-Breeding Bird*. Princeton: Princeton University Press.

Wursig, B., and M. Wursig. 1977. The photographic determination of group size, composition, and stability of coastal porpoises (*Tursiops truncatus*). *Science* 198: 755–56.

Wyman, R. L., and J. A. Ward. 1973. The development of behavior in the cichlid fish *Etroplus maculatus* Bloch. *Z. Tierpsychol.* 33:461–91.

Wynne-Edwards, V. C. 1962. *Animal Dispersion in Relation to Social Behavior*. Edinburgh: Oliver and Boyd.

———. 1986. *Evolution Through Group Selection*. Boston: Blackwell Scientific Publications.

Young, W. C., R. W. Goy, and C. H. Phoenix. 1964. Hormones and sexual behavior. *Science* 143:212–18.

Zahavi, A. 1974. Communal nesting by the Arabian babbler: A case of individual selection. *Ibis* 116:84–87.

———. 1975. Mate selection—A selection for a handicap. *J. Theor. Biol.* 53:205–14.

Zahl, P. A. 1963. The mystery of the monarch butterfly. *Nat. Geogr.* 123:588–98.

Zarrow, M. X. 1961. Gestation. In *Sex and Internal Secretions*, ed. W. C. Young. Baltimore: Williams & Wilkins.

Zarrow, M. X., R. Gandelman, and V. H. Denenberg. 1971. Prolactin: Is it an essential hormone for maternal behavior in the mammal? *Horm. Behav.* 2:343–54.

Zeveloff, S. I., and M. S. Boyce. 1980. Parental investment and mating systems in mammals. *Evolution* 34:973–82.

Zigmond, R. E., F. Nottebohm, and D. W. Pfaff. 1973. Androgen-concentrating cells in the midbrain of a songbird. *Science* 179:1005–6.

Zimen, E. 1972. *Vergleichende Verhaltungsbeobachtungen an Wölfen und Königspudeln*. Munich: Piper.

Zolman, J. F. 1982. Ontogeny of learning. *Perspectives in Ethology* 5:275–324.

Zucker, I., B. Rusak, and R. G. King. 1976. Neural bases for circadian rhythms in rodent behavior. *Adv. Psychobiol.* 3:36–74.

Zuk, M. 1984. A charming resistance to parasites. *Nat. Hist.* 4 (84):28–34.

Zumpe, D., and R. P. Michael. 1968. The clutching reaction and orgasm in the female rhesus monkey (*Macaca mulatta*). *J. Endocr.* 40:117–23.

CREDITS

454

Figure 15-7 Adapted from data by G. Kramer in *Proceedings of the X International Ornithologist's Congress*, p. 269–280, as presented by S. T. Emlen, "Migration, Orientation, and Navigation," *Avian Biology*, 5:152. Copyright © 1975 Academic Press, Orlando, FL. Reprinted by permission.

Figure 15-8 From S. T. Emlen, "Migrating Orientation in the Indigo Bunting. I. Evidence for Use of Celestial Cues," *Auk*, 84:309–342. Copyright © 1967 by the American Ornithologists' Union. Reprinted by permission.

Figure 15-9 From W. T. Keeton, "The Orientation and Navigational Basis of Homing in Birds," *Advances in the Study of Behavior*, 5:47–132, 1974. Copyright © 1974 by Academic Press, Orlando, FL. Reprinted by permission.

Figure 15-10 From S. T. Emlen, "Migratory Orientation in the Indigo Bunting, *Passerina cayanea*," in *Auk*, 84:463–489. Copyright © 1967 by the American Ornithologists' Union. Reprinted by permission.

Figure 15-13 From A. Carr, "Adaptive Aspects of the Scheduled Travel of Chelonia," in *Animal Orientation and Navigation*, p. 43, edited by R. M. Storm. Copyright © 1967 by Oregon State University Press, Corvallis, OR. Reprinted by permission.

Figure 16-3 From Robert H. MacArthur and Edward O. Wilson, *The Theory of Island Biogeography*. Copyright © 1967 by Princeton University Press, Princeton, NJ. Reprinted by permission.

Figure 16-4 From J. M. Diamond, "Distributional Ecology of New Guinea Birds," in *Science*, 179(1973):759–769. Copyright © 1973 by the American Association for the Advancement of Science. Reprinted by permission.

Figure 16-7 From D. L. Cheney and R. M. Seyfarth, "Nonrandom Dispersal in Free-ranging Vervet Monkeys: Social and Genetic Consequences," *American Naturalist*, 122:392–412. Copyright © 1983 by the University of Chicago Press, Chicago, IL. Reprinted by permission.

Figure 16-8 Reprinted with permission from: Anne E. Pusey and C. Packer, "The Evolution of Sex-Based Dispersal in Lions," in *Behaviour*, 101 (1987) pp. 275–310, E.J. Brill, Publishers, Leiden.

Figure 16-9 From J. Brodeur and J. N. McNeil, "Seasonal Microhabitat Selection by an Endoparasitoid through Adaptive Modification of Host Behavior," in *Science*, 244:226–228. Copyright © 1989 by the American Association for the Advancement of Science. Reprinted by permission.

Figure 16-10 Reprinted by permission from: Peter F. Sale from "A Suggested Mechanism for Habitat Selection by the Juvenile Manini *Acanthurus triostegus sandvicensis* Streets" in *Behaviour*, 35 (1970), pp. 27–44 by E.J. Brill, Leiden.

Figure 16-11 From L. Partridge, "Habitat Selection," in *Behavioural Ecology: An Evolutionary Approach*, edited by J. R. Krebs and N. B. Davies. Copyright 1978 by Blackwell Scientific Publications Ltd, Oxford, England. Reprinted by permission.

Figure 17-1 From Bruce A. Colvin, *Owl Foraging Behavior and Secondary Poisoning Hazard from Rodenticide Use on Farms*. Ph.D. dissertation, Bowling Green State University, 1984. Reprinted by permission.

Figure 17-4 Redrawn from J. R. Krebs and R. H. McCleery, "Optimization in Ecology," in *Behavioural Ecology: An Evolutionary Approach*, 2d edition, edited by J. R. Krebs and N. B. Davies. Copyright © 1984 by Blackwell Scientific Publications Ltd, Oxford, England. Reprinted by permission.

Figure 17-5 Reprinted by permission from *Nature*, 268:137–139. Copyright © 1977 Macmillan Magazines Ltd.

Figure 17-6 From D. W. Stephens and J. R. Krebs, *Foraging Theory*. Copyright © 1986 by Princeton University Press, Princeton, NJ. Reprinted by permission.

Figure 17-15 From N. B. Davies and A. I. Houston, "Territory Economics," in *Behavioural Ecology: An Evolutionary Approach*, 2d edition, edited by J. R. Krebs and N. B. Davies. Copyright © 1984 by Blackwell Scientific Publications Ltd, Oxford, England. Reprinted by permission.

Figure 17-21 From T. Caracao and L. L. Wolf, "Ecological Determinants of Group Sizes of Foraging Lions," *American Naturalist*, 109:343–352. Copyright © 1975 by University of Chicago Press, Chicago, IL. Reprinted by permission.

Figure 17-22 From C. Packer, "The Ecology of Sociality in Felids," *Ecological Aspects of Social Evolution*, edited by D. I. Rubenstein and R. W. Wrangham. Copyright © 1986 Princeton University Press, Princeton, NJ. Reprinted by permission.

Quote, pages 353–354 From H. Kruuk. *The Spotted Hyena: A Study of Predation and Social Behavior*. Copyright © 1972 by the University of Chicago Press, Chicago, IL. Reprinted by permission.

Figure 17-24 Excerpt from *The Wolf* by L. David Mech, copyright © 1970 by L. David Mech. Used by permission of Doubleday, a division of Bantam Doubleday Dell Publishing Group, Inc.

Figure 18-1 Reprinted with permission from *The Canadian Entomologist*, Volume 91, p. 302.

Figure 18-2 From M. J. Taitt, "The Effect of Extra Food on Small Rodent Populations: I. Deermice (*Peromyscus maniculatus*) in *Journal of Animal Ecology*, 50:111–124. Copyright © 1981 by Blackwell Scientific Publications Ltd, Oxford, England. Reprinted by permission.

Figure 18-5 From J. R. Krebs, "Territory and Breeding Density in the Great Tit, *Parus major L.*," in *Ecology* 52:2–22. Copyright © 1971 by the Ecological Society of America, Tempe, AZ. Reprinted by permission.

Figure 18-7 From Stephen Vessey, "Effects of Chlorpromazine on Aggression in Laboratory Populations of Wild House Mice," *Ecology*, 48:367–376. Copyright © 1967 by the Ecological Society of America, Tempe, AZ. Reprinted by permission.

Figure 19-1 Reprinted by permission from: N. Tinbergen, "Comparative Studies of the Behavior of Gulls (*Laridae*): A Progress Report," in *Behaviour*, 15 (1959) pp. 1–70, E.J. Brill, Leiden.

Figure 19-4 Redrawn from G. F. Van Tests, "A Comparative Study of Some Social Communication Patterns in the *Pelecaniformes*," in *Ornithological Monographs* 2:1–88. Copyright © 1965. Used by permission of the American Ornithologists' Union, Washington, D.C.

Text, pages 386–387 From E. L. Kessel, "The Mating Activities of Balloon Flies," in *Systematic Zoology*, 4:97–104. Copyright © 1955 *Systematic Zoology*. Reprinted by permission.

Figure 19-7 From N. E. Collias and E. C. Collias, "Evolutionary Trends in Nest Building by the Weaverbirds (*Ploceidae*)" in *Proceedings of the XIII International Ornithologists' Congress*, 518–530. Copyright © 1963 by the American Ornithologists' Union. Reprinted by permission.

Figure 19-9 From N. E. Collias and E. C. Collias, "Evolutionary Trends in Nest Building by the Weaverbirds (*Ploceidae*)" in *Proceedings of the XIII International Ornithologists' Congress*, 518–530. Copyright © 1963 by the American Ornithologists' Union. Reprinted by permission.

Figure 19-11 From D. Hunsaker, "Ethological Isolating Mechanisms in the *Sceloporus torquatus* Group of Lizards," in *Evolution*, 16:62–74. Copyright © 1962. Reprinted by permission.

Figure 19-12 From N. G. Smith, "Evolution of Some Artic Gulls (*Larus*): An Experimental Study of Isolating Mechanisms," in *Ornithological Monographs* 4:1–199. Copyright © 1966. Used by permission of the American Ornithologists' Union, Washington, D.C.

Figure 19-13 From M. J. Littlejohn, "Premating Isolation in the *Hyla ewingi* Complex (*Anura: Hylidae*), in *Evolution*, 19:234–243. Copyright © 1965. Reprinted by permission.

Figure 20-7 Drawing by Stephen T. Emlen based on J. L. Brown, "Helpers: Effects of Experimental Removal on Reproductive Success," in *Science*, 215:421–422, 22 January 1982. Copyright © by the American Association for the Advancement of Science. Reprinted by permission.

Figure 20-9 From J. C. Bednarz, "Cooperative Hunting Among Hawks," in *Science*, 239:1526, 1988. Copyright © 1988 by American Association for the Advancement of Science. Reprinted by permission.

NAME INDEX

SUBJECT INDEX